CRYOGENIC FUNDAMENTALS

P

CRYOGENIC FUNDAMENTALS

G. G. Haselden

Department of Chemical Engineering
University of Leeds
Leeds, England

 1971

ACADEMIC PRESS · London and New York

ACADEMIC PRESS INC. (LONDON) LTD.

Berkeley Square House
Berkeley Square,
London, W1X 6BA

U.S. Edition published by

ACADEMIC PRESS INC.

111 Fifth Avenue,
New York, New York 10003

Library of Congress Catalog Card Number: 74-153521
ISBN: 0 12 330550 0

Printed in Great Britain by
ABERDEEN UNIVERSITY PRESS

Contributors

J. BLACKFORD, *Lucas Industrial Equipment, Liverpool, England.*

J. A. CATTERALL, *National Physical Laboratory, Teddington, England.*

J. A. CLARK, *Department of Mechanical Engineering, University of Michigan, Ann Arbor, U.S.A.*

A. H. COCKETT, *British Oxygen Co. Ltd., London, England.*

F. J. EDESKUTY, *Los Alamos Scientific Laboratory, University of California, Los Alamos, New Mexico, U.S.A.*

R. M. GIBBONS, *The Gas Council London Research Station, Fulham, London, England.*

P. HALFORD, *Petrocarbon Developments, Wythenshawe, Manchester, England.*

R. HANCOX, *Culham Laboratory, Culham, Abingdon, England.*

G. G. HASELDEN, *Department of Chemical Engineering, University of Leeds, Leeds, England.*

N. KURTI, *Department of Physics, University of Oxford, Clarendon Laboratory, Oxford, England.*

W. MOLNAR, *BOC-Airco Crygenic Plant Ltd., Edmonton, London, England.*

R. REIDER, *Los Alamos Scientific Laboratory, University of California, Los Alamos, New Mexico, U.S.A.*

R. V. SMITH, *Cryogenics Division, Institute of Basic Standards, National Bureau of Standards, Boulder, Colorado, U.S.A.*

D. H. TANTAM, *BOC-Airco Cryogenic Plant Ltd., Edmonton, London, England.*

R. M. THOROGOOD, *Air Products Ltd., New Malden, England.*

H. TOWNSLEY, *BOC-Airco Cryogenic Plant Ltd., Edmonton, London, England.*

D. A. WIGLEY, *Department of Mechanical Engineering, University of Southampton, Southampton, England.*

K. D. WILLIAMSON, JR., *Los Alamos Scientific Laboratory, University of California, Los Alamos, New Mexico, U.S.A.*

K. WILSON, *Air Products Ltd., New Malden, England.*

Preface

During the war years (1939–45) in the Chemical Technology Department of Imperial College, London, the late Sir Alfred Egerton drew together a team of applied scientists to investigate the potentialities of liquid methane, and particularly its use as a substitute motor fuel. Dr. Martin Ruhemann played an important part in the project, and at the same time, his pioneering book "The Separation of Gases" was published by Oxford University Press.

In many ways, the present book is a direct consequence of these happenings. I was privileged, early in 1944, to be invited to join Egerton's team and, ever since, I have continued to be fascinated by low temperature phenomena and their uses. Ruhemann's book became my guide. Indeed such was its vision and quality that only at the end of the 1950's was the need felt for an additional English text in this field. The significant developments which had occurred in the U.S.A., many of them at the N.B.S. Boulder Laboratory, were brought together by its Director, the late Dr. R. B. Scott, in his classic text "Cryogenic Engineering". It marked a new era in the growth of applied low temperatures, with the emphasis shifting from industrial gas liquidation and separation to the lower temperatures involved in the use of liquid hydrogen for various large-scale nuclear applications and for rocketry, and the early exploration of superconducting devices at liquid helium temperatures.

During the 1960's, there was a tendency for the two temperature regimes of cryogenics to grow apart, and most of the new texts appearing in that period concentrated in temperatures below about 20°K. Nevertheless, significant studies were being made at higher temperatures with tonnage oxygen plants achieving sizes of up to 1000 tons per day, very large ethylene separation plants being installed in every major industrial country, and liquid natural gas becoming a significant world fuel.

The present book aims to restore the unity of the subject by taking the full temperature spectrum below about 150°K as its territory. It aims to give guidance to the experimentalist and the plant designer. A comprehensive coverage is attempted of all the special factors that arise at low temperatures, whether in relation to the process, the materials and methods of construction required for it, its control, or its safety requirements. The only process of separation which is specifically covered is adsorption, since it is employed more intensively in association with cryogenic plant than in any other field.

vii

The decision was made not to include distillation because, by and large, no new principle is involved in its use at cryogenic temperatures.

I wish to acknowledge the help and encouragement of Dr. Ruhemann, Professor Kurti, Dr. J. B. Gardner (BOC-Airco), and Dr. Bascombe Birmingham (N.B.S. Boulder Laboratory) in deciding the contents of the book, and finding the best authors.

Leeds   GEOFFREY G. HASELDEN
June, 1971

Acknowledgements

Some of the illustrations contained in this book are reproduced from other publications, and permission to do so is gratefully acknowledged.

In Chapter 2 figures are taken from "Technology and Uses of Liquid Hydrogen" (R.B. Stott, W. H. Denton and C. M. Nicholls, eds.) published by Pergamon Press Ltd. Several other figures are included by courtesy of N. V. Philips' Gloeilampenfabrieken of Eindhoven, Netherlands. Other acknowledgements are included at appropriate points in the text.

Contents

Chapter 1. Introduction

N. KURTI

Chapter 2. Refrigeration and Liquefaction Cycles

G. G. HASELDEN

Chapter 3. Heat Transfer

J. A. CLARK AND R. M. THOROGOOD

Chapter 4. Insulation

W. MOLNAR

Chapter 5. Fluid Dynamics

R. V. SMITH

Chapter 6. Materials of Construction and Techniques of Fabrication

D. A. WIGLEY AND P. HALFORD

Chapter 7. Adsorption

K. WILSON

Chapter 8. Expanders and Pumps

J. BLACKFORD, P. HALFORD AND D. H. TANTAM

Chapter 9. Superconductivity

R. HANCOX AND J. A. CATTERALL

Chapter 10. Instrumentation

H. TOWNSLEY AND A. H. COCKETT

Chapter 11. Safety

F. J. EDESKUTY, R. REIDER AND K. D. WILLIAMSON, JR.

Chapter 12. Thermophysical Data for Cryogenic Materials

R. M. GIBBONS

Chapter 1

Introduction

N. Kurti

Clarendon Laboratory, University of Oxford, Oxford, England

I. General

Cryogenics, i.e. the technology and art of producing low temperatures and, in a wider sense, the study of phenomena, techniques and concepts occurring at or pertaining to low temperatures, has grown spectacularly since about 1950. Before discussing the practical reasons for this development one might look briefly at the conceptual reasons which make the production and maintenance of low temperatures so fundamentally different from the achievement of high temperatures. The answer lies in the second law of thermodynamics which in fact postulates the asymmetry of the temperature scale and the fundamental irreversibility of certain physical and chemical processes. The subject of this book shows clearly this consequence of the second law. Whereas high temperatures, i.e. temperatures higher than those of the surroundings, can occur spontaneously, and in fact very often do, low temperatures must be generated artificially which usually means the expenditure of free energy, which may be mechanical or electrical or chemical—to mention only its most important forms. Should this assertion need illustration one only has to remember that while damaging fires often get started and do *spread*

1

during winter, pipes burst by frost during summer would be regarded with suspicion or revered as miracles.

Before reviewing the various aspects of low temperatures which make their use so rewarding in fundamental research—chiefly in physics—and in many fields of applied science and engineering, ranging from biology and medicine to chemical, electrical and communication engineering—a brief historical survey seems to be justified.

II. Historical Survey

Since this book is concerned chiefly with temperatures below 120°K (i.e. generally below the boiling point of methane which is 112°K) we may take as a starting point 1877 when almost simultaneously but independently Cailletet in France and Pictet in Switzerland succeeded in liquefying oxygen. It was left to the Polish scientists Olszewski and Wroblewski to produce six years later, in 1883, not just a quickly vanishing mist of liquid oxygen, but a stable liquid whose properties could be studied. These experiments were done before the invention of the vacuum flask by Dewar. To reduce the obnoxious effects of heat influx the tube containing the boiling oxygen was shielded by liquid ethylene which in turn was protected by forcing the cold ethylene vapour to pass through an annular space surrounding the ethylene container. It is interesting to remark that some of the most modern storage vessels for liquid hydrogen and liquid helium are based on a combination of vacuum insulation, highly reflective surfaces and an effective use of the enthalpy of the escaping vapour, and are an embodiment of both the Wroblewski and the Dewar principles.

Thus, within a span of less than six years, not only were the constituents of atmospheric air liquefied, but, more significantly, the three basic processes for gas liquefaction and refrigeration were demonstrated in practice. Cailletet's experiment in which the gas expanded in a cylinder doing work against a piston is the forerunner of refrigeration relying on isentropic expansion. Pictet allowed compressed oxygen cooled initially to about −100°C to expand through a nozzle producing the mist of liquid by virtue of the Joule–Thomson effect—gas liquefaction by isenthalpic expansion. Finally, Olszewski and Wroblewski laid the foundations of the cascade liquefaction method in which gases with successively lower critical points are first liquefied under pressure and then evaporated under atmospheric or subatmospheric pressure.

The next important developments took place round the turn of the century with the large scale liquefaction of air by Linde in Germany in 1895 using the Joule–Thomson isenthalpic expansion effect and by Claude in 1902 using isentropic expansion in a reciprocating engine. The foundations of a vast industry for the separation by distillation, first of oxygen and nitrogen from the atmosphere, and then of hydrocarbons were thus created. It is important to realize that it was the heat exchanger and heat regenerator principle which enabled the small cooling effect of the isenthalpic expansion or of the isentropic expansion through the relatively small pressure ratios used by Claude, to bridge the large temperature gap of more than 3 to 1 from room temperature to the boiling point of oxygen (90°K), and the cooling processes to be operated continuously.

The next important step in the progress of cryogenics was the liquefaction of hydrogen by Dewar in 1898. Since the critical point of hydrogen is well below the lowest temperature attainable by means of condensed nitrogen or oxygen, Dewar used the Joule–Thomson effect applying it to hydrogen precooled by liquid air. Moreover, Dewar, largely thanks to the vacuum vessel named after him was able to determine several physical properties of many substances down to the temperature of liquid and solid hydrogen.

Helium, the most "permanent" of all gases, i.e. the one having the lowest boiling point, was first liquefied in 1908 by Kamerlingh Onnes in Leiden in a Joule–Thomson liquefier using liquid hydrogen for pre-cooling. This event marked the end of the first phase of the history of cryogenics, a period largely devoted to the development of gas liquefaction and of some of the basic low temperature techniques. While the liquefaction of helium was a great achievement, Kamerlingh Onnes' importance in cryogenics was of a more fundamental nature. By building up at the University of Leiden a Physics Laboratory (now named after him) devoted almost entirely to the study of phenomena at low temperatures he established a world centre which for many decades remained unrivalled and which set an example and a standard for many subsequent low temperature laboratories.

The second phase of the history of cryogenics may be taken as the period from, say from 1908, the liquefaction of helium, until the early 1950s. It was a period during which cryogenics in its wider sense was becoming more and more widespread, as research groups sprung up throughout the world with the specific aim of studying the properties of matter at very low temperatures. Several large low temperature laboratories were established in Canada, U.S.S.R., Germany, U.S.A. and England; they had fair sized liquefiers and could provide liquid hydrogen and liquid helium—in "bulk", as it were—for many experiments running simultaneously. All helium liquefiers of that period were using the Joule–Thomson effect with one exception: Kapitza in Cambridge succeeded in liquefying helium by employing an expansion engine without the use of liquid hydrogen pre-cooling.

Another important step during this period was the development of small— almost miniature—liquefiers which incorporated the measuring apparatus or equipment so that small quantities of liquid helium or liquid hydrogen, often only a few cm³, were sufficient for experiments of several hours. Simon, in 1925, conceived and built his desorption refrigerator in which the cooling produced by the desorption of helium from charcoal resulted in the bridging of the gap between the temperature of condensed hydrogen and the critical point of helium, thus permitting the liquefaction of small but useful quantities of helium. Miniature Joule–Thomson liquefiers for hydrogen and for helium were developed by Ruhemann in 1930 and finally in 1932 came Simon's single expansion helium liquefier, similar in conception to Cailletet's method but which, starting from the temperature of solid hydrogen, could produce sizeable quantities of liquid helium with relatively simple means.

An entirely new temperature range was opened up during this same phase by the magnetic cooling method, proposed independently by Giauque and by Debye in 1926 and first put into practice in 1933 by Giauque and McDougall in Berkeley, California, and by De Haas, Wiersma and Kramers in Leiden. This method, the magnetic analogue of the Cailletet or Simon process, consists in the isentropic demagnetization of a suitable paramagnetic salt which has previously been magnetized at the temperature of liquid helium and it enabled temperatures down to the order of $10^{-3}\,°K$ to be reached.

Finally, this same period saw a spectacular growth of the gas separation industry and, in particular, the setting up of plants for the production of oxygen in tonnage quantities.

A new era of cryogenics was ushered in in the late 1940s by the commercial development of hydrogen and helium liquefiers. The production of these liquids which till then was regarded almost as an art, practised only in highly specialized laboratories, became within a few years a well engineered, reliable process. In the case of helium this development was due first and foremost to Collins' expansion machine manufactured and marketed by A. D. Little Inc. Liquid helium became available to non-specialized laboratories and this had so great an effect on the spread of research at low temperatures that the jocular reference of the pre-1950 period of cryogenics as B.C. (before Collins) is both fitting and felicitous. In recent years several automatic helium liquefiers have come on the market and this fact, together with the easy availability of commercially supplied liquid helium, means that liquid helium temperatures are now just as accessible to non-specialized laboratories as were liquid air temperatures fifty years ago.

The main impetus for the construction of large hydrogen liquefaction plants came from interest in the production of heavy water by the distillation of liquid hydrogen and the use of liquid hydrogen as a rocket fuel, and both these developments, being of an industrial character, brought with them an increasing interest in the low temperature properties of a host of materials and devices.

This historical review shows the impressive growth of cryogenics over nearly 100 years, and it would be interesting to estimate the growth rate of cryogenics in monetary terms over this period. This however does not seem to be possible and we have to content ourselves with an analysis of recent years which puts the annual growth rate at between 10% and 15%.

III. The Principal Uses and Applications of Low Temperatures

What are the reasons for the upsurge in the interest for cryogenics and its practical applications since the 1940s? High temperatures, between a few hundred and a few thousand degrees Celsius are the normal environments for many industrial processes and one is tempted to reason that the very conditions and properties which make high temperatures so desirable would militate against the use of low temperatures. This is true up to a point. Thus the use of high temperatures for the conversion of chemical energy into mechanical or electrical energy in a heat engine has no low temperature counterpart. Similarly, in many industrial processes high temperatures are used to facilitate chemical reactions be it through change of phase—a fluid is a better medium for a reaction than a solid—through increased reaction velocity or because of the shift of chemical equilibrium. But low temperatures have advantages in other respects. A liquid may be preferable to a gas in some circumstances; low reaction velocity may be preferable to high re-action velocity. In both basic research and technology the high molecular order associated with low temperatures is of great importance. Finally there is the increased efficiency of some processes at low temperatures, often con-nected with enhanced molecular order. The applications of low temperatures will now be briefly described along the lines just indicated. The discussion will not be restricted to well established cases; potentialities, often amount-ing to nothing more than hopeful speculations will also be discussed—mainly to stimulate further thought.

A. SOME DESIRABLE PROPERTIES OF CRYOGENIC LIQUIDS

The separation of liquid mixtures by distillation has been practised for more than 1,000 years. Since the composition of a liquid is usually different from that of the vapour in equilibrium with it transfer through the liquid-vapour phase-boundary is accompanied by a change in concentration. Moreover, evaporation even at high rates is almost an equilibrium process and this makes high throughputs under near-ideal conditions possible. It is therefore not surprising that once the abundantly occurring gas mixtures, in particular air, were liquefied the same distillation process should have been applied to them.

For several decades after Linde had set up his first air separation plant in 1895 the resulting oxygen, although produced as liquid, used to be shipped as compressed gas, and it was not until the 1930s that the transport and storage of oxygen in liquid form became established. With the advent of more effi-cient and cheaper insulated containers this method has become widespread, e.g. for the transport of natural gas and of hydrogen. In fact the use of hydrogen as a rocket fuel would be unthinkable without its availability in liquid form, and the setting-up of tonnage liquefaction plants for hydrogen and the development of reliable and safe transport and storage of liquid hydrogen constitutes one of the major advances in modern cryogenic technology.

It may well be that because of the pollution associated with the use of petrol as fuel for motor cars, methane might replace it. It would be a great challenge to the cryogenic industry to prove that the distribution of liquid methane could be made just as safe and reliable, even on a small scale, as the distribution of petrol through filling stations.

Ultimately we may even see motor cars powered by methane–air or hydrogen–oxygen fuel cells which would need the development of retail distribution of liquid methane or hydrogen, a suggestion which in the 1970s looks less fantastic than did in the 1930s the idea of manufacturing and shipping tons of liquid hydrogen.

Finally the bulk-supply of helium in liquid form is becoming important especially in countries with no indigenous sources of helium. For the U.K. and presumably for some other European countries this may become essential if helium is to be used as heat transfer agent in advanced high temperature nuclear power station. A rational integrated bulk-supply-system for helium for all industrial and research needs may then be desirable and this might also provide all the *liquid* helium used for low temperature work, i.e. in 1970 in the U.K. about 30,000 kg corresponding to 240,000 l of liquid or 6·5 million ft^3 of gas at normal temperature and pressure. The gas would afterwards be returned to central purification plants to be reprocessed for use in nuclear reactors and for other purposes.

It should also be remembered that in evaporating and reheating to room-temperature a liquefied gas, the mechanical and electrical energy that could be gained if the process were carried out near-reversibly is not negligible. To give an idea: the evaporation of 10^6 tons of liquid natural gas (LNG) per year—this is the order of magnitude of the throughput of the Canvey Island LNG depot—would produce 10 MW of electrical power.

B. COOLING AS A MEANS OF SLOWING DOWN REACTION VELOCITIES

This is perhaps the oldest application of low temperatures and was originally the chief stimulus for mechanical refrigeration since the preservation of food by cooling or freezing or deep-freezing is based on the slowing down of chemical reactions. Since about 1950 the interest in long-term preservation of biological material and of simple living organisms has been steadily increasing. Long-term preservation, i.e. with virtually no change during storage lasting for months or years, involves complete freezing and cooling to temperatures well below those of ordinary refrigerators and deep-freezers and relies usually on liquid nitrogen. However for the biological material not to sustain, during freezing, physical damage due to the formation of ice-crystals or chemical damage due to concentration changes, it is often necessary to perfuse it with some protective agent such as ethylene glycol or dimethylsulphoxide. In this way it has been possible to preserve in a reversible fashion a wide variety of simple biological material such as white blood cells, sperm, bone marrow and cartilage. Great efforts are being made to extend these techniques to more complex systems and eventually to whole organs. If this were successful "banks" for kidneys, livers, etc., could be established and the task of matching donor and recipient in transplant operations would become much easier.

While it has been possible by cooling to just below 0°C to arrest all meta-bolism and to keep certain mammals, e.g. hamsters and galagos as it were in "suspended animation", biochemical reactions were only slowed down but not arrested. These experiments however brilliant and exciting in themselves should therefore not be regarded as heralding in an era of reversible suspen-sion of life for long periods. It is worth emphasizing this point in view of recent reports about deep-freeze preservation of corpses in the hope of eventual resuscitation. Since the time elapsed between death and deep-freezing in the reported cases amounted to hours the irreversible brain damage —to say nothing of the physical and chemical damage due to freezing—rules out the possibility of eventual revival. It may just be that if a technique were found to perfuse *before death*, preferably while the subject is in good health, the entire body with a protective agent which could be removed during thawing ultimate revival might become possible but at present such speculations belong to the realm of science fiction. One might laugh off such wild claims were it not for the fact that they are liable to raise false hopes resulting in bitter and painful disappointments.

The lengthening of the time-constant of physical and in particular atomic processes also comes under the same heading of the effect of temperature on reaction velocities. A long relaxation time of energy exchange is equivalent to a sharpening of energy levels and hence narrower spectral lines range. This explains the widespread use of low temperatures both in optical and micro-wave spectroscopy to obtain higher resolution and thus to observe effects which otherwise might be indiscernible. Long relaxation times also make it possible to introduce the notion of thermally insulated "systems" for the description of certain properties. For instance in a dielectric crystal the system of elementary magnets associated with the atomic nuclei may be regarded as thermally insulated from the system of lattice vibration even though the two systems are not separated in space.

Such an *isolated* system of elementary magnets has the remarkable property that its energy cannot increase indefinitely. In an external magnetic field the energy of an individual magnetic dipole can vary only between a lower limit corresponding to it pointing parallel to the field and an upper limit corresponding to an antiparallel orientation. At absolute zero all dipoles are in their lowest energy state pointing in the direction of the field. The system is magnetized to saturation and the dipoles are in a perfectly ordered state corresponding to zero entropy. As energy is gradually introduced into the system more and more dipoles will be forced to take up a direction of higher

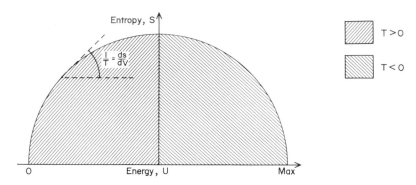

Fig. 1.1. Entropy as a function of energy for a system having a bounded energy spectrum.

energy, i.e. at an angle to the field: the temperature rises. Finally when the orientations are completely random, with equal number of dipoles pointing in all directions permitted by quantum theory, the system may be said to have reached its maximum possible molecular disorder and the corresponding temperature is infinite. It is thus possible to bring such a system having a bounded energy spectrum to $T = \infty$ with a *finite* amount of energy.

If the system is forced to absorb more energy than is needed to bring it to $T = \infty$, the additional energy can only be taken up by more dipoles pointing *opposite* the magnetic field. In other words the system which at $T = 0$ was magnetized to saturation and at $T = \infty$ had no resulting magnetic moment is now magnetized opposite to the direction of the external field corresponding to a *negative* absolute temperature which, it must be remembered, is *higher* than $T = \infty$. The highest energy that the system can assume corresponds to all dipoles pointing antiparallel to the field. Once again the system is in a perfect state of order; its entropy is zero and we have an absolute zero corresponding to the negative temperature range.

Fig. 1.1 illustrates the thermodynamic behaviour of such a system and shows in particular that absolute zero, unlike $T = \infty$, is a two-valued singularity of the temperature scale. Negative absolute temperatures which were first realized in 1951 represents a useful concept in nuclear spin and electron spin resonance and, more specially, in maser and laser physics. Although they have no place in conventional cryogenics a brief discussion of the concept of negative absolute temperatures is justified since it shows an important departure from the law of equipartition of energy which postulates a proportionality between absolute temperature and the mean energy of the particles of a system.

C. THE CREATION OF MOLECULAR ORDER

Low temperatures, decreasing entropy and increasing molecular order go hand in hand, and the study of phenomena that accompany the transition into an ordered state are among the most important and certainly most thrilling applications of low temperatures in physics. The third law of thermodynamics postulates that at absolute zero—which is unattainable—molecular order must be perfect. Often the transition into the ordered state is a gradual process, as for instance is the case of the lattice vibrations in a crystalline solid or of the conduction electrons in many metals. In other cases this transition is accompanied by marked changes in physical properties. The self-ordering of elementary magnets which gives rise to ferromagnetism is a well known example occuring at ordinary or elevated temperatures. Nevertheless low temperatures have helped our understanding of ferromagnetism since it enabled these studies to be extended to other metals besides iron, cobalt and nickel, such as the rare earth metals which have Curie points below room temperature. Some of this work may also have practical consequences since the saturation magnetization of certain rare earth metals and alloys is larger than that of Fe and Co alloys and thus may permit the construction of considerably smaller electromagnetic machinery—a fact which may have some importance for certain applications.

There are two phenomena, both of them marking the transition into the ordered state, that are wholly associated with very low temperatures. One of these is superconduction, the *complete* disappearance of electrical resistance in certain metals and alloys. Discovered in 1911 superconduction remained for nearly fifty years the object of mainly academic interest. Its study contributed much to our understanding of the interactions that govern the behaviour of electrons in metals and, more particularly, it showed that ordering need not be geometrical, i.e. positional or orientational, but could occur with respect to velocities or momenta resulting in such remarkable phenomena as the loss of electrical resistance. Since the discovery in 1960 of alloys which retain their zero resistance even under high current densities and in strong magnetic fields superconduction has become potentially the most important industrial application of very low temperatures, i.e. those attainable by liquid helium.

Another ordering phenomenon, in many ways akin to superconduction occurs in liquid helium itself, which, however far it is cooled under its vapour, does not solidify. Now it is difficult to conceive geometrical order in a liquid and indeed, similarly to what happens in superconduction, ordering sets in at a well defined temperature with respect to the momenta. The result is the so-called helium II modification or superfluid helium with vanishing viscosity, a many millionfold enhanced thermal conductivity and several other fascinating properties unencountered anywhere else.

D. THE REDUCTION OR ELIMINATION OF WASTEFUL DEGRADATION OF ENERGY

The transmission of energy or the performance of useful work usually involves the concerted motion of particles—e.g. the electrons in the case of an electric current, molecules in case of the flow of a fluid. Random molecular motion or thermal agitation hinders such an ordered process and the result is a degradation of energy into heat—electrical resistance or viscous dissipation. By reducing temperature and thereby thermal agitation one can hope to reduce inefficiency and waste and the question is whether if the process is carried out at low temperature instead of at ordinary temperature the resulting gain is larger than the increased costs. Joule heating is probably the most important example of the reduction of wasteful energy dissipation by cooling. Since the resistivity of pure metals drops sharply with temperature—the resistivity of commercially produced pure aluminium at $20°K$ is about one thousandth of its room temperature value—the possible gain appears very large at first sight. However, this albeit greatly reduced heat output must now be removed artificially, while, if it had occurred at ambient temperature it could have been, at least in principle, dissipated by natural convection or conduction, and therefore at no cost. The ratio of the consumption power to heat removal rate for an ideal heat engine is $(T_0/T - 1)$ where T is the temperature of the electrical conductor and T_0 is the ambient temperature. With $T_0 = 300°K$ and $T = 20°K$ one finds that the overall power consumption, i.e. Joule loss plus refrigerating power, when using an aluminium conductor at $20°K$ is only $1·5\%$ of what it would be at room temperature. This is a very idealized picture since it does not take into account the actual efficiency of the refrigerators and the capital cost of the complex low temperature installations, but nevertheless it indicates the great energy savings that can be effected by working at low temperatures.

Joule losses in electrical conductors could in principle be completely eliminated by the use of superconductors, but a substantial refrigerating power would still be needed to neutralize heat influx from the outside and in the case of a.c. the energy dissipation occurring especially at high current densities.

There are several areas of heavy electrical engineering in which low temperatures could be very useful. The most important in principle—though perhaps not in practice—is power transmission, where at present something like one part in 10^5 of the power transmitted is wasted per km. Although according to our present knowledge underground cables cannot compete under normal conditions with overhead powerlines there are cases when undergrounding is necessary or highly desirable. Leaving aside undergrounding for aesthetic reasons, often demanded by those who while drooling over examples of fifteenth century technology, e.g. windmills, refuse to see any beauty in the manifestations of our twentieth century civilization, we are faced with the problem of built-up areas which must be supplied by underground cables. With the increased demand for electric power it is no longer practicable to have transmission cables simply buried in the ground allowing the heat to seep out into the surrounding soil; forced cooling will have to be adopted more and more, the heat removed from the cable being dumped at suitable places into the atmosphere or some other heat sink. Once the need for forced cooling is admitted the question is at what temperature it should occur so as to provide in the most economical transmission.

The answer depends on many factors, e.g. the cost of refrigeration—to be discussed in detail in Section IV—the variation in the cost of the actual cable, questions of reliability, maintenance, etc. The problem is further complicated by the fact that estimates have to be made on the basis of the relatively new cryogenic technology with barely fifteen years of research and development behind it and without being able to guess whether increased demand over the next years may result in a dramatic downward plunge in the cost of cryogenic equipment. Estimates made in 1970 indicate that the two main contenders for low temperature power transmission are aluminium at about 80°K and a superconductor, e.g. niobium, at 4°K, with aluminium at 20°K occupying third place. Beryllium has interesting possibilities since its resistivity at 80°K is only $\frac{1}{65}$ of its room temperature value; on the other hand its high cost (at present), its brittleness and its toxicity are strong arguments against its use.

Less hopeful according to present indications is the use of superconductors in transformers. On the other hand, aluminium-wound transformers cooled by liquid hydrogen to 20°K might be advantageous and an experimental 100 kW transformer has recently been built and tested in France by L'Air Liquide and Alsthom.

It seems that as regards motors and generators running at low temperatures the interest is entirely in superconductors. The emphasis here is not so much on energy saving but rather on the other advantages accruing from the use of superconductors, viz. high torque at low speeds, and greatly reduced dimension for a given power. A 3,000 horsepower homopolar motor, having a superconducting stator and a room temperature copper disc for rotor has been designed and built by International Research and Development (IRD)

Ltd. in England and if the tests it is undergoing in industrial surroundings are successful there is likely to be a growing market for such machines with power of 10,000 h.p. or more in several industrial fields, e.g. in rolling mills.

IV. The Cost of Refrigeration

To be able to decide whether a process could be carried out more economically at low temperatures than at ambient temperature the cost of refrigeration must be known. At the time of writing, refrigerators for temperatures below 25°K are still mainly tailor made to suit the requirements of the individual customer. The exceptions are helium liquefiers which during the decade 1960–1970 have been supplied in large numbers to research laboratories —although with the improvement in the supplies of liquid helium in bulk this tendency seems to be on the decline.

The efficiency of refrigerators expressed as a fraction of the ideal thermodynamic or Carnot efficiency depends on many factors, the chief ones being the working temperature, T, and the refrigerating performance or heat removal rate, q. The power required to run the refrigerator, w_r, is then given by

$$w_r = q\left(\frac{T_0}{T} - 1\right)\frac{1}{\eta},$$

where η is the efficiency. One finds from a survey of existing installations that for $T = 4.5°K$ η varies between 0·1 and 0·3, for $T = 20°K$, between 0·2 and 0·4 and for $T = 80°K$ between 0·4 and 0·5. Figure 1.2 shows the power consumption per unit refrigeration performance w_r/q. These curves enable us to estimate the saving in total power consumption w_t, e.g. when running an electric cable at low instead of at ambient temperature. We shall assume a power dissipation at ordinary temperature of $q_{300} = 100$ kW, which roughly corresponds to a 10 km long 1,000 mVA transmission line. Since $w_t = w_r + q$ and $q = q_{300}(\rho/\rho_{300})$, where ρ and ρ_{300} are the resistivities at T and at $T_0 = 300°K$ respectively we have

$$\frac{w_t}{q_{300}} = \frac{\rho}{\rho_{300}}\left(1 + \frac{w_r}{q}\right)$$

For beryllium at 80°K, $\rho/\rho_{300} = 0.013$, and since according to Fig. 1.2 $w_r/q = 7$ when $q = 0.013 \times 100 = 1.3$ kW, we obtain $w_t/q_{300} = 0.1$.

For aluminium at 20°K, $\rho/\rho_{300} = 0.15$, $q = 60$ kW, $w_r/q = 6$, hence $w_t/q_{300} = 0.1$.

All this shows that by running a cable at low temperature, e.g. a beryllium cable at 80°K or an aluminium cable at 20°K net power losses could be reduced by as much as 90%. This calculation neglects the unavoidable heat-leaks from the outside into the core of the cable. Imperfect thermal insulation will make the saving less. More important is the effect of the capital cost of the refrigeration plant. Estimates of these costs as a function of working temperature and size are even more speculative than those for efficiencies. In particular much depends on the amount of controls and automation required and on the specification for the length of continuous running. The curves of Fig. 1.3 claim no accuracy and are given more as an illustration.

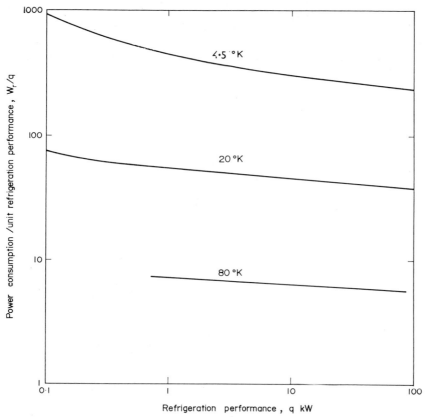

Fig. 1.2. Power consumption of cryogenic refrigerators related to size and operating temperatures.

The dependence of capital cost C on size is similar for 80°K and 20°K being roughly proportional to $q^{-\frac{1}{3}}$, while for refrigeration at 4·5°K the cost varies more like $q^{-0.6}$. The following three relations (C in £/W, q in kW) may be found useful for a rough assessment. The error, in short, may be as much as a factor 2 either way.

$$T = 80°K \quad \log_{10} C = -0.33 \log_{10} q + 0.5 \pm 0.3$$
$$T = 20°K \quad \log_{10} C = -0.33 \log_{10} q + 1.45 \pm 0.3$$
$$T = 4.5°K \quad \log_{10} C = -0.6 \quad \log_{10} q + 2.25 \pm 0.3$$

With the help of the graphs of Fig. 1.2 and 1.3, it is possible to estimate the cost of refrigeration R expressed in £ per kWh at various temperatures and various heat removal rates. R is made up of running costs and capital depreciation (based on a life of 10 years and interest at 8% per annum) but does not include the distribution of the cooling along the cable. By making some assumptions about the refrigeration performances q one is likely to want at various temperatures, e.g. 0·5 kW, 2 kW and 10 kW at 4·5°K, 20°K and 80°K, respectively, one can establish the relation between R (£/kWh) and T. It is given by

$$\log_{10} R = -1·5 \log_{10} T + 1·6.$$

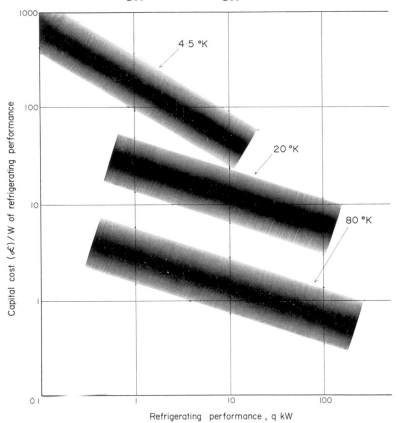

Fig. 1.3. Approximate capital cost of cryogenic refrigerators as a function of size and operating temperature.

Because of the many arbitrary assumptions about refrigeration performances, period of amortization, interest rates, maintenance, and because of the uncertainty of the graphs on which this relation is based it would be pointless even to try to assess the likely margin of error.

According to this relation, the cost of refrigeration of an aluminium cable at 80°K would be about the same as the saving in power dissipation compared with a room temperature cable. But as said before the comparison should be made with a forced-cooled room temperature cable and if the capital cost and running cost of the room temperature cooling plant are taken into account the balance may well tip in favour of the low temperature cable.

V. Conclusion

We have seen that over the last four decades the use of low temperatures have penetrated many scientific, technological and industrial activities. The use of liquid air or nitrogen in practically every scientific laboratory is so commonplace that it is hardly worth mentioning; but for reasons explained in Sections IIIB and IIIC increasing use is made of helium in many fields of research. Even if the actual investigation does not call for low temperatures, ancillary apparatus based on superconductivity may need a reliable and preferably cheap supply of liquid helium. Most important in this connection are superconducting magnets ranging from the small laboratory models weighing barely 1 kg to the large magnets weighing tens of tons used in or planned for high energy physics, e.g. for bubble chambers or particle accelerators.

The potentialities of low temperatures in industry are great but are as yet mostly unrealized. The future of the large scale transmission of electricity at low temperature, of superconducting electric motors and generators, of improvements in the technology and manufacture of superconducting materials seems to be in a vicious circle with the necessary cryogenic developments. Not until there is a likelihood of a large and developing market for its products will cryogenic industry be prepared to invest the money necessary to bring helium-refrigerators to the level of reliability and compactness of, say, refrigerators for deep-freezers. The same consideration applies to manufacturers of superconducting materials. On the other hand potential users of low temperature processes are chary to base ambitious developments on a new technology which to them seems barely to have reached adolescence.

To break this vicious circle a new attitude of mind regarding cryogenics is called for. First of all it must be realized that the wasteful irreversible dissipation of energy is not ordained by any law of nature, that it can be reduced and even made to vanish by various means and in particular by low temperatures. Nor is there any reason why the additional cost of producing and maintaining the low temperature environment should not be more than compensated by the increased efficiency due to cooling.

But, more generally, it should be emphasized that in view of the advanced state of the cryogenics industry and of its potentialities, a low temperature environment for any industrial process should no longer be regarded as a last resort but as a serious contender worthy of detailed economic analysis and forecasting.

Chapter 2

Refrigeration and Liquefaction Cycles

G. G. HASELDEN

Department of Chemical Engineering, University of Leeds, Leeds, England

I. Introduction

In designing or assessing any cooling system, including one which involves liquefaction, it is important to separate the duty and the cycle used to perform it. The need arises because thermodynamic irreversibility, or power wastage, may occur not only in the cooling cycle itself, but also in matching the cooling cycle to the duty. All such irreversibilities will lead to excess energy requirement, and also, in some cases, to additional investment in plant.

Thus a duty involving condensation of a pure gas (at constant temperature) which uses a single phase refrigerant stream accepting sensible heat through a temperature range (for instance, cold gas from an expansion engine) is bound to involve a high entropy gain in the heat exchanger employed, due to the mis-match of the temperature profiles. Therefore the power required will be well above the thermodynamic minimum although the expansion engine itself may have a very high efficiency. The separation of duty and cooling cycle is particularly necessary in liquefiers in which the fluid being liquefied is also the refrigerant.

The selection of the ideal refrigerator for a specified duty will depend on a number of factors in addition to power consumption. Capital cost, reliability, safety and ease of control will generally enter into the decision, and any of these factors may dominate in special circumstances. The capital cost will be important in most cases, but whereas in many forms of plant increased efficiency is only achieved by additional complexity and cost, in low temperature processes this may not be so. In cooling systems the compressor with its prime-mover is often the most expensive plant item. Reduced power consumption will generally mean a smaller compressor and cheaper motor, and these savings may substantially off-set extra costs elsewhere. The extent to which the power consumption dominates both capital and running costs increases with the scale of operation. Excessive complexity must be avoided, not only because it increases capital costs, but because it conflicts with reliability.

Cooling cycles using gases at atmospheric pressure, or below, will generally be more expensive than those working at pressures of, say, 5 atm and above. The reasons are at least two-fold. High pressure plants will be smaller for the same duty since the working media are denser. Thus pipes and vessels which are strong enough to withstand normal wear and tear (i.e. will not be damaged in transit, convenient for jointing and rigid structurally) will almost always be strong enough to contain pressures of at least 5 atm. The exceptions are large vessels such as storage tanks, or vessels having flat surfaces. For pressures above 5 atm the amount of metal required will not normally increase in proportion to the pressure. Secondly, heat transfer will usually be assisted by the use of high pressure, and therefore a smaller area of transfer surface will be required for the same duty. For similar reasons cycles using fluids which are partly in the liquid phase, will generally be more compact than those involving only gases, but there are exceptions which will be mentioned.

II. Refrigeration Duties and Minimum Work

A. ISOTHERMAL DUTIES

The simplest cooling duty is that involving the removal of heat at constant temperature, as in the condensation of a saturated gas—say atmospheric pressure nitrogen gas at $77°K$ ($-196°C$). The thermodynamic minimum work of cooling is found very readily from a Carnot cycle operating between this temperature level and a heat sink, which will be taken as cooling water at $303°K$ ($30°C$).

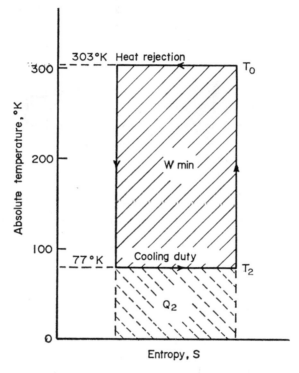

Fig. 2.1. Carnot cycle for a constant temperature cooling duty (for instance, condensing saturated nitrogen vapour).

Fig. 2.1 shows the temperature–entropy diagram for the Carnot cycle. The area within the rectangular cycle diagram represents the theoretical work requirement W_{min}, whilst the area below the $77°K$ heat acceptance line, down to the $T = 0$ axis, represents the cooling duty Q_2. If the heat rejected is Q_0, then by the First Law

$$Q_0 = Q_2 + W_{min}.$$

2

Thus the total heat rejected to the heat sink is given by the total area of the diagram below the heat rejection line.

From the geometry of the diagram it is evident that

$$\frac{Q_2}{W_{min}} = \frac{T_2}{T_0 - T_2} = \beta_{max} \tag{2.1}$$

The ratio Q_2/W is defined as the coefficient of performance β. Equation 2.1 gives the minimum work requirement, or the maximum coefficient of performance, for any isothermal duty. For condensing nitrogen at atmospheric pressure it is seen that $\beta_{max} = 77/(303 - 77) = 0.34$.

Corresponding values for other fluids are:

$$\text{Freon-12 (b.p. } 243°\text{K) } \beta_{max} = 4.05$$
$$\text{Ethylene (b.p. } 169°\text{K) } \beta_{max} = 1.26$$
$$\text{Hydrogen (b.p. } 20.4°\text{K) } \beta_{max} = 0.072$$
$$\text{Helium (b.p. } 4.2°\text{K) } \beta_{max} = 0.014.$$

These figures illustrate the rapid rise in the power requirement for refrigeration as the temperature level goes down. Thus at normal refrigeration levels (243°K or −30°C) the expenditure of 1 kW could potentially produce more than 4 kW of cooling whilst at nitrogen temperatures the potential cooling is only $\frac{1}{3}$ kW, and in the helium region only 14 W. In practice it is found, as will be shown later, that the efficiency of refrigerators (related to Carnot performance) normally decreases as the temperature differential between heat absorption and rejection increases, hence practical refrigeration performances diminish even more rapidly as the temperature goes down.

B. COOLING THROUGH A TEMPERATURE RANGE

Frequently the required cooling duty is that of taking a given quantity of a specified material and reducing its temperature from an initial temperature T_1 to a final temperature T_2. This may be done batchwise or in a flow system (in which case the quantity is related to unit time). If the material has a specific heat at constant pressure C_p, which may for instance be a linear function of temperature over the specified temperature range, $C_p = a + bT$, then the cooling duty is given by

$$Q = \Delta H = \int_{T_1}^{T_2} C_p \, dT = \int_{T_1}^{T_2} (a + bT) \, dT = a(T_2 - T_1) + \frac{b}{2}(T_2^2 - T_1^2) \tag{2.2}$$

As shown in Fig. 2.2, which is a T–S diagram for the process material, the minimum work of cooling may be found by considering that every increment of cooling, dT, has its own incremental Carnot cycle, hence

$$W_{min} = \int_{T_1}^{T_2} \left(\frac{T_0 - T}{T}\right) C_p \, dT = T_0 \Delta S - \Delta H \tag{2.3}$$

which, for the case of C_p being a linear function of temperature, yields

$$W_{min} = aT_0 \log_e\left(\frac{T_2}{T_1}\right) - (a - bT_0)(T_2 - T_1) - \frac{b}{2}(T_2^2 - T_1^2) \qquad (2.4)$$

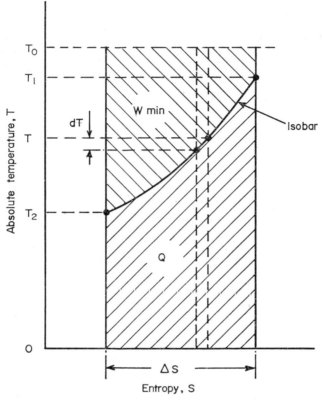

Fig. 2.2. Cooling duty and work requirement for non-isothermal heat removal.

The shaded area below the isobar joining T_1 and T_2 down to the $T = 0$ axis represents the total cooling duty $Q(= \Delta H)$, whilst the area above the isobar up to $T = T_0$ gives W_{min}. This analysis can be applied to any isobaric process whether the working substance is solid, liquid or gaseous, or of mixed phase. It should be noted that even when C_p is constant the isobar on the T–S diagram is not straight, and this fact is particularly relevant at very low temperatures.

C. MINIMUM WORK OF LIQUEFACTION

The minimum work of liquefaction is best demonstrated by reference to the T–S diagram for the given fluid, see Fig. 2.3. For a typical gas, initially at atmospheric pressure and ambient temperature A, say 293°K (20°C), it is apparent that liquefaction involves firstly cooling the gas to its dew point B, and then removing latent heat isothermally to the bubble point C. Thus the total cooling duty H is made up of two components, as is also the minimum work of liquefaction (given by $W_{min} = T_0 \Delta S - \Delta H$ for each stage).

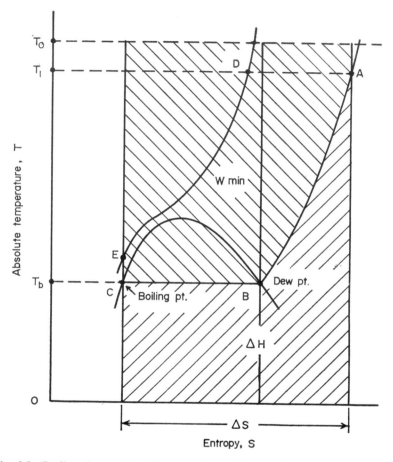

Fig. 2.3. Cooling duty and minimum work requirement for a liquefaction process. ABC—Atmospheric pressure isobar. DE—High pressure isobar.

Values of these separate and composite quantities are presented for a number of gases in Table 2.I, T_1 being 293°K and T_0 being 303°K in each case. It will be seen that for ethylene and methane the minimum work of condensation is high compared to that for removing sensible heat from the gas, whilst for hydrogen they are nearly equal and for helium the reverse is the case. Clearly this factor should have a considerable bearing on liquefaction cycle design.

Further, reference to a $T–S$ diagram for a typical cryogen (Fig. 2.3) shows that liquefaction may be achieved by two alternative routes. The route thus far described keeps the fluid at atmospheric pressure and involves the sensible and latent heat stages. However, if the feed gas is first compressed to D at a super-critical pressure (in theory this can be done isothermally and reversibly) then the cooling path down a super-critical isobar will involve no discontinuity, since there will be no discernible phase change. Instead, for the real fluid, the specific heat of the high pressure gas will have become uniformly higher throughout the whole cooling range (as shown by the closer spacing of the isenthalps as they intersect the isobar). If cooling is stopped at E when the entropy of the compressed gas falls to the entropy of the saturated liquid, then a final engine expansion may (reversibly) convert the cooled gas to saturated liquid.

TABLE 2.I. Heat Removal and Minimum Work of Liquefaction for a Number of Gases

Gas	Normal boiling point °K	ΔH kJ/kg			Minimum work kJ/kg		
		Cooling	Latent heat	Total	Cooling	Condensation	Total
Ethylene	169·4	163	485	648	75	396	470
Methane	111·8	397	512	908	277	870	1,147
Oxygen	90·2	197	213	410	133	498	631
Nitrogen	77·3	224	199	423	197	580	777
Neon	27·3	275	86	361	482	859	1,341
Hydrogen	20·4	3,413	434	3,847	6,100	6,090	12,190
Helium[4]	4·2	1,509	21	1,530	6,901	1,488	8,389

This alternative route is of great interest in liquefier design since it involves a simplified refrigeration (or cooling) step with no latent heat removal. Thus by using a super-critical feed the refrigeration duty is more easily matched to that generated over a temperature range by an engine or throttle expansion system. In practical systems the final expansion is carried out in a valve rather than an engine, but the resulting energy loss is not serious in most cases.

III. Cooling Methods

A. INTRODUCTION

Cooling methods are basically of two types. The first involves inducing a change of state in the working substance at ambient temperature such that its enthalpy is reduced, and heat is rejected to a heat sink. Then the change of state is reversed at low temperature and heat is absorbed so as to produce refrigeration. Alternatively, the refrigerant is made to perform work at low temperature, when the cooling produced will normally be about equal to the work done. Where the working substance is a fluid the process can normally be made to take place continuously. A further common feature of almost all these methods is the use of a heat exchanger to extend the temperature range between heat acceptance in the refrigerator and its rejection to the heat sink.

B. CONTINUOUS FLOW SYSTEMS

1. *Engine Expansion*

When a compressed gas is made to expand within an engine, which may be of the positive displacement type (reciprocating, for instance), or may harness a change of momentum (turbine), work is generated and the gas cools down.

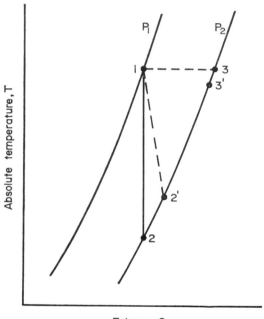

Fig. 2.4. The cooling effect produced by engine expansion of gas from a pressure P_1, to a pressure P_2.

Referring to Fig. 2.4, which shows two isobars on a T–S diagram, if the gas expands adiabatically and reversibly it will follow the path 1–2. The process will be isentropic and the work generated per unit mass of gas for a continuous flow or cyclic system will be ΔH. The useful cooling produced will normally be equal to the heat accepted by the expanded gas as it warms along the P_2 isobar from 2→3 (given by the area beneath 2–3 down to the $T = 0$ axis. For the case of a perfect gas only, and when $T_3 = T_1$, the cooling duty will equal the expansion work, since $H_3 = H_1$.

Magnitudes of temperature drops for an ideal engine, and a gas of average $\gamma = 1\cdot4$, are shown in Table 2.II. In each case the gas starts at 293°K and expands to atmospheric pressure. For any practical engine operating between the same pressure limits the state of the gas after expansion will be given by a point 2' whose temperature is higher than T_2, and the work output $\Delta H_{2'-1}$ will be lower. The efficiency of the engine expansion should be defined by the ratio $\Delta H_{2'-1}/\Delta H_{2-1}$ and not by the corresponding temperature ratio, since for real gases an engine with zero work output (zero efficiency) will still, in general, give a temperature change. The expansion is still assumed adiabatic but is it not reversible, and the path cannot be represented on a T–S diagram. Also in practical systems it will not be possible to have a useful cooling duty extending to T_3 since a temperature head will be required in a heat exchanger, hence the useful cooling duty will be restricted to $\Delta H_{3'-2'}$. The construction and operating characteristics of various expansion engines are described in Chapter 8.

TABLE 2.II. Temperature Drop for Ideal Engine Expansion to Atmospheric Pressure from a Range of Starting Pressures at 293°K, assuming $\gamma = 1\cdot4$

Initial pressure bar	Final temperature °K	Temperature drop °K
2	240	53
5	184	109
10	152	141
50	96	197
100	78	215

2. Throttle Expansion

When a gas expands adiabatically without performing any external work (throttle expansion), it will generally experience a temperature change, called the Joule–Thomson effect. The terminal conditions for a throttle expansion, though *not* the path between them, will lie on an isenthalp. By inspecting the slopes of the isenthalps of the given fluid on a T–S diagram the regions in which a throttle expansion gives a useful cooling effect are readily apparent, see Fig. 12.20. At high temperatures, say $>3T_C$, the isenthalps are rather flat, and a specific value of pressure exists above which expansion will lead to a temperature rise, and below which there will be a temperature drop. As the starting pressure falls to near atmospheric the temperature drop becomes vanishingly small. At lower absolute temperatures the isenthalps become more peaked, and near the critical region very large temperature changes can be achieved by throttle expansion. This behaviour continues into the 2-phase region. The locus of the pressure values at which the Joule–Thomson coefficient $(\partial T/\partial P)_H$, is zero defines the inversion curve shown in Fig. 2.5.

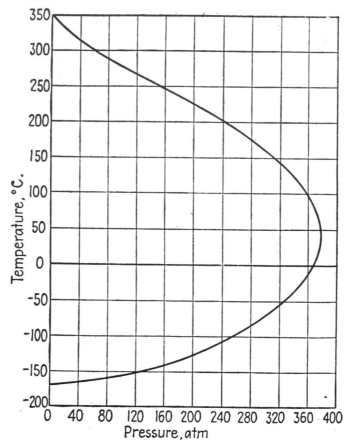

Fig. 2.5. The inversion curve for nitrogen.

A positive value of the coefficient implies cooling, and a negative value heating, on throttle expansion. From the law of corresponding states, which for simple gases, may be assumed adequate for estimation purposes, it follows that a generalized inversion curve based on reduced temperatures and pressures can be drawn.

It is useful to have an order-of-magnitude appreciation of Joule–Thomson cooling. Thus for ethylene at a pressure of 50 bar and 293°K expansion to atmospheric pressure will produce a temperature drop of 99°K, whilst for nitrogen the corresponding value is 11°K, and helium gives a temperature rise of 3°K.

Except in cases where throttle expansion enters the 2-phase region the product is cold gas, and the refrigerating effect corresponds to the heat it will absorb whilst returning to the pre-expansion temperature—as with engine expansion.

3. *Vapour compression*

The vapour compression cycle is used in most domestic refrigerators. Although it is dependent on throttle expansion it is useful to differentiate it from the normal Joule–Thomson effect. It relies basically on the fact that the vapour pressure of a fluid varies with temperature, and therefore a boiling liquid can be made to absorb heat at one temperature and reject heat at a second higher temperature by compressing and condensing the vapour thus generated. It differs from Joule–Thomson cooling in being essentially isothermal.

The simple vapour compression cycle is shown in Fig. 2.6 and the corresponding *T–S* diagram in Fig. 2.7. In the latter it is apparent that the refrigeration duty is given by the area under 4–1 whilst the heat rejected is given by the area under 2–3. Hence the net work of compression is given by the difference between these areas. The throttle expansion and the practical compression process are both indicated by broken lines, since neither process can be represented on a thermodynamic diagram.

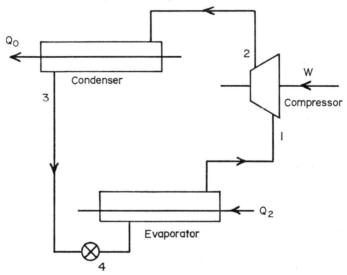

Fig. 2.6. The vapour compression cycle.

Calculations for the cycle are best made by reference to a pressure–enthalpy chart or table for the chosen refrigerant. Firstly, the evaporator and condenser pressures are established from the lowest process temperature required and the outlet temperature of the coolant, allowing appropriate ΔT's

2*

for heat exchange in both cases. Then the rate of refrigerant circulation is found by dividing $(H_3 - H_1)$ into the cooling duty. Finally the work of compression is found by multiplying $(H_2 - H_1)$ by the refrigerant circulation rate.

The vapour compression cycle is suited to the removal of heat at constant temperature, and this is also true of heat rejection. The cycle can be idealized by replacing the throttle valve with an engine expansion 3′–4′ and by compressing wet vapour from 1′ to 2′ isentropically. The idealized cycle is then equivalent to a Carnot cycle. In practice, even with throttle expansion and vapour superheat after compression, efficiencies of 60–85 % related to Carnot are achieved. Hence the complication of engine expansion, and the hazard of wet compression, are not attempted.

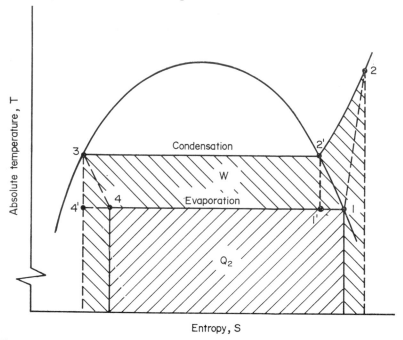

Fig. 2.7. Temperature–entropy diagram for the vapour compression cycle, showing the duty Q_2 and the work requirement W.

In addition to being efficient, especially for isothermal duties, the vapour compression cycle is generally low in capital cost. The rates of heat transfer in boiling and condensation are very high compared to those achievable with gases, thus requiring small heat exchangers. The latent heats of refrigerants are high compared with sensible heats, thus giving low flow rates and small pipe sizes.

To adapt the vapour compression cycle to duties involving a temperature range compression can be staged and a number of evaporators employed at different pressures (and temperatures). Alternatively the refrigerant may take the form of a non-aerotropic mixture. In the co-current evaporation of such a mixture, although the total composition remains constant, the boiling point of the liquid solution will continuously rise from the bubble point to the dew point. A similar temperature gradient will tend to exist in the condenser. The use of the mixed refrigerant cycle will be illustrated in relation to natural gas liquefaction.

The temperature range usefully achievable with any pure refrigerant system is determined by its thermodynamic properties. From the T–S diagram Fig. 2.7, it is seen that as the condenser pressure approaches the critical the losses associated with compression and throttling become large, and the overall cycle efficiency therefore diminishes rapidly. It is primarily for this reason that carbon dioxide is now no longer favoured as a refrigerant. Also the evaporator pressure is not normally allowed to fall much below atmospheric, partly due to the risk of ingress of air, and partly because this causes the plant to be large. To reach low temperatures it is necessary to employ a cascade system in which the evaporating first-stage refrigerant is used to condense the lower boiling second-stage refrigerant.

It is worth noting that the absorption refrigeration cycle is simply a variant of the vapour compression cycle in which the refrigerant is conveyed from the region of low pressure to the region of high pressure by dissolving it and pumping the solvent. For a high solubility system the volume of solvent pumped is far less than the volume of vapour, and hence the work required is much lower. In its place there is a need for a substantial quantity of heat to regenerate the refrigerant from solution at the higher pressure. There is more scope than is generally realized for the use of absorption refrigeration systems for first-stage refrigeration levels down to say $-60°C$.

4. *Vortex Tube*

The vortex tube was apparently invented by Ranque in 1933, although it received little attention until publicized in a paper by Hilsch (1947). A sketch of the Hilsch design is given in Fig. 2.8. Compressed gas, for instance dry compressed air, is injected tangentially into a tube through a carefully profiled nozzle, shown at the centre of the diagram. On the near side of this nozzle the tube extends for a distance of at least 50 diameters before being closed by an adjustable valve. On the other side, at a distance of only about $\frac{1}{2}$ diameter from the centre line of the nozzle, the tube is sealed by a flat diaphragm containing a concentric hole of approximately half the diameter of the tube.

With the valve at the near end of tube fully open the device acts as an ejector, a small amount of air being drawn in through the diaphragm, which having mixed with the drive air, discharges through the valve. As the valve is

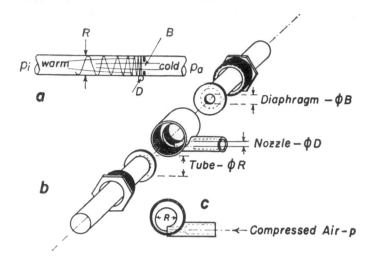

Fig. 2.8. The vortex or Hilsch tube. (*a*) Shows a section through the tube, illustrating the flow process (*b*) Shows the parts of an actual tube construction (*c*) Shows a section through the inlet nozzle.

progressively closed the flow through the diaphragm reverses and the discharged air is found to be significantly colder than the drive air, whilst that leaving through the valve is found to be hotter.

The range of results Hilsch obtained using a tube of 4·6 mm diameter and for four different diaphragm sizes is given in Fig. 2.9. It is seen that best

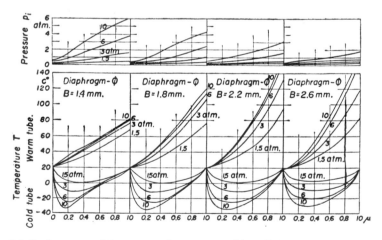

Fig. 2.9. Performance of the vortex tube, as reported by Hilsch (1947). The feed air is at 20°C, the tube diameter is 4·6 mm. The parameter for the various curves is *P*, the drive pressure. P_c is the recovery pressure for the warm air, the cold air discharges to atmosphere. *B* is the diaphragm diameter and μ the fraction of the feed air leaving as cold product.

performance is obtained with a diaphragm orifice diameter of 2·2 mm when, for feed air at 10 atm, about 30% of the feed air ($\mu = 0·3$) can be withdrawn at a temperature below $-30°C$—a cooling effect in excess of 50°C. The upper curves show the pressure recovery in the warm product air.

The explanation given was that the nozzle creates a high centrifugal field in the neighbourhood of the diaphragm. The feed air enters at the perimeter, and that part which leaves through the diaphragm must diffuse through the centrifugal field. The air will tend to conserve angular momentum and therefore will try to rotate at a progressively increasing angular speed as it moves inwards. However, the influence of viscosity will oppose this acceleration and will cause all the air to rotate at nearly the same angular velocity. Hence the air diffusing to the centre will give up kinetic energy to the outer layers, so becoming cooled. The outer layers of rotating gas, having an enhanced velocity, will move along the tube, and the kinetic energy will be converted progressively into heat by turbulent dissipation.

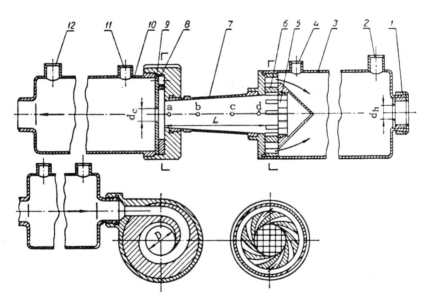

Fig. 2.10. Advanced vortex tube described by Metenin (1964). It has a toroidal inlet nozzle and a tapered tube which terminates in a diffuser.

This simple explanation was soon found to be inadequate. For instance it ignored the fact, as did a number of later more sophisticated versions, that the air leaving the nozzle was already at a temperature well below that achieved at the diaphragm. Moreover when a significant cooling effect was being achieved it was normally accompanied by a noise like that of boiling, i.e. a shock wave phenomenon was present.

Spasmodically since Hilsch, a large number of investigators have struggled with the vortex tube to improve and explain its performance. A more recent unit (Metenin, 1964), shown in Fig. 2.10 is claimed to produce a temperature drop of 65°C with 30% of the feed air (pressure $4 \cdot 9 \times 10^5$ N/m²) as cold product. A fairly complete theory has been provided by van Deemter (1952).

Measurements of the cooling effect produced by a vortex tube operating at low temperatures have been reported by Gulyaev (1966). He supplied the device with precooled helium gas at temperatures down to 80°K, and showed that, as with engine expansion, the temperature drop decreased linearly with the absolute temperature of the feed.

The temperature drop achieved in the best vortex tubes is about 50% of that achievable by isentropic expansion. However when allowance is made for the fact that only 30% of the feed air is cooled, and there is no power recovery, the overall thermodynamic efficiency is found to be less than 10%.

Thus it appears unlikely that the vortex tube will ever be attractive other than in duties where the amount of refrigeration required is small, and its low efficiency is offset by convenience, or by the availability of a cheap supply of dry compressed gas. Otherwise it will always be excelled by a turbine or other expansion engine which converts all the feed gas to cold product at a lower temperature, and gives a bonus of work output as well.

5. Peltier effect

The Peltier effect, though known since about 1860, did not justify attention as a significant cooling method until the 1950s when semiconductors became available. These materials displayed thermoelectric powers with a magnitude 2–3 orders of magnitude higher than for metals, and in certain cases the ratio of the electrical to the thermal conductivity was also more favourable. The development of thermoelectric refrigeration has been described by Goldsmid (1964).

Consider a simple circuit made up of conductors a and b joined together, one of the conductors being interrupted to insert a battery delivering a current I around the circuit. If the circuit is thermally insulated except at the junctions then in general a small quantity of heat q will be rejected at one junction, and an equal amount absorbed at the other. The differential Peltier coefficient π_{ab} is defined as

$$\pi_{ab} = \frac{q}{I} \qquad (2.5)$$

If the battery is replaced by a potentiometer, and the circuit used as a thermocouple with a temperature difference ΔT between the junctions, then the thermoelectric power α_{ab} (also described as the differential Seebeck coefficient) is given by

$$\alpha_{ab} = \frac{V}{\Delta T} \tag{2.6}$$

where V is the voltage generated. By irreversible thermodynamics it may be shown that

$$\pi_{ab} = \alpha_{ab}T \tag{2.7}$$

Since the differential Seebeck coefficient is far easier to determine than the Peltier coefficient the former is more widely used in design.

It is also useful to be able to define an absolute Seebeck coefficient for the conductors separately so that the effect of combining different pairs can be predicted from the properties of the components. To do this it must first be noted that, by the Third Law, the differential Seebeck coefficient between any two conductors at $0°K$ is zero. Thus it is reasonable to take the absolute coefficient as zero for all materials at $0°K$. Then it is necessary to select a reference substance against which refer all other conductors. Very pure lead has been chosen for this purpose, and its absolute Seebeck coefficient has been determined from near $0°K$ to room temperature. In the range $0–18°K$ this was done by forming a thermocouple between high purity lead and a Nb_3Sn superconductor. All superconductors have a zero Seebeck coefficient below their critical temperature. The relationship between the absolute and differential Seebeck coefficients is

$$\alpha_{ab} = \alpha_a - \alpha_b \tag{2.8}$$

A thermoelectric cooling element may be constructed by joining the ends of equal length rods of the two semi-conductors, laid side-by-side, to a copper plate which is to conduct the cooling duty. The remote ends of the semi-conductors are separately placed in thermal communication with a heat sink, and also electrically connected to the poles of a d.c. source. If a current I flows in the correct direction then the copper plate will start to cool as a quantity of heat q is extracted from it. However the effective cooling is reduced by two factors, by thermal conduction through the semi-conductors to the heat sink, and by ohmic heating within them—about half this heat being conducted to each junction for each semi-conductor. If the copper plate is thermally isolated from the surroundings then its temperature will drop until the potential cooling effect is exactly balanced by thermal conduction and ohmic heating. For good semi-conductors, with optimal geometry, the temperature differential may be as high as $70°K$, but at this condition the coefficient of performance will be zero.

For a good cooling element it is obviously desirable to use materials which have a high electrical conductivity but a low thermal conductivity. In metals, since the conduction electrons are responsible for the transport of both heat and electrical charge, the ratio of the two conductivities at any temperature is essentially the same, and this is expressed in the Wiedemann–Franz Law. In semi-conductors the transport processes are far more complex. There is a tendency for thermal transport to be facilitated by lattice vibrations, causing an adverse swing in the conduction ratio. However in metal alloys the thermal resistance can be significantly increased, without a proportionate penalty in electrical resistance, and further advantage can be attained by careful crystal alignment through zone melting.

For refrigeration duty at not much below room temperature the best materials are solid solutions of bismuth telluride in antimony telluride and bismuth selenide. These materials are dense ($7{\cdot}86$ g/cm^3) and difficult to machine. The optimum Seebeck coefficient is about ± 200 V$^\circ$K^{-1}. Small changes of composition lead to the materials being either n-type or p-type semi-conductors, and couples are formed by using alternate limbs of each. The best p-type alloy has the composition $Bi_{0.5}Sb_{1.5}Te_3$ and the best n-type alloy has the composition $Bi_2Te_{2.7}Se_{0.3}$.

Newton (1967) describes a thermoelectric refrigerator using alloy elements of the above type, assembled in complex structures. For an operating ΔT of 35°C he quotes a coefficient of performance of 0·66—which is less than 10% of Carnot performance, though comparable with that of an ammonia absorption system. Goldsmid (1964) gives reasons why the prospects for substantially better performance are not encouraging.

Fully codified methods for designing cooling elements, including nomograms, are presented by Birkholz (1961).

Thermoelectric cooling elements may be assembled in cascades to achieve lower temperatures. A 3-stage cascade has achieved a temperature of 150°K.

Smith and Wolfe (1962) showed that effective cooling can be achieved down to 77°K, but attempts to use the Peltier effect at liquid helium temperatures have so far been abortive (Blatt, 1962).

With present materials the useful temperature differentials are rather too low to make thermoelectric cooling exciting for small-scale duties even except in some instrumentation applications.

The thermodynamic efficiency is too low to justify large-scale applications.

6. Dilution Cycles

The dilution cycle, first hinted at by London in 1951, and translated into a practical cycle by London, Clarke and Mendoza (1962), provides effective cooling at absolute temperatures about 100 times lower than can be achieved by low pressure vaporization of helium. With normal helium (He^4) the onset of superfluidity and film-flow limits the normally achievable temperature to about 1°K. Liquid He^3, which has a lower normal boiling point (3·2°K) and does not exhibit superfluidity, will maintain a bath temperature down to about $0.3°K$ (at a pressure of 0.13 N/m²). It is not practicable to go to lower temperatures due to the very high specific volume of the vapour to be pumped.

The dilution cycle operates by harnessing the entropy increase which occurs when He^3 transfers from a concentrated solution in He^4 as solvent to a very weak solution, across a liquid-liquid interface. This is possible because at temperatures below $0.87°K$ solutions of He^3 in He^4 exist in two liquid phases with the concentrated He^3 phase being less dense and therefore on top. The phase transition (from upper to lower) is similar to evaporation, and in fact below about $0.5°K$ He^4 behaves, in respect of the He^3 atoms, as a mechanical vacuum in that the He^3 atoms experience no drag force from the surrounding He^4. However, the He^3 quasi-particles in the He^4 atmosphere have an effective mass 2.4 times that of single He^3 atoms. Another singular feature of the dilute phase is that the osmotic pressure of He^3 does not extrapolate to zero at $0°K$ but to a finite value of approximately 17 mmHg. Hence a 6.3% solution would be stable with respect to the pure isotopes at absolute zero. Thus in contrast to the normal process of evaporation, where the rapid decrease of vapour pressure with temperature sets a finite mechanical limit (due to pipe and pump sizes) on the minimum achievable temperature, no such direct limit exists in the dilution process. As will be seen second order effects limit the approach to absolute zero in this case.

The special properties of helium relevant to an understanding of the dilution refrigerator are presented by J. F. Allen (1966), and J. C. Wheatley (1969).

Fig. 2.11(a) shows the simplified flow diagram of a typical continuously operating dilution cycle. A gas containing about 85% He^3, 15% He^4 is admitted to a vacuum pump at a pressure of about 5N/m² and is compressed to about 4×10^3 N/m². After external cooling it is passed to a heat exchanger, followed by helium baths at $4.2°K$ and $1°K$. It then passes through a throttle device to sustain the required pressure drop so that the He^3 rejects heat under optimal conditions in the still. As the temperature drops below $0.87°K$ the feed proceeds to separate into two liquid phases, the greater part comprising a low density phase rich in He^3, and the lesser part a higher density phase rich in He^4. On entering the mixing vessel the liquid feed tends to accumulate in the top part of the chamber. The lower part of the chamber communicates via the heat exchanger with the lower part of the still, this pipeline being filled with superfluid He^4 through which the He^3 can diffuse without resistance. Since nearly pure He^3 vapour is being driven from solution in the still further He^3 is drawn from the weak phase at the base of the mixing chamber through the heat exchanger to replace it. This in turn causes He^3 to pass across the phase boundary in the mixing chamber, and in so doing to increase in entropy and to absorb heat. Thus refrigeration is continuously generated at the phase boundary in the mixing chamber.

One aspect of the thermodynamic behaviour of the equilibrium He^3–He^4 mixtures is shown in Fig. 2.11(b). If, at constant temperature, He^3 passes from the concentrated to the dilute phase it must absorb heat to increase its enthalpy (essentially due to a reduction of osmotic pressure, or an increase of osmotic enthalpy). Since the scales in this diagram are logarithmic it is apparent that this enthalpy step will fall off rapidly as the temperature goes down. In fact at 50m°K it has a value of 0·2 J/mol of He^3 transferred, and at

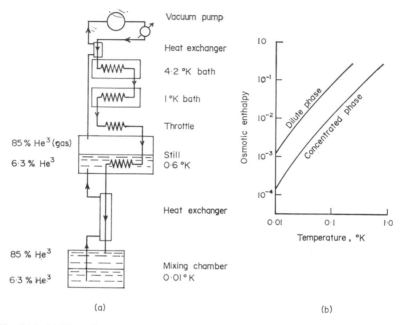

Fig. 2.11. (a) Flow diagram of a simplified helium dilution cycle. (b) Osmotic enthalpies for equilibrium He^3–He^4 mixtures.

5 m°K it is 2×10^{-3} J/mol. This behaviour compares unfavourably with that of the latent heat of evaporation which remains essentially constant with diminishing temperature. The equivalent of the throttle expansion (irreversible isenthalpic change) in the normal refrigerator is given by a horizontal line on Fig. 2.11(b) drawn between the two curves. This shows the temperature drop which would be experienced by He^3 passing from the concentrated to the dilute phase in a thermally isolated system. The two temperatures have a nearly constant ratio of 2·8 : 1. The presence of the heat exchanger between the still and mixing chamber allows this temperature drop to be made cumulative.

The practical limitations of the Dilution Cycle will be discussed in Section IVᴇ.

1. *Joule Expansion*

If a valve sealing a vessel charged with compressed gas is cracked open, so that a proportion of the gas escapes, the gas which remains will have performed displacement work and hence will have become substantially cooled. In the absence of heat transfer from the wall, and of frictional energy dissipation in the gas, the temperature drop will be identical with that for isentropic engine expansion. Thus expansion from an initial pressure P_1, to a final pressure P_2 will produce a cooling effect

$$\frac{T_2}{T_1} = \left(\frac{P_2}{P_1}\right)^{(\gamma-1)/\gamma} \tag{2.9}$$

The fraction of the feed gas remaining in the vessel, f, is given by

$$f = \left(\frac{P_2}{P_1}\right)^{\frac{1}{\gamma}} \tag{2.10}$$

Thus the mass of cold gas will be substantially less than the starting amount, no useful work will have been generated, and the cold fluid will not be easily accessible for use. Nevertheless the Joule expansion has been incorporated in several ingenious ways into liquefiers and refrigerators. Of these the Gifford–McMahon and the pulse tube refrigerators will be described in later sections.

2. *Adiabatic Demagnetization*

The use of energy transformations in the solid phase to provide cooling at temperatures below 1°K was predicted simultaneously but independently by Debye and Giauque in 1926. Both recognized that the temperature drop produced in a paramagnetic material by first aligning the material in a strong magnetic field and then, after thermally isolating it, removing the field, could be especially significant at helium temperatures due to the low specific heats of solids in this region. Experimental verification had to wait until Giauque and MacDougall (1933) achieved a temperature of 0·25°K by this means. Suitable paramagnetic materials include chromium and iron alums. For producing very low temperatures dilute salts are preferred in which the paramagnetic ions are separated from each other by non-magnetic atoms so that there is minimal interaction between neighbouring magnetic ions. Thus chromium alum may be combined with a larger mass of potassium alum, the concentration of the magnetic chromium ions being restricted to a level such that each ion is surrounded by more than forty non-magnetic atoms.

It is difficult to achieve good thermal contact between the paramagnetic salt and its container in order that the heat liberated by magnetization can be dispersed rapidly, and so that test cells may be cooled by the salt. One method is to produce a thick slurry of the salt in aqueous glycerine. The remaining problem is that of minimizing the ohmic heating produced by eddy currents which, due to the high magnetic fields employed, may be large. The achievement of efficient cooling by adiabatic demagnetization is well summarized by Ambler and Hudson (1955).

IV. Liquefaction Cycles (Illustrated by Reference to Nitrogen)

A. THROTTLE EXPANSION CYCLES

1. *Simple Linde (or Hampson) Cycle*

The first continuously operating liquefier for air (or nitrogen) was developed by Carl von Linde in 1895, and depended on throttle expansion alone.

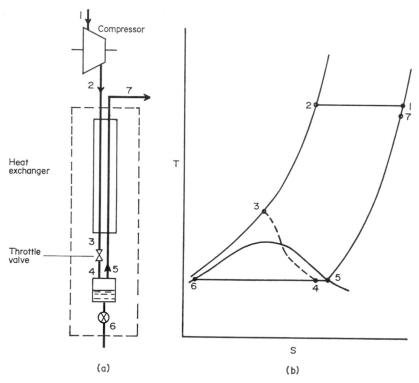

Fig. 2.12. (*a*) Flow diagram for a simple Linde liquifier. (*b*) *T–S* diagram for the same cycle.

The flow circuit and corresponding *T–S* diagram are shown in Fig. 2.12. It uses a heat exchanger to make the throttle expansion (3–4) cumulative, and by compressing the feed gas to a super critical pressure avoids the necessity for removing latent heat at low temperature. If the fraction of the feed gas liquefied is ε, then by an energy balance over the dotted rectangle, ignoring heat inleak,

$$H_2 = \varepsilon H_6 + (1-\varepsilon)H_7$$

or $$\varepsilon = (H_7-H_2)/(H_7-H_6) \tag{2.11}$$

Thus no liquid can be produced unless the enthalpy of the outlet gas is higher than that of the feed gas, and this despite the fact that its temperature must be lower. Paradoxically therefore the liquefaction performance is dominated by the degree of non-ideality of the gas at the warm end of the heat exchanger. Maximum liquefaction would be achieved by feeding the gas at the pressure corresponding to the maximum in the isenthalp at the heat sink temperature. Thus for nitrogen entering the heat exchanger at 300°K a pressure of about 350 bar would give the highest value of ε, though for minimum power consumption per unit mass of liquid product the optimum pressure will be somewhat lower.

Inspection of the slopes of the isenthalps on the T–S diagram for nitrogen (Fig. 12.20) will show that the lower the feed gas temperature at 200 atm the more non-ideal the gas, and hence the higher the liquid yield.

Conversely as the feed temperature goes up ε diminishes, and must become zero before $T = 600°K$.

In calculating the performance of the simple Linde cycle, as with other cycles to be described, the following standard assumptions will be made:

Feed gas	Nitrogen at 300°K
Heat exchanger	Minimum temperature approach of 5°K
Compressor	70% isothermal efficiency

Frictional pressure drops and heat inleak will be ignored.

Thus assuming a feed pressure of 200 bar, and noting that the minimum temperature approach will occur at the warm end of the heat exchanger, the following performance is calculated

$$\varepsilon = 6\cdot42\% \qquad W_L = 10{,}630 \text{ kJ/kg} \qquad \eta = 7\cdot3\%$$

(where W_L is the work of liquefaction for 1 kg liquid nitrogen and η is the thermodynamic efficiency).

2. Linde Double Expansion Cycle

In the case of the simple Linde cycle it was shown that the available cooling was determined by the difference of enthalpy of the streams at the warm end of the heat exchanger. However it is also apparent, from the slope of the isenthalps on the T–S diagram in the 300°K region (Fig. 12.20), that compression up to about 20 bar produces negligible enthalpy change, since the greater part of the non-ideality occurs between 20 and 200 bar. Thus virtually the same cooling effect could be obtained by circulating the gas only between these higher pressures whilst the work of compression will be reduced in the approximate proportion log 200/20 to log 200/1 (i.e. 1 : 2·3).

This saving is exploited in the Linde double expansion cycle, a version of which, with its corresponding T–S diagram, is shown in Fig. 2.13.

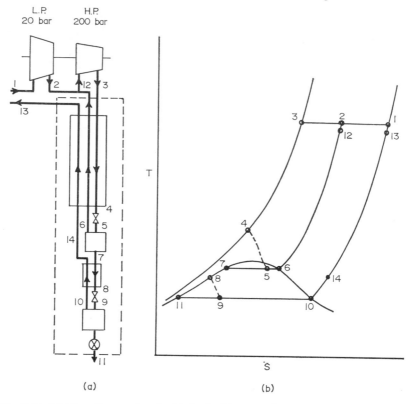

Fig. 2.13. (a) Flow diagram for the Linde double expansion cycle. (b) T–S diagram for the same cycle.

The overall heat balance gives, per unit mass of h.p. feed

$$H_3 = \varepsilon H_{11} + (1-x)H_{12} + (x-\varepsilon)H_{13}$$

where x is the fraction going forward after the first expansion and ε is the fraction of the feed liquefied.

A further heat balance on the fraction x going forward gives

$$xH_7 = \varepsilon H_{11} + (x-\varepsilon)H_{14}.$$

Using an intermediate pressure of 20 bar, and other assumptions as previously, all the H's may be specified, hence the equations can be solved for x and ε. The following results are obtained for nitrogen

$$x = 10\cdot16\%, \quad \varepsilon = 6\cdot35\%, \quad W_L = 5{,}250 \text{ kJ/kg}, \quad \eta = 14\cdot8\%.$$

This cycle can be further improved by employing an auxiliary vapour compression cycle (using, for instance, ammonia) to provide most of the cooling down to say $-40°C$. This has the effect of depressing the temperature at the warm end of the main heat exchanger, and therefore employing a region of the T–S diagram which is more non-ideal (a larger enthalpy difference for the same pressure ratios). Even allowing for the extra work required for the auxiliary refrigerator an efficiency increase of about 7% is attainable.

B. ENGINE EXPANSION CYCLES

1. Claude Cycle

In 1902 Georges Claude developed the first successful expansion engine cycle for liquefying air. As shown in Chapter 8 the problem centred on finding a means of sealing and lubricating the moving pistons at temperatures well below those at which all normal lubricants solidify. Claude eventually found

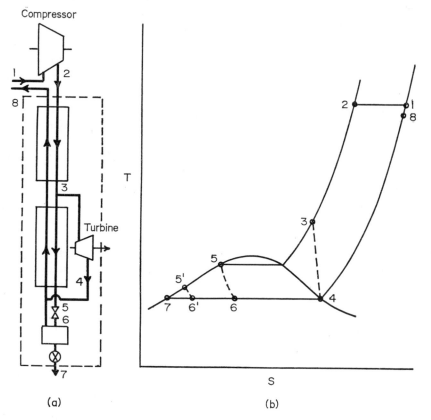

Fig. 2.14. (a) Flow diagram for the Claude cycle. (b) T–S diagram for the same cycle.

that certain leathers, carefully degreased, retained sufficient low temperature flexibility.

His cycle is illustrated in Fig. 2.14. The feed air (or nitrogen in the present case) is compressed to only about 20 bar—partly at least because a 20:1 expansion ratio is as much as can reasonably be handled in a single stage expander. Since this pressure is sub-critical the cooling duty 3→5 involves a latent heat step, which is poorly matched to the refrigeration output of the engine. Hence although improved thermodynamic efficiency may be expected due to the engine, it will be partially off-set by irreversibilities of heat exchange. Moreover in the low temperature heat exchanger it is no longer likely that the temperature pinch (i.e. the point of closest temperature approach between the two streams) will occur at one end, in fact it can only be found by plotting the temperature profiles.

Two cases will be considered. In (a) the compressed nitrogen is cooled and condensed to the bubble point 5 before undergoing throttle expansion. In (b) the condensed liquid is further cooled to 5'. In both cases it will be assumed that the effluent gas from the expansion engine is at the dew point 4. This is thermodynamically desirable for cases where the feed gas is at a sub-critical pressure, and the resulting cooling duty is concentrated at low temperatures. It would pose very difficult, if not insurmountable, problems for a recipro-cating expansion engine since liquid could condense in the engine during part of the stroke. The expansion duty could, however, be handled by a turbine.

The nitrogen feed pressure will be taken as 20 bar and the expander effi-ciency (taken as the fraction of the enthalpy drop for an isentropic expansion) will be taken as 80%, with the useful work recovery being 70% of theoretical. Thus the temperature of the nitrogen entering the engine is calculated to be 169·4°K and the useful work recovery 2,160 kJ/kg mol. The calculation procedure again involves making an energy balance for the streams entering and leaving the dotted rectangle, but this time the work generated by the engine must be included.

The temperature profiles in the low temperature heat exchanger for cases (a) and (b) are shown in Fig. 2.15, from which it can be calculated that approximately 3·39 kg of return gas are required for each kg of gas cooled in (a) and 54·1 kg in (b). Since however the recovery of liquid per unit quantity of expanded liquid is far greater in (b) than (a), the total masses of gas to be compressed per kg of liquid product is slightly greater in (a) (7·08 kg) than in (b) (6.92 kg). Taking into account work recovery in the expander the following performances are calculated:

Case (a)   Case (b)

$W_L = 2,478$ kJ/kg, $\eta = 31·4\%$ $W_L = 2,244$ kJ/kg $\eta = 34·6\%$

These efficiencies are appreciably higher than the practical performance figures quoted by Ruhemann (1949). The explanation lies in the fact that his figures relate to the use of a reciprocating expander whose outlet temperature was of the order of 100°K (rather than 77°K) and whose efficiency was probably lower than the turbine value used in the above calculations.

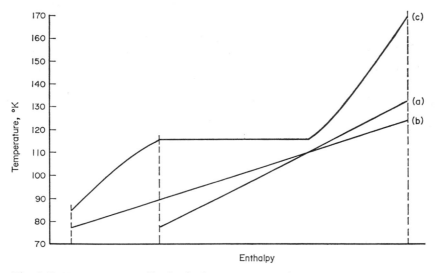

Fig. 2.15. Temperature profiles in the low temperature heat exchanger of the Claude cycle. (*a*) gives the isobar for the case when the feed air is cooled only to the bubble point, and (*b*) gives the corresponding isobar when the feed air is condensed and subcooled. Curve (*c*) is the 20 bar feed air curve.

2. *Heylandt Cycle*

If the feed gas pressure in the Claude cycle is raised above the critical then it is no longer desirable to provide for the expander to exhaust at the dew point temperature. This arises from the fact that the cooling load is now distributed almost uniformly, whilst the cooling performance of the expander (and its work output) increases with increasing inlet temperature.

These factors were exploited by Heylandt, who found that for a feed pressure of about 200 bar the optimum expander inlet temperature is about equal to the ambient temperature. Not only did this eliminate one heat exchanger but it also greatly eased the design of the expander which could be a reciprocating machine employing a low freezing-point lubricant. Thus many of the difficulties of the Claude cycle were avoided.

The flow sheet of the Heylandt cycle is shown in Fig. 2.16. It will be seen that it combines the best features of engine and throttle expansions. The high feed gas pressure distributes the cooling load, so that it is well matched to the refrigeration output of the expander as shown in Fig. 2.17. Throttle expansion is used only at the lowest temperature for liquid production.

The exhaust pressure from the expander is largely controlled by the allowable expansion ration in the engine. In many applications this cycle is used in conjunction with air distillation to generate liquid oxygen and nitrogen products. Then it is desirable to arrange for the engine exhaust to be at the pressure of the lower column of the Linde double column (about 5 bar) so this pressure will be used in the subsequent calculation.

The nitrogen feed pressure is taken as 200 bar. Since the expansion ratio in the engine is 40 : 1 the thermodynamic efficiency is taken as 75 % (compared to 80 % for the turbine in the Claude cycle), and the work recovery 65 %. The engine exhaust temperature is calculated to be 138°K. Of the high

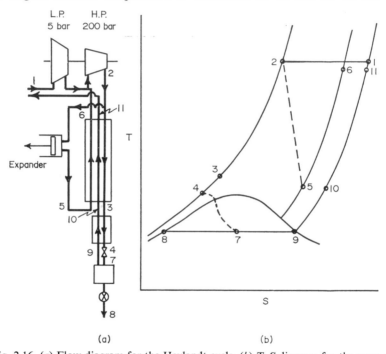

(a)   (b)

Fig. 2.16. (a) Flow diagram for the Heylandt cycle. (b) T–S diagram for the same cycle.

pressure nitrogen cooled in the low temperature heat exchanger approximately 44 % liquefies. Also, to satisfy the heat balance in the higher temperature exchanger, approximately 1·2 moles of nitrogen are expanded in the engine for each mole being cooled. The nett work requirement is calculated to be

$$W_L = 2,496 \text{ kJ/kg}, \quad \eta = 31 \cdot 2 \%.$$

The work recovery from the engine is approximately 12 % of the total work of compression. It is at first sight surprising that the efficiency calculated for the Heylandt cycle is lower than that for the Claude cycle. The explanation lies in the optimal use in the Claude cycle of a turbine exhausting at saturation temperature. As mentioned previously this was not practically possible in the reciprocating expanders developed by Claude. Also although the throttle expansion employed in the Heylandt cycle traverses a region in which the isenthalps have a high slope (i.e. high non-ideality) the cooling is nevertheless far inferior to that achieved in an engine. In the calculated Claude cycle a higher proportion of the feed gas is passed to the expander, and this fact more than off-sets the inferior heat exchanger match.

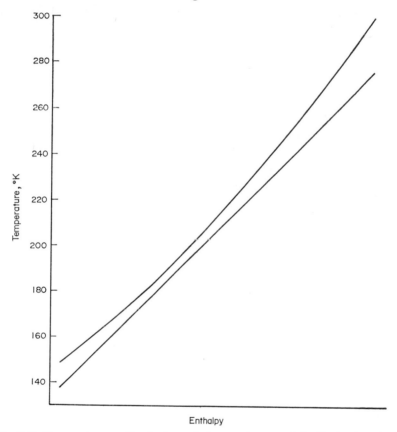

Fig. 2.17. Temperature profiles in the main heat exchanger of the Heylandt cycle. The upper curve is for the feed air and the lower curve represents the sum of the two return streams.

3. Stirling Cycle

The heat engine invented by Stirling in 1816 was virtually rediscovered by the Philips Company of Eindhoven, Netherlands, in the early 1940s as a convenient mobile power unit. It was subsequently developed quite brilliantly

as a low temperature refrigerator by Dr. J. W. L. Kohler (1954, 1960) and
co-workers of the same Company.

The principle of the machine is illustrated in Fig. 2.18 and a diagrammatic
section through one version of the machine used for air liquefaction is shown
in Fig. 2.19. Referring first to Fig. 2.18 it is seen that the expansion cylinder
on the left is separated from the compression cylinder on the right by a
packed bed (thermal regenerator). Starting with the expander piston at top
dead centre and the compressor piston at bottom dead centre the working
fluid (normally hydrogen) will be at its lowest pressure of about 20 bar, and
will exist mostly in the compression cylinder. The compression piston then

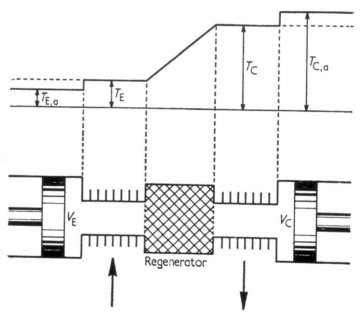

Fig. 2.18. Diagram illustrating the principle of the gas refrigerating machine based on the
Stirling engine. The expansion cylinder is on the left and the compression cylinder on the
right. The upper diagram shows the temperature distribution.

moves inwards until the hydrogen pressure has increased to about 45 bar,
with a consequent rise in temperature. At this stage the expander piston also
starts to move, but more slowly than the compressor piston, hence the
hydrogen is progressively displaced through the first heat exchange section
(where it rejects heat to cooling water) then through the regenerator (where
it is further cooled), through a second heat exchange passage (with no
further heat transfer) into the expander. During this displacement phase the
hydrogen pressure remains essentially constant. If the absolute temperature
of the hydrogen on the low temperature side of the regenerator is only say

one-third that on the inlet side, whilst the displacement volumes of the two piston and cylinder assemblies are the same, then the expander piston will have traversed only about one-third of its stroke by the time the compression cylinder reaches top dead centre. If the expander piston then continues to move the hydrogen will expand to a lower pressure and temperature with the generation of work. The cycle is then completed with the displacement of the

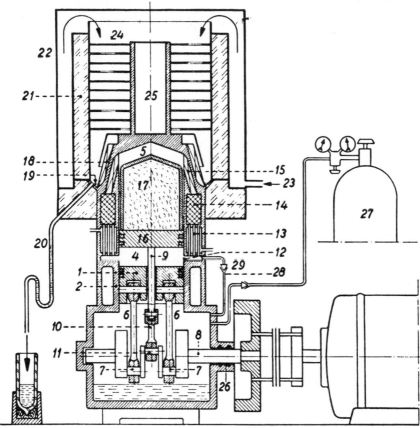

FIG. 2.19. Simplified cross-section through a Philips gas refrigerating machine employed as an air liquefier.

cold hydrogen back into the compression cylinder as both pistons return to their starting positions. During displacement the cold gas first performs its refrigeration duty as it picks up heat in the low temperature heat exchange section, and is then restored to approximately room temperature in the regenerator.

Referring now to the actual embodiment of this cycle in the machine shown in Fig. 2.19, it is seen that the compression volume comprises the space (4) between the lower piston (1) and the upper "displacer" (16). The regenerator (14) is housed in an annular cavity around the lower end of the expansion chamber which itself occupies the volume (5) between the displacer and the cylinder head. The insulated crown on the displacer enables its sealing rings to operate at normal temperatures, thus eliminating the need for any rubbing surfaces at low temperatures. The high temperature heat exchanger (13) comprises a bundle of narrow water-cooled tubes placed below the regenerator, whilst the low temperature heat exchanger (18) takes the form of fins machined into the cylinder head. The crankshaft (8) provides for the piston and displacer to reciprocate with simple harmonic motion 90° out of phase.

The resulting motion is a close approximation to the ideal described above. It is one of the virtues of this machine that it contains no valves.

The cycle analysis of the gas expansion machine is complex even when assumptions are made that the working fluid is ideal and all components operate will negligible driving forces. The difficulty arises from the fact that there are five main elements, the compression and expansion cylinders, two heat exchangers and a regenerator, containing between them a fixed mass of gas. As a first approximation, the mean temperatures of each of these elements may be assumed constant. The volumes of the compression and expansion cavities change with simple harmonic motion, but with phase displacement, the volumes of the other elements remaining fixed. The instantaneous pressure, which may be assumed constant throughout the system, will thus be determined by the distribution of the total gas charge between the fixed and varying volumes at their respective mean temperatures. In this way the pressure cycle of the machine may be built up, and hence the cooling duty and nett work requirement. Details of this calculation, together with procedures for accounting for some of the irreversibilities arising in the practical machine are given by Köhler (1954, 1960). It is impossible to represent the cycle on a T–S diagram because not all the gas within the machine goes through the same cyclic process.

The performance of the Stirling cycle is highly dependent on the thermal efficiency of the regenerator. When, for instance, the machine is used for air liquefaction the temperature difference across the regenerator is more than 200°C, yet its length is only about 5 cm. The described forms of construction involve either stacked annular stampings of very fine copper gauze, or a random mat of fine copper wire with the wire lying generally perpendicular to the flow direction. It is also important that the voidage of the regenerator, and its pressure drop, should be as low as possible. It is claimed with the Philips regenerator that these conditions are satisfied whilst giving a mean temperature difference at any level of only 2°C between the streams flowing in the alternate directions.

The complex head (25) on the Philips machine comprises an insulated heat exchanger in which, in air liquefaction, water and carbon dioxide impurities can be frozen out over an operational period of many hours, so that the actual condensing surface (18) remains essentially clean.

In larger machines using this cycle described by Dros (1964) the crank drive has been replaced by a hydraulic transmission, and the ring seal on the displacer with a rolling diaphram. By these and other small changes the efficiency has been further improved and the risk of oil contamination has been eliminated.

The Carnot efficiencies of both the laboratory and industrial size Philips machines are plotted as a function of refrigeration temperature in Fig. 2.20 for the case of heat rejection at 300°K. At liquid nitrogen temperature (77°K) it is seen that the efficiencies are approximately 33% and 42%, respectively, for isothermal refrigeration. These efficiencies are remarkably high for the Stirling cycle, and imply component efficiencies well in excess of 90%.

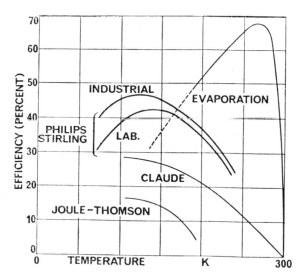

Fig. 2.20. The performance of the Philips gas expansion refrigerator related to the Carnot cycle, including a comparison with other cooling cycles.

If a single Philips machine is used as a nitrogen liquefier then all the heat for both cooling and liquefying the nitrogen must be removed at the lowest temperature. Hence the total work requirement for the laboratory machine is calculated to be 3,744 kJ/kg, and that for the larger machine 2,940 kJ/kg. Thus as nitrogen liquefiers the overall efficiencies are 20·8% and 26·4% respectively. It is apparent that these machines are much more efficient as isothermal refrigerators than as liquefiers. Some improvement could be achieved in a large liquefier installation by using an additional smaller machine to precool the feed nitrogen to say 170°K.

4. *The Bell–Coleman or Brayton Cycle*

The Bell–Coleman cycle, shown in Fig. 2.21(*a*) closely resembles the Stirling cycle but uses a heat exchanger in place of the regenerator and physically separates the compressor and expander. It enables centrifugal equipment

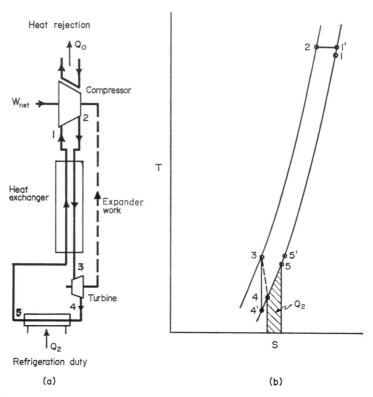

Fig. 2.21. (*a*) The Bell–Coleman or Brayton cycle employing an expansion turbine. (*b*) The *T–S* diagram for this cycle, using hydrogen at 5 and 10 bar, for refrigeration at liquid nitrogen temperature (77°K).

to be used, thus rendering it suitable for large duties. Moreover the greater geometric freedom allows the refrigerant to be taken to the duty rather than the reverse. Also the compressor can be staged and water cooled so that it approaches isothermal behaviour whilst the expander is normally insulated to make it nearly adiabatic. This cycle has been analysed fairly fully in relation to methane liquefaction by Barber and Haselden (1957).

Best performance is gained with the Bell–Coleman cycle when the working fluid behaves as an ideal gas. Thus for nitrogen liquefaction it would be appropriate to use either hydrogen or helium. The form of the T–S diagram for a nitrogen liquefier using hydrogen as the working fluid, and upper and lower pressures of 10 and 5 bar, respectively, is shown in Fig. 2.21(b). Ideally the compression would be isothermal along the path $1'$–2 and the expansion isentropic along 3–$4'$. Then cooling could be available whilst the hydrogen warmed from $4'$ to $5'$, the refrigeration duty being given by the area under the isobar between these two points. For this ideal case the nett work requirement would be the area enclosed within the diagram $1'$ 234$'$. In practice the expansion is likely, at best, to be only 85% efficient terminating at a temperature indicated by 4, and the need for a finite temperature difference for heat exchange will limit the refrigeration duty to the point 5. Hence the real duty will be given by the area Q_2, whilst the nett work can no longer be represented, though it will be substantially greater than before.

Theoretically the cycle could be further idealized by employing isothermal engine expansion from 3 to $5'$. In this case the available refrigeration duty would be far greater, as also would be the recovered work. Unfortunately no one has yet devised a practical engine which allows heat transfer to the expanding fluid. The proportions of the T–S diagram show that the loss arising from isentropic rather than isothermal expansion is minimized by operating the cycle with only a small pressure differential. In contrast the losses arising from the finite T for heat exchange are minimized by using a large pressure differential, so that 5–$5'$ is a small proportion of 4–$5'$. Hence there is an optimal design for each duty which minimizes the sum of the various loss terms. This cycle is in fact so sensitive to component efficiencies that even with 90% efficiencies for the compressor and expander the cycle efficiency as a nitrogen liquefier is unlikely to be more than 25%.

C. VAPOUR COMPRESSION CYCLES

1. Cascade Liquefiers

The very early liquefiers originated by Pictet, and developed by Kamerlingh Onnes, employed vapour compression cycles in cascade, the evaporator of the first serving to remove heat from the condenser of the second, and so on. Figure 2.22 shows a possible cascade for nitrogen liquefaction which employs the three auxiliary refrigerants ammonia, ethylene and methane. This circuit is in fact a rather simple one; improved power economy can be gained by precooling all liquids before throttling with additional exchangers. A cascade of this type is described by Lyon (1963). A practical installation must also include provision for storing all refrigerants at shut-down, and systems for detecting refrigerant leaks and counteracting them.

3

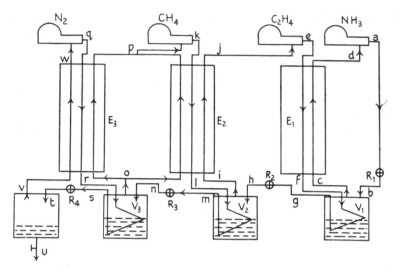

Fig. 2.22. A possible cascade cycle for liquefying nitrogen.

TABLE 2.III. Flow and Energy Analysis for the Cascade Cycle

m (kg)	No.	p (atm)	$T(°C)$	f	h (kcal/kg)	Work required kJ/kg liquid nitrogen
NH_3 0·331	a	10·2	25		128·0	
	b	1·0	−34	0·196	128·0	107
	c	1·0	−34	1·000	390·8	
	d	1·0	17	—	417·7	
C_2H_4 0·963	e	19·0	25	—	74·2	
	f	19·0	−7	—	64·9	
	g	19·0	−31	0·000	−25·5	222
	h	1·0	−104	0·341	−25·5	
	i	1·0	−104	1·000	44·0	
	j	1·0	17	—	79·3	
CH_4 0·660	k	24·7	25	—	220·4	
	l	24·7	−75·9	—	165·7	
(0·636 through E_3,	m	24·7	−101·0	0·000	64·3	319
0·024 through E_2)	n	1·0	−161·6	0·447	64·3	
	o	1·0	−161·6	1·000	132·6	
	p	1·0	17	—	221·0	
N_2 1·770	q	18·6	25	—	108·9	
	r	18·6	−158·6	0·872	52·6	
	s	18·6	−168·6	0·000	27·1	
	t	1·0	−195·8	0·435	27·1	492
1·000	u	1·0	−195·8	0·000	6·7	
0·770	v	1·0	−195·8	1·000	53·65	
	w	1·0	17	—	110·0	

The descriptions of all the flow streams in the cascade cycle shown in Fig. 2.22 are presented in Table 2.III taken from Ruhemann (1949). The final column gives the work required by an ideal compressor for each stage, the total being 1,140 kJ/kg liquid nitrogen. If a compressor efficiency of 70% is assumed then the actual work requirement becomes 1,630 kJ/kg liquid nitrogen, corresponding to an efficiency of 47·6%. This calculation makes no allowances for pressure drops in pipelines and heat exchangers, nor for any heat in-leak, hence practical efficiencies will be slightly lower. If additional heat exchangers are used to sub-cool the liquids before throttling then the resulting entropy gains (and lower losses) can be reduced and the overall efficiency raised by between 5–10%.

V. Practical Liquefiers

A. NATURAL GAS OR METHANE (Barber and Haselden, 1957)

The first large scale natural gas liquefier was built in Cleveland, U.S.A. in 1939 as part of a peak-load sharing installation. It employed a conventional cascade system with ammonia and ethylene as first and second stage refrigerants (Clark and Miller, 1940 and 1941).

A similar plant, designed by Dresser Industries Ltd., of Dallas, Texas, was installed on the Saratov natural gas pipeline near to Moscow in 1948. It was designed to liquefy about 5,000 m³/hr of gas, and employed five gas-engine driven compressors each of 600 h.p., of which one was used for ammonia and two each for ethylene and methane.

A large barge-mounted liquefier built in 1953 for an abortive scheme for transporting liquefied natural gas (LNG) up to river Mississippi from the Mexican Gulf to Chicago, but later used for supplying liquid for the first trans-atlantic LNG shipping trials to Canvey Island, England, used a novel expansion turbine cycle developed by Morrison (1954). The feed gas, after purification was compressed to approximately 70 bar, precooled in a heat exchanger to about −60°C and then expanded down to near atmospheric pressure in a turbine. The exhaust from the turbine was designed to contain between 10 and 20% of liquid to be separated, the residual gas being returned through the other pass of the heat exchanger. It was to be expected that certain mechanical problems would accompany expansion to this degree of wetness, also there would be a tendency for part of the precipitated liquid to exist as a fog, and so be difficult to separate. Moreover, the cycle involved high inherent thermodynamic losses, and thus had a high power consumption. Its main virtue was that of simplicity.

The large liquefier built in Arzew, Algeria, in 1963 reverted to the cascade cycle. In fact, due to the relative cheapness of fuel, a rather simple cycle with minimal cold recovery was chosen (see Fig. 2.23).

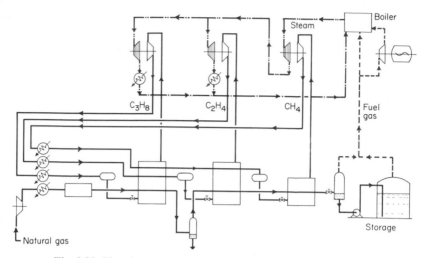

Fig. 2.23. Flowsheet of the Arzew Natural Gas Liquefaction Plant.

The use of mixed refrigerants to extend the cooling range of a single stage refrigerator had been pioneered for absorption refrigerators by Maiuri (1939) and for vapour compression cycles by Ruhemann (1947). The idea of extending this concept to multicomponent mixtures for natural gas liquefaction appears to have been grasped almost simultaneously in U.S.A., U.S.S.R.

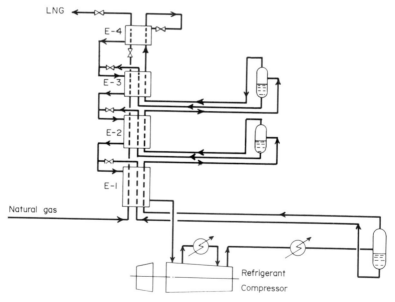

Fig. 2.24. Mixed refrigerant, single pressure, natural gas liquefaction process.

(Kleemenko, 1959), France (Grenier, 1967; Salama and Eyre, 1967; Darradeua, 1969) and probably other countries also.

The new cycle of this type chosen by Air Products and Chemicals Inc. (Allentown, U.S.A.) for the Esso plant in Libya is shown schematically in Fig. 2.24. The feed gas at a pressure of 620 lb/in² absolute is cooled continuously in a heat exchanger until it is both condensed and subcooled. To increase the heat rejection from the feed gas (and hence the fraction remaining as liquid after final expansion) the feed may be expanded to an intermediate pressure before flowing through the lowest temperature zone of the exchanger.

The multicomponent refrigerant (MCR) consists of a mixture of hydrocarbons, which may be extracted in a separate plant from the feed natural gas, carefully blended with nitrogen to give the optimum composition. After compression in a single large centrifugal machine, with intercooling, the gas is passed to a water-cooled condenser. This unit is designed for co-current flow of gas and condensate with good mass transfer between the two. At exit about 25% of the feed has condensed, and closely approaches phase equilibria with the residual gas. The condensate and residual gas are separately passed through the first stage of the main heat exchanger so that the condensate sub-cools whilst the gas mixture undergoes further co-current condensation. The sub-cooled condensates is now expanded into the shell of the heat exchanger where it evaporates in co-current flow with diluent gas from the upper part of the exchanger. The refrigeration available from the combined stream is sufficient not only to condense the next stage refrigerant but also to cool the natural gas feed.

The partially condensed residual refrigerant stream is passed to a phase separator and the gas and liquid streams separately passed to the next stage of the main heat exchanger. In this way the temperature is progressively lowered and each stage is made to support the cooling requirements of the next lower one until finally the refrigerant gas feed is totally condensed, sub-cooled, and fed after throttling into the cold end of the heat exchanger.

For a typical natural gas mixture the shape of the cooling curve is shown in Fig. 2.25. To match this duty together with that of cooling, condensing, and sub-cooling the various fractions of the refrigerant mixture by the progressive evaporation of the different liquid refrigerant fractions, is obviously a task of great complexity. Furthermore, if the power requirement is to be minimized (which will also keep down the capital cost of the compressor and its prime mover) the mean temperature differences have to be kept uniformly low throughout the heat exchanger. To achieve this end it is necessary to use a refrigerant mixture of several components and to calculate the feed composition by an extensive optimizing procedure. Even then it cannot be expected that practical performance will exactly duplicate design and therefore provisions should be incorporated to modify the mixed refrigerant composition during operation to minimize temperature pinches.

The heat transfer conditions involve partial condensation and partial evaporation under conditions of progressively varying flow rates and in situations in which maldistribution of the phases can easily occur. When these factors are combined with the thermodynamic ones, it is apparent that a design problem of quite exceptional sophistication is posed. The challenge is justified by the potential capital savings inherent in a single stream plant. Thus, although the power requirement is marginally greater than for a well-designed classical cascade system, it is far cheaper to purchase one large

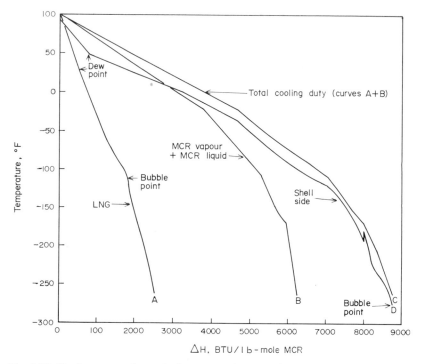

Fig. 2.25. Cooling curves for a single-pressure MCR cycle for liquefying natural gas.

centrifugal compressor than two or three smaller machines having a slightly lower aggregate power. Similarly, one installed large heat exchanger is significantly cheaper than an assembly of individually piped smaller units.

For the Esso plant the refrigerant compressors have available approximately 160,000 b.h.p. per train. Each train contains two parallel heat exchangers; each exchanger having a shell size of 12 ft 8 in. diameter by 190 ft high, and containing more than 400 miles of aluminium tubing. The two completed heat exchangers during transit are shown in Fig. 2.26.

For very much smaller duties, for instance, the reliquefaction of natural gas on ocean-going tankers, the large Stirling cycle machines are probably justified by their convenience and reliability. Vogelhuber and Parish (1968)

Fig. 2.26. Two heat exchangers, belonging to the Esso Libyan LNG plant, shown in transit.

described machines for this application, and claimed that they would normally have economic advantages over other machines for liquefaction duties of less than 40,000 m³/day.

Especially if the boil-off gas is prevented from warming up before being reliquefied the cooling characteristic of the Stirling cycle will be ideally matched to the duty, and an overall thermodynamic efficiency in excess of 40% can be achieved.

In the long term the competition between the pure and mixed refrigerant cascades is likely to continue. Whilst the thermodynamic match is potentially better with the mixed refrigerant system this is off-set in practice by the impossibility of maintaining equilibria between the phases during the boiling and condensation processes. The rates of heat transfer are bound to be less for the mixed refrigerants not only because of the additional mass diffusional resistances, which must be present in both phases, but also because of mal-distribution effects. For instance if in the shell of the main exchanger the liquid flow on one side is 5% greater than on the other, whilst the vapour flowrate is everywhere uniform, then the whole equilibrium pattern is disturbed. This could be countered only by dropping the evaporation pressure appreciably and hence increasing compressor power.

The pure refrigerant cascade can be simplified, with no loss of efficiency, by the use of cold compression (Haselden, 1967). Since the density of the refrigerants going forward to the compressors is higher in this case the compressors may be more compact, and therefore potentially cheaper. Moreover since virtually all the heat transfer occurring in the revised cycle accompanies either boiling or condensation the heat transfer coefficients will all be high, and therefore the areas of the remaining heat exchangers will be comparatively low. Finally, by designing the refrigerant circuits to have minimum liquid hold-up the problem of refrigerant storage and replenishment is kept in bounds.

B. OXYGEN AND NITROGEN

A comparison of nitrogen liquefiers has already been presented, and the thermodynamic picture with oxygen is very similar. On a small scale the convenience and comparatively high efficiency of the Stirling cycle machines makes them most attractive. Versions of this machine are available which incorporate not only automatic provisions for eliminating moisture and carbon dioxide from the feed air, but also a distillation column to separate a liquid nitrogen product.

On a large scale the Heylandt cycle held sway over many years and is still used. However the reciprocating expander (see Chapter 8), operating with a feed air pressure of about 200 bar, inevitably requires appreciable maintenance, and there is a self-evident risk of lubricating oil, or its decomposition products, entering the plant from either the high pressure compressor or the expander. Also the demand for both liquid oxygen and liquid nitrogen from many air separation plants now frequently exceeds that which can be generated by operating the Heylandt cycle on the feed air.

To meet the need for additional liquid supplementary circuits are added which work either at high or intermediate (10 to 20 bar) pressures. At the lower pressure centrifugal compressors may be used together with expansion turbines. These circuits also include auxiliary vapour compression circuits operating at say 210°K to enable the turbine to give maximum benefit at low temperature. Undoubtedly a lower power consumption could be achieved by using a cascade system, but this has been avoided not only for reasons of complexity but because methane is the only readily available refrigerant to bridge the temperature interval between about 110° and 170°K. However since the methane would normally be used to condense nitrogen under pressure it would seem that the explosion hazard associated with having methane in the neighbourhood of an oxygen plant could be designed out.

As the proportion of liquid products from an air separation plant increases, through the use of auxiliary liquefiers, new difficulties arise both in the column and in the clean-up system. Distillation is impaired because liquid nitrogen drawn off as product is no longer available as reflux. The re-vaporization of carbon dioxide in the reversing heat exchangers is rendered difficult due to the reduced volume of reject impure nitrogen. If, in addition, a pure argon side-stream is required the plant becomes almost uncontrollable.

C. HYDROGEN

1. *Special Factors*

Since the inversion temperature of hydrogen (204°K) is well below ambient it is impossible to liquefy hydrogen by use of throttle expansion alone. Hence all hydrogen liquefiers must employ either a secondary refrigerant (normally

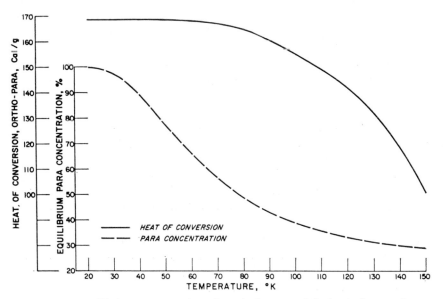

Fig. 2.27. Equilibrium concentration of parahydrogen and the heat of conversion.

liquid nitrogen) or one or more expansion engines. Further problems arise due to the ortho–para conversion. Hydrogen gas at room temperature is a

3*

mixture of two isotropic forms differing in the orientation of the two nuclear spins in the diatomic molecule. The equilibrium concentration is 75% ortho, 25% parahydrogen. As the mixture is cooled the equilibrium para concentration increases as shown in Fig. 2.27, whilst there is an associated heat of reaction, also shown. At its normal boiling point the equilibrium conversion to parahydrogen is virtually complete. In the absence of a catalyst the conversion process takes several weeks to approach completion. Therefore, if liquid hydrogen is to be used within a few days, it will normally be best to isolate it from materials having catalytic activity, and then its change of composition will be relatively small. However, if it is to be stored, then the hydrogen should be contacted with catalyst during cooling and liquefaction. Suitable materials include chromium oxide on an alumina support, or a granular iron hydroxide gel. The catalyst must be activated by intense drying. The design of these catalytic reactors has been described by Singleton and Lapin (1965). The minimum work of conversion is achieved only if the equilibrium shown in Fig. 2.27 is maintained throughout the cooling process, and the heat of conversion reversibly removed at every temperature level. An extra power consumption of 2,260 kJ/kg is incurred for the ideal conversion process, giving a total power requirement for liquefaction to parahydrogen of 14,450 kJ/kg. In practice the best that can be done is to contact hydrogen with catalyst at three or four temperature levels with heat removal at each.

Two remaining factors are important with all hydrogen liquefiers. The first is the need to incorporate means for removing trace impurities such as nitrogen, oxygen and methane from the feed gas. The purification system normally comprises active charcoal and silica gel adsorbers working at liquid nitrogen temperature. The second is the need for extreme safety precautions to guard against explosion hazards. An extensive review of hydrogen liquefiers has been presented by Scott, Denton and Nicholls (1964).

2. *The Boulder Liquefier* (Scott, 1959)

The large hydrogen liquefier at the N.B.S. Cryogenic Engineering Laboratory in Boulder, Colorado, is a good example of the Hampson type liquefiers which have been custom-designed at many famous low temperature laboratories.

The basic flowsheet is shown in Fig. 2.28. The feed gas is compressed to about 100 bar, heat-exchanged against outgoing nitrogen and hydrogen gas, and cooled in a liquid nitrogen bath at 66°K (about 0·2 bar). The gas is then further cooled against outlet hydrogen gas before being throttle expanded. It would be possible to improve the power consumption by employing a first nitrogen bath at atmospheric pressure and by using a high pressure recirculation stage for the hydrogen, but the additional complication is not justified in a laboratory unit.

The full flow circuit of the liquefier also contains a refrigerated drier and adsorber for purifying the feed gas. The form of the liquefier stage is shown in Fig. 2.29. When producing 95% parahydrogen the hydrogen compressors take in 1,300 m³/hr of gas and the liquid output is 230 l/hr. The liquid nitrogen consumption is 340 l/hr.

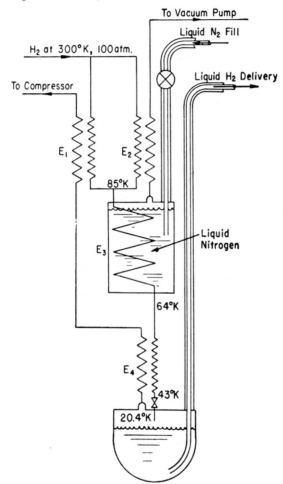

Fig. 2.28. Simplified flow diagram of a large laboratory hydrogen liquefier.

3. Proprietory Engine Expansion Liquefiers

Two forms of hydrogen liquefier based on engine expansion cycles are available commercially. The first is an adaptation of the Collins Helium liquefier and will be described under that heading. The second is a 2-stage version of the Philips gas refrigerator.

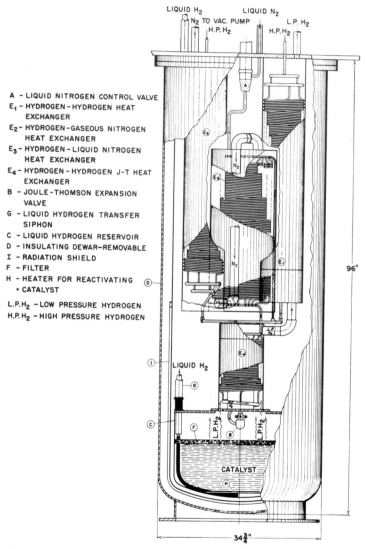

Fig. 2.29. Cut away assembly of the hydrogen liquefier vessel at the National Bureau Standards Cryogenic Engineering Laboratory, Boulder, Colorado, U.S.A.

The principle of this machine is illustrated in Fig. 2.30, from which it is seen to be an ingenious but logically staged version of the Stirling Cycle in which the first stage refrigerator at about 80°K acts as heat-sink for the smaller second stage unit. Helium gas is the working fluid for both stages. Particular problems arise with the low temperature regenerator since the specific heats of most solids become vanishingly small as the temperature of liquid hydrogen (20°K) is approached. The most suitable material appears to be lead. A photograph of the Philips hydrogen liquefier is shown in Fig. 2.31. If this unit is supplied with bottled gas at 8 bar the declared production rate is 5·5 l/hr at a power consumption of 11 kW. The unit achieves the rated output in 40 min starting from warm, and it includes many automatic control and safety features.

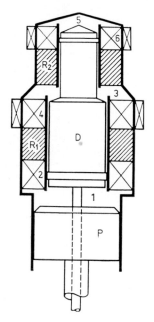

Fig. 2.30. Schematic cross-section of a 2-stage gas expansion refrigerator. Compression takes place in 1 and expansion in both 3 and 5 simultaneously.

4. Large Hydrogen Liquefiers (Tonnage Productions)

For use in large space vehicles hydrogen must be liquefied on the scale of many tons per day. Representative of this class is the 27 ton/day plant built by Air Products and Chemical Inc. for the U.S. Air Force. The flowsheet of the liquefier is shown schematically in Fig. 2.32. The feed and recycle hydrogen gas streams are restricted to quite separate circuits in order to simplify the purification systems, and to restrict the use of catalytic converters to the feed stream. Refrigeration is provided at 278°K by a Freon system (not shown) then by nitrogen gas from an expansion turbine. Liquid nitrogen cooling is applied firstly at atmospheric pressure and then at about 0·14 bar. A single turbine expands hydrogen gas at a pressure of about 7 bar to provide refrigeration at 36°K. The feed to this turbine comprises in part the L.P. hydrogen recycle stream, the remainder coming from the intermediate pressure liquid hydrogen bath, which itself provides cooling at 30°K. Final cooling is provided by a bath of liquid hydrogen boiling at essentially atmospheric pressure.

Fig. 2.31. The Philips hydrogen liquefier, based on the 2-stage gas expansion machine.

Thus below 36°K the system is dependent upon throttle expansion. It is seen that converters are provided at six different temperature levels. In the design of plants of this type a difficult balance must be struck between efficiency on the one hand and simplicity on the other. Clearly the power consumption could be substantially reduced by using more turbines to bridge the cooling range from 60°K to 20°K, but the risk of breakdowns tends to be more than a linear function of the number of mechanical components in use.

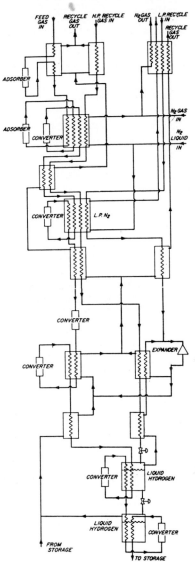

Fig. 2.32. Simplified flowsheet for a large hydrogen liquefier.

D. HELIUM (Kropshot *et al.*, 1968)

1. *General Considerations*

Reference back to Table 2.1 shows that by far the greatest cooling duty in liquefying helium is associated with cooling the gas to its dew point, and that the theoretical work to condense the gas is relatively trivial in comparison.

Also the temperature–entropy diagram for helium shows that the upper inversion temperature occurs at about 40°K, whilst to obtain significant cooling by throttle expansion requires precooling to below 20°K. These factors underline the credit due to Kammerlingh Onnes for the vision and skill which enabled him to first liquefy helium at Leiden in 1908. His apparatus used a pre-cooling bath of liquid hydrogen pumped down to a temperature of about 14°K.

Croft (1961) makes the point that since the density of liquid helium is only about 8 times that of its saturated vapour, it is sensible to consider the minimum *net* work of liquefaction taking account of the work potentially recoverable in reheating the displaced gas. On this basis the net work required to accumulate 1 kg of liquid helium whilst displacing an equal volume of saturated gas at atmospheric pressure is 5.93×10^3 kJ rather than 8.39×10^3 kJ as quoted in Table 2.I.

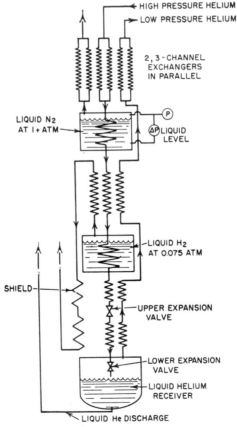

Fig. 2.33. Flowsheet of Hampson type helium liquefier installed at the N.B.S. Cryogenic Engineering Laboratory, Boulder, Colorado, U.S.A.

A related issue, though depending additionally on the fact that specific heats of constructional metals become negligibly small below liquid hydrogen temperature, is the efficiency of a simple expansion liquefier for helium. Pickard and Simon (1948) were the first to high-light this fact and to show that by pre-cooling with pumped liquid hydrogen a high pressure vessel charged with high pressure helium gas, insulating it with a vacuum, and then allowing the helium to expand down to atmospheric pressure, liquefaction can occur to an extent of more than 50% of the vessel capacity.

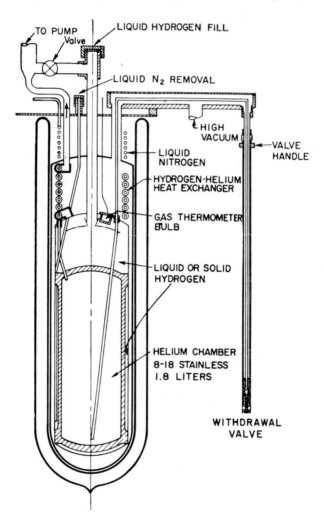

Fig. 2.34. A Simon-type liquefier for helium designed by Gifford, and suitable for decanting the product.

2. Throttling Helium Liquefier

The Hampson type liquefier, using nitrogen and hydrogen pre-cooling, installed in the N.B.S. Cryogenic Engineering Laboratory has the flow-sheet shown in Fig. 2.33. It takes helium gas at a pressure of 30 bar, and at approximately $1 \cdot 5$ m^3/hr, and produces 15 l/hr of product for a consumption of 24 l/hr of liquid hydrogen. The helium is expanded in two stages to optimize heat rejection before the final throttle valve.

3. Simon Liquefier

A version of the Simon-type liquefier, designed for the withdrawal of liquid product, has been refined by Gifford and is illustrated in Fig. 2.34. All parts of the unit are first cooled with liquid nitrogen before feeding liquid hydrogen. During the ensuing helium feed period the ingoing gas is cooled by heat exchange against the outgoing hydrogen gas. Maximum pre-cooling is achieved by allowing the outgoing hydrogen to by-pass the heat exchanger during the final period so that the helium chamber, and its gaseous contents, are cooled to solid hydrogen temperature (below 14°K). For a helium feed pressure of 100 bar, and pre-cooling to 12°K, the yield of liquid helium is approximately 50% of the chamber volume. Virtually all of this liquid can be transferred without loss to a secondary apparatus, since the cold expansion gas which precedes it may be used for pre-cooling purposes.

4. Collins Helium Liquefier (1952)

The Collins Liquefier (Fig. 2.35) reflects the fact that helium liquefaction is mainly concerned with the removal of sensible heat from the feed gas from room temperature down to the dew point, and that this is ideally accomplished by the engine expansion of successive fractions of helium gas over consecutive temperature ranges. Depending on the gas pressure, and the availability of liquid nitrogen, this could require between three and five expansion engines, but in practice only two or three are used. Liquefaction is achieved finally by throttle expansion of cold gas.

The success of the Collins Liquefier is also related to the excellence of its mechanical design as exemplified by the expansion engines and the heat exchanger. Collins developed very efficient, reciprocating expanders, using nitrided steel pistons and cylinders, with the piston rods in tension. This combination allows the use of very small clearances and gas lubrication, whilst the long piston rods minimize heat inleak. The heat exchanger was accommodated in an annular channel formed between parallel conical surfaces, the outer one comprising one wall of the surrounding vacuum jacket. The heat exchanger tubes have secondary surface in the form of ribbon spirals brazed edge-on.

The form of the vacuum jacket is simple, with no gas connections within it, and therefore it could be made reliably. The various components of the

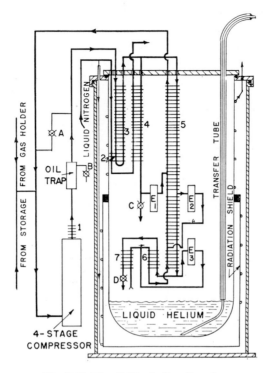

Fig. 2.35. The Collins helium liquefier.

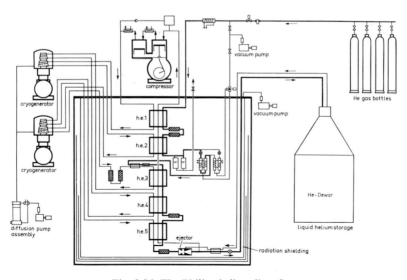

Fig. 2.36. The Philips helium liquefier.

liquefier are surrounded by an atmosphere of gaseous helium at atmospheric pressure, the vertical position of each component being deliberately chosen to create and sustain a steady temperature gradient from room temperature at the top to 4°K at the bottom. In so far as this was achieved natural convection currents in the helium gas were absent, and therefore very little heat was

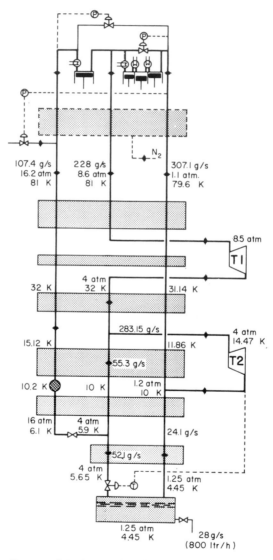

Fig. 2.37. Flow diagram of the large helium liquefier built by Sulzer and installed in Otis, Kansas, U.S.A.

conducted down to the base of the liquefier. Extreme gas tightness of the expansion engines was not crucial.

5. *Philips Helium Liquefier*

The Philips helium liquefier, shown in Fig. 2.36, employs two 2-stage gas expansion refrigerators for precooling a separate helium stream which itself generates the lowest temperature for liquefaction by throttle expansion. The multiple precooling levels, each achieved with quite high efficiency, obviously suit the thermodynamic requirements of the process. An interesting additional feature is the use of an ejector for part of the throttle expansion (Rietdijk,

Fig. 2.38. Photograph of the Otis helium liquefier.

1966). In this application the ejector allows the liquid helium product to be generated at atmospheric pressure whilst boosting the pressure of the return gas stream to a level more than sufficient to overcome the back-pressure of the five heat exchangers. Thus the gas enters the recycle compressor at a pressure above atmospheric, which saves power and increases throughput. There are parallels between this system and the Linde Double Expansion Liquefier, but the use of the ejector avoids the complication of the low pressure stage of the compressor.

6. *Sulzer Helium Liquefier*

The Sulzer liquefier, described by Trepp (1966), exemplifies the approach required in large units where nitrogen sub-cooling can be supplemented by multiple turbine expanders. The flow diagram, and operating conditions, for the 800 l/hr unit installed in Otis, Kansas, are shown in Fig. 2.37. The greater part of the feed gas is compressed to only 8·9 bar, and after pre-cooling is expanded in two stages through turbines. The lowest stage exhausts at 10°K, which thus fixes the temperature level (10·2°K) of the remaining part of the feed gas as it enters the Hampson liquefier stage. Reference to the helium T–S diagram will show that the enthalpy maximum at 10·2°K is at approximately 16 bar, which determines the liquefier stream pressure. This pressure is, however, too high for a liquid-producing expansion, so Trepp employs throttle expansion to an intermediate pressure of 4 bar, after which about half the gas goes to the low-stage turbine whilst the other half flows through the throttle valve. A very satisfactory integration of 2-stage turbine expansion with 2-stage throttling is thereby achieved.

A photograph of this plant is shown in Fig. 2.38. The isothermal compression work is approximately 450 kW and the production rate is somewhat in excess of 800 l/hr.

VI. Cryogenic Refrigerators

A. GAS EXPANSION ENGINES

The cooling cycles described thus far have been related to the duty of cooling and liquefying gases. There are many duties such as those of maintaining electronic devices at a constant low temperature, counteracting heat inleaks to super-conducting magnets, or removing the energy input to bubble chambers, where refrigeration alone is called for. Generally this will be at a single constant temperature, though it may be useful to have refrigeration available at one or more intermediate temperatures to service radiation shields.

In many instances totally enclosed gas-expansion units, such as the Philips machine, will provide the best answer, not only for reasons of convenience but also because their high potential efficiencies are most fully realized in essentially isothermal duties.

The performance of a large single stage machine is illustrated in Fig. 2.39, from which it is apparent that it retains an acceptable efficiency for cooling duties down to about 60°K. Two-stage machines are also available in various sizes. Verbeek (1969) has described a large machine with a refrigeration capacity of approximately 1·7 kW at 20°K and 1 kW at 16°K. Gifford and others have been responsible for developing simpler, less efficient, units for very small duties, and these will be described.

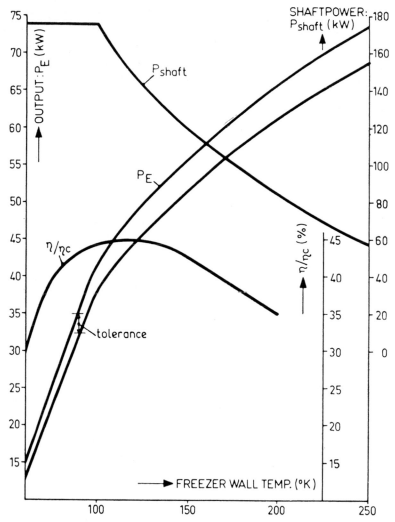

Fig. 2.39. Refrigeration performance of a large single-stage gas expansion machine using hydrogen as the working fluid (Philips PPG-2500 unit).

Only where the duties are very large or where the nature of the installation to be cooled requires that cooling fluid be widely distributed, will other systems justify consideration.

B. NORMAL EXPANDER CYCLES

Most forms of liquefaction cycle are directly adaptable as refrigerators and most will show an increase of efficiency, compared to liquefaction duty, due to the balanced flows in heat exchangers. The characteristics of pure fluid cascades, including those involving cold compression (Haselden, 1967), will generally give them the lowest power consumption—where the temperature level and scale of operation justifies their use. Otherwise a closed flow system involving one or more expansion engines, acting as pre-cooling stages for a final throttle expansion, will normally generate the required refrigeration, the operating fluid being chosen to have the refrigeration temperature within its boiling range. These systems may incorporate vapour-compression first-stage cooling.

Illustrations of helium refrigerators employing expanders followed by throttle expansion are furnished by Baldus and Sellmaier (1964), Winters and Snow (1965), and Carter *et al.* (1969), whilst Trepp (1969) reviews the factors controlling efficiency and costs of large refrigerators.

Where refrigeration is required at temperatures below the λ-point of helium it is desirable to use He^3 which has a lower boiling point than He^4 and involves no complications of superfluidity. Baldus and Sellmaier (1967) describe such a refrigerator which employs liquid nitrogen for pre-cooling and a single Doll–Eder valveless piston expander for obtaining 30 W refrigeration at $1\cdot8°K$. The power consumption was 76 kW and the nitrogen useage about $8\cdot1$ kg/hr. The scarcity and high cost of He^3 dictates that apparatus shall be designed for minimum hold-up, and with every possible precaution against leakage.

C. GIFFORD–MCMAHON REFRIGERATOR

A method of harnessing the Joule effect in a cyclically operating refrigerator of low mechanical complexity was invented by McMahon and Gifford (1959). It requires a source of pure compressed gas, a cylinder containing a closely fitting displacer, a regenerator and two valves (Fig. 2.40.)

The starting point of the cycle may be taken when the displacer is at the bottom position with the upper volume 1, and the regenerator, filled with low pressure gas. The inlet valve opens to admit high pressure gas to volume 1 and to the regenerator. The temperature of the gas in volume 1 rises due to isentropic compression (Joule heating). Depending on the degree of mixing all the gas in 1 at this stage may be at a more or less uniformly high temperature, or there may be a layer of gas at a higher constant temperature surmounted by a second layer having a temperature gradient within it. With the inlet valve still open the displacer is now moved to its top position, thus displacing the gas at constant volume and pressure from 1 to 2 through the regenerator. If the bottom end of the regenerator is already cold then the gas will contract as it flows into 2 and additional high pressure gas will be drawn in. All the gas will now exist in the regenerator and volume 2. The inlet valve is closed and the exhaust valve opened. The gas in volume 2 undergoes Joule expansion and cools, that remaining being at the lowest achieved temperature. The gas leaving through the exhaust valve will be hotter than the feed gas since it will have picked up the Joule heating effect in the regenerator. With the exhaust valve still open the displacer is moved to its bottom position, thus displacing the cold gas through the regenerator partly back into volume 1 and partly to waste, to complete the cycle.

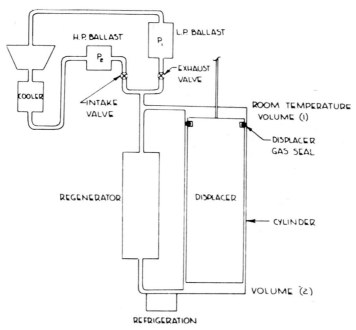

Fig. 2.40. Flow diagram of the Gifford–McMahon refrigerator.

The regenerator has the effect of accumulating the refrigeration produced in volume 2, hence its temperature will progressively decrease until the refrigerating effect is exactly counter-balanced by the cold losses (giving the lowest temperature which the device can achieve), or it produces a nett refrigerating effect at some higher temperature.

It will be seen that at no stage is there a pressure drop across the displacer apart from the small flow resistance of the regenerator. Hence there is no requirement for a very close fit between the displacer and the cylinder, and the seal can be at the warm end. Typically the displacer is moved at a frequency of only 50 to 150 cycles/min, hence its speed, and the force required to move it, are low. In early machines a crank system was employed but later the feed gas was itself used as the propulsive medium via a small diameter piston and cylinder assembly mounted above the main cylinder (Gifford, 1965).

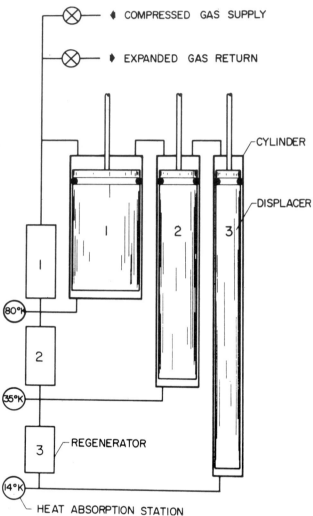

Fig. 2.41. Flow diagram of the 3-stage Gifford–McMahon refrigerator.

Whilst any pure gas may be used in the device, it is best to use one which approximates ideal gas behaviour and has physical properties which facilitate high heat transfer and low pressure drop in the regenerator. Gifford used helium and a pressure ratio of between 1·5 and 2·5 to obtain cooling down to 23°K from a gas supply at 300°K. The exhaust gas may be recycled through a compressor, as in Fig. 2.40.

An additional advantage of the cycle is that it may be cascaded readily to form 2- or 3-stage machines, as shown in Fig. 2.41. In this 3-stage version it is seen that the flow system is still very simple with single inlet and exhaust valves. The three displacers are moved together.

The analysis of the cycle (Gifford, 1965) is complicated by the fact that, as with the Stirling cycle, not all the gas goes through the same sequence. For this reason it is impossible to represent the cycle as a single trace on a $T–S$ diagram.

From the First Law it is evident that for a perfectly insulated system the refrigeration effect, at steady state conditions, will be equal to the difference in enthalpy between the gas entering through the inlet valve and that leaving through the exhaust valve. If the regenerator is ideal then the nett heat rejected will be equal to the Joule heating in volume 1, and in this sense the refrigerating effect is governed by what happens at the warm end of the regenerator.

The total refrigeration per stroke for an idealized unit which has a regenerator of zero free volume, no flow resistance and perfect heat exchange, is given by Gifford (1965) as $(pi-p_0)V$, where pi and p_0 are the inlet and outlet pressures and V the free volume of the cylinder. He also investigated the magnitudes of the various losses in a practical machine. For a machine operating with helium between pressures of 14·5 and 1 bar producing refrigeration at 80°K from supply gas at 300°K he measured an actual refrigerator duty equal to 68% of the ideal for a regenerator efficiency of 98·2% (corresponding to a mean ΔT of 3°C).

Gifford and McMahon (1959) described a variant of this machine in which the upper gas volume was eliminated and the displacer or piston was connected to a crank so that work could be generated. Starting with the piston in its lowest position the high pressure gas is supplied to the lowest point of the cylinder, through the regenerator. The admission stroke then commences and feed gas does work on the lower face of the piston. At an appropriate intermediate point the inlet valve is closed and the feed gas expands isentropically, doing more work on the piston. At top dead centre the cylinder is filled with gas at its lowest temperature. The exhaust valve now opens and the cold gas is expelled at constant pressure through the regenerator as the piston descends. The nett cooling effect is now equal to the work generated by the piston less heat inleak and regenerator losses. This type of refrigerator is particularly suited to miniaturization.

Both the above machines could be made more efficient, at the price of increasing mechanical complexity, by replacing the external gas supply (or compressor) and control valves by a piston and cylinder assembly in closed circuit. At this point, however, it becomes almost indistinguishable from the Stirling machine.

D. PULSE TUBE

When a gas at room temperature is admitted to one end of a tube, closed at the far end, so that the pressure in the tube is raised, there will be a tendency for a temperature gradient to be established within part of the tube. This gradient will be most pronounced if the gas enters with plug flow, without turbulent mixing in the tube, and with minimal heat transfer to the wall. Then all the gas initially within the tube will have undergone isentropic compression, its temperature T being given by the usual relationship.

$$\frac{T}{T_0} = \left(\frac{P}{P_0}\right)^{(\gamma-1)/\gamma}$$

where the subscript 0 implies the initial condition. This gas will have been displaced towards the closed end of the tube. Between this isothermal, increased temperature, region and the gas at the open end of the tube, which will still be at T_0, there will be a temperature gradient.

If heat is now rejected in the region of the closed end of the tube to restore the gas temperature to near T_0, and the pressure suddenly released through the open end of the tube, the gas will expand by a near-isentropic process back to its original pressure and reoccupy most of the tube. This gas will be at a temperature below T_0, and therefore will be capable of performing a refrigeration duty.

The method of harnessing this effect cyclically to produce a heat-pumping or refrigeration action was discovered more or less simultaneously by Gifford and Longsworth (1964) and Ruhemann, and is called the Pulse Tube refrigerator.

Gifford's design is shown schematically in Fig. 2.42, together with the temperature profiles established under pseudo steady-state conditions. Feed gas from the inlet valve is cooled in the regenerator and passes unchanged through the no. 1 heat exchanger into the pulse tube. The rapid compression raises the temperature profile from the dotted curve to the full curve, also displacing the gas into the no. 2 heat exchanger. Compression is followed by a quiescent period at near constant pressure during which heat is rejected to this exchanger, and also to the wall of the tube. The outlet valve is then opened and the pressure rapidly released. The temperature profile drops to the lower value and gas is displaced through no. 1 heat exchanger. All but the initial mass of gas passing through this exchanger will have a temperature lower than T_c, and therefore refrigeration is performed as the exit gas warms to a value near to T_c. The gas leaves the warm end of the regenerator at a temperature only a few degrees below the supply temperature. The nett refrigeration, for a well insulated system, is clearly equal to the heat rejected in no. 2 heat exchanger less the small enthalpy difference between the inlet and outlet streams at the warm end of the regenerator.

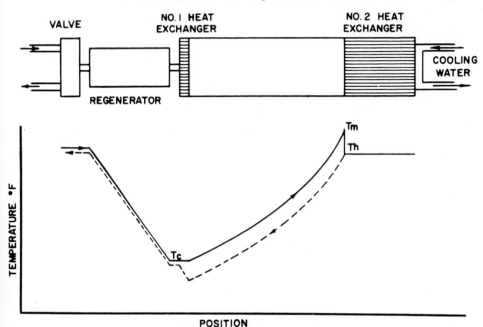

Fig. 2.42. The simple pulse tube refrigerator.

Contrary to his initial expectations Gifford found that it was unnecessary for some of the gas to undergo complete displacement from one heat exchanger to the other in order to get effective heat pumping. Typically he found that with a pulse tube 150 mm long and 19 mm diameter, operating with helium cycling between 20 and 10·5 bar at 42 cycles/min, a refrigeration

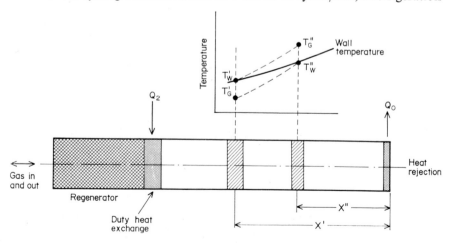

Fig. 2.43. The heat-pumping action of the pulse tube.

temperature of $-92°C$ could be attained. Due mainly to its higher value of γ helium gave a substantially better performance than air.

The definitive description of the underlying heat pumping mechanism was presented by Gifford and Longsworth (1965). They appreciated the crucial contribution of heat transfer to the wall of the pulse tube. A modified version of their theory may be presented in the following terms.

Referring to Fig. 2.43 a small disc element of gas is compressed and displaced from position X' to X'' (measured from the closed end of the tube). It is assumed that the disc of gas was initially in thermal equilibrium with the wall at temperature T'_w. If the movement is sufficiently rapid for heat transfer during transit to be negligible, then it will undergo isentropic compression from T'_w to T''_G. During the quiescent period the gas will reject heat to the wall; in the limit achieving the wall temperature T''_w. Now rapid isentropic expansion occurs and the element of gas finds itself back essentially at the starting position. Its temperature is now T'_G as a result of isentropic expansion. The cycle is completed by a second quiescent period during which the gas warms up to T'_w, absorbing heat from the wall. In this way heat is pumped along the wall against the temperature gradient.

Clearly any heat transfer which occurs during the compression or expansion stages will also contribute to the heat pumping process, but since the temperature heads for heat exchange will be smaller it will be less effective. Since refrigeration performance is directly determined by the amount of heat transfer which occurs, optimum design requires that the quiescent periods should be sufficient to allow say 75% of the available heat transfer to occur between the gas and adjacent wall, whilst compression and expansion periods should be as short as possible consistent with minimal turbulence.

A full analysis is further complicated, as pointed out by Gifford (1965), by the gradient of velocity which inevitably occurs within the gas near to the tube wall. As the surface of the tube is approached the gas displacement will fall off to zero. Gifford claims that this surface effect is advantageous because it will tend to increase the temperature gradient along the tube. Viewed in a larger context it will be seen that the existence of the laminar sub-layer will create an unfavourable radial temperature profile, and the nett effect will be disadvantageous.

Longsworth (1966) has investigated the effect of geometrical factors and gas properties on pulse tube performance. To optimize the design of the pulse tube refrigerator the following criteria are significant.

Pressure. The mean pressure should be as high as possible since the higher the heat capacity per unit volume of the displaced gas the higher will be the refrigerating duty of a given size of tube.

Tube diameter. For a given average gas displacement per cycle (related to tube length) the larger the diameter the closer the approach to plug flow, but the longer the path length for heat conduction to the wall. Also longitudinal heat conduction will increase in relation to radial conduction. Longsworth (1966) finds that heat pumping is characterized by the Fourier Number, N_F, where $N_F = \alpha/nD^2$ and α is the thermal diffusivity of the gas, n the cycle frequency, and D the tube diameter.

Tube length. The longer the tube the smaller will be the losses due to back conduction along the wall of the tube and axially through the gas. At the same time the gas velocity, for a fixed cycle time, will increase and with it the departures from plug flow. Longsworth claims that heat pumping rate is directly proportional to tube length, other things being equal.

Cycle time. Increased frequency will tend towards increased refrigeration capacity, but it will be off-set by diminished efficiency of heat transfer.

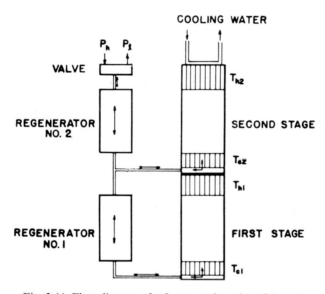

Fig. 2.44. Flow diagram of a 2-stage pulse tube refrigerator.

The pulse tube analysis given by Colangelo *et al.* (1967) does not include many of the factors mentioned above. Gifford and Kyanka (1966) discussed the problems of linking the pulse tube to a piston and cylinder assembly, so that the expansion work could be recovered. He showed that potentially the pulse tube is capable of high thermodynamic efficiency. Gifford (1964) also showed how the pulse tube could be cascaded for the production of lower temperatures. Figure 2.44 shows the arrangement of a 2-stage version with which he achieved a refrigeration temperature of about 140°K, whilst with a 3-stage unit he obtained about 100°K.

E. HELIUM DILUTION REFRIGERATOR AND RELATED UNITS

The principles of the Dilution cycle have been presented briefly in Section IIIB. In practice there are many other factors which limit performance and which dominate design. These are mainly concentrated in the lowest temperature region of the cycle, and will be discussed in relation to Fig. 2.45.

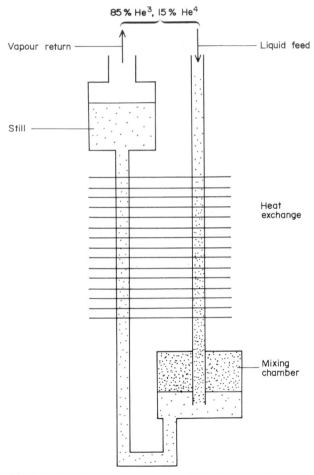

Fig. 2.45. Low temperature section of the dilution refrigerator.

As the liquid feed is cooled in the heat exchanger it passes the temperature threshold (about $0.87°K$) at which two liquid phases begin to separate out, and the separation is progressive. Data on the equilibrium properties of the phases is given by D. S. Betts (1968) and Conte (1970). It is desirable that the smaller volume of superfluid He^4 should separate out as a dispersion of droplets to be carried forward by the liquid He^3. If the He^4 forms a continuous thread or film through the tube or liquid passage then a "fountain" flow opposed to the direction of liquid flow will occur.

When the feed liquid enters the mixing chamber the liquid He^3 will have a high viscosity—approximating that of castor oil at room temperature—and therefore it is desirable that it should be injected well down in the vessel in a way which will promote stirring. Transfer of the He^3 atoms occurs across the liquid-liquid interface producing refrigeration. This transfer will continue so long as there is an osmotic pressure gradient in the columns of superfluid helium leading from the base of the mixing chambers to the still. The rate of transfer is limited by the viscosity of the He^3 atoms interacting with each other, although there is no mechanical drag exerted by the He^4. The temperature of the fluids in the mixing chamber will go down until the refrigeration produced is counterbalanced by heat inleak down the connecting tubes (and their contents) and from the surroundings.

Additional factors which govern the flow of heat and mass include the following:

Kapitza Boundary Resistance to heat transfer. At very low temperatures an additional thermal resistance occurs between liquid helium and any bounding metal surface (see Clement and Frederking, 1966). This insulating effect, termed the Kapitza boundary resistance, varies as T^{-3} and tends to dominate any heat transfer processes below about $0.1°K$. Thus this effect limits the operation of the heat exchanges and also can result in the temperature of the outer wall of the mixing chamber being nearer to the temperature of the feed liquid than of the colder dilute phase. The latter aspect is important both in determining the temperature achieved by the dilution refrigerator and in contacting a test cell with the dilute phase. One of the ways of minimizing the heat exchanges problem is to flow both liquid streams through plugs of sintered copper placed in thermal contact by bonding them on either side of the dividing wall. To minimize heat transfer in the longitudinal direction the total heat exchanger is constructed from a series of short sintered copper units separated by lengths of capillary.

Flow inversion. An early dilution refrigeration in which the heat exchanger comprised two capillaries 0.25 mm bore and 1 m long twisted together and wound round a stainless steel tube of 5.1 mm outer diameter and surrounded by a further tube of 8.25 mm bore, with the dilute phase flowing through the remitting annulus was found not to work. This exchanger was mounted with the warm end uppermost. It was deduced that the gradient of He^3 concentration (due to the viscosity of the He^3 itself) caused the dilute phase in the upper parts of the annulus to be more dense than that at the bottom thus inducing a natural convection flow which negated the osmotic flow. It has been shown by Abel and Wheatley (1968) and others that the flow passages for the dilute

4

phase must be substantially less than 1 mm inner diameter if flow inversion is to be prevented.

Viscous heating. The work done by the diffusing He³ atoms in viscous interaction in passing from the mixing chamber to the still generates heat, part of which is conducted back to the mixing chamber. If the passage way is made wider to reduce the viscous drag then heat conduction through the fluid will become excessive. Abel and Wheatley estimate that the conjunction of these two processes limits the ultimate temperature of continuous dilution refrigeration to about 5 m°K.

Efficient still heating. Heat must be added to the still in such a way that the vapour generated contains the maximum concentration of He³, and the minimum quantity of heat is conducted back to the dilution chamber. Orifices and other devices may be used with advantage to minimize film flow into the exit vapour line. In addition the heater should be in the closest possible thermal contact with the still liquid near to the liquid-vapour interface. It is not satisfactory to wrap the heater around the lower part of the still since this can result in a higher concentration of He³ in this region of the still than at the liquid surface.

A well engineered dilution refrigeration (see Fig. 2.46) can maintain temperatures down to about 15 m°K. The same principle can be exploited in a single-shot device where, due to the fact that heat is not being added continuously with the feed liquid, a lower temperature of 4–5 m°K is obtainable.

H. London (1969) has proposed a further development of the dilution refrigerator which will extend cooling temperature further or increase its capacity in the lower range by effectively making the dilute He³ phase undergo adiabatic expansion. The expansion cylinder consists of a vessel with a volume several times that of the mixing chamber. By the rapid opening and closing of a valve, some of the dilute phase from the base of the mixing chamber is admitted. The vessel is then connected to a reservoir of superfluid He⁴ (for instance the still of the dilution refrigerator) through a superleak and capillary. He⁴ then flows into the vessel causing the feed liquid to expand in volume with a resulting temperature drop. London has shown how to operate this system cyclically.

An alternative method of achieving lower temperature which uses the anomalous properties of solid He³ has been proposed by Pomeranchuk (1950) and developed by Wheatley (1969). As far as is known pure liquid He³ will not solidify even at 0°K without the imposition of a pressure of about 34 bar. The melting curve has a minimum at 0·3°K at which the required pressure is 29 bar. Below 0·3°K the solid is more highly disordered than the liquid, and has a higher entropy. The resultant solid entropy is $R \ln 2$ per mole, whilst the liquid entropy is about 4·8 $RT/°K$, R being the Universal Gas Constant. At 20 m°K the molar liquid entropy is only $\frac{1}{7}$ that of the solid, and therefore

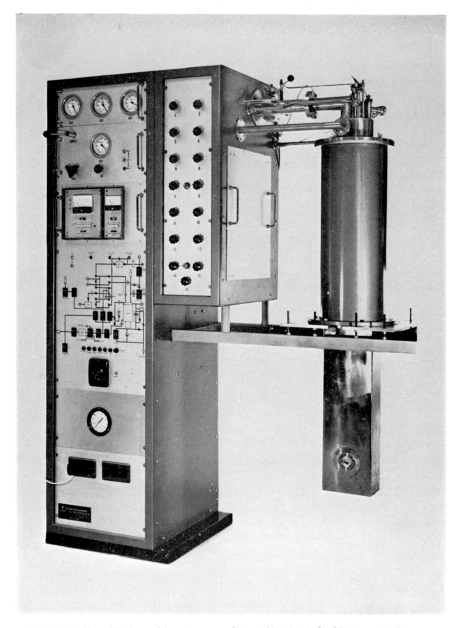

Fig. 2.46. Helium dilution refrigerator manufactured by the Oxford Instrument Company.

adiabatic solidification brought about by pressure will be accompanied by a substantial fall in temperature.

The rate of heat absorption is given by an expression of the following form

$$Q \approx n_3 RT\left(\ln 2 - \frac{b}{T^2} - \gamma T\right) \qquad (2.12)$$

where n_3 is the rate of conversion of He^3 from liquid to solid.

Two main factors limit the cooling achievable by Compressional Cooling. The first is that of transmitting the pressure change into the whole mass of liquid H^3. If this is attempted in the normal way by passing He^3 through a connecting tube to the refrigerator vessel then solidification will occur first in this tube and will result in its blockage. The second is the high viscosity of liquid He^3 in the millidegree temperature range. The resulting viscous dissipation of energy accompanying compression can nearly cancel out the cooling effect of crystallization. One solution to these problems developed by Wheatley has been to contain the pre-cooled liquid He^3 in a flexible capsule which is then compressed by an external atmosphere of superfluid He^4. Cooling to about 2 m°K is claimed by this means.

F. MAGNETIC REFRIGERATOR

A semi-continuous refrigerator for temperatures below 1°K employing the magnetic cycling for a paramagnetic salt has been described by Danut *et al.* (1954). The form of this refrigeration, and its *T–S* cycle are given in Fig. 2.47.

The paramagnetic salt (15 g of $FeNH_4(SO_4)_2 . 12 H_2O$) is contained in a thin-walled stainless steel cannister shown at the centre of the diagram. In good thermal contact with the cannister are two thermal switches, each comprising a short length of superconductor with its own coil. At the prevailing temperature the superconductivity of either heat switch may be destroyed by imposing on it a sufficient magnetic field from the surrounding coil.

At the beginning of the cycle the cannister is thermally connected to a pumped helium bath at temperature $T°$ through switch S_1 whilst switch S_2 is "opened" by energizing its coil. The salt is in a low field H_A, and at a temperature $T_s°$ slightly above that of the helium (see Fig. 2.47(b), point A). The field strength is raised to H_B and heat is continuously rejected to the helium bath to maintain the salt temperature $T_s°$. The thermal switch S_1 is now opened and the external field is reduced to H_C, so that the condition of the salt changes from B to C, the increased randomness of the magnetic ions being compensated by a drop in temperature, the entropy remaining constant. The thermal switch S_2 is now closed and heat is absorbed isothermally by the salt at T_s whilst the field strength is further reduced from H_c to H_D. A small temperature driving force is required so that the effective refrigerator temperature is T_L. The cycle is completed by opening S_2 and restoring the field strength to H_A.

The minimum temperature is achieved by delaying closure of the lower switch until the field is reduced to H_D, when a temperature corresponding to the point C^1 would be reached but then the duty would be a non-isothermal one along the curve C^1D. The cooling in this case, given by the shaded area under C^1D would be very much less than in the isothermal case (area under CD).

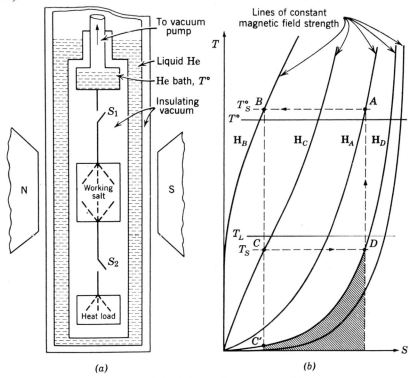

(a)   (b)

Fig. 2.47. Principle of the cyclic magnetic refrigerator.

The identified cycle ABCD is, of course, equivalent in performance to a Carnot cycle. Eddy current and heat transfer losses will lend to a practical performance which is considerably inferior.

G. HELIUM VORTEX REFRIGERATOR

The helium vortex technique was developed by F. A. Staas of the Philips Research Laboratory, Eindhoven, in 1970. It relies on the special flow characteristics of superfluid He⁴ to generate a heat pumping action, and is capable of giving continuous refrigeration in milliwatt quantities in the range 1·3 to 0·7°K.

The principle of the vortex refrigerator is illustrated in Fig. 2.48. It comprises the pump B and the refrigerator A, both connected to the base of a liquid helium bath at about $1.5°K$ (3–4 mmHg absolute pressure). The bath liquid enters the super-leak S_2 which consists of a porous plug of very fine powder. Effectively only the superfluid helium can pass through the superleak whilst the normal helium is retained in the bath. Below S_2 is a small heater

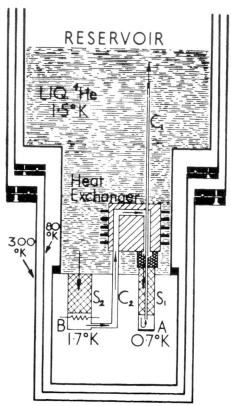

Fig. 2.48. Flow system for the Helium vortex refrigerator.

which raises the temperature of the helium and therefore reduces the concentration of superfluid helium. Thus a concentration gradient is established across the super-leak which causes a positive flow of superfluid helium across S_2 and raises the pressure in B. A flow is thereby induced in the capillary C_2, and a pressurized stream of liquid helium, restored by heat exchange to approximately $1.5°K$ is supplied to the top of the other super-leak S_1. Frictionless flow of helium II again occurs through the porous plug into the refrigerator vessel A and up through the capillary C_1. The later capillary is sized so that the critical flow velocity of superfluid helium is exceeded and hence normal liquid helium is carried out with it. The entropy and temperature of A is thereby reduced until a thermal balance is created between the heat flow into A and the loss of heat due to vortex trapping of normal helium in the discharge through C.

In the unit developed by the Philips Company the refrigerating capacity is 10 mW at $0.8°K$ for a unit with overall dimensions of 5 cm diameter and 15 cm high. Maintenance of the main bath at a temperature of $1.5°K$ means that the required vacuum equipment is relatively simple.

VII. References

Abel, W. R. and Wheatley, J. C. (1968). *Physics Letters* 27a, 599–610.
Ambler, E. and Hudson, R. P. (1955). *Rep. Prog. in Physics* 18, 251.
Allen, J. F. (1966). "Superfluid Helium". Academic Press, London and New York.
Baldus, W. and Sellmaier, A. (1964). *Int. Adv. in Cryogenic Eng.* 10, 13.
Baldus, W. and Sellmeier, A. (1967). *Int. Adv. in Cryogenic Eng.* 13, 434–440.
Barber, W. R. and Haselder, G. G. (1957). *Trans. Inst. Chem. Eng.* 35, 77–86.
Betts, D. S. (1968). *Contemporary Phys.* 9, 97–114.
Birkholz, U. (1961). Kaltechnik, 13, 335–339.
Blatt, F. J. (1962). *Phil. Mag.* 7, 715–722.
Carter, C. N. and Lewis, K. G., Maddock, B. J. and Noe, J. A. (1969). "Proceedings of the Low Temperatures and Electric Power Conference, London", pp. 31–43. Published Int. Inst. of Refrigeration, Paris.
Clark, J. A. and Miller, R. W. (1940). *Oil and Gas J.* 39, 48–53.
Clark, J. A. and Miller, R. W. (1941). *Oil and Gas J.* 39, 57–63.
Colangels, J. W., Fitzpatrick, E. E., Rea, S. N. and Smith, J. L. (1967). *Int. Adv. in Cryogenic Eng.* 13, 494–503.
Collins, S. C. (1952). *Science* 116, 289–299.
Coute, R. R. (1970). "Elements de Cryogenie." Masson et Cie, Paris, pp. 248–268.
Croft, J. (1961). *Progress in Cryogenics* 3, 1–22.
Darradena, M. (1969). "Proceedings Int. Conference on Liquefied Natural Gas, London", pp. 240–266. Inst. Mech. Eng., London.
Dros, A. A. (1964). *Int. Adv. In Cryogenic Eng.* 10, 7–12.
Gianque, W. F. and MacDougall, D. P. (1933). *Phys. Rev.* 43, 768–775.
Gifford, W. E. and McMahon, H. O. (1959). *Int. Adv. in Cryogenic Eng.* 5, 368–372.
Gifford, W. E. and Longsworth, R. C. (1964). *Int. Adv. in Cryogenic Eng.* 10, 69–79.
Gifford, W. E. (1965). *Int. Adv. in Cryogenic Eng.* 11, 152–159.
Gifford, W. E. and Longsworth, R. C. (1965). *Int. Adv. in Cryogenic Eng.* 11, 171–9.
Gifford, W. E. and Kyanka, G. H. (1966). *Int. Adv. in Cryogenic Eng.* 12, 619–630.
Goldsmid, H. J. (1964). "Thermoelectric Refrigeration." Heywood, London.
Grenier, M. R. (1967). "Proceedings XII Int. Congress of Refrigeration, Madrid", 1, paper 1.35, 7 pp.
Gulyaev, A. I. (1966). *Int. Chem. Eng.* 6, 300–302.
Haselden, G. G. (1967). "Proceedings XII Int. Congress of Refrigeration, Madrid."
Heer, C. V., Barnes, C. B. and Daunt, J. G. (1954). *Rev. Sci. Inst.* 25, 1088–1092.
Hilsch, J. (1947). *Rev. Sci. Inst.* 18, 108–113.
Kleemenko, A. P. (1959). "Proceedings X Int. Congress of Refrigeration, Copenhagen", 1, 34–39. Pergamon Press, London.
Kohler, J. W. L. (1960). *Progress in Cryogenics* 2, 43–67.

Kohler, J. W. L. and Jonkers, C. O. (1954). *Philips Tech. Rev.* **16**, 69–78; 105–115.

Kropschot, R. H., Birmingham, B. W. and Mann, D. B. (1968). "Technology of Liquid Helium." N.B.S. Monograph III, Washington.

Kurti, N. and Simon, F. E. (1935). *Proc. Roy. Soc.* (*London*) **A149**, 152–170.

London, H. (1969). British Patent Specification No. 1, 117, 246.

London, H. Clarke, G. R. and Mendoza, E. (1962). *Phys. Rev.* **128**, 1992–1999.

Longsworth, R. C. (1966). *Int. Adv. in Cryogenic Eng.* **12**, 608–618.

Lyon, D. N. (1963). *In* "Cryogenic Technology" (Vance, R. W., ed.), p. 43. John Wiley and Sons. Inc., New York.

Maiuri, G. (1939). British Patent Nos. 494, 693 and 530, 044.

McMahon, H. O. and Gifford, W. E. (1959). *Int. Adv. In Cryogenic Eng.* **5**, 354–367.

Metenin, V. 1. (1964). *Int. Chem. Eng.* **4**, 461–464.

Morrison, A. (1954). *Power Eng.* **58**, 93–100.

Newton, A. B. (1967). "Proceedings XII Int. Congress of Refrigeration, Madrid", **2**, pp. 1079–1089.

Pickard, G. L. and Simon, F. E. (1948). *Proc. Phys. Soc.* **60**, 405–502.

Pomeranchuk, I. (1950). *Zh. Eksp. Teor. Fiz.* **20**, 919–925.

Rietdijk, J. A. (1966). "Liquid Helium Technology." Int. Inst. of Refrigeration, Commission 1, Boulder, U.S.A., pp. 241–248. Int. Inst. Refrigeration, Paris.

Rubemann, M. (1947). British Patent Nos. 643, 886.

Rubemann, M. (1949). "The Separation of Gases", 2nd Edn. Clarendon Press, Oxford.

Salama, C. and Eyre, D. V. (1967). *Chem. Eng. Prog.* **63**, 62–67.

Scott, R. B. (1959). "Cryogenic Engineering." D. Van Nostrand and Co. Princetown.

Scott, R. B., Denton, W. H. and Nicholls, C. M. (1964). "Technology and Uses of Liquid Hydrogen." Pergamon Press, Oxford.

Singleton, A. H. and Lapin, A. (1965). *Int. Adv. in Cryogenic Eng.* **11**, 613–617.

Smith, G. E. and Wolfe, R. (1962). *Appl. Phys.* **33**, 841–850.

Trepp, C. (1966). "Liquid Helium Technology." Int. Inst. of Refrigeration, Commission 1, Boulder, U.S.A., pp. 215–226. Int. Inst. of Refrigeration, Paris.

Trepp, C. (1969). "Proceeding of the Low Temperatures and Electric Power Conference, London", pp. 31–43. Int. Inst. of Refrigeration, Paris.

van Deemter, J. J. (1952). *Applied Sci. Res. A* **3**, 174–179.

Vogelhuber, W. W. and Parish, H. C. (1968). "Proceedings 1st Int. Conference on LNG, Chicago", paper 28, 12 pp. Inst. Gas Technology, Chicago.

Wheatley, J. C. (1969). *Int. Adv. Cryogenic Eng.* **15**, 415–421.

Winters, A. R. and Snow, W. A. (1965). *Cryogenic Eng.* **11**, 116–125.

Chapter 3

Heat Transfer

J. A. CLARK

*Department of Mechanical Engineering, University of Michigan,
Ann Arbor, Michigan, U.S.A.*

and

R. M. THOROGOOD

Air Products, Ltd, New Malden, Surrey, England

I. Introduction

Cryogenic processes have created several unique problems in heat transfer. The handling and transporting of cryogenic fluids at very low temperatures have necessitated the development of specialized insulating methods and design techniques. The high energy requirement of refrigeration at low temperatures has emphasized the need for efficient heat exchanger designs with close temperature approaches, and consequently for improved accuracy of design. The occurrence of boiling and condensing heat transfer regimes and of two phase flow in cryogenic fluids is very frequent. Thermal conduction within the supporting structure of cryogenic containers requires special design attention if heat gain is to be minimized. The condensation of water vapour, carbon dioxide and other contaminants on cooled surfaces is another problem of importance in cryogenic systems.

At low temperatures the physical properties of many substances are significantly temperature dependent. This fact has made it necessary to consider the question of variable properties in the analysis and calculation of heat transfer processes. Important cases are changing thermal conductivities at low temperatures and varying convective phenomena in the region of the critical point.

A significant characteristic of most cryogens is that they behave as "classical" fluids. That is, their physical behaviour follows the well-established principles of mechanics and thermodynamics and they obey the laws of similarity. Thus, the fundamental concepts of fluid behaviour based on the classical laws of physics may be expected to apply. Accordingly, convective heat transfer correlations can be formulated in terms of such well-known dimensionless quantities as the Nusselt, Reynolds, Prandtl and Grashof numbers, and the length-to-equivalent-diameter ratio.

An important exception to classical behaviour at very low temperatures is liquid helium (He^2) at temperatures below $2 \cdot 19°K$, and of certain other "electron gases" in solids (Lane, 1962).

Several useful sources of references on the subject of cryogenic heat transfer, follow. Each year the proceedings of the Cryogenic Engineering Conference, held in the United States, is published under the title "Advances in Cryogenic Engineering" (Plenum Press). The publication contains review papers on all aspects of cryogenic engineering. The journal *Cryogenics* has been published in England by Heywood and Co., Ltd., London, since 1960.

Useful textbooks include: "Cryogenic Engineering" by Scott (1959), "Technology and Uses of Liquid Hydrogen", edited by Scott, Denton and Nicholls (1964), "Applied Cryogenic Engineering", edited by Vance and Duke (1962) and "Cryogenic Technology", edited by Vance (1964).

The publication of the various commissions in the "Proceedings of the International Congress of Refrigeration" frequently contain papers on cryogenic heat transfer. Other sources in the United States of occasional papers and reports on cryogenic heat transfer are the "Transactions of the American Society of Mechanical Engineers" (ASME), especially its *Journal of Heat Transfer (Series C)*, *The American Institute of Chemical Engineers Journal (AIChE)* and the *Journal of the American Institute of Aeronautics and Astronautics (AIAA)*.

II. Basic Design Concepts

The design of all but the simplest heat transfer equipment requires two consecutive stages. The first stage defines the basic duty required of the equipment in terms of process, flows and temperatures. This may be expressed in non-dimensional form, and allows the designer to assess the practicability of the process in relation to the heat exchanger design. The second stage is the transformation of the non-dimensional design into practical equipment. This requires determination of fluid properties, calculation of heat transfer coefficients and pressure drops, and optimization of the heat transfer surface area and its disposition in relation to the mechanical design. Interaction may occur between these first and second phases of design in accordance with the experience and skill of the designer.

A. NON-DIMENSIONAL RATING OF HEAT EXCHANGERS

The difficulty for interchange of heat between fluids may be characterized in terms of their relative temperatures and heat capacities. A commonly encountered non-dimensional characteristic is the number of transfer units (N_T) which is expressed in terms of exchanger entry and exit temperatures. This is defined for the cold stream in a two stream exchanger by:

$$N_{T_c} = \int_{T_{ci}}^{T_c} \frac{dT_c}{T_h - T_c} \tag{3.1}$$

Subscripts c and h denote cold and hot streams, respectively, and i the inlet condition.

For heat exchange between streams of constant heat capacity as shown in Fig. 3.1, the N_T may be calculated from:

$$N_{Tc} = \frac{T_{co} - T_{ci}}{\Delta T_{lm}} ; \qquad N_{Th} = \frac{T_{hi} - T_{ho}}{\Delta T_{lm}} . \tag{3.2}$$

The logarithmic mean temperature difference, ΔT_{lm}, is defined by:

$$\Delta T_{lm} = \frac{(T_{hi} - T_{co}) - (T_{ho} - T_{ci})}{\ln (T_{hi} - T_{co})/(T_{ho} - T_{ci})} . \tag{3.3}$$

The general case of heat exchange between fluids with varying specific heats requires integration along the cooling curves (temperature versus stream enthalpy relationship). This may be carried out conveniently by reduction of the cooling curves to several straight line sections. Multiple fluid exchangers must be reduced to pseudo two-stream systems by separately summing the respective cooling and warming fluids.

The N_T concept may be applied strictly only to systems without phase change. It is also of value for comparison of process designs in systems where temperature-dependent phase changes occur (multicomponent systems). It is of no value for boiling or condensing single component fluids; these systems are generally assessed by temperature difference alone. Its particular value is in giving an indication of the required length of a heat exchanger and thus indicating whether a process design is feasible.

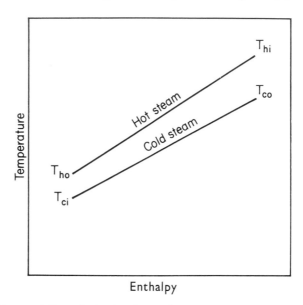

Fig. 3.1. Representation of a heat exchanger by a temperature versus enthalpy plot.

The efficiency or effectiveness of a heat exchanger is the ratio of the temperature change experienced by the stream of lower heat capacity to the temperature change it would have undergone in an exchanger of infinite area. Thus, for fluids of constant specific heat where the warm stream has the lower heat capacity, the effectiveness, E', is defined by:

$$E' = \frac{T_{hi} - T_{ho}}{T_{hi} - T_{ci}} \tag{3.4}$$

Exchanger effectiveness, heat capacity ratio and N_T have been used by Kays and London (1964) and Domingos (1969) to analyse the thermal behaviour of assemblies of exchangers.

B. SURFACE AREA REQUIREMENTS

The local heat flux, dq/dA is related to the local temperature driving force in accordance with the expression:

$$\frac{dq}{dA} = h\Delta T \tag{3.5}$$

The constant of proportionality, h, is termed the heat transfer coefficient and has the dimensions W/cm^2 °C or B.T.U./hr ft^2 °F. In many situations the same quantity of heat will pass successively through several media, then h is defined separately for each medium and suffixed accordingly. Heat flux through solid media is determined directly in terms of thermal conductivity and thickness.

$$\frac{dq}{dA} = \frac{k}{s}\Delta T \tag{3.6}$$

Heat transfer, under steady state conditions, through several media in combination is denoted by the overall coefficient U.

For the commonly encountered situation of heat transfer between two fluids across a wall, the overall heat transfer is composed of at least three terms in series: a film convection component on the cold side; a wall conduction component; and a film convection component on the hot side. The reciprocal of the overall heat transfer coefficient (termed the overall thermal resistance) is determined from the series addition of the component thermal resistances: Thus:

$$\frac{1}{U_c} = \frac{1}{h_c} + \frac{s}{(A_w/A_c)k_w} + \frac{1}{(A_h/A_c)h_h} \tag{3.7}$$

U_c is based upon unit area of wall surface on the cold stream side. s is the wall thickness and A_w the average wall area.

The terms of equation 3.7 are representative of a situation in which the heat transfer surface is classified as primary (i.e. without fins) and clean. For surfaces which are subject to fouling an additional term is frequently added to the equation as a fouling resistance. Where secondary heat transfer surface such as finning is utilized then the appropriate convective term(s) must be modified. Thus the hot side term would become, for example:

$$\frac{1}{[(A_{h1} + \eta A_{h2})/A_c]h_h} \tag{3.8}$$

where A_{h1} represents primary surface, A_{h2} the secondary, or fin, surface and η is the thermal efficiency of the secondary surface.

The total heat flux across a heat exchanger may be expressed:

$$\int dq = q = UA\Delta T_m \tag{3.9}$$

where ΔT_m is the mean effective temperature difference and U is the mean value of the overall heat transfer coefficient. For process design purposes the product UA $(= q/\Delta T_m)$ is a useful measure of the process requirements for the heat exchanger. The experienced designer can readily assess the feasibility of a process cycle by an examination of the calculated values of UA. The mean effective temperature difference depends not only on the specified cooling curves and temperature approach but also on the flow geometry of the exchanger. For countercurrent or parallel flow operation ΔT_m is equal to the logarithmic mean temperature difference ΔT_{lm}. The solution is more complex for cross-flow exchangers but has been treated by Nusselt (1930). When the specific heat of either a cold or hot stream is variable with temperature or the value of U changes rapidly it is necessary to calculate the exchanger by subdivision into sections which may each be approximated by linear cooling curves.

III. Conduction Heat Transfer

Probably the principal new problems associated with conduction heat transfer at cryogenic temperatures are those of variable thermal properties and low temperature insulation, as indicated in the Introduction. In addition, the principles of conduction heat transfer and diffusion phenomena have been recently applied to some important applications in cryogenics, namely, interfacial processes in single and multicomponent systems. These will be discussed later in this section.

A. CONDUCTION IN SOLIDS†

The differential equation governing the steady and unsteady conduction of heat in isotropic substances having variable thermal properties and including the influence of internal heat generation is

$$\frac{\partial T}{\partial t} = \frac{1}{\rho C p}\left[\frac{\partial}{\partial x}\left(k\frac{\partial T}{\partial x}\right) + \frac{\partial}{\partial y}\left(k\frac{\partial T}{\partial y}\right) + \frac{\partial}{\partial z}\left(k\frac{\partial T}{\partial z}\right)\right] + \frac{P''_s(x, y, z, t)}{\rho c p} \tag{3.10}$$

This equation is written in cartesian coordinates for convenience only. The properties k, ρ, cp are usually temperature dependent for a given substance. Typical low temperature variations in k and c_v (for solids c_v and c_p are essentially the same) are shown in Figs. 12.4 to 12.7. Equation 3.10 is non-linear owing to the temperature dependency in k, ρ and c_p and accordingly is very difficult to solve. A simple transformation (Clark, 1969), however, can be used to aid in the solution. A new variable E is defined as

$$E = \int_{T_R}^{T} k(T)\, dT, \tag{3.11}$$

† The emphasis in this section will be on the influence of temperature dependent physical properties. Heat conduction analysis and calculation for uniform properties is well documented (Arpaci, 1966; Carslaw and Jaeger, 1959; Scheider, 1963; Kreith, 1965).

where T_R is an arbitrary reference temperature. Hence, we may formulate the space and time derivatives of T in equation 3.10 as

$$\frac{\partial T}{\partial t} = \frac{1}{k(T)}\left(\frac{\partial E}{\partial t}\right) \tag{3.12}$$

and

$$\frac{\partial}{\partial x}\left(k\frac{\partial T}{\partial x}\right) = \frac{\partial^2 E}{\partial x^2}, \text{ etc.} \tag{3.13}$$

With these transformations solutions to heat conduction problems are sought in terms of the function E. Since E is a unique function of T by equation 3.11 the temperature distribution $T(x,y,z,t)$ is determined from the function $E(x,y,z,t)$. Thus, equation 3.10 becomes

$$\frac{\partial E}{\partial t} = a(T)\left[\frac{\partial^2 E}{\partial x^2} + \frac{\partial^2 E}{\partial y^2} + \frac{\partial^2 E}{\partial z^2}\right] + a(T)p_s''(x, y, z, t), \tag{3.14}$$

where $a(T)$ is the variable thermal diffusivity, defined as $k(T)/\rho(T)C_p(T)$, and is a function of temperature. Equation 3.14 is still non-linear since both $a(T)$ and E are functions of temperature. However, it has been put into a more useful form than equation 3.10. In case $a(T)$ is a severe function of temperature it will be necessary to employ numerical methods to the integration of equation 3.14, something which will be discussed later. For numerical formulations equation 3.14 is a more convenient and useful representation than equation 3.10.

There are certain practical circumstances in which $a(T)$ is much less variable with temperature than is $k(T)$. This is a consequence of similar temperature variations of $k(T)$ and $C_p(T)$ in some regions such that their ratio remains essentially independent of temperature. For these conditions $a(T)$ may be approximated by a constant value, designated as a^*. Equation 3.14, for no internal heat generation, then becomes,

$$\frac{\partial E}{\partial t} = a^*\left[\frac{\partial^2 E}{\partial x^2} + \frac{\partial^2 E}{\partial y^2} + \frac{\partial^2 E}{\partial z^2}\right] \tag{3.15}$$

This result will be recognized as the classical linear diffusion equation for which a vast number of analytical solutions have been generated for various boundary conditions (Arpaci, 1966; Carslaw and Jaeger, 1959). For a one-dimensional case equation 3.15 is written

$$\frac{\partial E}{\partial t} = a^*\left(\frac{\partial^2 E}{\partial x^2}\right) \tag{3.16}$$

Further, for a convective heat transfer at $x = 0$ and $x = 2L$ (Fig. 3.2) with an ambient fluid at T_∞, the typical convective boundary condition is (Clark, 1969)

$$\frac{\partial E(0, t)}{\partial x} = \frac{h}{k_\infty}[E(0, t) - E_\infty] \tag{3.17}$$

where, $k_\infty(T_\infty - T_R) = E_\infty$. The initial condition would be written $E(x, 0) = E_i$. Hence, we may write equation 3.16 and its boundary condition equation 3.17 in the following form,

$$\frac{\partial E^*(x^*, F_0)}{\partial F_0} = \frac{\partial^2 E^*(x^*, F_0)}{\partial x^{*2}} \tag{3.18}$$

and

$$\frac{\partial E^*(0, F_0)}{\partial x^*} = B_i[E^*(0, F_0)] \tag{3.19}$$

where

$$E^* = \frac{E(x, t) - E_\infty}{E_i - E_\infty}, \qquad B_i = \frac{hL}{k_\infty}$$

$$F_0 = \frac{a^* t}{L^2}, \qquad x^* = \frac{x}{L}. \tag{3.20}$$

in which L is a characteristic length.

A comparison of equations 3.18 and 3.19 for variable properties with the corresponding differential equation and boundary conditions for uniform properties shows them to be identical. Hence, for the same geometrical shape and initial condition their solutions also will be identical. This means that the various heat conduction charts formulated on the basis of uniform properties, such as the extensive tabulation of Schneider (1963), can be employed for the solution of a variable property problem in terms of the function E. Multi-dimensional problems also may be solved by the product method for those geometric shapes for which analytical solutions are available.

The analytical solution to equations 3.18 and 3.19 for a slab is given in Fig. 3.2 in chart form as $(1-E^*)$ for variable properties of k, ρ, and c_p for the case of constant thermal diffusivity a^*.

Analytical solutions for steady-state heat conduction problems having variable properties may be obtained from solutions to equation 3.14 with $(\partial E/\partial t)$ set equal to zero. Thus, the governing equation is,

$$\frac{\partial^2 E}{\partial x^2} + \frac{\partial^2 E}{\partial y^2} + \frac{\partial^2 E}{\partial z^2} + p_s''(x, y, z) = 0. \tag{3.21}$$

For a convective heat transfer to a wetting fluid at T_∞ the boundary condition would be of the type given in equation 3.17. In the absence of internal heat generation, we have then

$$\frac{\partial^2 E}{\partial x^2} + \frac{\partial^2 E}{\partial y^2} + \frac{\partial^2 E}{\partial z^2} = 0. \tag{3.22}$$

This is the classical Laplace equation well-known in field theory and in the study of diffusion phenomena. As may be noted the variable property

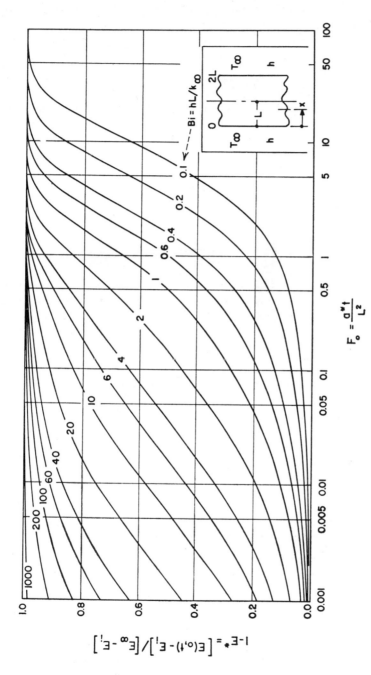

Fig. 3.2. Temperature response of a slab, $0 \leqslant x \leqslant L$ after sudden exposure to a uniform temperature convective ambient T_∞ at $x = 0$ (Schneider, 1963).

5

problem in this case involves only variations in $k(T)$ and not in the thermal diffusivity $a(T)$.

Analytical solutions to equations 3.21 or 3.22 are rather broadly available in the standard literature (Clark, 1971; Arpaci, 1966; Kreith, 1965) for a wide variety of problems. Such solutions also may be found by analogical techniques such as the analogue field plotter. For these reasons they will not be reproduced here. The main difference in the present formulations from those

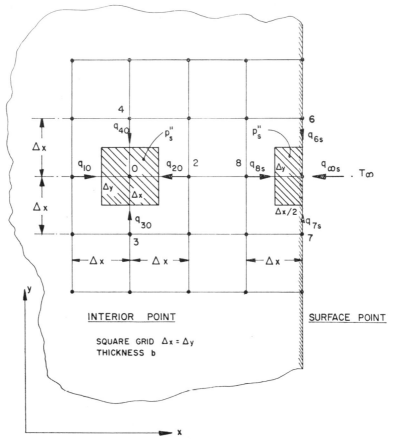

Fig. 3.3. Finite-difference network for a typical interior and surface point using a square grid $\varDelta x = \varDelta y$.

currently available rests in the treatment of the function E, equation 3.11, to determine the temperature distribution rather than the temperature itself. This is the principal distinction of the variable property problem.

For those cases of unsteady heat conduction in which the property variation is very great, and the constant thermal diffusivity approximation of equation

3.15 cannot be made satisfactorily it, will be necessary to solve equation 3.14 by numerical methods using a digital computer. In this discussion we shall study the case without internal heat generation, $p''_s(x,y,z,t) = 0$ as this function, even with spatial and time variations, is a simple additive term in the numerical formulation which does not complicate the calculation in any special way. The most convenient form for a numerical calculation is to arrange the equations in an explicit formulation (Barakat and Clark, 1966). This enables a marching-type solution and avoids the time-consuming iterative computer procedures of the implicit-type formulations. The development to be given here will be for the two-dimensional region shown in Fig. 3.3 which has a convective heat transfer at the wetted boundary. A detailed development of this form of the equations is given in Kreith (1965) and McAdams (1954). For the interior point O, using a square grid, the value of the function E_0 at the $(n+1)$ time interval may be computed in terms of the functions at the adjacent lattice points at the $(n$th$)$ time interval (Clark, 1969) as,

$$E_0^{n+1} = \frac{E_1^n + E_2^n + E_3^n + E_4^n + (M_0 - 4)E_0^n}{M_0} \tag{3.23}$$

where

$$M_0 = \frac{(\Delta x)^2}{a(T_0)\Delta t} \tag{3.24}$$

For variable properties $a(T_0)$ will vary with time and space and it will be necessary to up-date the calculation of M_0 at the start of each new sequence of calculations. This may require adjustment in the time interval Δt. The numerical value of M_0 is related to the stability of the calculation and will be discussed below.

For the surface point S the value of the function E_s at the $(n+1)$ time interval is given (Clark, 1969) as

$$E^{n+1} = \frac{2E_8^n + E_6^n + E_7^n + 2N_s E_\infty^n + [M_s - (2N_s + 4)]E_s^n}{M_s} \tag{3.25}$$

where

$$M_s = \frac{(\Delta x)^2}{a(T_s)\Delta t} \tag{3.26}$$

$$N_s = \frac{h\Delta x}{k^*} \tag{3.27}$$

$$k^* = \frac{1}{T_\infty - T_s^n} \int_{T_s^n}^{T\infty} k(T)\, dT \tag{3.28}$$

and,

$$E_\infty^n = \int_{T_R}^{T\infty} k(T)\, dT \tag{3.29}$$

Equation 3.25 is basically a form of a boundary condition. In this case it allows for convective heat transfer at the boundary surface which is described by the heat transfer coefficient h. The great power of the numerical solution may be seen by the fact that, using equations 3.25–3.29, provision may be made for both spatially as well as timewise variations in h, something impossible with analytic forms of the solution. This flexibility of the numerical solution makes it generally the most useful in practical engineering applications. The stability of the numerical calculation requires that,

$$M_s \geqslant 2N_s + 4, \tag{3.30}$$
$$M_0 \geqslant M_s, \tag{3.31}$$

for constant grid size. Because $a(T_s)$ and to a lesser extent k^* will vary during the calculation owing to changes in T_s, the value of M_s must be up-dated at each step to check the compliance with the stability criterion, equation 3.30. Should changes in M_s become necessary it is probably best done by altering the size of the time interval, Δt, rather than the grid size, Δx.

Equations 3.23 and 3.25 will be recognized as "marching" type explicit formulations. Because of the restrictions on the time interval, Δt, this kind of formulation can require considerable computer time to complete a calculation. Its advantage, however, rests with its "explicit" form. Another type of "explicit" formulation suitable for this kind of a problem and which is unconditionally stable without restriction on the size of Δt is given by Barakat and Clark, (1966). Owing to space limitations is will not be outlined here. The calculation is complete when the computed values of E are related to their corresponding values of T, using equation 3.11.

Steady heat conduction problems also usually require the use of numerical methods when the geometry of a region or the boundary conditions are not simple. The general formulation of the equations for the steady case corresponding to the system in Fig. 3.3 are given as follows (Clark, 1969; Kreith, 1965) for the interior and surface points,

$$E_1 + E_2 + E_3 + E_4 - 4E_0 = R_0, \tag{3.32}$$

and,

$$2E_8 + E_6 + E_7 + 2N_s E_\infty - [N_s + 2]E_s = R_s. \tag{3.33}$$

The solution of equations 3.32 and 3.33 is usually accomplished by the "relaxation" method (Arpaci, 1966) or by iterative procedures on a digital computer. The desired value of R_0 and R_s, the residuals, is zero but as a practical matter they are reduced to as small a value (positive or negative) as required by the demands of accuracy of the problem. The quantities N_s, E_∞ and E_s are defined above.

B. INTERFACIAL PHENOMENA

Interfacial transfer of heat and mass is intimately associated with both pressurization and stratification phenomena in cryogenic vessels containing co-existent liquid and vapour phases. It is, of course, the conditions at the liquid-vapour interface which couple the simultaneous transport processes in the liquid and gas phases. A detailed summary of this problem is given by Clark (1969).

Mass transfer by condensation or evaporation at a vapour-liquid interface depends on the relative rates of heat transfer from each phase at the interface. Should heat transfer from the vapour dominate that to the liquid, evaporation will occur at the interface; if the opposite is true, the vapour will condense; if the respective heat transfer rates are the same, neither evaporation nor condensation occurs and the interface remains stationary. These circumstances will exist generally. For physical systems having convective action in both phases adjacent to the interface there is little known at present for predicting the interfacial transport of heat and mass. Clark *et al.* (1967) treat the subject for cryogenic containers.

A particularly useful case is that which corresponds to the conditions of zero net evaporation or condensation. A simple formulation may be used as a criterion for judging in a particular instance which process, condensation or evaporation, may be expected to occur. The interfacial temperature for this case is found by Yang *et al.* (1965) to be

$$\left(\frac{T_s - T'_\infty}{T''_\infty - T'_\infty}\right)_0 = A'(0)$$

$$= \frac{1}{1 + \sqrt{[(k\rho c_p)'/(k\rho c_p)'']}}$$

(3.34)

It is shown (Yang *et al.*, 1965) that if $(T_s - T'_\infty)/(T''_\infty - T'_\infty)$ is greater than $A'(0)$ then interfacial condensation occurs, whereas if this temperature ratio is less than $A'(0)$, then interfacial evaporation occurs. The usefulness of this criterion is that it provides a simple expression in terms of known system parameters and the thermal property ratio $(k\rho c_p)'/(k\rho c_p)''$. This expression is shown in Fig. 3.4 along with representative values of the thermal property ratio for O_2, N_2, H_2 and H_2O at 1 and 3 atm, saturated conditions. From this result it may be observed that interfacial evaporation may reasonably be expected in the pressurization of sub-cooled liquid hydrogen, whereas much larger temperature differences, $T''_\infty - T'_\infty$, are required to cause evaporation at liquid nitrogen, oxygen or water interfaces. In these latter systems condensation may more often prevail. Experimental data for both liquid hydrogen systems (Liebenberg and Edeskuty, 1965), where vaporization was reported, and liquid nitrogen systems (Bailey and Churgay, 1962) where condensation was reported, are included in Fig. 3.4. The relative positions of these data points on the figure confirm the prediction of equation 3.34.

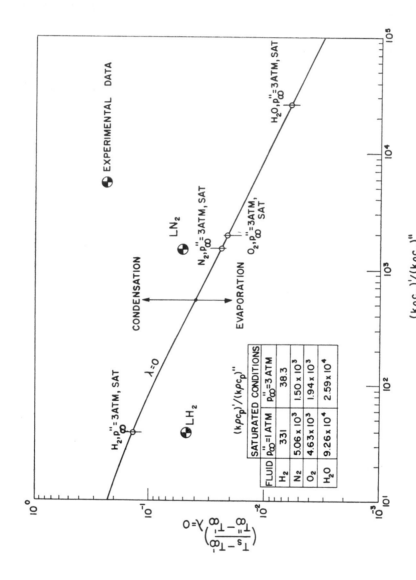

Fig. 3.4. Interfacial temperature for no condensation nor evaporation, $\lambda = 0$.

IV. Forced-Convection Processes

In general, single-phase forced convection heat transfer processes for cryogens may be described by the same scaling parameters as found useful for other substances. The exceptions to this rule include all transport phenomena in the region of the critical point. Here variations in the values of cp, μ, k, ρ and β with temperature and pressure are so great that correlation equations developed for "constant" property conditions are invalid. It is currently felt that heat transfer processes in the region of the critical point will be successfully described when the general problem of convective heat transfer in a system of severely variable properties in adequately solved. Some work has been accomplished along this line (Hess and Kunz, 1965; Szetela, 1962) for liquid hydrogen and will be discussed later. Another hypothesis tendered to explain near-critical point behaviour is to ascribe it to a pseudo-boiling phenomena (Hendricks et al., 1962). The frictional pressure drop for duct flow can be fairly reliably predicted using standard friction factor–Reynolds number correlations, except in the region of the critical point. The total pressure drop in many cases, however, is made up mostly of momentum changes resulting from density variations on heating or cooling. These can be computed using property data (Hendricks et al., 1962).

The most useful results to date for design purposes are the generalized empirical and semi-empirical correlations. These include correlations formulated from studies conducted on both cryogenic and non-cryogenic substances (Vance, 1964), Richards et al. (1961) have summarized some results for cryogenic substances. Compact heat exchangers have been sized for cryogenic application (Wright, 1960; Kahl, 1962; Fleming, 1967) using the extensive tabulation of heat transfer and friction data of Kays and London (Kays, 1964). For single phase flow the standard heat transfer and pressure-drop correlations have been employed for design (Kahl, 1962; Pope et al., 1960; Kroeger, 1967). Other heat exchanger design procedures for cryogenic application are reported by Bartlit and Williamson (1966), Hargis and Stokes (1965) and Kroeger (1967).

Single-phase convective heat transfer is characterized by the type of flow, laminar, transition or turbulent and the manner in which the flow is obtained, forced convection or natural convection. Heat transfer correlations depend on these characterizations as well as on the geometry of the flow, external, internal (duct flow) or separated. The principal scaling parameters are the Nusselt number $Nu = hD\hat{e}/k$, the Reynolds number $Re = \rho VD\hat{e}/\mu$, the Grashof number $Gr = D_e^5\rho^2 g\beta\Delta T_0/\mu^2$, the Prandtl number $Pr = cp\mu/k$ and the length to diameter ratio $L/D\hat{e}$. Where variable property effects are important these parameters are either evaluated at a particular temperature, such as the film temperature $T_f = (T_w + T_b)/2$, or an additional parameter consisting of (T_w/T_b) or (v_w/v_b) may be included in the correlation.

A. FLOWS WITH MODERATE PROPERTY VARIATION

Except for the superfluid condition most cryogens behave in a "classical" manner when their thermodynamic state is well removed from the critical state. For gases, this will always be true whenever an ideal gas equation of state describes the p, v, T relationship and the transport properties (cp, μ, k) vary only slightly with temperature. Under these circumstances the heat transfer correlations may be written for a moderate property variation corresponding to low to moderate differences in temperature between the surface and that of the fluid bulk.

1. *Laminar Flow*

Laminar flow exists when disturbances decay when introduced into the flow. Generally, for flow inside of ducts this corresponds to a condition of a Reynolds number, based on diameter, less than approximately 2,000. In this instance the Reynolds number is defined as

$$Re = \frac{\rho V D_e}{\mu} = \frac{D_e G}{\mu} \tag{3.35}$$

Here D_e is the equivalent diameter defined by

$$D_e = 4\frac{\text{wetted flow area, } A_f}{\text{wetted perimeter, } p} = 4\left(\frac{A_f}{p}\right) \tag{3.36}$$

which provides for consideration of ducts having non-circular cross section.

Heat transfer data for laminar flow inside ducts having uniform wall temperature is correlated by the following two equations for h based on the arithmetic mean temperature difference. The subscript b indicates that physical properties are evaluated at mixed mean fluid temperature.

(a) $$Re \cdot Pr (L/D_e) \geqslant 10$$

$$\frac{h D_e}{k_b} = 1 \cdot 86 \left[\frac{4}{\pi} \cdot \frac{A_f}{D_e^2} \cdot \frac{Re \cdot Pr}{L/D_e}\right]_b^{\frac{1}{3}} \left(\frac{\mu_b}{\mu_w}\right)^{0 \cdot 14} \tag{3.37}$$

(b) $$Re \cdot Pr (L/D_e) \leqslant 5$$

$$\frac{h D_e}{k_b} = \frac{1}{2}\left[\frac{Re \cdot Pr}{L/D_e}\right]_b \tag{3.38}$$

These equations are shown on Fig. 3.5 taken from McAdams (1954) and compared with some experimental data for air in circular ducts. The agreement is good at low values of $Re \cdot Pr/(L/D_e)$ and satisfactory at higher values where the data fall slightly above the curve for equation 3.37. This probably was caused by a superposed effect of free convection. The parabolic velocity profile correlation, equation 3.37 is shown for circular tubes for which $A_f/D_e^2 = \pi/4$.

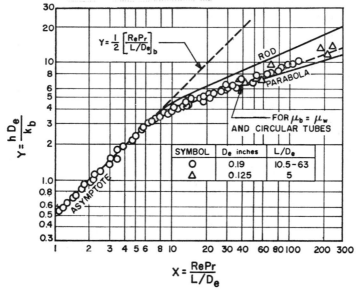

Fig. 3.5. Correlation of heat transfer data for laminar flow in ducts.

2. Turbulent Flow

Turbulent flow is found to exist when the flow Reynolds number is sufficiently large. For flow inside ducts a condition of fully turbulent flow occurs when the Reynolds number based on diameter D_e exceeds approximately 10,000. Between a Reynolds number of 2,000 and 10,000 the flow is in a transition state, being partly laminar and partly turbulent and characterized by some unsteadiness.

Heat transfer data for turbulent flow inside ducts is correlated ($\pm 25\%$) by the following equation for $1 \leqslant Pr \leqslant 120$ and $10,000 \leqslant Re \leqslant 500,000$,

$$\left(\frac{hD_e}{k}\right)_f = 0.023\left(\frac{D_e G}{\mu_f}\right)^{0.8} Pr_f^{\frac{1}{3}}\left[1+\left(\frac{D_e}{L}\right)^{0.7}\right] \tag{3.39}$$

The subscript f in this expression indicates that the physical properties are to evaluated at the film temperature which is the arithmetic mean of the surface and mixed mean fluid temperature.

Another correlation which may be used to compute heat transfer coefficients in turbulent flow ($\pm 25\%$) and having the same restrictions as equation 3.39, is written as follows:

$$\frac{h}{c_{pb}G}Pr_b^{\frac{2}{3}}\left(\frac{\mu_w}{\mu_b}\right)^{0.14} = 0.0205\left(\frac{D_e G}{\mu_b}\right)^{-0.2}\left[1+\left(\frac{D_e}{L}\right)^{0.7}\right] \tag{3.40}$$

5*

Except for the viscosity ratio correction factor, μ_w/μ_b, all physical properties are evaluated at the mixed mean bulk temperature, designated by the subscript b. A comparison of this expression with experimental data is shown in equivalent form in Fig. 3.6. The influence of L/D_e as well as the general interrelationship of turbulent, transition and laminar heat transfer data also is shown.

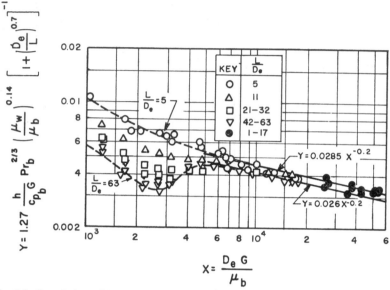

Fig. 3.6. Correlation of heat transfer data for laminar, transition and turbulent flow in ducts.

Thompson and Geery (1963) studied heat transfer to liquid hydrogen at super critical pressures (45·3–89·6 bar and inlet temperatures 57 to 86°R) for turbulent flow in a 0·194 in. internal diameter tube. Two regimes of flow were found characterized by the magnitude of the level of heat flux and wall–fluid temperature ratio. Except at low values of these quantities, the data were correlated (±30%) by

$$Nu_b = 0.0217 Re_b^{0.8} Pr_b^{0.4} \left(\frac{T_W}{T_b}\right)^{-0.34} \tag{3.41}$$

Pressure drop was satisfactorily correlated by the conventional friction factor for Newtonian fluids.

In low pressure regions where hydrogen has almost ideal gas behaviour, Hendricks et al. (1962) found that local heat transfer data could be correlated by

$$\frac{h_x D_e}{k_f} = 0.023 \left(\frac{\rho_f V D_e}{\mu_f}\right)^{0.8} \left(\frac{c_p \mu}{k}\right)_f^{0.4} \tag{3.42}$$

At higher pressures (12–50 atm) and for fluid temperatures in excess of 90°R, these authors find their local heat transfer data to correlate approximately as equation 3.42.

Hendricks *et al.* (1962) summarize turbulent convective heat transfer correlations for hydrogen at pressures above and below the critical pressure (12·53 bar) in the range 1–100 atm, but not too close to the critical temperature (60°R). This summary is given in Table 3.I and includes some results for helium as well.

A comparison of the various correlations in Table 3.I is shown in Fig. 3.7. Except for equation 3.43 they all predict essentially the same magnitude of result over a large range of T_w/T_b. Equation 3.43 diverges severely for values of T_w/T_b greater than about 4. The range of actual experimental values of T_w/T_b is not great and since many modern power and propulsion systems require design at higher values of T_w/T_b additional data would be welcome. Hendricks *et al.* (1962) recommend the use of equation 3.47 at large ratios of T_w/T_b as this correlation is formulated in terms of local properties and consequently takes some account of property variation.

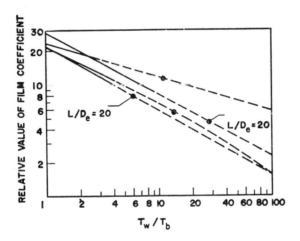

Fig. 3.7. Comparison of convective film correlations over a range of T_w/T_b. —, based on experiment ---, extrapolated.

At low pressures the pressure drop may be computed by the standard turbulent flow equation. For pressures above the critical pressure and temperatures in excess of 90°R the fluid friction appears to be similar to that observed with other gases in pipes of varying roughness. Wolf and McCarthy (1960) give values of the isothermal fluid friction comparable to results in the literature. However, it was noticed that the nonisothermal friction factors were less than the isothermal values at the same bulk Reynolds number.

TABLE 3.I. Summary of Heat Transfer Correlations for Regions Remote from the Critical Point (Hendricks *et al.*, 1962)

Fluids	Max $\dfrac{T_w}{T_b}$	Entrance bulk temp. range, °R	Correlation	
Hydrogen (McCarthy and Wolf, 1960)	2·8	360 to 560	$Nu_b = 0\cdot023\ Re_b^{0\cdot8} Pr_b^{0\cdot4} \left(\dfrac{T_w}{T_b}\right)^{-0\cdot3}$ (overall)	(3.43)
Hydrogen (Wolf and McCarthy, 1960)	9	135 to 515	$Nu_b = 0\cdot045\ Re_b^{0\cdot8} Pr_b^{0\cdot4} \left(\dfrac{T_w}{T_b}\right)^{-0\cdot55} \left(\dfrac{L}{D_e}\right)^{-0\cdot15}$ (overall)	(3.44)
Helium	11			
Hydrogen (Wright and Walters, 1959)	4	88 to 130	$Nu_b = 0\cdot021\ Re_b^{0\cdot8} Pr_b^{\frac{1}{3}} \left(\dfrac{T_w}{T_b}\right)^{-0\cdot575} \left(1 + \dfrac{6}{L/D_e}\right)$ (overall)	(3.45)
Helium (Taylor and Kirchgessner, 1959)	4	1500 max	$Nu_f = 0\cdot034\ Re_f^{0\cdot8} Pr_f^{0\cdot4} \left(\dfrac{L}{D_e}\right)^{-0\cdot1}$ (overall)	(3.46)
Unpublished hydrogen (Hendricks *et al.*, 1962)			$Nu_f = C Re_f^{0\cdot8} Pr_f^{0\cdot4},\quad C = 0\cdot021$ (overall)	(3.47)

The friction coefficient with helium at high wall temperatures was observed (Taylor and Kirchgessner, 1959) to be predictable by the Karman–Nikuradse relation for turbulent flow:

$$\frac{1}{\sqrt{f}} = 4 \log_{10} (Re\sqrt{f}) - 0{\cdot}4 \tag{3.48}$$

where f is the Fanning friction factor, defined as

$$f = \frac{\tau_w}{\rho V^2 / 2 g_0}, \tag{3.49}$$

and

$$\Delta p = 4f\left(\frac{L}{D_e}\right)\frac{\rho V^2}{2 g_0}. \tag{3.50}$$

McAdams (1954) recommends a modified form of equation 3.48 for flow in heated tubes having T_w/T_b up to 2·5. The result is a reduction in the friction factor for the higher values of T_w/T_b, as was observed by Wolf and McCarthy (1960). For compressible flows, a significant contribution of momentum change to the total pressure drop must be anticipated.

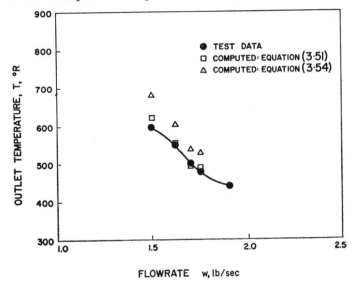

Fig. 3.8. Comparison of test data and computed performace predictions for a typical oxygen heat exchanger.

Hargis and Stokes (1965) report on the design of heat exchangers for the Saturn I and Saturn V pressurization systems. These units consist of a helical tube coil through which the gaseous pressurant (H_2, O_2 or He) is pumped. The heat exchanger coil is heated by combustion gases from the rocket exhaust which flows over the outside of the coil. A comparison of test data and computer performance predictions for an oxygen heat exchanger is shown in Fig. 3.8. The heat transfer correlation used for average tube side coefficient for oxygen which best fitted the data is

$$Nu_b = 0.023 Re_b^{0.8} Pr_b^{0.4} \left(\frac{T_w}{T_b}\right)^{-0.34} \phi_1 \phi_2, \qquad (3.51)$$

where ϕ_1 and ϕ_2 are correction factors, given below.

TABLE 3.II. (a) Oxygen Correction Factor ϕ_1 for equation 3.51

T_b °R	T_w °R		
	600	1000	1800
200	1·3	0·95	—
278	0·58	0·42	0·32
300	0·68	0·556	0·434
350	0·836	0·764	0·655
400	1·0	1·0	1·0
Over 400	1·0	1·0	1·0

(b) Correction Factor ϕ_2 for equation 3.51

For $L/D_e \leqslant 50$:

$$\phi_2 = \frac{1·48}{(L/D_e)^{0·1}} \qquad (3.52)$$

For $L/D_e \geqslant 50$:

$$\phi_2 = 1·0 \qquad (3.53)$$

Equation 3.51 is shown in Fig. 3.8 and agrees satisfactorily with the experimental data. As a demonstration of the need for property value correction Hargis and Stokes (1965) include a comparison with a simplified correlation sometimes used in heat transfer correlations. This equation also shown in Fig. 3.8 is

$$Nu_b = 0.023 Re_b^{0.8} . Pr_b^{0.4} \qquad (3.54)$$

This result compares much less favourably with the experimental data than does equation 3.51.

As may be observed equation 3.51 for oxygen at temperatures above 400°R and L/D_e in excess of 50 is essentially the same correlation found to predict the behaviour of supercritical hydrogen, equation 3.41 and equation 3.43.

A discussion of the application of these various correlations to hydrogen cooled rocket nozzles is given by Benser and Graham (1962). The cooling of large masses by ducted supercritical helium is reported by Koln (1965) and

Koln *et al.* (1966). To maintain superconducting magnets at low temperature (40°K) these investigators employed a forced circulation system of liquid helium at pressures above 100 atm in small tubing integral with the object to be cooled. Performance data but not heat transfer data are reported.

3. *Transition Flow*

Between the Reynolds numbers of 2,000 to 10,000, the flow in a duct is in a transition condition. This is a flow region in which the characteristics of both laminar and turbulent flow co-exist. There also is a tendency for instability in the flow pattern. Very little is known about this flow regime and no really satisfactory method or correlation exists for computing its heat transfer coefficients. A residual L/D_e influence is observed which is greatest at the lower range of Reynolds numbers and gradually diminishes at higher Reynolds numbers. To obtain heat transfer coefficients in the transition region it is recommended that the data of Fig. 3.6 be smoothed and the coefficients computed from the ordinate corresponding to the expected Reynolds number and the L/D_e.

4. *Flow Outside of Ducts*

An important configuration encountered frequently is flow external and normal to the axis of a duct. Heat transfer data for the case of round ducts are correlated Knudsen and Katz (1959) by the expression

$$\left(\frac{hD_0}{k}\right) = C \cdot \left(\frac{\rho_\infty V D_0}{\mu_f}\right)^n \tag{3.55}$$

the subscript f denotes the physical properties are evaluated at the film temperature. Both C and n are functions of the Reynolds and Prandtl numbers in accordance with Table 3.III below.

TABLE 3.III

$\dfrac{\rho_\infty V D_0}{\mu_f}$	n	C Gases	C Liquids
1–4	0·330	0·891	0·989 $Pr^{\frac{1}{3}}$
4–40	0·385	0·821	0·911 $Pr^{\frac{1}{3}}$
40–4,000	0·466	0·615	0·683 $Pr^{\frac{1}{3}}$
4,000–40,000	0·618	0·174	0·193 $Pr^{\frac{1}{3}}$
40,000–250,000	0·805	0·0239	0·0266 $Pr^{\frac{1}{3}}$

Experimental results for air are correlated closely by equation 3.55. Data also are available for non-circular ducts and spheres and may be found in McAdams (1954) and Knudsen and Katz (1959).

The flow around a bluff object is of a variable nature. It starts as a laminar condition at the forward stagnation point, undergoes transition to a turbulent boundary layer and finally separates from the surface in the downstream regions producing a highly turbulent wake. For this reason the flow is characterized by the Reynolds number only without reference to a laminar or turbulent description.

Data for air flow across banks of tubes have been correlated by Grimison (1937, 1938). The correlation is summarized by McAdams (1954). Pressure drop data are also correlated by Grimison.

A correlation for pressure drop in two phase flow across tube banks has been presented by Diehl (1957).

B. FLOW WITH LARGE PROPERTY VARIATION

The conditions corresponding to the limits of application of constant or variable property correlations—the ideal gaseous or liquid states, on one hand, and the critical or superfluid states on the other—are fairly easily

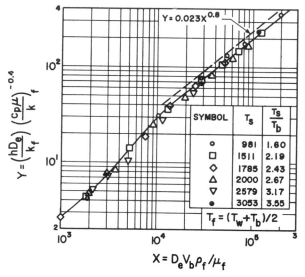

Fig. 3.9. Correlation of heat transfer data at inlet-air temperature of 531 to 533°R. Using modified film Reynolds number.

identified. Their intermediate or transition states, where fluid property variation with temperature and/or pressure becomes severe, are much less easily defined. No simple criterion exists. One is best guided by the magnitude of the variation in such properties as k, cp and v in the range of temperatures expected in a design. McAdams (1954) reports the data of Demon and Sams (1950) which indicates the influence of large temperature variations on heat transfer to air in turbulent flow. Some results are shown in Fig. 3.9 where the data are correlated in accordance with a modified form of equation

3.42. In this instance the Reynolds number is evaluated at the bulk temperature. The corresponding property variations for air suggests a limiting criterion for the application of "moderate" property variation correlations.

Hendricks et al. (1962) present another guideline in the form of a dimensionless temperature ratio, originally suggested by E. R. G. Eckert. This ratio indicates the proximity of the fluid state to the critical state as measured by the temperature, T_m, at which the specific heat, c_p, reaches its maximum.

An analysis of the velocity and temperature distributions, the wall shear stress and the heat transfer rate for the fully developed turbulent flow of supercritical hydrogen is given by Hess and Kunz (1965). Their analysis accounts for the influence of variable physical properties on the universal velocity and temperature profiles. Based on the method of Wiederecht and Sonnemann (1960), these authors have solved the momentum and energy transport equations for variation in the eddy diffusivities of heat and momentum as well as the thermodynamic and transport properties. They showed that the influence of variable properties is most significant in the turbulent core.

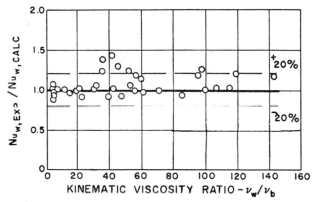

Fig. 3.10. Comparison between experimental and calculated Nusselt numbers using the proposed analysis.

It was found necessary to relate the viscous damping constant for the eddy diffusivity of momentum to the kinematic viscosity ratio, v_w/v_b, in order to describe the heat transfer to supercritical hydrogen. The specific form of the functional relationship to do this was determined by selectively fitting the results of the calculation to experimental data. As a consequence, however, reasonable agreement was found between the analytical calculations and the experimental heat transfer data of Hendricks, et al. (1962). The results are shown in Fig. 3.10 where the ratio of the experimental Nusselt number to the computed from the theory is plotted against the ratio v_w/v_b, the principal governing parameter for this system. As may be seen the comparison is within $\pm 20\%$, except for a few data in the range of v_w/v_b between 30 and 60.

As a further test of their analysis Hess and Kunz computed the local Nusselt numbers and the wall temperature distribution along the heated length of a tube and compared their results with the experimental hydrogen data of Hendricks *et al.* (1962). A typical results is shown in Fig. 3.11. Except in the entrance region ($x/D<30$) the comparison of computed and experimental results is favourable. The lack of agreement for x/D less than 30 is attributed to an extended influence of variable properties on the thermal entrance region. The unusual temperature distribution, which has a peak, had previously been found only in boiling experiments. The region downstream from the peak with the decreasing wall temperature appears to be predicted well by the analytical results based on the hypothesis of variable fluid properties.

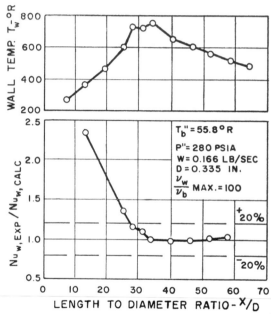

Fig. 3.11. Experimental data exhibiting a peak in the tube wall temperature

The analytic results obtained by Hess and Kunz, while providing some insight into the process, require extensive calculations using a digital computer in order to obtain useful numerical results. Hence, they are too complex for most practical design work. Because of this the following simpler formulation was obtained to fit the experimental data,

$$Nu_f = 0.0208 Re_f^{0.8} Pr_f^{0.4}\left(1+0.01457\frac{v_w}{v_b}\right) \qquad (3.56)$$

where all properties are evaluated at the film temperature, except that the bulk density is used to determine the velocity. The fit was generally within $\pm20\%$.

V. Natural Convection Processes

Natural convection describes a type of convective heat transfer process in which fluid motion is a direct consequence of buoyant and viscous forces in the fluid. The only requirements for flow is that the fluid have a temperature gradient and be in a force field such as a gravity or centrifugal field. The coupling of the fluid motion with the temperature distribution gives rise to complexities in the mathematical analysis of such processes. However, empirical correlations have been obtained which confirm those theoretical solutions that are available. It has been found that for both laminar and turbulent flow next to a heated (or cooled) surface of various geometry a correlation ($\pm 25\%$) of the following form is obtained:

$$\frac{hL}{k_f} = C \left[\frac{L^3 \rho f^2 g \beta_f \Delta T_0}{\mu_f^2} \cdot \left(\frac{c_p \mu}{k} \right)_f \right]^n \tag{3.57}$$

The term in the square brackets is sometimes called the Rayleigh number and is the product of the Grashof and Prandtl numbers. The subscript f indicates that all physical properties are to be evaluated at the film temperature.

For vertical plates and cylinders the constant C and exponent n of equation 3.57 are functions of the Grashof–Prandtl number product, $Gr \cdot Pr$, as shown in Table 3.IV below.

TABLE 3.IV

$Gr \cdot Pr$	C	n	Type of flow
10^4–10^9	0·59	$\frac{1}{4}$	laminar
10^9–10^{12}	0·13	$\frac{1}{3}$	turbulent

This expression has been found to confirm experimental measurements on liquid and gaseous nitrogen (Fenster et al., 1960; Merte and Clark, 1964). Partial comparison of this correlation with experimental results of liquid helium and liquid hydrogen indicate qualitative agreement, except for helium data at approximately half an atmosphere pressure (Richard et al., 1961).

Correlations of the form of equation 3.57 are available for other geometry as well. Some of these are given by the following expressions (McAdams, 1954):

Horizontal cylinders, $10^3 \leqslant Gr \cdot Pr \leqslant 10^9$

$$\frac{hD_0}{k_f} = 0·53 \left[\frac{D_0^3 \rho_f^2 g \beta_f \Delta T_0}{\mu_f^2} \cdot \left(\frac{c_p \mu}{k} \right)_f \right]^{\frac{1}{4}} \tag{3.58}$$

This equation with the constant slightly less than 0·53 correlates natural convection data from a 1 in. sphere for subcooled liquid nitrogen at 3 and 5 atm (Merte and Lewis, 1967).

Heated horizontal square plates facing upward and cooled plates facing downward:

Laminar flow, $10^5 \leqslant Gr \,.\, Pr \leqslant 2 \times 10^7$

$$\frac{hL}{k_f} = 0.54 \left[\frac{L^3 \rho f^2 g \beta_f \Delta T_0}{\mu_f^2} \,.\, \left(\frac{c_p \mu}{k} \right)_{\!f} \right]^{\frac{1}{4}} \tag{3.59}$$

Turbulent flow, $2 \times 10^7 \leqslant Gr \,.\, Pr \leqslant 3 \times 10^{10}$

$$\frac{hL}{k_f} = 0.14 \left[\frac{L^3 \rho f^2 g \beta_f \Delta T_0}{\mu_f^2} \,.\, \left(\frac{c_p \mu}{k} \right)_{\!f} \right]^{\frac{1}{3}} \tag{3.60}$$

This expression has been confirmed experimentally for unbounded plates in force-fields resulting from system accelerations up to 21 times standard gravity (Merte and Clark, 1961). Results from this investigation are given in Fig. 3.12 for both a bounded and unbounded plate. The results for the bounded plate fall significantly above the correlation for the bounded plate, equation 3.60, probably owing to increased eddy motion in the fluid. Similar results were found (Clark and Merte, 1963) for liquid nitrogen in the range a/g from 1 to 20. Natural convection data for liquid hydrogen on a horizontal flat plate showed favourable agreement with equations 3.59 and 3.60 (Clark et al., 1967).

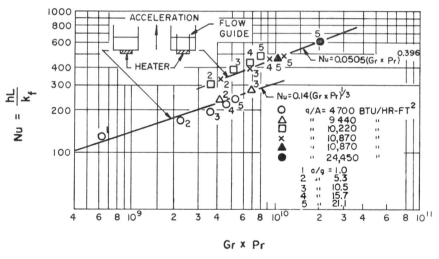

Fig. 3.12. Correlation of natural convection data for systems having acceleration normal to heated surface.

Heated square plates facing downward or cooled plate facing upward, $3 \times 10^5 \leqslant Gr \,.\, Pr \leqslant 3 \times 10^{10}$

$$\frac{hL}{k_f} = 0.27 \left[\frac{L^3 \rho f^2 g \beta_f \Delta T_0}{\mu_f^2} \,.\, \left(\frac{c_p \mu}{k} \right)_{\!f} \right]^{\frac{1}{4}} \tag{3.61}$$

The characteristic dimension, L, in these correlations is the edge of the plate. It will be noted that for the laminar condition the length dependency is weak, $L^{-\frac{1}{4}}$, and for the turbulent condition, exponent $\frac{1}{3}$, it is completely absent. For non-square geometry an estimate of a mean L is recommended for use in the foregoing expressions.

During the orbital flight test of the Saturn I B, vehicle AS–203, the above correlations were found to be satisfactory for the prediction of heat transfer to liquid hydrogen and oxygen in a low gravity (a/g as low as 8×10^{-4}) environment (Chrysler Corporation, 1967; Swalley *et al.*, 1966).

VI. Pressurized-Discharge Processes and Stratification Phenomena in Cryogenic Containers

A great deal of research, engineering design effort and testing has been devoted to the problem of the pressurized-discharge of a cryogenic liquid from a container. Most of this effort has been directed at the optimization of the propellant tank design, determination of the pressurant requirements and the selection of the operating parameters for large rocket vehicles. Similar problems arise, however, in other applications such as the pressurized-transfer and storage of cryogens from vessels in ground installations.

The problems of pressurization and stratification are considered in detail by Clark (1969). Although the results are general they have been used almost exclusively on cryogenic propellant systems for both ground and flight application such as the booster stage of the Saturn V moon rocket.

Considerations of propellant or storage tank design and performance characteristics and the interaction of these with other sub-systems requires an examination of several related processes. Of primary importance among these are:

(a) pressurization, including the calculation of the transient temperature, velocity and concentration profiles in the gas space and the flow rate and quantity of pressurant;

(b) liquid stratification, including the calculation of transient temperature, velocity and concentration distribution in the liquid; and

(c) interfacial phenomena, including the study and prediction of mass and heat transfer rates across gas–liquid and gas–solid interfaces.

Thermal stratification of a cryogenic liquid in a vessel results from external heat exchange and consequent non-equilibrium phenomena within the liquid. In both flight vehicles and storage containers the most dominant influence is heat transfer through the (vertical) side walls of the vessels. Other effects may be important as well, such as nuclear energy absorption within the liquid, heating from the bottom of the vessel and both heat and mass exchange between the liquid surface and the pressurant. The phenomena of thermal stratification is important to propellant tank design and operation as it influences the selection of venting devices, insulation, pumps and tank structure, among others. The pressure in the tank is directly related to the interface temperature which is established by the convective transport processes with the tank. In the case of liquid hydrogen a 1°R increase in the liquid–vapour interface temperature represents approximately a 3 p.s.i. increase in tank pressure. As a result of stratification the pressure in cryogenic vessels has been found to be significantly greater than that corresponding to the vapour pressure at bulk (mixed) liquid temperature.

Stratification is caused by the natural convective flow of heated liquid along the side-walls of a tank and into the upper regions near the liquid interface. Here it flows toward the centre of the tank, dispersing and mixing, and causing a downward penetration of heated liquid, the depth of which increases with time. This depth is known as the thermal stratification layer.

VII. Multi-Phase Processes

A. BOILING HEAT TRANSFER

Boiling will exist at a surface when the surface temperature exceeds the saturation or bubble point temperature by a few degrees. This superheat depends upon the type of fluid and surface and the system pressure. It decreases to near zero as the pressure approaches the critical pressure. The word boiling is used to describe the process of vapour bubble generation within a liquid and almost always occurs at a solid surface in heat transfer systems. Two general types of boiling systems exist: pool boiling, a process similar to natural convection, and forced convection boiling. For each of these systems the fluid can be sub-cooled and thus have no net vapour generation or it can be saturated in which case a net vapour generation will occur.

The boiling phenomenon itself is characterized by three regimes, shown in Fig. 3.13, namely, nucleate boiling, transition boiling and film boiling. In addition, two other unique phenomena are observed in boiling heat transfer. These are the conditions of maximum and minimum heat flux, also shown in Fig. 3.13, which separate the transition boiling regime from those of nucleate and film boiling. For systems such as nuclear reactors and electronic equipment in which the heat flux (q/A) is an independent variable the point of maximum heat flux is of utmost importance. Should an attempt be made to increase the power level (and hence q/A) of such systems beyond that of the maximum heat flux corresponding to a given set of circumstances, the surface-fluid saturation temperature difference would increase to that of film boiling at this heat flux. As may be observed from Fig. 3.13 this would result in heat transfer surface temperatures 1,000 to 2,000°F above the fluid saturation

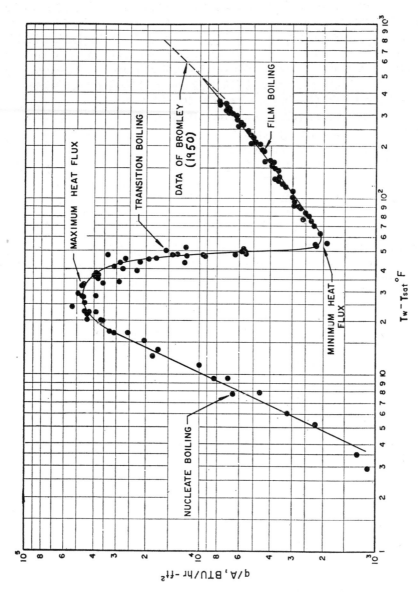

Fig. 3.13. Typical pool boiling characteristic curve.

temperature. For most fluid-surface combinations this would mean physical destruction of the surface. In cryogenic application this consequence is minimized owing to the low saturation temperatures but it still is to be avoided in most instances.

The literature of boiling heat transfer is very large. It will be possible in this section to describe only some of the significant results as related to cryogenic heat transfer. Excellent summaries of the subject have been prepared by Westwater (1956), Balzhiser, *et al.* (1961), Zuber and Fried (1961), Richards, Steward and Jacobs (1961), Giarratano and Smith (1965, 1966), Brentari, *et al.* (1965), Seader, *et al.* (1964), and Tong (1965), for both cryogenic and non-cryogenic application. In general, the properties of pool boiling including nucleate, transition and film boiling and the maximum and minimum heat flux can be computed from available correlations and some knowledge of the fluid-surface characteristics. This is less true with forced convection boiling although some progress has been made and will be discussed later.

1. *Pool Boiling*

a. Nucleate boiling. Nucleate boiling exists in the $(q/A) - \Delta T$ range from incipient boiling to the point of maximum heat flux. This type of boiling has probably received the greatest attention in terms of the total number of investigations owing to the relative simplicity of conducting measurements. Complete agreement is lacking among the results of the various investigators because of the important influence the kind and nature of the surface has on the process, an effect reported by Marto, *et al.* (1968) for liquid nitrogen. In addition to this variation in system geometry, method of taking data and uncertainties in measurement contribute to the general scatter of data. Incipient nucleate boiling data are reported for hydrogen by Coeling and Merte (1969).

This general effect is seen in Fig. 3.14 and 3.15 taken from Brentari and Smith (1965), where the experimental data for nucleate and film boiling for nitrogen and hydrogen are shown. Oxygen would behave similarly to nitrogen. Except for the data of Lyon *et al.* (1964) the width in the band of the data reflects the spread of each investigator's measurements on a given system. The spread in the data of Lyon *et al.* is a result of their study on a range of geometries, orientations and surfaces. For comparison, the correlation for nucleate boiling of Kutateladze (1959, equation 3.62) is included.†
The results for the maximum heat flux, the minimum heat flux and film boiling will be discussed later. In general Kutateladze's equation represents the data reasonably well. It should be pointed out that this also will be true

† Kutateladze (1959) gives two equations. His second equation is used by Seader, *et al.* (1964) and Zuber and Fried (1961). Each appears to give approximately the same results, although their $(q/A)-\Delta T$ relationship differ.

of several other correlating equations to be discussed below. Kutateladze's first correlation, originally derived for water and various organic liquids, is

$$\frac{h}{k_l}\left(\frac{g_0\sigma}{g\rho_l}\right)^{\frac{1}{2}} = 3\cdot25(10^{-4})\left(\frac{(q/A)c_{pl}\rho_l}{h_{fg}\rho_v k_1}\left(\frac{g_0\sigma}{g\rho_l}\right)^{\frac{1}{2}}\right)^{0\cdot6} \cdot$$

$$\left(g\left(\frac{\rho_l}{\mu_l}\right)^2\left(\frac{g_0\sigma}{g\rho_l}\right)^{\frac{3}{2}}\right)^{0\cdot125} \cdot \left(\frac{p}{(\sigma g \rho_l/g_c)^{\frac{1}{2}}}\right)^{0\cdot7} \quad (3.62)$$

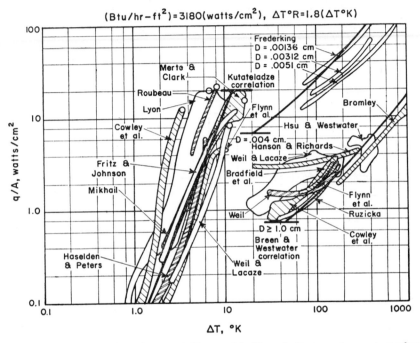

$(Btu/hr-ft^2) = 3180(watts/cm^2)$, $\Delta T°R = 1.8(\Delta T°K)$

Fig. 3.14. Experimental nucleate and film pool boiling of nitrogen at one atmosphere compared with the predictive correlations of Kutateladze and Breen and Westwater (1965).

From this result the $(q/A)-\Delta T$ relationship is,

$$(q/A) = f(p)\Delta T^{2\cdot5} \quad (3.63)$$

This equation has been computed by Brentari and Smith for a range of pressures as shown in Fig. 3.16 for N_2. The presence of g in these correlations is primarily for dimensional considerations; they cannot be used for values other than 32·2 ft/sec².

Seader, et al. (1964) have arranged several nucleate boiling correlations into the form,

$$\frac{L^*G^*}{\mu_l} = F(p)\left/\left(\frac{q/A}{\Delta T c_{pl}G^*}\right)^a Pr^b\right. \quad (3.64)$$

$(Btu/hr-ft^2)=3180(watts/cm^2)$, $\Delta T°R=1.8(\Delta T°K)$

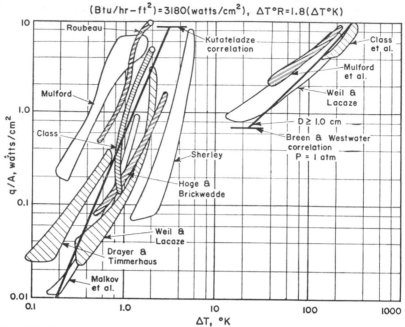

Fig. 3.15. Experimental nucleate and film pool boiling of hydrogen at one atmosphere compared with the predictive correlations of Kutateladze and Breen and Westwater (1965).

$(Btu/hr-ft^2)=3180(watts/cm^2)$, $\Delta T°R=1.8(\Delta T°K)$

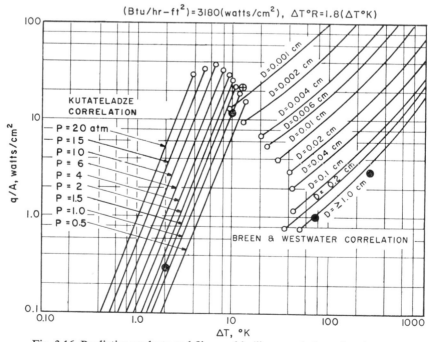

Fig. 3.16. Predictive nucleate and film pool boiling correlations for nitrogen.

which may be written in the form of a Reynolds number, Stanton number, Prandtl number correlation as

$$Re_B = \frac{F(p)}{St_B^a Pr^b} \tag{3.65}$$

It should be noted that a is the exponent on ΔT in the relation $(q/A) = f(p)\Delta T^a$. The value of a depends on the pressure and the nature of the surface.

They found that the correlations of Rohsenow (1952), Levy (1959) and Michenko (1960) were found to fit the nucleate boiling data for liquid nitrogen at one atmosphere from a polished copper sphere 1 in. in diameter (Clarke and Merte, 1963). These data are shown in Fig. 3.17. This fit was accomplished using a value of C_{sf} of 0·015 in Rohsenow's original equation. Zuber and Fried (1961) report good agreement between the correlations of Rohsenow (1952), Michenko (1960), Forster and Zuber (1955) and Labountzov (1960) and experimental data of liquid hydrogen in nucleate boiling from a smooth flat surface, (1960) in the pressure range from 0·8 to 5·1 atm. In this comparison a value of 0·0147 was used for C_{sf} in Rohsenow's correlation.

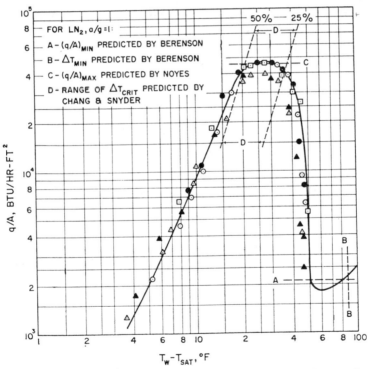

Fig. 3.17. Comparison of results for liquid nitrogen at 1 atm reduced manually and by digital computer. $a/g = 1$, 1 in. diameter sphere (Merte and Clark, 1963).

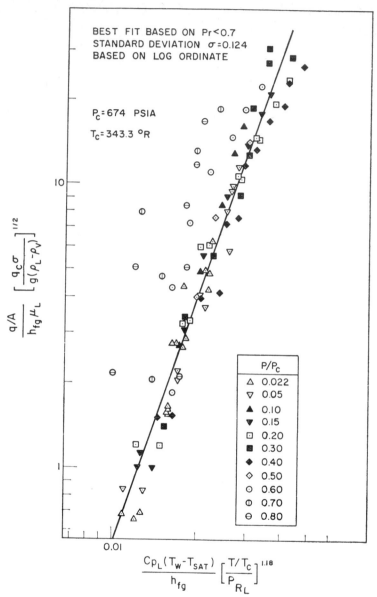

Fig. 3.18. Nucleate Boiling data for methane compared with proposed correlation (Sciance, 1967).

Nucleate boiling heat transfer has been studied for hydrogen, neon, nitrogen and argon by Bewilogua *et al.* (1966) and for neon by Astruc *et al.* (1967). Each of these papers reports an hysteresis effect in the transition from natural convection to nucleate boiling, an effect usually associated with systems in which contamination has been carefully avoided. A different nucleate boiling characteristic was obtained depending on whether the data were taken by increasing or decreasing the heat flux.

Nucleate boiling data for methane from an 0·811 in. diameter, 4 in. long gold plated cylinder at pressures from 1 atm to 41·2 atm (0·9 critical pressure) is reported by Sciance et al. (1967). A modified form of Rohsenow's correlation (1952) in which pressure dependence was introduced was found to correlate the data for pressures below 0·7, the critical pressure. This result is shown in Fig. 3.18 in comparison with the methane data.

b. *Maximum and minimum heat flux.* The upper limit of heat flux for nucleate boiling and the lower limit of heat flux for film boiling, Fig. 3.13 are each marked by unique physical states known as the maximum (or, first critical) heat flux and the minimum (or, second critical) heat flux. Each of these states is apparently characterized by a critical stability condition relating to the ability of liquid to maintain contact with the surface.

TABLE 3.V. Comparison of $(q/A)_{max}$ Correlations for Saturated Liquid Nitrogen (at Atmospheric Pressure and $a/g = 1$). The Experimental Data of Clark (1962) gave $47,000 \pm 1,000$ BTU/hr ft^2

Investigator	$(q/A)_{max}$, BTU/hr ft^2
Noyes (1963)	45,000
Chang and Snyder (1960)	56,500
Zuber (1958)	50,100
Kutateladze (1951)	61,000
Borishanskii (1956)	61,000
Chang (1963)	38,500–50,000
Moissis and Berenson (1963)	65,000

Kutateladze (1951) was probably the first to point out that the critical heat flux conditions was a matter of hydrodynamic stability. However, Zuber (1961), Chang (1959) and Berenson (1964) first applied the concepts of stability analysis to the process. Borishanskii (1956) extended the work of Kutateladze to include the influence of viscosity, which usually is not particularly significant. Other work include that of Zuber et al. (1959), Lienhard and Wong (1964), Zuber and Tribus (1958), Roshenow and Griffith (1956) and Chang (1963). A number of correlations for both maximum and minimum heat flux have appeared in the literature. Some of these can be represented by the functions

$$\frac{(q/A) \max}{h_{fg}\rho_v} = \Phi_1 \tag{3.66}$$

and

$$\frac{(q/A)\min}{h_{fg}\rho_{vf}} = \Phi_2 \tag{3.67}$$

In most cases the dependence on acceleration is given as $g^{\frac{1}{4}}$.

The functions Φ_1 and Φ_2 have been listed and compared by Seader *et al.* (1964). It should be noted that both Φ_1 and Φ_2 have the dimensions of a

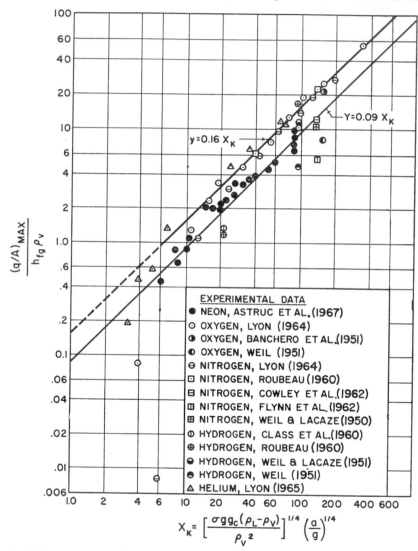

Fig. 3.19. Comparison of the maximum nucleate heat transfer fluxes with the Kutateladze maximum correlation.

velocity. Several of these correlations in Table 3.III are compared with experimental maximum heat flux data for liquid nitrogen (Merte and Clark, 1964) in Table 3.V.

Maximum heat flux data for the cryogenic liquids O_2, N_2, H_2, He and Ne taken by a large number of investigators on plates, cylinders and wires is shown in Fig. 3.19. Comparison is made with the equation of Kutateladze

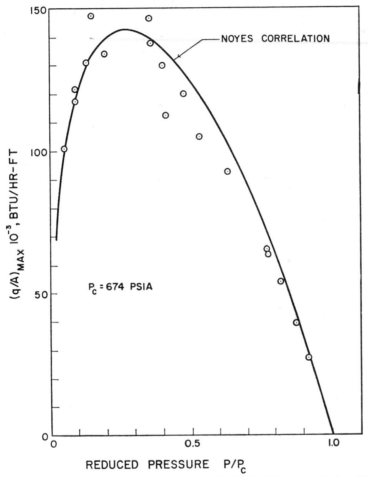

Fig. 3.20. Methane burnout heat flux compared with the Noyes correlation (1963).

(1951) but only fair to moderate agreement is found. Astruc *et al.* (1967) found that a best fit with their neon data was with a relation $y = 0.09\chi_k$, whereas the remainder of the data in Fig. 3.19, except for exceptionally low O_2 and N_2 points of Lyon, seemed to correlate best by $y = 0.16\chi_k$, the

Kutateladze equation. Interestingly, the approximate mean curve through all the data would be that of Zuber (1958), $y = (\pi/24)\chi_k$.

The maximum heat flux data of Sciance *et al.* (1947) for the boiling of methane outside a 0·811 in. diameter tube over a wide range of pressures is shown in Fig. 3.20 in comparison with the correlation of Noyes (1963). An excellent order of agreement is found over the full range of pressures. Wright and Colver (1968) present maximum heat flux data for saturated ethane and ethylene mixtures.

A comparison of Berenson's correlation (1964) for the minimum heat flux is shown in Table 3.VI with experimental liquid nitrogen data at various accelerations (a/g) from 1 to 0·001 (1964). The correspondence is generally favourable and tends to confirm the $(a/g^{\frac{1}{4}})$ dependence predicted by Kutateladze (1959), at least for an (a/g) above approximately 0·10.

TABLE 3.VI. Comparison of Experimental and Predicted Values of $(q/A)_{min}$. Merte and Clark (1964)

a/g	$(q/A)_{min}$-predicted	$(q/A)_{min}$-experimental
1·0	2,100	1,700–2,100
0·6	1,850	1,550
0·33	1,590	1,300–1,400
0·20	1,400	1,300
0·03	875⎫	
0·01	666⎭	870–1,100
0·003	491⎫	
0·001	374⎭	180–530

An interesting result is obtained for the ratio $(q/A)_{max}/(q/A)_{min}$ using the correlations of Zuber. This ratio may be written,

$$\frac{(q/A)_{max}}{(q/A)_{min}} = (0·845 - 1·436)\left(\frac{\rho_l}{\rho_v}\right)^{\frac{1}{4}} \tag{3.68}$$

The results for N_2 in equation 3.65 suggest the larger value of the constant best fits the data.

c. Film boiling. Film boiling can exist practically with cryogenic fluids because of their low saturation temperatures. In fact, it is possible to have a heat flux in film boiling greater than that of $(q/A)_{max}$ and still maintain sufficiently low surface temperatures to prevent melting. This may be seen especially for nitrogen in Fig. 3.14 and hydrogen in Fig. 3.15. Because film boiling presents a reasonably well defined physical model (as compared with nucleate boiling) it has been treated analytically. Bromley (1950) was one of the first to analyze this process and he modelled his study on the similarity to film condensation. Other such studies include those of Breen and Westwater

(1960), Frederking and Clark (1963), Berenson (1964), Chang (1959) and Hsu and Westwater (1960).

In practical systems the total rate of heat transfer from a surface in film boiling will include an important component of radiation because of the high surface temperatures. This effect is discussed by Brentari et al. (1965) who point out that if the surface temperature is below 400 to 425°K (260 to 306°F) the maximum error in neglecting radiation is less than 5% for film boiling of O_2, N_2, H_2, and He. The error becomes approximately 50% for a doubling of the surface temperature.

Bromley (1950) correlated film-boiling to liquid nitrogen on the outside of horizontal cylinder by

$$\frac{hD_0}{k_f} = 0\cdot62\left(\frac{D_0^3\rho_f(\rho_l-\rho_f)gh_{fg}}{k_f\mu_f\Delta T_{\text{sat}}}\left(1+0\cdot4\frac{c_{pf}\Delta T_{\text{sat}}}{h_{fg}}\right)^2\right)^{\frac{1}{4}} \qquad (3.69)$$

For fluids of low latent heat ($c_p\Delta T_{\text{sat}}/h_{fg} \gg 2$) such as helium and hydrogen, Frederking (1961) modified this result by the following

$$\frac{hD_0}{k_f} = 0\cdot522\left(\frac{1}{2}\frac{\rho_l}{\rho_f}\right)^{\frac{1}{4}}\left(\frac{D_0^3\rho_f^2g\beta_f\Delta T_{\text{sat}}}{\mu_f^2}\left(\frac{c_p\mu}{k}\right)_f\right)^{\frac{1}{4}} \qquad (3.70)$$

Merte and Clark (1962) obtained liquid nitrogen film boiling data from a 1 in. diameter sphere using the transient method. These data compared well with Bromley's results for a 0·350 in. cylinder. These data have been correlated by Frederking and Clark (1963) by

$$\frac{hD_0}{k_f} = 0\cdot14\left[\frac{D_0^3\rho f(\rho_l-\rho_f)g}{\mu_f^2}\left(\frac{c_p\mu}{k}\right)_f\frac{h_{fg}}{c_p\Delta T_{\text{sat}}}\left(1+0\cdot5\frac{c_p\Delta T_{\text{sat}}}{h_{fg}}\right)\right]^{\frac{1}{3}} \quad (3.71)$$

Equation 3.71 has also correlated the film boiling data of Banchero et al. (1951) and those of Merte and Lewis (1967).

The transition from laminar to turbulent vapour flow, and the influence of special geometries is reviewed by Clark (1969).

The film boiling data of Sciance et al. (1969) for methane is correlated in Fig. 3.21. The correlation of the data from an 0·811 in. diameter cylinder presented by these authors is

$$Nu^* = 0\cdot346\left[\frac{Ra^*h'_{fg}}{c_{p_v}\Delta T(T/T_c)^2}\right]^{0\cdot276}, \qquad (3.72)$$

where,

$$Nu^* = \frac{(q/A)}{k_{vf}\Delta T}\left[\frac{g_0\sigma}{g(\rho_l-\rho_v)}\right]^{\frac{1}{2}}\left(\frac{a}{g}\right)^{-\frac{1}{2}} \qquad (3.73)$$

$$Ra^* = \left(\frac{\lambda_c}{2\pi}\right)^3\frac{g\rho_{vf}(\rho_l-\rho_v)}{\mu_{vf}^2}\left(\frac{c_p\mu}{k}\right)_{vf}\left(\frac{a}{g}\right) \qquad (3.74)$$

Equation 3.72 is shown in Fig. 3.21 as the solid line.

6

These same experimental data can be correlated in the manner of Breen and Westwater (1962) by an equation similar to that of Berenson (1964) and obtain an equally satisfactory result. This may be done by fitting the data in Fig. 3.21 by an equation in which the exponent in equation 3.72 is changed from 0·276 to 0·250 and determining a new value for the constant. As a result the following equation is obtained

$$\frac{h\lambda_c^{\frac{1}{4}}}{F_1} = 0.745\left(\frac{T}{T_c}\right)^{-\frac{1}{4}} \tag{3.75}$$

Equation 3.75 is shown by the dotted line in Fig. 3.21.

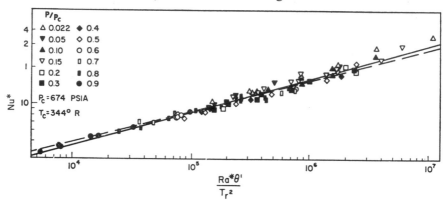

Fig. 3.21. Methane film boiling data compared with the proposed correlation (Sciance *et al.*, 1967).

2. *Forced Convection Boiling*

a. Sub-cooled, nucleate boiling. Nucleate boiling will occur in forced convection flow when the temperature of a heated surface exceeds the saturation temperature by a few degrees. When the bulk of the fluid is sub-cooled the bubbles which form at the surface condense in the liquid, either while still attached to the surface or after detaching and penetrating the liquid a short distance. Owing to the action of the bubbles themselves the degree of liquid mixing or turbulence is significantly enhanced near the surface and the heat transfer rate is significantly increased. The process mechanics then appear to be governed by both bubble induced flows and the flow of the bulk liquid as influenced by the presence of the bubbles. Accordingly, most nucleate boiling heat transfer correlations in forced convection boiling attempt to account for these two simultaneous effects.

At present there are very little forced convection, sub-cooled nucleate boiling data available for cryogenic liquids. Because of the limited amount of available cryogenic data, essentially all of the correlation methods to be discussed have been developed for non-cryogenic substances. However, it may be expected that they also will apply to cryogenic liquids. Forced convection boiling with cryogenic fluids is reviewed by Seader *et al.* (1964), Giarrantano and Smith (1966) and Brentari *et al.* (1965).

The simplest method which has been proposed is to use the pool boiling equations alone for forced convection boiling. Zuber and Fried (1961) in a review paper place the work of Kutateladze (1959), Michenko (1960), Gilmour (1958), Labountzov (1960) and Forster and Grief (1959) in this category. This method, however, seems hazardous in view of the demonstrated variation between some of the forced convection and pool boiling data, especially with regard to the influence of sub-cooling (Bergles and Rohsenow, 1964).

Rohsenow (1953) and Clark and Rohsenow (1954) have suggested that nucleate boiling data be predicted by a superposition of data for non-boiling forced convection and pool boiling. Thus, the heat flux in nucleate boiling, $(q/A)_{NB}$, would be written,

$$(q/A)_{NB} = h(T_w - T_l) + (q/A)_{PB}, \tag{3.76}$$

where $(q/A)_{PB}$ is the pool boiling heat flux corresponding to $(T_w - T_l)$, and h is computed from an appropriate non-boiling correlation, such as

$$\frac{h_x C_e}{r_f} = 0 \cdot 023 Re_f^{0 \cdot 8} Pr_f^{0 \cdot 4} \tag{3.77}$$

This method was employed on data for high pressure water at low velocities in tubes (Clark and Rohsenow, 1954) where it was observed that at high (fully developed nucleate boiling) heat flux, the flow velocity had no noticeable influence on the process. However, equation 3.76 should be used with discretion.

b. Saturated (film) boiling. Heat transfer to a flowing saturated liquid produces bubbles which do not condense but cause a net increase in the vapour fraction of the stream. This is called saturated boiling or, if the vapour fraction is sufficiently large, film boiling, especially if the vapour is concentrated near the surface. One important consequence of this is the large reduction in stream density which can occur. For flow inside of pipes and tubes the flow then accelerates, causing a momentum pressure-drop which in many instances becomes the predominant component of the total pressure-drop. Under these circumstances, the flow patterns in the stream are also very complex and difficult to define. These conditions contribute to the problem of correlating heat transfer data in this flow regime.

A nucleate boiling correlation designed for saturated liquids has been proposed by Chen (1963) and tested against data for water, methanol, cyclohexane, pentane, benzene and heptane. The average deviation of the data from Chen's equation is $\pm 11\%$ for a vapour quality range from 1% to 70%. For flow in a vertical tube in the annular or annular-mist flow regime,

Chen employed a weighted superposition hypothesis to account for the interaction between the vapour bubbles and the flowing liquid. Thus, the boiling heat flux is written

$$(q/A) = h(T_w - T_{sat})F + (q/A)_{FZ}S, \tag{3.78}$$

where h is computed from equation 3.42, or equivalent, and $(q/A)_{FZ}$ is the nucleate pool boiling heat flux using the Forster–Zuber (1955) correlation. F is a two-phase correction factor given in fig. 3.22 as a function of the parameter X_{tt}, where

$$X_{tt} = \left(\frac{1-x}{x}\right)^{0.9}\left(\frac{\mu_l}{\mu_v}\right)^{0.1}\left(\frac{\rho_v}{\rho_l}\right)^{0.5} \tag{3.79}$$

S is a factor which accounts for the influence of the flow in suppressing the growth of the vapour bubbles. This factor is given in Fig. 3.23, as a function of $Re_l F^{1.25}$, where

$$Re_l = \frac{DG(1-x)}{\mu_l} \tag{3.80}$$

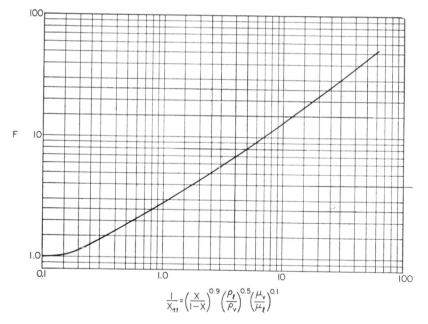

Fig. 3.22. Reynolds number factor, F, of Chen (1963).

Presently most of the available data is for liquid hydrogen. This includes the work of Hendricks *et al.* (1962), Wright and Walters (1959), Walters (1961), Chen (1963), Graham *et al.* (1961), Chi (1965) and Core *et al.* (1959). Burke and Rawdon (1965) and Laverty and Rohsenow (1965) have studied

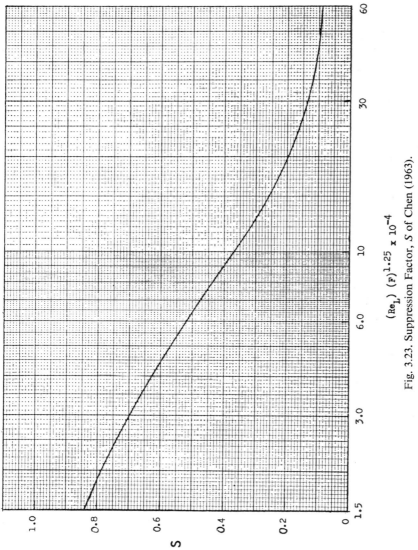

$(Re_L)\,(F)^{1.25} \times 10^{-4}$

Fig. 3.23. Suppression Factor, S of Chen (1963).

nitrogen in film boiling. A summary of film boiling correlating methods is given by Seader *et al.* (1964), Giarratano and Smith (1966) and Brentari *et al.* (1965).

Saturated and film boiling correlations have been formulated in terms of the Martinelli parameters X_{tt} and the vapour mass fraction (quality) x, where,

$$X_{tt} = \left(\frac{1-x}{x}\right)^{0.9}\left(\frac{\mu_l}{\mu_v}\right)^{0.1}\left(\frac{\rho_v}{\rho_l}\right)^{0.5} \tag{3.81}$$

These correlations are written in terms of the ratio

$$\frac{Nu, \exp}{Nu, \text{calc}},$$

where, Nu, exp is the experimentally determined Nusselt number and Nu calc. is a Nusselt number computed on the basis of a single phase correlation.

Hendricks *et al.* (1962) proposed that

$$\frac{Nu, \exp}{Nu, \text{calc}, f, \text{t.p.}} = f(X_{tt}) \tag{3.82}$$

where,

$$Nu, \exp = \frac{h \exp D}{k_{f,v}} \tag{3.83}$$

$$Nu, \text{calc}, f, \text{t.p.} = 0.023 Re_{f,m,t.p.}^{0.8} Pr_{f,v}^{0.4} \tag{3.84}$$

$$Re_{f,m,t.p.} = \frac{\rho_{f,m,t.p.} V_{\text{avg}} D}{\mu_{f,v}} \tag{3.85}$$

$$\rho_{f,m,t.p.} = \frac{1}{(x/\rho_f) + [(1-x)/\rho_l]} \tag{3.86}$$

$$V_{\text{avg}} = \frac{w}{\rho_b A_c} \tag{3.87}$$

$$\rho_b = \frac{1}{(x/\rho_v) + [(1-x)/\rho_l]} \tag{3.88}$$

The correlation of experimental data by equation 3.82 showed general agreement but considerable scatter. Because the experiments were conducted on hydrogen alone, a correlation was sought (Girrantano and Smith, 1966) for the Nusselt number ratio of equation 3.82 in terms of the local vapour quality x, but no significant improvement in the degree of correlation resulted.

An improvement in the correlation was obtained by Ellerbrook *et al.* (1962), by recognizing that the experimental data could be grouped into families of curves identified by constant "Boiling Number", B_0, where,

$$B_0 = \frac{(q/A)}{h_{fg}G_{mix}} \tag{3.89}$$

where,

$$G_{mix} = \frac{w}{A_c} \tag{3.90}$$

The separation of the data by boiling number is shown in Fig. 3.23. The boiling number may be interpreted physically as the ratio of the transverse mass velocity of the vapour formed at the wall, $(q/A)/h_{fg}$, to the axial mass velocity of the mixture, G_{mix}.

The boiling number is used to modify equation 3.82 to give

$$\frac{Nu, \exp}{Nu, \text{calc}, f, \text{t.p.}} B_0^{-0.4} = f(X_{tt}) \tag{3.91}$$

This form of the correlation is shown in Fig. 3.24 where it will be observed that the scatter is probably within the experimental error.

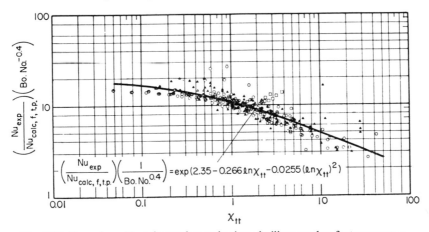

Fig. 3.24. Two-phase Nusselt number ratio times boiling number factor versus χ_{tt}.

c. Dispersed flow film boiling. Dispersed flow occurs at moderate to high quality conditions and is characterized by liquid droplets entrained in the vapour flow. Because of the absence of continuous direct contact between tube wall and entrained droplet, two paths for heat transfer exist simultaneously. Part of the heat flow is direct from the tube wall to droplet; this has the major importance at lower qualities. At high qualities the heat transfer route from tube wall to vapour and thence from vapour to drop assumes an equal or greater contribution. In this latter situation the vapour becomes superheated and a discrepancy may exist between the equilibrium quality of the stream (as determined by the quantity of heat transferred to it) and the actual quality (the mass fraction of fluid that has actually evaporated).

A series of studies of disperse flow film boiling of liquid nitrogen have been carried out at the Massachusetts Institute of Technology by Laverty and Rohsenow (1965), Forslund and Rohsenow (1968), and Hynek et al. (1962). Two regimes of disperse flow boiling have been observed by these workers at the same heat flux and flow conditions. These correspond to boiling with an annular film present in addition to entrained droplets (annular disperse flow, termed Type 1 boiling) and boiling with only an entrained dispersed liquid phase (Type 2).

The existence of either regime was determined by whether or not the wall surface temperature was maintained above the Leidenfrost temperature during the establishment of boiling conditions. The two boiling regimes are distinguished by a different degree of superheat in the vapour, and thus Type 2 boiling exhibits a higher wall temperature (lower heat transfer coefficient).

Laverty, Forslund, Hynek et al. developed a thermodynamic non-equilibrium model which accounted for the simultaneous heat transfer directly from wall to droplets, from wall to vapour, and vapour to droplets, and by stepwise integration along the tube were able to correlate the experimental data by fitting a value to the product of two constants, $K_1 K_2$, in the direct wall to droplet heat transfer correlation. However, the value of $K_1 K_2$ was found to vary substantially with the fluid being correlated. Thus, for nitrogen a value of 0·2 was found appropriate whilst for methane $K_1 K_2 = 2·0$ was found to give the best fit.

These studies of disperse boiling signal an area of caution for the heat exchanger designer in which he must differentiate carefully between actual and equilibrium conditions. On the basis of the equilibrium temperature versus enthalpy relationship, the heat flow may be given as

$$\left(\frac{q}{A}\right)_e = h_c(T_w - T_e) \tag{3.92}$$

Where the subscript e denotes thermodynamic equilibrium conditions. For the actual non-equilibrium case,

$$\left(\frac{q}{A}\right)_a = h_a(T_w - T_a) \tag{3.93}$$

where a denotes actual. From an enthalpy balance,

$$x_e h_{fg} = x_a(h_{fg} + c_{pv}(T_a - T_e)) \tag{3.94}$$

where x_e and x_a are the equilibrium and actual qualities. Thus,

$$\left(\frac{q}{A}\right)_a = h_a\left(T_w - T_e - \left(\frac{x_e}{x_a} - 1\right)\frac{h_{fg}}{c_{pv}}\right) \tag{3.95}$$

By comparison of equations 3.92 and 3.95, it is apparent that the actual heat flux may be below the value calculated from an assumed thermodynamic equilibrium.

B. CONDENSATION

The process of condensation, though physically complex, can often be calculated with precision due to the fact that almost all the resistance to transfer is normally concentrated in the condensate layer. The necessary conditions for a pure vapour are that the condensate spreads to form a uniform layer (this tends to be true with cryogenic liquids), and that the thickness of the layer is calculable in terms of the forces acting. Then the heat transfer coefficient is given by the thermal conductivity of the condensate divided by its average thickness. The condensate film thickness, in the simplest case, is determined by a balance between viscous forces and drainage due to gravity. For situations involving multi-component vapours a substantial resistance to mass and heat transfer may develop in the vapour phase also. At low pressures there may also be a significant resistance at the vapour–liquid interface.

1. *Laminar Film Condensation*

The elementary analysis of laminar film condensation on a vertical plate from a stagnant pure vapour was made by Nusselt (1916). He assumed a linear temperature profile across the film and a constant surface temperature.

He showed that at a distance z down a vertical surface from an initial point of zero film thickness, the film thickness is given by

$$\delta = \left(\frac{4k_l\mu_l\Delta Tz}{g\rho_l(\rho_l-\rho_v)h_{fg}} \right)^{\frac{1}{4}} \tag{3.96}$$

in which ΔT is the temperature difference from the saturation vapour temperature to the wall, and h'_{fg} is the average enthalpy change of the vapour in condensing to liquid and subcooling to the average liquid temperature ($h'_{fg} = h_{fg}+\frac{3}{8}c_l\Delta T$).

The local heat transfer coefficient is given by

$$h_z = \left(\frac{q}{A} \right) \bigg/ \Delta T = \frac{k_l}{\delta} \tag{3.97}$$

Thus:

$$h_z = \left(\frac{g\rho_l(\rho_l-\rho_v)k_l^3 h'_{fg}}{4z\mu_l\Delta T} \right)^{\frac{1}{4}} \tag{3.98}$$

The average coefficient determined by integrating down the length of the surface is obtained as

$$h = 0\cdot943\left(\frac{g\rho_l(\rho_l-\rho_v)k_l^3 h'_{fg}}{L\mu_l\Delta T} \right)^{\frac{1}{4}} \tag{3.99}$$

2. Condensation on Inclined Plates and Outside Horizontal Tubes

The equations above may be used directly for an inclined flat plate by the substitution for g of $g \sin \theta$, the component of gravity along the plate. θ is the angle to the horizontal plane.

For a horizontal tube $g \sin \theta$ varies continuously with the position around the tube. For the top tube of a bank of tubes, the average heat transfer coefficient is given by

$$h = 0{\cdot}728 \left(\frac{g \rho_l (\rho_l - \rho_v) k_l^3 h'_{fg}}{D_0 \mu_l \Delta T} \right)^{\frac{1}{4}} \tag{3.100}$$

For a bank of n horizontal tubes in which all the liquid from one tube drops to the next, the heat transfer coefficient averaged over all the tubes is

$$h = 0{\cdot}728 \left(\frac{g \rho_l (\rho_l - \rho_v) k_l^3 h'_{fg}}{n D_0 \mu_l \Delta T} \right)^{\frac{1}{4}} \tag{3.101}$$

However, equation 3.101 does not account for the fact that the liquid condensate falling between tubes is subcooled and will permit additional condensation. Chen (1961) proposed that additional condensation would equal the subcooling of the liquid leaving the tubes. He derived

$$h = 0{\cdot}728 \left(1 + \frac{0{\cdot}2 c_l \Delta T (n-1)}{h_{fg}} \right) \left(\frac{g \rho_l (\rho_l - \rho_v) k_l^3 h'_{fg}}{n D_0 \mu_l \Delta T} \right)^{\frac{1}{4}} \tag{3.102}$$

This agrees well with experimental data provided $(n-1) c_l \Delta T / h_{fg} < 2$.

3. Effect of Vapour Shear

Where the vapour flow rate is high such as for condensation in a plate-fin exchanger or in tubes, the effect of the vapour velocity is to produce a shear force at the liquid film surface which may advance or retard the condensate flow depending upon whether the vapour flow is cocurrent or countercurrent. In cocurrent flow, the effect of vapour shear is to reduce the condensate film thickness and thus to increase the heat transfer coefficient.

For laminar flow of the liquid film, an analytical solution for the heat transfer may be derived similarly to the Nusselt solution by addition of the vapour shear stress term in the force balance. The reader is referred to heat transfer texts (see, e.g. Rohsenow and Choi, 1961) for the derivation and presentation of the sets of equations representing this situation.

The transition from laminar to turbulent flow of the liquid film is a function of both the condensate flow rate and the vapour flow rate. For a freely falling film (shear stress, $\tau_v = 0$) the transition is observed at a film Reynolds number $= 1800$ where the film Reynolds number is defined as $4\Gamma/\mu$ and Γ is the mass flow rate of condensate per unit width of surface. The transition Reynolds number has been observed to decrease sharply with increasing vapour flow rate to a value of approximately 240 at high vapour velocity (Carpenter and Colburn, 1951).

An analysis for turbulent film flow condensation has been presented by Rohsenow et al. (1956) and the corresponding trial and error design calculation procedure is outlined by Rohsenow and Choi (1961).

A more convenient correlation for annular turbulent condensing flow where there is no stratification has been presented by Soliman et al. (1968). They obtained the following correlation for the liquid film heat transfer coefficient.

$$h = 0.036 \frac{k_l \rho_l^{\frac{1}{2}}}{\mu_l} Pr_l^{0.65} \tau_0^{\frac{1}{2}} \tag{3.103}$$

τ_0 is the shear stress in the laminar sublayer and is given by:

$$\tau_0 = \tau_F + \tau_A + \tau_H \tag{3.104}$$

The component shear stresses are given by Soliman as Friction:

$$\tau_F = \frac{A}{S}\left(-\frac{dp_f}{dz}\right) \tag{3.105}$$

Momentum:

$$\tau_H = g_z\left(\frac{1-\alpha}{4}\right)D_0(\rho_l - \rho_v) \tag{3.106}$$

Gravity:

$$\tau_A = \frac{G^2}{4\rho_v}\left(D_0\frac{dx}{dz}\right)\left(a_1\left(\frac{\rho_v}{\rho_l}\right)^{\frac{1}{3}} + a_2\left(\frac{\rho_v}{\rho_l}\right)^{\frac{2}{3}} \cdots + a_5\left(\frac{\rho_v}{\rho_l}\right)^{\frac{5}{3}}\right) \tag{3.107}$$

$$a_1 = 2x - 1 - \beta x$$

$$a_2 = 2(1-x)$$

$$a_3 = 2(1-x-\beta+\beta x)$$

$$a_4 = \frac{1}{x} - 3 + 2x$$

$$a_5 = \beta\left(2 - \frac{1}{x} - x\right)$$

In the above equations, A is the cross-sectional area for flow, S is the internal perimeter of the flow channel. dP_f/dz is the frictional pressure gradient from the Lockhart–Martinelli correlation (1949). Y is the vapour volume void fraction from the correlation of Zivi (1964). x is the "quality" or vapour phase mass fraction compared to the total flow. The constant β is given as 1·25 for turbulent flow in the film.

The Soliman et al. correlation is derived for a range of experimental data with vapour velocities from 6 to 300 m/sec, Pr from 1 to 10 and x from 0·99 to 0·03, and with data for both vertical and horizontal flow. It is however derived for annular flow and cannot be considered applicable where the ratio ρ_v/ρ_l departs significantly from zero, say at $\rho_v/\rho_l > 0\cdot125$ or it is known from a flow regime map that flow is not annular.

4. Empirical Forced Convection Correlations

In order to determine the flow regime boundaries for condensing flow it is recommended that the methods proposed by Wallis (1969) are used. Of particular importance is the transition from annular to slug flow for upward vertical vapour flow. This transition range is expressed by the equations

$$j_v^* = 0\cdot4+0\cdot6j_e^*$$

and

$$j_v^* = 0\cdot9+0\cdot6j_e^* \tag{3.108}$$

which have already been presented in Section III B.4.b. Condensation heat transfer coefficients in the slug and bubble flow regimes are best predicted from empirical correlations of the homogenous two-phase flow type as used similarly for boiling. Typical of these is the Boyko–Kruzhilin correlation (1967):

$$h = 0\cdot021\frac{k_l}{D_0}Re_{lo}^{0\cdot8}Pr_l^{0\cdot43}\left(\frac{\rho_l}{\rho_{\text{hom}}}\right)^{0\cdot5} \tag{3.109}$$

in which Re_{lo} is the Reynolds number for the total flow with properties determined for the liquid phase. ρ_{hom} is the homogenous two phase density:

$$\frac{1}{\rho_{\text{hom}}} = \frac{x}{\rho_v}+\frac{(1-x)}{\rho_l} \tag{3.110}$$

5. Horizontal Flow Inside Tubes

For non-stratified flow in horizontal tubes in which the condensate film is uniformly distributed around the tube surface, the correlation of Soliman et al. (1968) may be applied as above. The determination of the flow regime boundary is best determined from the map by Baker (1954).

Where stratification is occurring the vapour shear forces do not predominate and the form of correlation required is similar to the Nusselt equation for the outside of a horizontal tube.

Thus,

$$h = 0\cdot725\Omega_1\left(\frac{g\rho_l(\rho_l-\rho_v)k_i^3h_{fg}'}{\mu_lD_0\Delta T}\right)^{\frac{1}{4}} \tag{3.111}$$

in which Chato (1962) gives Ω_1 equal to $0\cdot78$ an$_{fg} = h_{fg}+0\cdot68c_l\Delta T$.

6. Condensation with Non-Condensable Gases Present

The effect of a non-condensable gas present in the vapour phase is to generate a resistance to the diffusion of the vapour molecules towards the condensate surface. In order for there to be a net diffusion of vapour to the interface a concentration driving force exists and thereby a reduction in the vapour phase partial pressure towards the interface. Thus, the interface saturation temperature is reduced to a temperature intermediate between the bulk vapour and the surface, and the temperature driving force across the condensate film is reduced when compared to the comparable pure vapour situation.

The heat flux is given by:

$$\frac{dq}{dA} = h_f(T_w - T_i) = h_g(T_i - T_g) + K_g(\rho_i - \rho_g)h_{fg} \tag{3.112}$$

in which h_f is the condensate film heat transfer coefficient, h_g is the single phase gas heat transfer coefficient, K_g is the gas phase mass transfer coefficient for the condensing component, and subscripts w, i, and g denote wall, interface and bulk gas respectively.

Let $\Delta T_f = T_i - T_w$ then as a first approximation,

$$\frac{\Delta T_f}{T_w - T_g} = \frac{h_g + K_g(d\rho^0/dT)h_{fg}}{h_f + h_g + K_g(d\rho^0/dT)h_{fg}} \tag{3.113}$$

Where $d p^0/dT$ is the variation of the condensate vapour pressure with temperature. K_g and h_g are usually calculated as the convective coefficients for the vapour/gas phase mixture flowing alone.

C. INJECTION COOLING

Another process which has found important application in cryogenic systems is the cooling of a cryogen by the injection of a non-condensible gas of low solubility. One significant application of this has been the injection of gaseous helium into the liquid-oxygen suction lines of large rocket boosters during pre-launch operations. This provides the necessary subcooling by the evaporation of the liquid into the helium bubbles to prevent pump cavitation and engine failure at start-up (Randolph and Vaniman, 1961; Marshall Space Flight Center, 1961). This process has been analysed in depth in recent years. However, in the interest of brevity the available papers on the subject will be cited only.

Larsen *et al.* (1963) analysed the process of gas injection including the effects of liquid evaporation, gas solubility, gas enthalpy flux, ambient heating and liquid displacement due to the presence of gas bubbles in the system. The analytical model is space-wise lumped and the results compare favourably with measurements on vertical columns of liquid oxygen cooled by the injection of gaseous helium and nitrogen. The production of solid argon by the injection of gaseous helium into liquid argon has been observed in experiments reported by Lytle and Stoner (1965). The cooling of liquid hydrogen by helium gas injection has been studied by Schmidt (1963). A detailed analytical investigation of the dynamics of single bubbles injected in a liquid is reported by Arpaci *et al.* (1965). Reasonable agreement with the theory was obtained from high speed motion pictures of the dynamics of nitrogen and helium bubbles injected into water.

D. FROST FORMATION

The problem of frost formation on cooled surfaces is not well defined although it is commonly encountered on cryogenic systems in an atmospheric ambient. Should cryogenic substances be stored in atmospheres other than the terrestrial other kinds of frost can be expected.

The prediction of frost formation in atmospheric air is complicated by several factors. For surface temperatures below the temperature of liquid air the overriding layers of water and carbon dioxide frost have a liquid air substrate. This results in an instability of the growth of the frost causing it to coat a surface in patches. The shearing action of gravity and aerodynamic drag forces cause a similar destruction of the frost layers. The mechanical structure of frost is largely unknown and its physical properties are not well tabulated. Frost formation from normal ambient air involves a complex process of simultaneous transient heat and mass transfer of at least two or three components (H_2O, CO_2, air) and two or three phases.

A study of frost formation inside of tubes in forced convection is reported by Chen and Rohsenow (1964). They attribute the increased heat transfer and pressure drop under conditions of frost to the roughness of the frost surface. Smith *et al.* (1964) analysed the problem of frost formation on cooled surfaces in forced convection and give results concerning the thermal conductivity of frost. They conclude that the thermal conductivity is a function of the conduction and diffusion paths through the frost.

Barron and Han (1965) investigated the formation of frost on vertical flat surfaces in laminar and turbulent natural convection. They observed that a simultaneous mass transfer of water vapour from the moist ambient air to the frost layer increased the rate of heat transfer. This effect was not large, however, being about 5% for a mass fraction of water vapour of 0·10. Their heat transfer data were correlated within $\pm 20\%$ by formulations in terms of Nusselt, Grashof, Prandtl and Schmidt numbers and parameters which relate to the effects of mass transfer. On the other hand the predicted mass transfer rates were about ten times greater than those measured. This was attributed to the interference by frost particles in the boundary-layer to the diffusion of water vapour to the cold surface. The predominant component to the energy transport was the convective component and the influence of thermal diffusion and diffusion thermo effect was found to be completely negligible. Barron and Han measured the thermal conductivity of the frost and found it to be related to its mean density and mean temperature. Some of their results are given in Figs. 3.25 and 3.26.

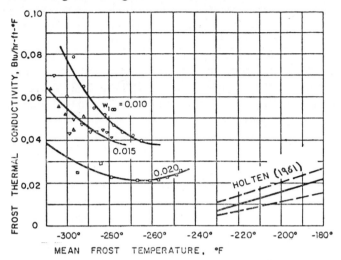

Fig. 3.25. Mean apparent frost thermal conductivity as a function of the mean frost temperature, $T_f = 0.5(T_w + T_s)$.

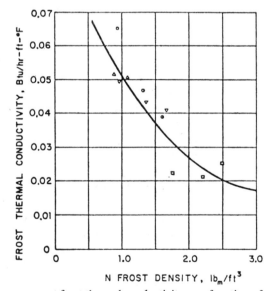

Fig. 3.26. Mean apparent frost thermal conductivity as a function of mean frost density.

Studies of frost formation also have been reported by Holten (1961), Van Gundy and Uglum (1962), Arthur D. Little, Inc. (1958), The National Bureau of Standards (1961) and Loper and Heatherly (1955).

Holten (1961) investigated the growth of frost on the outside of spherical aluminium containers in a free convection atmospheric ambient and found a definite dependence on the specific humidity. The rate of formation of frost was found to go through a maximum while its surface temperature increased. The fact that the latter did not reach 32°F after long times is attributed to the mechanical failure of the condensate which became excessive after 150 minutes. The data of Arthur D. Little, Inc. (1958) indicate k-values of frost to range from 0·02 B.T.U./hr ft °F at -275°F to 0·035 B.T.U./hr ft °F at -200°F, corresponding to frost thicknesses of 0·05 and 0·15 in., respectively. Only 20–40% of the water vapour that is predicted to diffuse to the cooled surface by mass transfer calculations is retained on the surface as frost, an observation similar to that of Barron and Han (1965).

Van Gundy and Uglum (1962) conducted experiments on the mechanics of frost on liquid hydrogen cooled surfaces at 20°K for the forced flow of moist air. They report an important influence of the condensation of liquid air. The highest heat flux is obtained when the condensed air can flow off the surface. A low ambient temperature and low humidity was found to inhibit frost formation.

VIII. Radiation

The principal problem of radiation heat transfer at cryogenic temperatures is the determination of the radiation properties of surfaces. Gaseous radiation is less troublesome since most of the substances which remain as gases at low temperatures are not significant radiators nor absorbers. Radiation enclosure calculation methods such as those of Hottel (1954) and Gebhart (1961) are valid under these conditions. However, low temperature systems have introduced a new consideration into the treatment of radiation heat transfer. This is the effect of condensed gases on the radiation properties of cold surfaces. Such condensate layers build up complex systems known as cryo-deposits on the cold substrate. Much of the work to date has been for systems of H_2O and CO_2, the common cryodeposits from a terrestrial ambient. In general this is an extension of the problem of frost and much yet remains to be learned about these systems. Knowledge of the radiation properties of the cryodeposits is important in connection with studies of cryopumping, simulation of the solar space environment and the storage of cryogenic fluids.

Kropshot (1962) gives some values of the total hemispherical emissivities for some common metals at low temperatures. His results are included in Table 3.VII. Other tabulations of the emissivities of various materials with different surface conditions as a function of temperature include the work of Fulk and Reynolds (1957).

The emissive properties of vacuum-deposited metallic coatings on poly-ester film for low temperature service has been investigated by Ruccia and

Hinckley (1967). These highly reflective surfaces are employed in multi-layer, vacuum insulation where, except for aluminium, savings in weight and cost as well as increases in strength are obtained by depositing copper, gold and silver on a polyester substrate. An important factor to be determined is the influence of thickness of the metallic deposit on the radiative properties. Another is the effect of environmental conditions of temperature, humidity and contaminants on the stability and adhesion of the metal film.

TABLE 3.VII. Selected Minimum Total Emissivities †

Surface	Surface temperature, °K			
	4	20	77	300
Copper	0·0050		0·008	0·018
Gold			0·01	0·02
Silver	0·0044		0·008	0·02
Aluminium	0·011		0·018	0·03
Magnesium				0·07
Chromium			0·08	0·08
Nickel			0·022	0·04
Rhodium			0·078	
Lead	0·012		0·036	0·05
Tin	0·012		0·013	0·05
Zinc			0·026	0·05
Brass	0·018			0·035
Stainless steel, 18–8			0·048	0·08
50 Pb 50 Sn solder			0·032	
Glass, paints, carbon				>0·9
Silver plate on copper		0·013	0·017	
Nickel plate on copper		0·027	0·033	

† These are actually absorptivities for radiation from a source at 300°K. Normal and hemispherical values are included indiscriminately. Data taken from Kropschot (1962).

The results of Ruccia and Hinckley (1967) for the emittance of metallic deposits of aluminium, gold, silver, silicon monoxide protective coating on silver, copper and silicon monoxide protective coating on copper are given in Fig. 3.27. The substrate consisted of ¼-mil (0·00025 in.) DuPont, type A, polyester film. As the thickness of the deposited metal layer increases the emittance decreases until an asymptote is reached at about 750–1,000 Å for silver and aluminium and 1,500 Å for gold. The lowest emittance is found for silver with copper, gold and aluminium following next in order of increasing emittance for any thickness. The protective coatings of SiO on silver and copper degrade their emittance by about 40%. The authors also investigated the influence of environment on emissivity. Except for the 95% relative humidity environment the aluminium coating showed good stability. Gold

7

appears to be less stable in all environments while silver and copper were the least stable.

Cunningham and Young (1963) have studied the absorptance of a CO_2 cryodeposit on various substrates at 77°K (139°R). Their measurements gave results for the bare surface which agree well with literature values for the black and polished surface and a weighted calculation based on area for the other surfaces. The influence of cryodeposit thickness is to change the absorptance significantly for small thickness but to have essentially no effect for thicknesses greater than about 0·8 mm. For substrates of low absorptivity the effect of the deposit is to increase the absorptivity whereas the opposite effect is observed for high absorptivity substrates. Even for relatively large

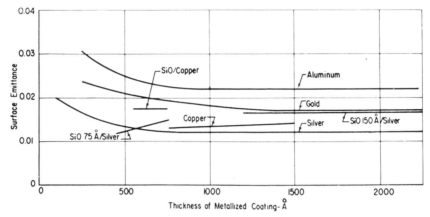

Fig. 3.27. Emittance of vacuum-metallized polyester film at 553°R for various metal coating materials and thicknesses (Ruccia and Hinckley, 1967).

thickness of CO_2 the absorptivity of the cryodeposit on the various surfaces did not approach a common value. Apparently for a thickness in excess of about 0·8 mm the frost absorbs all radiation in the absorption band thus having no further influence on the radiating characteristics for greater thicknesses.

The absorptance of water vapour and carbon dioxide cryodeposits at 77°K (139°R) for both solar and 290°K (522°R) black body radiation is reported by Caren et al. (1964). The substrates consisted of polished aluminium and aluminium coated with a black epoxy paint (cat-a-lac flat black). The absorptance of an H_2O cryodeposit on these substrates is shown in Fig. 3.28 as a function of deposit thickness for black radiation at 522°R. The influence of the H_2O deposit is to increase the absorptance very significantly for the low absorptance substrate. The effect is much less for the painted surface. Agreement with data for H_2O given by Moore (1962) is good. The absorptance data for a CO_2 deposit on the same substrates for black radiation at 522°R is similar. Agreement with the results of Moore (1962) is reasonably good but poor with the data of Cunningham and Young (1963). This variation is attributed to differences in the physical nature of the deposits.

A solar source was simulated by using filtered radiation from a mercury–xenon lamp. In this case the absorptance decreases with H_2O deposit thickness from nearly 1·0 for an essentially black substrate to 0·5 for a deposit thickness of 3 mm.

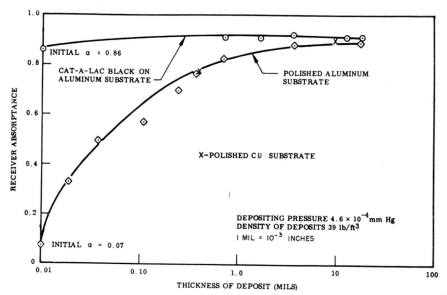

Fig. 3.28. Absorptance of H_2O cryodeposit for room temperature blackbody radiation (Caken et al., 1964).

An analysis of the thermal transport processes in a cryodeposit subject to radiation at the vacuum interface is given by McConnell (1966). His results compare favourably with the experimental measurements of Caren et al. (1964) for H_2O on a reflecting substrate. Tien and Cravalho (1967) survey recent advances in the study of the thermal radiation properties of solids at cryogenic temperatures. For radiative transport between solids they discuss the influence or non-grey surfaces (wavelength and temperature dependent properties), non-equilibrium fields and the effect of small surface spacing, a factor of particular significance at cryogenic temperatures. This latter is important because of the concentration of low temperature thermal radiation at long wavelengths. Theoretical studies are needed to provide calculation procedures for the transport of radiation between surfaces whose spacing is of the same order or less than the wavelength at the maximum radiative heat flux.

IX. Selection and Design of Heat Exchangers

Cryogenic heat exchangers range in magnitude from units of up to 60 m in length for processing natural gas (Fig. 3.29), to miniature coil wound exchangers of 6 mm diameter manufactured from finned hypodermic tubing, used for cooling electronic equipment (Fig. 3.30). Exchangers also vary in complexity from simple rock filled regenerators used in some air separation plants to the multi-stream exchangers with simultaneous laminar, turbulent, boiling and condensing flows which are encountered in light hydrocarbon process plants.

A. SELECTION

The selection of an exchanger for a particular duty is normally determined by three criteria: process design requirements, mechanical design limitations and economic considerations. The principal types available are plate and fin extended surface, tube-in-shell (both straight tube and coil wound), and paired tube exchangers, and regenerators.

1. *Process Design Requirements*

a. N_T. N_T may be defined for an individual stream using equation 3.1 but based on the difference in temperature between the bulk stream and the heat transfer surface.

Consider an idealized heat exchanger for two fluids in turbulent convective counterflow with parallel cooling curves, i.e. temperature against heat capacity profiles. The component N_T may then be written:

$$N'_{Tc} = \frac{T_{ci} - T_{co}}{\Delta T_{wc}} \; ; \qquad N'_{Th} = \frac{T_{ci} - T_{co}}{\Delta T_{hw}} \qquad (3.114)$$

and

$$\frac{1}{N_{Tc}} = \frac{1}{N'_{Tc}} + \frac{1}{N'_{Th}} \qquad (3.115)$$

ΔT_{wc} and ΔT_{hw} are the temperature differences (constant) between the heat transfer surface and cold stream and the hot stream and the surface respectively.

The overall heat transferred between streams may be written (equation 3.9).

$$q = (wc_p)_c(T_{ci} - T_{co}) = h_c A \Delta T_{wc} \qquad (3.116)$$

Thus,

$$N'_{Tc} = \frac{h_c A}{(wc_p)_c} \qquad (3.117)$$

For heat transfer in tubes and channels where the fluid flow is fully developed (i.e. with a length to diameter ratio in excess of approximately 20), the heat transfer coefficient may be expressed conveniently for the present

Fig. 3.29. Tube bundle and tube plate for a large coiled tube in shell heat exchanger. (Courtesy of Air Products and Chemicals Inc.).

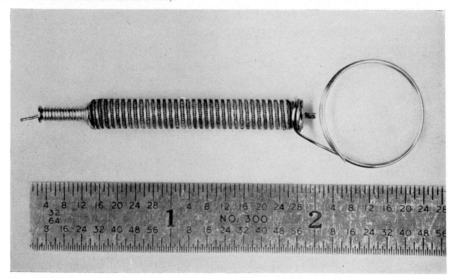

Fig. 3.30 Tube bundle for a miniature heat exchanger—part of a Joule Thompson nitrogen cycle refrigerator. (Courtesy of Air Products and Chemicals Inc.).

purpose by the Colburn correlation. The heat transfer coefficient is thus expressed solely in terms of physical properties, mass flow rate, and the heat transfer j factor (j_T).

$$h_c = G_{cp} j_T Pr^{-\frac{2}{3}}$$ (3.118)

j_T is a weak function of Reynolds number and is analogous to the momentum friction factor and mass transfer j factor. The exact form of j_T may be determined by comparison of equation 3.118 with the more complex correlations given for turbulent convective heat transfer.

Thus, from equations 3.117 and 3.118 for a channel of length L and diameter D,

$$N'_{Tc} = 4 j_T Pr^{-\frac{2}{3}} \frac{L}{D}$$ (3.119)

In turbulent flow j_T varies slowly with the stream Reynolds number ($\propto Re^{-0.2}$) in the same manner as the friction factor. Thus, the maximum values of N'_T are determined primarily by the L/D ratio of the flow channels.

Typically, the individual stream N'_T as determined by current economic considerations will have values of $0.025\, L/D$ to $0.02\, L/D$ and for an exchanger with balanced thermal resistances the overall N_T will have values from $0.05\, L/D$ to $0.033\, L/D$.

It is apparent therefore that for gas/gas heat exchange a conventional straight tube-in-shell exchanger is limited to small values of N_T (2 in. diameter tubes in a 20 ft long shell will give $N_T = 5$ approximately). A standard length (12 ft) brazed aluminium plate-fin exchanger in air separation service has a typical N_T of about 35. Larger values of N_T may be achieved in a single exchanger by the use of small diameter coiled tubes.

b. *Number of streams.* Heat exchange between multiple fluid streams is most readily accomplished in a plate-fin exchanger. Up to eight fluid streams may be accommodated, with intermediate withdrawal of any stream at intermediate positions along the exchanger.

It is also relatively straightforward to accommodate a number of co-current fluid streams in the tube circuits of a tube-in-shell exchanger provided that they are exchanging heat with a single countercurrent shell-side fluid. Efficient heat exchange between fluids flowing countercurrent in the tube circuits generally requires that the tubes are paired, i.e. the tubes of different circuits are thermally linked to give a direct conduction path. This may be accomplished by soldering together copper tubes (Fig. 3.31).

Regenerators, which operate by the alternate warming and cooling of a a high thermal capacity solid medium by hot and cold fluids, are restricted to heat exchange between two fluids only unless there are buried coils for additional fluids or additional regenerators.

c. Pressure drop. The choice of exchanger is not generally restricted by pressure drop considerations. Pressure drop requirements can be met by the provision of adequate cross-sections for fluid flow. A requirement for very low pressure drops may necessitate a large exchanger diameter; thus influencing the relative costs of different exchangers.

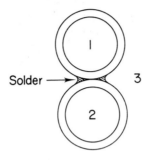

Fig. 3.31. Cross-section of a paired tube assembly.

2. Mechanical Design Limitations

a. Pressure. The maximum operating pressure that can presently (1970) be achieved in brazed aluminium plate-fin exchangers, designed in accordance with A.S.M.E. code, and operating at steady pressure loading is approximately 50 bar. The major factors determining operating pressure are the stresses occurring within the corrugations and their brazed connection with the parting sheets, and at the junction of the header tanks to the core. The latter in particular limits the maximum size of the blocks for high pressure. For example, at 45 bar operating pressure the present maximum block width is 0·65 m. Examples of other maximum sizes for various operating pressures are given in Table 3.VIII.

TABLE 3.VIII. Maximum Design Pressure Related to Block Size of Brazed Aluminium Plate–fin Exchangers (1970)

Maximum design pressure		Width of passages		Stack height of passages	
bar	lb/in² g	m	in	m	in
52	755	0·5	19·7	0·8	31·5
45	652	0·65	25·6	0·92	36·0
38	550	0·77	30·5	0·92	36·0
31	450	0·90	35·4	0·92	36·0

Heat exchangers used in the reversing service of air separation plants are subject to approximately one million pressure reversals over a life of fifteen to twenty years. This arduous duty reduces the maximum allowable pressure with brazed aluminium exchangers to a value of 10 to 11 bar.

Exchangers for pressures exceeding 50 bar are commonly of tube-in-shell design with the high pressure fluid inside the tubes. Very high pressure exchangers (200 bar) generally use copper tubes with brass or stainless steel tube sheets and soldered or brazed joints, although permissible operating pressures for aluminium tube exchangers are increasing with improved techniques for mechanical sealing of tubes into tube sheets.

b. *Materials*. Metals for cryogenic usage must retain their ductility over the full operating temperature range (see chapter 6). Plate-fin exchangers are manufactured from alternate layers of corrugated and plane aluminium sheets dip brazed in a molten salt bath. Aluminium tubes are used in exchangers in preference to copper from cost considerations, however, copper retains the advantage for very high pressures or for paired-tube exchangers due to the strength and reliability of jointing by silver brazing. Alloy steels are used primarily in exchanger shells and tube sheets. Low percentage nickel steels are used in olefine plants where they are economies due to the higher temperature level, and where the fire resistance of steels is favoured.

Aluminium is not suitable for service when alkali is contained in the process fluid and especially in the presence of chlorides. Such situations occur in nitrogen wash plants for the manufacture of ammonia synthesis gas, where the deposition of nitrogen oxide resins requires periodic washing by alkali, and in air separation plants in heavily contaminated industrial atmospheres, unless special precautions are taken to cleanse the process fluids.

c. *Mechanical design*. Brazed aluminium plate-fin exchangers are presently (1970) limited in size to maximum dimensions of 1·2 m (4 ft) square cross-section by 7·3 m (24 ft) length. These limitations are determined by the size of the brazing bath and the accumulation of tolerances in assembly. Although larger exchangers may become available the useful maximum block cross-section is likely to be limited effectively by pressure considerations.

The maximum diameter of wound coil exchangers is limited by the support technique employed between adjacent layers of tubes. Simple frictional support is adequate for small exchangers. Exchangers having a diameter of up to about 4 m have been manufactured, Fig. 3.29, using special interlayer supports. This size corresponds to the maximum width of load which can be shipped by normal means. The maximum exchanger length is determined by manufacturing facilities and especially the distance between mandrel support points on the winding lathe. Single bundles exceeding 15 m in length have been built, whilst exchangers 60 m long containing multiple tube bundles have been constructed within a single shell.

d. *Economic considerations*. Given that more than one acceptable design of heat exchanger remains after considering the various process and mechanical design parameters, then the final selection is made from economic considerations. These must include initial capital cost, effective life of exchanger to determine replacement cost, operating cost, and delivery period in relation to overall project.

The lowest cost per unit area of heat exchanger surface for a single exchanger element is generally obtained from brazed aluminium plate-fin exchangers. However, where very large flows are involved, the cost of manifolding multiple units in parallel may render the use of large coil wound exchangers more economic. For two fluid gas-phase service regenerators packed with aluminium ribbon, metallic spheres or pebbles may still occasionally be justified although they have been increasingly supplanted by aluminium plate-fin exchangers.

Replacement and maintenance cost is not an important factor for cryogenic systems since corrosion may generally be ignored. However, the usage of brazed aluminium exchangers in large scale air separation applications has not yet extended beyond eight to ten years of continuous service and some degree of exchanger repair and replacement has already been found necessary. The primary sources of trouble have been insufficient attention to the mechanical design of supports and piping, which has lead to fatigue cracking, the weakening of brazed joints by local welding, and occasionally from corrosion by condensate from industrially polluted air.

An operating expense is created by the pressure loss during fluid flow through the exchanger circuits. Pressure loss in the low pressure circuits is especially significant and may be costed in terms of compression power requirement.

B. BRAZED ALUMINIUM PLATE-FIN EXCHANGERS

1. General Assembly

The typical form of the plate-fin exchanger is illustrated by Figs. 3.32 and 3.33. The development, design and construction of these exchangers have been described by A. G. Lenfestey (1961). The basic elements of the construction are shown in Fig. 3.34. Alternate layers of primary surface (parting sheet) and secondary surface (corrugated fin) are assembled as required to provide the necessary flow circuits. Limitations to maximum exchanger size and the number of fluid streams have already been indicated. Each layer of primary and secondary surface is termed a passage and is bounded by a side bar of rectangular or channel section to retain the internal block pressure. The various forms of corrugation are shown in Fig. 3.35. Straight plain or perforated fins are used primarily in boiling or condensing duties and to minimize the possibility of solid deposits accumulating. This is necessary, for example, in the oxygen circuit of air separation plant reboilers where hydrocarbon accumulations would constitute a hazard. Herringbone and multi-entry fins are available to give maximum fluid turbulence and hence increased

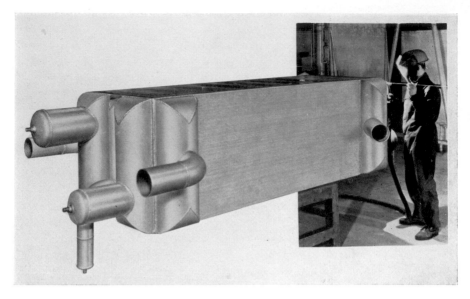

Fig. 3.32. A single plate-fin heat exchanger block for an air separation plant. (Courtesy of Marston Excelsior Limited).

Fig. 3.33. A bank of plate-fin heat exchangers assembled in the cold box frame of an air separation plant. (Courtesy of Marston Excelsior Ltd).

heat transfer especially in gas service. Multi-entry fins are used extensively in reversing exchangers of air separation plants where, in addition to maximum heat transfer surface, the fin surface also serves to retain deposited ice and solid carbon dioxide.

Fig. 3.34. The basic elements of a plate-fin heat exchanger. (Courtesy of Marston Excelsior Ltd).

The various flow geometries which are utilized have been amply illustrated by previous authors (Kays and London, 1964; Lenfestey, 1961). The most commonly used flow arrangement for cryogenic service (see Fig. 3.36) is the countercurrent flow formation which permits the achievement of 95 to 98% exchanger effectiveness. Cross-flow and cross-countercurrent flow arrangements are utilized principally for duties exchanging heat between gas and liquid streams and for liquefiers where a large throughput of low pressure gas at low pressure drop is required.

Each fluid stream is introduced into the exchanger from the process piping through header tanks and thence into each of the exchanger passages through distributor sections which are formed integrally within the passages. The distributor may take various forms (see Fig. 3.37) to suit the requirements of the header arrangements. The distributor corrugations are of the same

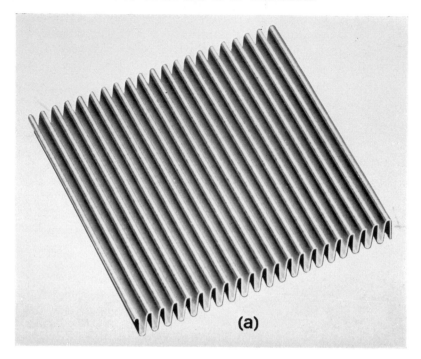

(a)

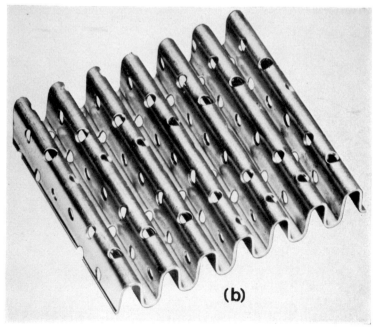

(b)

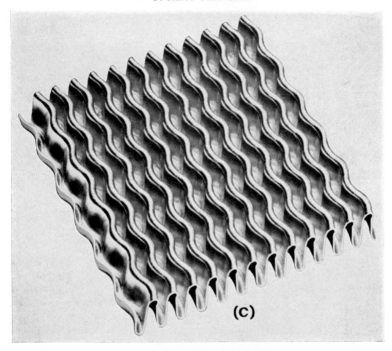

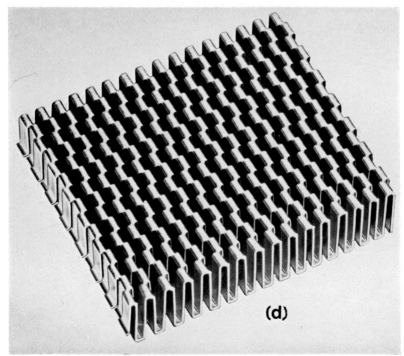

Fig. 3.35. Types of fin corrugation for plate–fin heat exchanger. (Courtesy of Marston Excelsior Limited.) (*a*) Plain (*b*) Plain Perforated (*c*) Herringbone. (*d*) Multi–entry.

Fig. 3.36. Passage arrangements (courtesy of Marston Excelsior Ltd. (*a*) Counterflow. (*b*) Cross-flow.

height as the basic secondary surface but are coarser and of thicker perforated metal. The distributor does not generally contribute significantly to the heat transfer and it is common practice to refer to the effective length of the exchanger as being the overall passage length less the length occupied by the distributors. However, recent work by Kays *et al.* (1968) presents a means for determining the heat transfer effectiveness of a counterflow heat exchanger with cross-flow distributors.

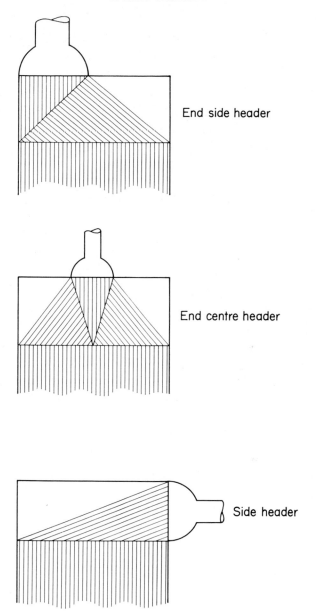

End side header

End centre header

Side header

Fig. 3.37. Types of flow distributor.

In areas of the passage where corrugations are not required for flow purposes, dummy corrugations are inserted to preserve the mechanical integrity of the assembly.

2. *Calculation of* UA

The overall UA of a plate-fin exchanger is calculated from the individual hA values for the component fluid streams by:

$$UA = \frac{\sum(hA)_h \sum(hA)_c}{\sum(hA)_h + \sum(hA)_c} \tag{3.120}$$

$\sum(hA)_h$ and $\sum(hA)_c$ are the summations of the component hA of the warm and cold streams, respectively.

The effective area for heat transfer of an individual fluid stream comprises its respective primary surface and secondary fin surface with appropriate corrections for the efficiency of the fin and the geometrical arrangement of the passages. The latter is generally termed the banking of the passages. For example, if the number of cold passages is double the number of warm passages, the most effective passage arrangement requires two cold passages to be arranged immediately adjacent to one another and separated from the next pair of cold passages by a single warm passage. Other passage banking arrangements are used to provide suitable uniform passage distribution.

For a single banked passage arrangement the effective area for heat transfer is given by

$$A = A_1 + A_2 \eta_2 \tag{3.121}$$

where subscript 1 denotes primary surface (parting shot) and subscript 2 denotes secondary surface (fin).

The fin efficiency of the secondary surface, η_2, is given by equation 3.6 for a straight fin with constant cross-section as:

$$\eta_2 = \frac{\tanh es}{es} \tag{3.122}$$

$$e^2 = \frac{2h}{k_2 d_2} \tag{3.123}$$

where h is the local heat transfer coefficient, d_2 is the fin thickness and k_2 is the fin thermal conductivity. s is generally assumed to be half of the fin height.

A double banked passage arrangement (two passages of the same fluid stream adjacent) has a reduced effective surface area for the two passages given by:

$$A = \frac{A_1}{2} + A_2 \eta_2 + \frac{A_1}{2} \eta_{2t} \tag{3.124}$$

The fin efficiency η_2 is calculated for a value of s equal to the fin height of a single passage. Half of the primary surface is assumed to act as secondary surface, modified by the fin tip efficiency η_{2t}. The fin tip efficiency is calculated from

$$\eta_{2t} = \frac{1}{\cosh es} \tag{3.125}$$

e and s are determined as for the secondary surface and thus η_{2t} is an approximate correction corresponding to the temperature of the tip of the secondary fin as if acting alone.

The heat transfer coefficient for single-phase flow is generally expressed in terms of the Colburn correlation

$$h = j_T G c_p Pr^{-\frac{2}{3}} = G c_p St \tag{3.126}$$

The Colburn factor j_T separates the effect of the fluid properties on the heat transfer coefficient and correlates the heat transfer coefficient as a direct function of Reynolds number. Curves of j_T as a function of Reynolds number are available from the heat exchanger manufacturers for their various surface geometries. Some data are also presented by Kays and London (1964). The typical form of data is shown in Fig. 3.28 from Lenfestey (1961) together with the corresponding friction factor for pressure drop. The values of j_T vary with the length/diameter ratio of the passage opening in the laminar and transition flow regimes in the manner shown in Fig. 3.6.

Few studies of two-phase heat transfer in plate-fin heat exchangers are known to have been made, and no general correlations are available for the prediction of two-phase heat transfer coefficients. Three procedures are commonly adopted for the estimation of two-phase coefficients.

Condensation is generally carried out on plain corrugations when the correlations presented later are applicable.

Little is known about boiling coefficients in the narrow channels existing between fin corrugations. An approximate assessment may be made from forced convection two-phase flow correlations, and from nucleate boiling correlations. In the absence of definitive experimental data a typical boiling coefficient of 0.3–0.4 W/cm^2°C (500–700 B.T.U./hr ft^2 °F) is frequently assumed.

In many instances involving multicomponent systems, it is more convenient to assume the single-phase coefficient for all liquid or all gas flow corresponding to the total two-phase flow. This will result in a conservative heat exchanger design except possibly in the high vapour fraction region (mist flow regime) of a boiling system.

3. *Pressure Drop*

The overall pressure loss is made up from component losses in the exchanger inlet and exit nozzles, header tanks and distributors and in the principal heat transfer surface. The major loss is generally concentrated in the heat transfer

8

surface and is expressed in terms of a friction factor, values of which are presented by the manufacturers for their various fin geometries.

$$\Delta P = 4 . f_T . \frac{L}{D_e} \frac{G^2}{2\rho g_0} \qquad (3.127)$$

Lenfestey (1961) has shown how the block geometry may be determined without change of heat transfer rating for a required pressure drop. Thus,

$$\frac{\Delta P}{hA} = \frac{f_T}{cj_T} \frac{G^2}{2\rho g_0} \qquad (3.128)$$

where c is a constant for a given heat transfer rating. The ratio f_T/j_T varies only slightly with Reynolds number for a given corrugation (see Fig. 3.38). Thus G may be determined for a required ΔP, and if the length of block is changed inversely as the change in j_T with G, the heat transfer rating is not affected.

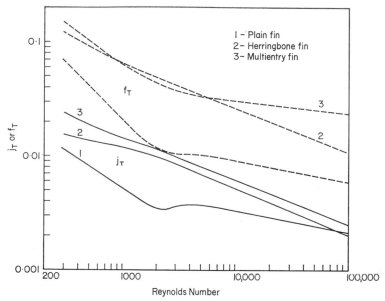

Fig. 3.38. Typical form of j_T and f_T factors for plate–fin exchanger corrugation.

Proprietary methods for determining pressure drop in the distributors and headers are given by the manufacturers.

Pressure drop data for two-phase flow are not available for plate–fin exchangers, an assumption of pressure drop equal to that for the total flow as vapour is commonly made. Some refinement of the two-phase pressure drop may be made by recourse to the standard correlations for two-phase flow in tubes, e.g. Chenoweth and Martin (1955) and Martinelli and Lockhart (1949) as described in Chapter 5.

4. Factors Requiring Special Consideration

Because of the very high heat transfer effectiveness demanded of plate–fin exchangers in cryogenic service, special consideration must be given to a number of other aspects of the fluid mechanics.

a. Flow distribution. Maldistribution of flow between passages or across the width of a given passage may lead to a serious loss of exchanger performance. Consider, for example, an exchanger split into two equal parallel elements. One of the two counterflowing fluid streams (having equal heat capacities) is maldistributed by 10% from the mean flow. Then for an ideal design temperature approach of 3°C in 100°C, the actual temperature approach is increased to 5°C. Thus, the exchanger UA has been reduced to 60% of its design value. An analysis of the effects of maldistribution upon counterflow exchangers has been presented by Fleming (1966). He showed that loss of performance is greatly reduced by intermediate remixing of fluid streams.

The manufacturers give careful attention to the distribution characteristics of their heat exchangers by design to close tolerances, standardized manufacturing techniques, and by subsequent flow testing. The latter is used for matching of series exchanger pairs in multiple exchanger banks to ensure the minimum of maldistribution. The extra work entailed for close flow matching increases the exchanger cost, and it is necessary therefore for the process designer to be able to specify the largest possible tolerances commensurate with design duty.

Design for two-phase flow distribution present a particularly difficult problem. An effective solution is generally provided by segregation of the phases and separate distribution. Distribution in the two-phase regime may be achieved in the dispersed flow regimes or in any regime by axisymmetric division of the fluid. Exchangers with multiple fluid streams require careful consideration of the optimum distribution of the passages to give the most uniform matching of the cooling curves (under all operating conditions). The best result is usually obtained by even distribution of the passages for each fluid circuit.

b. Flow instability. Problems due to flow instability occur primarily in ascending two-phase flow, either boiling or condensing. One type of instability is flow oscillation between parallel channels allowing intermittent liquid accumulation and discharge. For this there exists a critical minimum flow velocity at which all parallel flow channels in a heat exchanger will exhibit similar flow patterns. The optimum economic mass flow velocity is lower for a plate–fin exchanger than for other types and flow instability frequently becomes a limiting design parameter. The following criteria proposed by Wallis (1969) are useful as a guide to determine minimum acceptable upward flow velocities.

Transition from slug to annular flow occurs between:

$$j_v^* = 0.4 + 0.6 j_l^*$$

and

$$j_v^* = 0.9 + 0.6 j_l^* \qquad (3.108)$$

where

$$j_v^* = j_v \left[\frac{\rho_v}{(\rho_l - \rho_v) g D_e} \right]^{\frac{1}{2}}$$

$$j_l^* = j_e \left[\frac{\rho_l}{(\rho_l - \rho_v) g D_e} \right]^{\frac{1}{2}}$$

j_v and j_l are volumetric vapour and liquid flow rate per unit area based on the total cross-sectional area for flow.

C. COILED TUBE HEAT EXCHANGERS

1. *General Design Procedure*

The configuration of coiled tube-in-shell exchangers is illustrated in Figs. 3.29 and 3.30. The ensuing paragraphs will outline a design method and the particular problems that result from the coiled tube geometry. Two types of exchanger are considered:

(i) Shell side flow countercurrent to all fluid streams. Heat flow is thus direct between the shell fluid and all other fluids, and bundles of single tubes provide efficient heat transfer.

(ii) Co-current flow of one or more tube-side streams with the shell flow. These streams are generally being warmed against other tube-side fluids in countercurrent flow (cooling streams). The heat flow path between tube-side heating and cooling streams is thus indirect via the shell-side fluid. This inefficiency is overcome by directly coupling (pairing) the counter-flow tube circuits to provide a direct conduction path. This is often accomplished by soldering together copper tubes.

The design method will be treated as a series of steps.

a. Preliminary estimate. An initial estimate is made of the tube length and number of tubes from guessed overall coefficients using equation 3.129 applied to each fluid stream.

$$h_i A_i = \frac{q_i}{\Delta T_m} \qquad (3.129)$$

in which i denotes the fluid stream.

Tube diameter and wall thickness are selected, commensurate with exchanger type and pressure, from previous experience. To facilitate subsequent calculations the exchanger may be divided into segments corresponding to straight line approximations of the cooling curve and to variations from single-phase to two-phase flow in any stream.

b. Heat transfer calculation. The tube-side heat transfer coefficients are calculated using the methods given for forced convection flow (see Section IV). A mass flow velocity is assumed for the shell-side together with tube pitch and layer to layer spacer diameter. Shell-side coefficients are calculated using the correlations for flow across tube banks given in Section IV.4.

For exchangers without paired tubes, the overall coefficients are then calculated for each circuit and thus the required tube lengths. Where there is a poor match between tube lengths of different circuits then the numbers of tubes in each circuit should be reassigned and the above steps repeated. From the final set of tube lengths, the longest circuit should be selected as the value for exchanger design, and the appropriate design margin added.

Paired tube exchangers require a calculation of the heat transfer surface effectiveness using the methods of Section III.A in order to complete the calculation of overall coefficients and tube lengths. It is convenient at this stage to calculate tube circuit pressure drops using the methods given in Chapter 5. Where necessary the number of tubes must be adjusted to meet process design requirements. This is also a convenient point to evaluate flow stability of two-phase circuits.

c. Bundle geometry calculation. The minimum cross-sectional area of the shell side for fluid flow between tubes is calculated from the selected velocity assuming an in-line tube arrangement. The mandrel diameter is selected having regard to the mechanical design requirements of bending load during bundle winding and the minimum bending radius of the tubes. The number of tube layers (and bundle diameter) may then be calculated from the previously assumed layer spacer diameter. Allowances should be made for obstruction of the cross-sectional area by the spacers. The average tube spacing in each layer may then be calculated and thus the winding angle and bundle length. The shell-side pressure drop is determined using the methods given in Chapter 5. Any necessary corrections to the pressure drop are made by adjustment of the shell mass-flow velocity and repeating the above calculations. Finally, the number of tubes in each layer is determined to the nearest integral value and the tube winding pitch is adjusted to match.

Complete optimization of a coiled tube exchanger design is a complex problem involving a number of variables, viz., tube and shell flow velocities, tube diameter, tube pitch and layer spacer diameter. The optimum solution will give the minimum operating cost including capital depreciation.

2. Heat Transfer with Paired Tubes

Coil winding requirements generally dictate that the paired coil tubes are of equal outside diameter. Consider the pair of tubes shown in Fig. 3.31.

Let subscripts 1, 2, 3 denote a warming stream, a cooling stream, and the shell side warming stream respectively. Heat transfer between streams 2 and 3 takes place directly across the surface of tube 2 and by the fin contribution of tube 1. The effective area for heat transfer from 2 to 3 per unit of outside tube surface is given by

$$A_{23} = 1 - f + \eta(1 - f) \tag{3.130}$$

where f is the fraction of the tube perimeters in direct contact and is the fin efficiency of tube 1. The heat transfer coefficient for tube circuit 2 must be referred to the effective outside surface.

$$h_{23} = h_2 \frac{A_2}{A_{23}} \tag{3.131}$$

Hence the stream to tube surface temperature differences may be determined from the heat fluxes and coefficients:

$$\frac{\Delta T_{2w}}{\Delta T_{w3}} = \frac{q_2 h_3}{q_3 h_{23}} = r \tag{3.132}$$

and

$$\Delta T_{2w} = \frac{r \Delta T_{23}}{1 + r} \tag{3.133}$$

The required surface area (tube length) for tube circuit 2 is then determined by:

$$A_2 = \frac{q_2}{h_2 \Delta T_{2w}} \tag{3.134}$$

Similarly, for heat transfer between streams 2 and 1, the effective surface for heat transfer per unit of inside tube surface is given by

$$A_{21} = f + (1 - f)\eta \tag{3.135}$$

and

$$h_{21} = \frac{h_2 A_2}{A_{21}} \tag{3.136}$$

and hence the required surface area for tube circuit 1 may be determined.

Values of f for soldered copper tubes vary from 0·1 to 0·2. An exact determination of the fin efficiency is complex—values from 0·65 to 0·8 are representative for normal tube diameters and wall thickness.

The reader is referred also to an analysis of coiled tube exchangers with paired tubes by Kao (1965).

3. Special Considerations

a. Bypassing. In contrast to the integrally brazed structure of the aluminium exchanger, a serious loss of performance may occur in the coiled tube exchanger due to bypassing of the shell flow. Consider, for example, an exchanger with an overall temperature change of 100°F and a uniform temperature difference of 3°F. A by-pass of 1 % of the shell side flow would require an increase of 33 % in the surface area to match the performance with zero by-pass.

b. *Distribution.* Three distinct problems of flow distribution may be identified for a coiled tube exchanger.

(i) Tube distribution. Where two or more tube circuits exist, it is important to obtain an even radial distribution of tubes such that each tube layer contains an approximately equal proportion from each circuit.

(ii) Two-phase distribution in tube circuits. It is exceedingly difficult to achieve a uniform distribution of a two-phase fluid into the inlet of a tube circuit. Whenever possible this should be avoided by changes to the process cycle design or by proper selection of the break point between exchangers in series. It is particularly crucial to avoid this problem in two-phase multicomponent systems where not only is the heat duty of each circuit affected, but the temperature versus enthalpy relationship is also changed.

(iii) Two-phase distribution of the shell flow. Two-phase distribution of the shell flow is met commonly in cryogenic hydrocarbon processing plants. The gas flow is self distributing due to the uniform bundle geometry and the flow resistance induced by the tubes. It is important, however, to distribute the liquid phase uniformly. This is particularly true in the radial direction, since the liquid flow is not effectively redistributed between the layers of tubes in a vertical heat exchanger. The devices used for distribution are generally considered proprietary by cryogenic equipment manufacturers.

c. *Flow instability.* In a similar manner to brazed aluminium exchangers, flow instability occurs primarily in ascending two-phase flow. For flow in the tubes the j_v^* versus j_l^* criteria given previously are applicable but should be used with caution since the effect of tube inclination is unaccounted for. The slug to annular flow transition occurs at a higher value of j_v^* for an inclined tube; some data have been presented by Wallis (1969). Two-phase flow on the shell side has received scant attention but some studies of flow in nuclear reactor rod clusters have indicated that the effective diameter in equations 3.108 should reflect the housing (shell) rather than the tube diameter.

D. REGENERATORS

1. *Applications*

A regenerator exchanges heat between two fluids by the alternate accumulation of heat from a hot fluid into the mass of the regenerator and subsequent transfer into a cold second fluid. In the fixed-bed regenerator as used in cryogenic applications the hot and cold fluids are passed alternatively in opposite directions through a vessel containing the regenerator packing. Two or more vessels are used to obtain a continuous heat transfer operation. Two principal cryogenic applications exist; in air separation plant where stone or corrugated metal packings are used, and in reversed Stirling cycle refrigerators using compact wire mesh or small diameter spherical metallic packings.

The principal advantages of the regenerator are the cheapness of the heat transfer surface, the high surface area per unit volume of packing and low pressure drop. But two major disadvantages have restricted their use; the intercontamination of the fluid streams by mixing during fluid interchange, and the difficulty of regenerator design to involve three or more fluids. These problems do not exist in the reversed Stirling engine application since it involves only a single working fluid. In air separation applications, however, the regenerator has been largely superceded by brazed aluminium exchangers because of the trend to multiproduct plants. Regenerators with heat transfer coils embedded in the packing are used but these are susceptible to mechanical failure. Other difficulties have been dust formation by bed movement in stone packed regenerators and the substantial air losses from the regenerator void volume at switchover.

2. Regenerator Theory

a. Solution of the heat transfer equation. The theory of the regenerator is complex due to the additional time dependence of the heat transfer. The heat transfer is described by the following partial differential equations from a heat balance over an element of regenerator length.

$$h\frac{A}{L}(T_p - T_g)\, dL = Gc_g\frac{dT_g}{dL}\, dL \qquad (3.137)$$

$$h\frac{A}{L}(T_g - T_p)\, dL = M_p c_p\frac{dT_p}{dt}\, dL \qquad (3.138)$$

h is the overall heat transfer coefficient between the gas and packing interior, A and M the total area and mass of the packing and subscripts g and p denote gas and packing respectively.

The derivation and solution of these equations involve a number of simplifying assumptions. These are as follows:

The packing conductivity is zero parallel to flow.

The gas mass-flows and specific heats are constant with time and position.

The overall heat transfer coefficient, and packing density and specific heat are constant.

Wall effects are insignificant.

The gas flows enter the regenerator at constant temperature.

The gas hold-up is negligible.

Analytical solutions to equations 3.137 and 3.138 have been presented by a number of authors, the more notable of these being Hausen (1950), Nusselt (1927), Iliffe (1948), Tipler (1947), and Schalkwijk (1959). In order to account accurately for the effect of unbalanced flows, nonlinearity of the temperature profiles and varying packing specific heat with temperature, numerical solutions requiring the use of a digital computer have been presented by Lambertson (1958), Willmott (1962), and Rios (1967). Valuable summaries of the above solutions have been given by McDonald (1967) and Bretherton (1970).

For fixed bed regenerators, the solutions to equations 3.137 and 3.138 are presented in terms of reduced dimensionless parameters, some of these being comparable to the expressions for the rating of heat exchangers given in Section IIA.

Thus at a particular moment in time equation 3.137 may be integrated.

$$\int_{L=0}^{L=L} \frac{dT_g}{T_p - T_g} = \frac{hA}{Gc_g} = \Lambda \tag{3.139}$$

And for a particular distance along the regenerator equation 3.138 may be integrated.

$$\int_{t=0}^{t=P} \frac{dT_p}{T_g - T_p} = \frac{hAP}{M_p c_p} = \pi \tag{3.140}$$

Where P is the time period of operation between reversals. Λ is the reduced length of the regenerator and is a measure of the number of transfer units on a length basis. π is the reduced period and is a measure of the number of transfer units on a time basis.

b. *Efficiency.* The thermal efficiency of a regenerator, η, is defined as the ratio of actual enthalpy transfer to the maximum possible with infinite heat transfer area.

Separate efficiencies are defined for the cooling and heating periods.

Cooling period:

$$\eta_c = \frac{T_{gcom} - T_{gci}}{T_{ghi} - T_{gci}} \tag{3.141}$$

Heating period:

$$\eta_h = \frac{T_{ghi} - T_{ghom}}{T_{ghi} - T_{gci}} \tag{3.142}$$

Subscripts c and h refer to cold and hot gas streams, i and o to inlet and outlet conditions, m to time mean temperature.

8*

Regenerators are said to be in "equilibrium" when the enthalpy transfers to and from the packing during the heating and cooling periods are equal. Then,

$$\eta_c = \eta_h \frac{G_h c_{gh} P_h}{G_c c_{gc} P_c} \tag{3.143}$$

A "balanced" regenerator has

$$\Lambda_h = \Lambda_c$$

and

$$\pi_h = \pi_c$$

and

$$\delta_c = \eta_h.$$

For cryogenic applications of practical importance the equilibrium case only need be considered and for air separation and Stirling engine regenerators the balanced condition is approximately true.

The maximum thermal efficiency of a regenerator is obtained as the period time tends to zero. For a balanced regenerator the maximum efficiency is given by

$$\eta = \frac{\Lambda}{\Lambda + 2} \tag{3.144}$$

c. *Heat transfer coefficient.* The overall heat transfer coefficient between the gas and the interior of the packing has two components, the convective coefficient in the gas phase, h_g, and a conduction coefficient in the packing, h_p.

$$\frac{1}{h} = \frac{1}{h_g} + \frac{1}{h_p} \tag{3.145}$$

The packing coefficient has been considered by Hausen (1950) who correlated a dimensionless penetration factor to the period time and thermal diffusivity as

$$\phi = \frac{k_p}{D_p h_p} = 0 \cdot 1 - 0 \cdot 00143 \frac{D_p^2}{\alpha' P} \tag{3.146}$$

where D_p is a characteristic dimension of the packing, and for a sphere is given by Hausen as equal to $\frac{2}{3}$ diameter, α' is the thermal diffusivity of the packing. Generally, the packing coefficient is much larger than the convective coefficient. Thus, the overall coefficient is principally determined by the gas convection coefficient.

Convective heat transfer in packed beds has been extensively investigated. Although primarily derived for ambient or warmer temperatures, the correlations which have been developed are generally applicable to cryogenic systems. It is not appropriate in this text to review these correlations in detail, but the following are of value for initial evaluations.

The work of about 250 investigators has been reviewed by Barker (1965), and their results correlated in terms of the Colburn factor versus Reynolds number. A general agreement within a factor of about 2 was noted for $100 < Re < 100,000$.

Data for air through gravel beds, which may be compared to stone packings, have been correlated in a dimensional form by Löf and Hawley (1948) for $50 < Re < 500$ by

$$h_g = 0 \cdot 79 \left(\frac{G}{D'_p} \right)^{0 \cdot 7} \tag{3.147}$$

in which D'_p is the particle diameter, and G is based on the total bed cross-sectional area.

Meck (1962) has presented a correlation for the range $900 < Re < 4000$ which also extrapolates reasonably to lower values of Reynolds number. This was based on measurements with packed beds of 0.394 in. diameter steel spheres.

$$h_g = 0 \cdot 47 \frac{Gc_g}{e} \left(\frac{GD_h}{\mu e} \right)^{-0 \cdot 38} \left(\frac{c_g \mu}{k_g} \right)^{-0 \cdot 67} \tag{3.148}$$

This correlation utilizes a Reynolds number based upon the hydraulic diameter defined as:

$$D_h = 4 \times \frac{\text{void volume of packing}}{\text{surface area of packing}}$$

and a mass flow rate G/e which accounts for the void volume of the packing, e.

3. Regenerator Packing

a. *Stirling engine regenerators.* The key requirements for a regenerator packing are to have a high heat capacity per unit volume and a large surface area per unit volume. Further, the thermal conductivity in the direction of fluid flow should be small. These requirements are met in practice by the use of packed metallic beds. Typically, in a 2-stage Stirling engine the warmer stage uses a copper wool packing with the strands perpendicular to the direction of flow. The colder stage uses lead shot of about 0·01 in. diameter.

b. *Air separation plant regenerators.* Whilst the ideal theoretical requirements for the regenerator are no different from the Stirling engine, the type of packing is substantially modified by the larger scale and the cost of the packing. Two packing types have been commonly used—quartz pebbles and aluminium ribbon (Fränkl packing).

Quartz pebbles have found most favour in recent plants because of their low cost. Typically, ocean rounded quartz pebbles of from 0·25–0·725 in. diameter are used with a size tolerance of $\pm 20\%$. Natural grading of pebbles occurs on some beaches when a water wash is the only processing required before use in a regenerator.

The typical Fränkl packing is manufactured from 0·2 to 0·5 mm thick aluminium ribbon about 20 mm wide with diagonal corrugations of 0·6 to 2·3 mm depth. A pair of ribbons with the corrugations opposed are wound spirally on a former to form a flat pancake. The pancakes are stacked to form the regenerator.

4. *Special Considerations for Regenerator Design*

In addition to the design of the regenerator packing to obtain the required heat transfer and thermal capacity characteristics, the designer must also consider the effect of various non-idealities and practical limitations.

a. Axial conductivity. The thermal efficiency of a regenerator is reduced when the packing has a high effective thermal conductivity in the axial direction. This effect may be significant in short bed regenerator designs. The effective axial conductivity is composed of a number of separate contributions:

(i) Heat transfer through the packing by point contact or radiation or by convection through the stagnant fluid layer adjacent to the point contact.

(ii) Molecular diffusion in the gas phase as for a stationary gas.

(iii) Turbulent or eddy diffusion as the gas flows through the packing.

Thus it cannot be simply calculated and must generally be determined by experiment. The data of some experimenters has been summarized by Bretherton (1970).

The calculation of the effect of the axial conductivity upon regenerator performance generally requires recourse to numerical analysis. A typical approach is presented by Handley and Heggs (1969).

In practice, the effect of axial conduction is minimized by the use of discontinuous packings and with low thermal conductivity materials for the containing vessel, e.g. stainless steel.

b. Erosion of the packing. The regenerator packing is subjected to large forces during flow reversal. This is particularly a problem with brittle packings such as stone beds, where movement of the bed may lead to fracture or dusting of the pebbles. This results in further difficulties with quartz dust and fragments being entrained into the cryogenic plant where it may cause erosion of components, such as valves, or blockages. Also in the case of regenerators with buried tube coils (as used in air separation plants) erosion of the tube may be a serious problem.

A successful design requires attention to three factors:

(i) The mass velocity of the upward flowing fluid should be maintained below fluidization velocity. In air separation plants, the critical situation occurs in a regenerator with warm end upwards where very high velocities may be experienced with the high pressure air being expelled during flow reversal. If this condition cannot be met then the bed must be positively retained by grids.

(ii) The flow upset during reversal may be minimized by proper control of sequencing and opening speeds of the changeover valves.

(iii) The regenerator bed must be properly packed during construction.

c. *Flow distribution.* Uniform flow distribution is of paramount importance in the regenerator because of the common use of high N_T values. The distribution is primarily controlled by the regular grading of the packing but particular care must be exercised with short bed designs where entrance and exit effects may be comparable to the bed pressure drop.

X. Helium II

The common helium isotope (He[4]) is unique among the various substances in that it has two known and distinctly different liquid phases. These are indicated in the phase diagram for helium in Fig. 3.39. The liquid phase

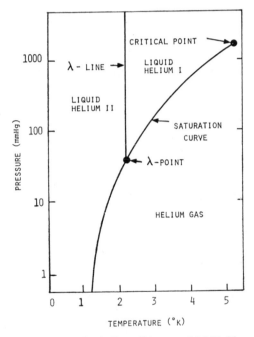

Fig. 3.39. Phase diagram for helium (Rivers and McFadden, 1966).

which exists in the temperature range from about 2·19°K to the critical point at 5·2°K behaves in a classical manner as do ordinary substances and the gaseous phase of helium. However, at temperature below approximately 2·19°K liquid helium undergoes a remarkable transformation. Within a fraction of a degree below 2·19°K the heat conducting ability of the liquid increases in an astonishing manner by a factor of 10⁷. Also, its heat capacity increases by about a factor of 6 in this temperature region. Because of these characteristics this phase of the liquid is identified as HeII and sometimes called a "superfluid". The classical liquid phase is known as HeI. An interesting and potentially significant aspect of HeII from a technological view point is the fact that it remains as a liquid down to the lowest attainable temperatures. HeII may be solidified if it is subjected to a sufficiently high pressure. The shape of the heat capacity curve has given the name "Lambda" to the line of transition between HeI and HeII.

The properties of the "superfluid" HeII have long been of great interest to physicists. Much of the known results have been summarized by Lane (1962) and Matheson (1966). Some of the extraordinary behaviour of HeII, described by Lane, include its superfluidity, the mechano–caloric effect and thermo–mechanical effect, sometimes known also as the fountain effect. In order to explain these observed phenomena, HeII was postulated to be made up of two fluids. One fluid, known as the "superfluid", is without viscosity, entropy or heat capacity while the second fluid has all the properties of a normal substance. It is called the "normal fluid". These two fluids mix in all proportions with HeII consisting of all normal fluid at the λ-point and all superfluid at the absolute zero. These assumptions are largely unproved but they have helped to "explain" certain observed behaviour of He^2. Obviously, this substance presents a challenge which requires a departure from classical concepts to obtain a quantitative description of its properties.

One of these properties is its extremely great heat conducting ability. The definition of a "thermal conductivity" for HeII, is principally a convenience for comparative purposes as HeII does not follow the Fourier concept in the usual sense. If a thermal conductivity is computed for this substance it will be found to be highly temperature sensitive, be a function of the thermodynamic state and be influenced by the geometry and temperature gradient in its system (Holridge and McFadden, 1966). Another factor in this is the fact that thermal transport ("conduction") in an apparently "stagnant" (zero net mass flow) HeII is not a consequence of a strictly diffusive process. Because of the presence of the superfluid within the mass of HeII powerful convective flow are established by the temperature differential. The large apparent thermal conductivity is a result of these internal flows. Nevertheless, there is useful value in formulating the thermal conduction characteristic in terms of an apparent thermal conductivity.

In a presentation of a review of this subject Clement and Frederking (1966) give the following expression for the apparent thermal conductivity of HeII,

$$k_{App} = C(T)T\mu_n^{\frac{1}{3}}\rho^{\frac{2}{3}}S^{\frac{4}{3}}\left(\frac{\Delta T}{L}\right)^{-\frac{2}{3}} \tag{3.149}$$

where μ_n is the absolute viscosity of the normal fluid, ρ is the total fluid density, S is the liquid entropy per unit mass and $C(T)$ is a function of temperature. The combined property function,

$$k_{App}\left(\frac{\Delta T}{L}\right)^{\frac{2}{3}} = C(T)T\mu_n^{\frac{1}{3}}\rho^{\frac{2}{3}}S^{\frac{4}{3}} \tag{3.150}$$

is given by Clement and Frederking (1966) and the corresponding values of k_{App} for various values of $(\Delta T/L)$ are given in Figure 3.40. The effectiveness of HeII as a "conductor" may be seen from Fig. 3.40. The maximum value of k_{App} shown is about 500 W/cm °K which may be compared with the thermal conductivity of room temperature copper, which is about 4 W/cm °K. For

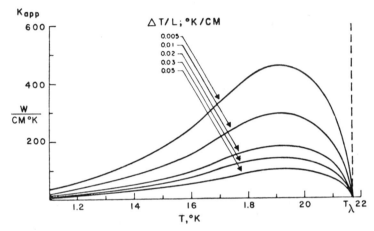

Fig. 3.40. Apparent thermal conductivity k_{app} of He II as a function of liquid temperature for constant temperature gradients (Clement and Frederking, 1956).

lower values of $(\Delta T/L)$ the k_{App} of HeII is even greater. Of particular interest is maximum in the property group $C(T)T\mu_n^{\frac{1}{3}}\rho^{\frac{2}{3}}S^{\frac{4}{3}}$ at approximately 1·9°K, as shown in Fig. 3.40, and the rapid reduction in k_{App} as the liquid temperature is either increased to the λ-point or reduced toward the absolute zero. This is a consequence of the liquid becoming completely a normal fluid at the λ-point with the thermal conductivity of liquid HeI, and completely a superfluid, but with no heat conducting ability, at 0°K. Because of the large magnitude of k_{App} HeII can sustain only very small temperature differentials. Thus, even during large rates of thermal transport the temperature of HeII appears uniform.

There is, however, another effect which becomes of importance under these circumstances and has a primary influence on the transport of heat from a solid surface to HeII. This is a boundary resistance known as the Kapitza resistance after its discoverer (Kapitza, 1941). At low heat flux rates to liquid HeII Kapitza observed what appeared to be a temperature discontinuity between the surface and the liquid. It is possible this is an effect to be found with many other fluid–solid systems as well but owing to the large apparent conductivity of HeII the temperature discontinuity is more evident in this case. The existence of this phenomenon is supported by theoretical considerations of Khalatnikov (1952) who assumed a radiative transport and showed the Kapitza effect to depend on T^3 and the acoustic and elastic properties of both the solid and the liquid. The temperature difference under these circumstances may be expressed in terms of a contact heat transfer coefficient (the reciprocal of a resistance) as

$$h = \frac{(q/A)}{\Delta T} \tag{3.151}$$

A limiting value to this coefficient, h_0, corresponding to ΔT equal to zero, may be shown to be written (Clement and Frederking, 1966).

$$h_0 = 4\sigma_K T^3, \tag{3.152}$$

and for larger ΔT but under the restraint that $\Delta T < T$, the coefficient h may be formulated in terms of h_0 as

$$\frac{h}{h_0} = f\left(\frac{\Delta T}{T}\right) = 1 + \frac{3}{2}\left(\frac{\Delta T}{T}\right) + \left(\frac{\Delta T}{T}\right)^2 + \frac{1}{4}\left(\frac{\Delta T}{T}\right)^3 \tag{3.153}$$

In equation 3.152, σ_K is a parameter which depends on the particular properties of the solid and liquid. In most cases it is necessary to determine σ_K empirically as the theory of Khalatnikov, while correctly predicting a solid–fluid property dependence, does not predict its observed magnitude (Madsen and McFadden, 1968). The temperature dependence is reasonably borne out by experiments with measured exponents ranging between 2·6 and 4·2. Thus, equation 3.153 may be written

$$h = 4\sigma_K T^3 f\left(\frac{\Delta T}{T}\right) \tag{3.154}$$

and the corresponding heat flux, $(q/A)_K$, in the range where the Kapitza resistance is controlling becomes

$$(q/A)_K = 4\sigma_K T^3 \Delta T f\left(\frac{\Delta T}{T}\right) \tag{3.155}$$

For a Pt–HeII interface Kapitza's data (1941) for $\Delta T \ll T$ may be expressed as Clement and Frederking (1966) as

$$h_0 = 0.065 T^3, \tag{3.156*}$$

and

$$(q/A)_{K,Pt} = 0.065 T^3 \Delta T \, f\left(\frac{\Delta T}{T}\right) \tag{3.157†}$$

---

* The units are W/cm−°K and T is °K.
† The units are W/cm² and both T and ΔT are °K.

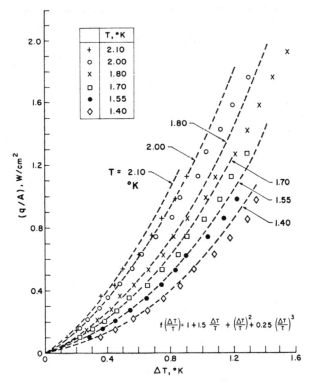

Fig. 3.41. Comparison of experimental heat flow density q/A with semiempirical correlation $(q/A)_{K,Ag} = 0.082\,f\Delta T\,T^3$.

These general types of relationships have been confirmed in several experiments. Clement and Frederking (1966) have studied the heat transfer from a silver surface to HeII in a short tube and for conditions of $\Delta T < T$ find that $(q/A)_{K,Ag}$ may be expressed as

$$(q/A)_{K,Ag} = 0.082\,T^3 \Delta T\, f\left(\frac{\Delta T}{T}\right) \text{ watts/cm}^2 \qquad (3.158)$$

Their data are shown in Fig. 3.41. Agreement between equation 3.158 and

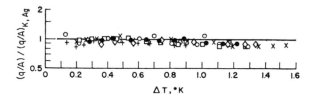

Fig. 3.42. Ration $(q/A)/(q/A)K,Ag$ for data of Fig. 3.31.

the data, especially for low values of $\Delta T/T$ is very good. A correlation of the measured (q/A) and that computed by the T^3-relation, equation 3.158, is given in Fig. 3.42. The comparison is favourable for low ΔT but a departure is observed at higher ΔT. It seems probable the reason for this is that at higher ΔT the surface temperature exceeds the λ-point temperature and HeII no longer is in contact with the solid. This would have the effect of introducing a thin, very low conducting film of HeI between the solid surface and the HeII in the liquid bulk. If this is the case then the measured (q/A) would fall below that predicted by the Kapitza-type theory, which is what is observed. At much larger ΔT a vapour film of HeI would be expected to form at the surface. Under these circumstances it is possible that two films, one of HeI vapour and the second HeI liquid, could co-exist in a HeII system in film boiling for pressures above that of the λ-point. In this circumstance the temperature discontinuity would probably be shifted to the liquid HeI–liquid HeII interface.

Thus, for computing heat transfer rates to liquid HeII a T^3-relationship of the form of equation 3.155 can be employed in which the constant σ_K must be determined for each solid–liquid combination. As a guideline this relationship may be used for ΔT up to the point at which the surface temperature exceeds the λ-point temperature of HeII. For conditions of higher surface temperature a HeII system goes into other modes of heat transfer. These could involve some vapour formation (boiling) of a HeI film or a film boiling condition of HeI vapour in contact with a bulk liquid of HeII. Because of its large heat transport characteristic bubbles or nucleate boiling have never been observed in HeII. As a guide to engineering calculations in these regions it is probably best to use the available experimental data, most of which is for natural convection processes from small, laboratory type surfaces. Some of these data will be discussed later.

Further evidence of the T^3-relationship for low (q/A) and low ΔT is provided by Irey *et al.* (1965). The test section for these data was a horizontal cylinder of soft-glass, 1·88 mm in diameter and heat transfer was by natural convection from the outside surface of the cylinder to a HeII bath.* The data for this combination appear to fit the relationship $h = 0 \cdot 139T^3$ very well. Clement and Frederking (1966) have summarized the results of several investigators for heat transfer to liquid HeII from a number of different surfaces for conditions in which the Kapitza effect controls the process. A linear relationship between h_0 and T^3 is indicated by most of the data, thus supporting equation 3.152. A low value of h_0 will produce the greater boundary ΔT for a given heat flux.

---

* In a later paper, Holdredge and McFadden (1966) suggest that some of the data of Irey, *et al.* (1965) are probably in error. These are data above the Kapitza range and below a ΔT of about 80°K.

A comparison of natural convection heat transfer data from a 1·79 mm diameter soft-lead glass tube to saturated and unsaturated ("sub-cooled") liquid HeII with the Kapitza resistance theory of Khalatnikov (1952) is given by Madsen and McFadden (1968). Good agreement is found for the low ΔT

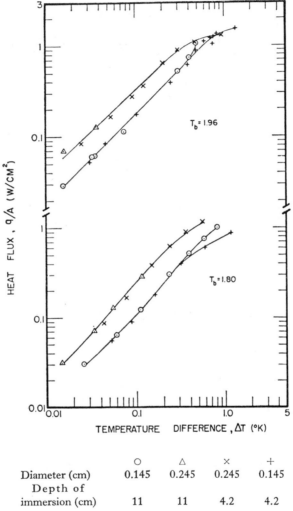

	○	△	×	+
Diameter (cm)	0.145	0.245	0.245	0.145
Depth of immersion (cm)	11	11	4.2	4.2

Fig. 3.43. Non-film boiling heat flux as a function of temperature difference (Holdredge and McFadden, 1966).

range but as the temperature difference is increased there apparently is an additional thermal resistance being created in the system.

Holdredge and McFadden (1966) studied the Kapitza resistance in a saturated HeII system with a test section similar to that of Irey *et al.* (1965).

They examined the influence of depth of immersion but found it not to be an important factor. Since the influence of depth is to establish the pressure level at the test section these results suggest that no vapour phase of HeI exists at the heated surface in these low ΔT regions. An effect of test section size is unobserved as well in this range of ΔT. The experimental data seem to depart from the Khalatnikov theory at a ΔT of approximately $0 \cdot 10°$K.

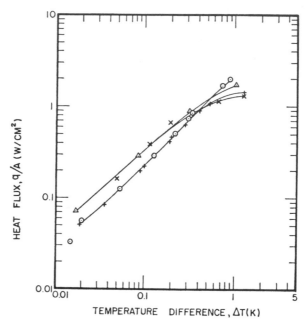

Fig. 3.44. Non-film boiling heat flux as a function of temperature difference for a $2 \cdot 1$ K helium bath (Holdredge and McFadden, 1966).

Above the ΔT corresponding to the range of the predominance of the Kapitza effect heat is transferred to liquid HeII by a transport mechanism which is largely unknown. It is called non-film boiling but does not appear to be similar to nucleate boiling. Data for this heat transfer process are given in Figs. 3.43 and 3.44 taken from Holdredge and McFadden (1966) and Madsen and McFadden (1968). In each case the test section is a soft-glass cylinder horizontally oriented in a bath of HeII. The data in these figures are for a saturated liquid with various bulk temperatures, depths and test section size. In these data an influence of size is observed at all bath temperatures suggesting a controlling effect of buoyant forces. Depth, or pressure, does not appear to be significant until fairly large ΔT are obtained. This is thought to be the point for the onset of film boiling and the appearance at the surface of the gaseous phase of HeI. The surface temperature increases with heat flux very significantly at this point and the phenomenon of a maximum heat flux is observed. The effect of "sub-cooling" on the non-film boiling was also examined. The maximum heat flux is in the range 2–4 W/cm² which may be compared to a corresponding value of about 1 W/cm² for HeI.

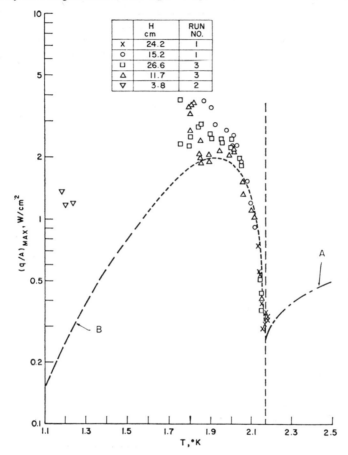

Fig. 3.45. Maximum heat flux at various depths of immersion versus temperature (Clement and Frederking, 1966). A—$(q/A)_{max}$, HeI. Lyon (1964); horizontal surface. B—$(q/A)_{max} = 0.148 \, C(T)TS^{\frac{4}{3}}\rho^{\frac{2}{3}}\mu_n^{\frac{1}{3}}$.

For a heat flux above a certain critical maximum the influence of the superfluid is destroyed, at least locally, in the vicinity of the heated surface. This critical maximum heat flux is similar to the maximum nucleate boiling heat flux but its mechanism is probably quite different. Clement and Frederking (1966) investigated this phenomenon for heat transfer to HeII from a horizontal surface through a short tube. In this study the transport processes were of a more bounded type than those cited above in which a test surface was totally surrounded by liquid HeII. The low heat flux-low ΔT data for this bounded system were discussed above and given in Fig. 3.41. The measured values of $(q/A)_{max}$ are given in Fig. 3.45 for various liquid depths H of the test surface below the HeII bath surface. There appears to be very little effect

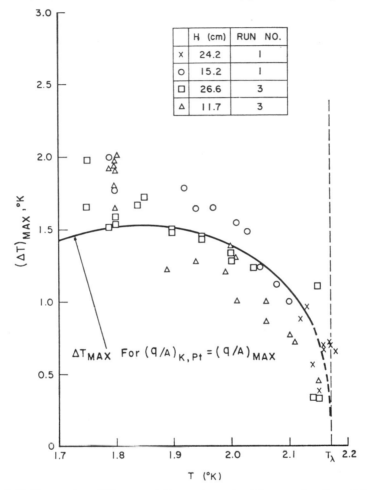

Fig. 3.46. Temperature differences at the maximum heat flux as a function of the bath temperature (Clement and Frederking, 1966).

of liquid depth (pressure) on $(q/A)_{max}$. Owing to limitations of pump capacity the pressure was difficult to stabilize at the lower temperatures. In the vicinity of the λ-point, however, the data are very accurate. The magnitudes of $(q/A)_{max}$ are seen to increase from $0 \cdot 10$ W/cm^2 at the λ-point to approximately 2–4 W/cm^2 at about $1 \cdot 9°$K, the point or maximum apparent thermal conductivity for HeII. The results near the λ-point coincide very well with the low pressure maximum heat flux data of Lyon (1965) for HeI. Clement and Frederking (1966) have fitted a curve to their data which follows the general characteristic of the heat flux corresponding to the apparent thermal conductivity of HeII. The equation, which fits the data well near the λ-point, is

$$(q/A)_{max} = 0\cdot148C(T)T\mu_n^{\frac{1}{3}}\rho^{\frac{2}{3}}S^{\frac{4}{3}} \tag{3.159}$$

The influence of depth of immersion (pressure) on the temperature difference at the maximum heat flux was found to be negligible. The temperature difference corresponding to the maximum heat flux is summarized in Fig. 3.38 as a function of bath temperature. The accuracy is limited at the lower temperatures owing to restricted pumping capacities in the experimental apparatus. A lower bound to ΔT_{max} is shown in Fig. 3.46 which is that ΔT from equation 3.155, the Kapitza relationship, at which the corresponding heat flux, $(q/A)_K$, is equal to the maximum heat flux, $(q/A)_{max}$, from equation 3.159. There is quite a bit of scatter in the data. It should be noted that these ΔT are about the same as those determined for the maximum heat flux in HeI. Clement and Frederking (1966) report $(q/A)_{max}$ values for HeII as high as 10 W/cm² but do not give the specific data. Reference is probably made to the small wire data such as that of Frederking (1961).

Maximum heat flux data of the order of 10 W/cm² for a horizontal platinum wire 15 μ diameter is reported by Rinderer and Haenseler (1959) for liquid HeII at 1·4°K. The magnitude of $(q/A)_{max}$ was found to depend on liquid height above the test surface.

Lemieux and Leonard (1968) investigated the maximum heat flux for a horizontal 76·2 μ diameter 90% platinum–10% rhodium wire. Depths of immersion below the liquid–vapour interface ranged from 5 to 70 cm. In this way both the pressure at the heated surface and the heat flow path length were varied. A large influence of depth of immersion was found. Near the λ-point, for a depth of 30 cm, $(q/A)_{max}$ increases very rapidly with bulk temperature reaching a maximum of about 15 W/cm² at about 2·0°K. The maximum values are considerably greater for this small diameter wire than for the larger unbounded cylinders (Holridge and McFadden, 1966; Madsen and McFadden, 1968) and the flat (bounded) plate. The smoothed data taken as a function of depth are given in Fig. 3.47. The influence of depth vanishes at the λ-point. Whilst there is agreement between the various authors who used heated wires, the results for the cylinder ($D = 0\cdot245$ cm) of Holdredge and McFadden (1966) differs appreciably. Apparently, there is a very significant diameter or size influence on $(q/A)_{max}$, perhaps similar to that observed in film boiling for small diameter cylinders by Breen and Westwater (1962).

Lemieux and Leonard (1968) report that on reduction of heat flux from a film boiling condition a minimum heat flux is reached, less than $(q/A)_{max}$, at which the system no longer can support the vapour film. For lower heat flux the heat transport reverts to that of the superfluid.

At heat flux rates in excess of $(q/A)_{max}$ a HeII system goes over into film boiling controlled by ordinary fluid and thermal phenomena. Under these conditions the superfluidity effects are destroyed at the heated surface and are replaced with those circumstances which govern the natural convection of a vapour layer in film boiling (1966). The liquid HeII is lifted off the surface but becomes an effective heat sink for the vapour. Since the greater thermal resistance resides in the vapour film the unusually great heat transport ability of the superfluid is no longer effective in promoting heat transfer.

Rivers and McFadden (1966) have analyzed the transport processes in laminar film boiling in liquid HeII for a flat plate and cylinder. Their results are given for both liquid and vapour films of HeI between the heated surface and the liquid HeII. The process is formulated in terms of the usual transport equations for laminar boundary layer flow but a temperature and velocity discontinuity is introduced at the film-HeII interface. The Nusselt number is determined in terms of the Grashof and Prandtl numbers and two parameters, the "interface enthalpy", H_i and the "interface heat flux", Q_b. The analytical results for the cylinder are given in Fig. 3.48 for films of liquid HeI ($Pr = 0.4$) and helium gas ($Pr = 0.7$). The results for the flat plate follow the same pattern. The following definitions are used

$$\overline{Nu} = \frac{\bar{h}D}{k_f} \tag{3.160}$$

where $D = r$ for a cylinder and L for a plate,

$$Gr = \frac{gD^3\rho_f(\rho_b-\rho_f)}{\mu_f^2} \tag{3.161}$$

$$H_i = \frac{(\Delta h)_i}{c_{pf}(\Delta T)_f} \tag{3.162}$$

where $(\Delta h)_i$ = enthalpy change at the He^2 interface, and

$$Q_b = \frac{D(q/A)_b}{k_f(\Delta T)_f} \tag{3.163}$$

in which $(q/A)_b$ is an "interface heat flux", a quantity required to compute Q_b but about which little is presently known. From these results Rivers and McFadden identify three regions. For $Q_bGr^{-\frac{1}{4}} < 0.01$, the heat transfer follows ordinary film boiling characteristics in which the Nusselt number is a function of the Grashof and Prandtl numbers and, of course, H_i. In this region the convective processes dominate. Above $Q_bGr^{-\frac{1}{4}}$ of about 10, the process is described by the asymptote $\overline{Nu} = Q_b$. Under these circumstances, the wall heat flux $(q/A)_w$ is identical with that at the interface, $(q/A)_b$. Conductive mechanisms govern the phenomena in this region. For intermediate

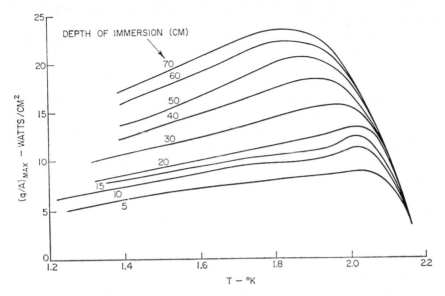

Fig. 3.47. Maximum heat flux as a function of bath temperature and depth of immersion (Lemieux and Leonard, 1968).

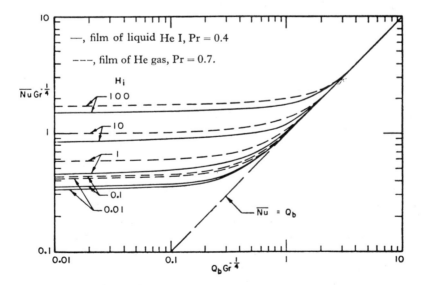

Fig. 3.48. Calculated results illustrating average heat transfer coefficient ($Nu = h_c r/k_f$) for a horizontal circular cylinder (Rivers and McFadden, 1966).

values of $Q_b Gr^{-\frac{1}{4}}$ a transition region exists in which both convection and conduction effects coexist.

Except in the "convection" region, $Q_b Gr^{-\frac{1}{4}} < 0.01$, it is necessary to determine $(q/A)_b$ in order to make calculations using the results in Fig. 3.48. This is the key step in the calculation procedure for the transition region. The "interface heat flux", $(q/A)_b$, is fixed by the state of the liquid HeII and is independent of the processes in the film. Accordingly, it has the role of a "heat sink" and enters the process as a boundary condition. Its magnitude is thought to depend on the bulk temperature of the liquid HeII and the interfacial temperature discontinuity. However, its *a priori* determination has not yet been resolved nor have any quantitative formulations been established for its calculation. Fortunately, in the high ΔT range where $Q_b Gr^{-\frac{1}{4}}$ is less than 0.01, $(q/A)_b$ is much less than $(q/A)_w$, the wall heat flux, and thus it does not influence the process significantly. In this range, however, the classical film boiling correlations are not valid owing principally to the unusual temperature discontinuity and "heat sink", $(q/A)_b$, boundary condition at the HeII interface. There is presently a need to clarify the role of $(q/A)_b$ and to achieve an explicit expression for its calculation.

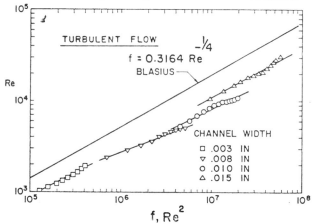

Fig. 3.49. Dimensionless flow rate (Reynolds number) versus dimensionless pressure gradient (Rinderer and Haenseler, 1959).

Rivers and McFadden (1966) report some experimental data of Holdredge for the case of film boiling from a horizontal cylinder in which the liquid HeII is separated from the heated surface by a film of HeI vapour. A comparison of the predicted and measured Nusselt numbers for primarily a "convective" condition in which $(q/A)_b$ has a subdued role is generally favourable for the range of variables considered.

A study of isothermal, turbulent flow of liquid HeII at 1·5°K in wide channels (0·003 to 0·015 in.) having an R.M.S. roughness of about 0·0001 in.

is reported by Frederking and Schweikle (1965). Their results indicated that under these flow conditions HeII behaved similar to ordinary fluids and could be correlated in terms of the usual similarity parameters. Pressure drop data correlated in terms of Reynolds number and friction factor are shown in Fig. 3·49 in comparison with the Blasius formulation for turbulent flow. Although the general nature of the correlation is evident, the magnitude of the friction factor for a given Reynolds number is considerably greater than that predicted by the Blasius expression for smooth surfaces. This increase is significantly greater than would normally be expected for roughened surfaces in the experimental ranges of relative roughness (hydraulic radius/ R.M.S. roughness = 30 to 150) and Reynolds number. The increase in friction factor at Reynolds number as low as 10^3 as a consequence of roughness alone is quite unusual and may be a result of other factors. In Fig. 3.49 the Reynolds number is defined as

$$Re = \frac{\rho VC}{\mu_n} \tag{3.164}$$

where ρ is the total liquid density, μ_n the absolute viscosity of the normal fluid and D the hydraulic diameter. The friction factor is defined from

$$\Delta p = f\left(\frac{L}{D}\right)\frac{\rho V^2}{2g_0} \tag{3.165}$$

Heat transfer data for zero net mass flow systems also were plotted using similarity parameters but less success was achieved.

XI. Nomenclature

A	Area, ft² or cm²
a	Acceleration, ft/sec²
a, a^*	Thermal diffusivity, ft²/hr
A_f	Flow area, ft²
B	Function for λ_c, see equation (3.20)
Bo	Boiling number, $(q/A)/h_{fg}G_{\mathrm{mix}}$
c_p, c_v	Heat capacity, B.T.U./lb$_m$°F, J/mol °K
c_0	Concentration, lb$_m$/ft³
D', D''	Mass diffusivity, ft²/hr
D_e	Equivalent diameter, ft
D_h	Hydraulic diameter
D_0	Outside tube diameter, ft
d	Thickness of heat transfer surface for plate–fin exchangers
E	k(T) integral, see equation (311.), B.T.U./hr ft
E'	Heat exchanger effectiveness
E_∞	Value of E at T_∞, B.T.U./hr ft
E_i	Value of E at T_i, B.T.U./hr ft
E^*	Defined by equation (3.20)

e	Void volume or fin efficiency parameter (equation 3.148)
Fo	Fourier number, a^*t/L^2
f	friction factor, or a fraction in equation 3.155
G	Mass velocity, $lb_m/hr\ ft^2$
G^*	See Table 3.X
G_{mix}	See equation (3.90)
Gr	Grashof number
g_0	Conversion factor, 32·2 $(lb_m/lb_f)\ (ft/sec^2)$
g, g_z	Gravitional acceleration, ft/sec^2, m/sec^2
H	Liquid height, ft, Fig. 3.46
H_i	See equation (3.181)
h	Heat transfer coefficient, B.T.U./hr ft^2 °F, W/cm² °K
h_x	Local heat transfer coefficient
h_0	Kapitza effect heat transfer coefficient, equation (3.152)
h_g	Gas space heat transfer coefficient
h_{fg}	Latent heat, B.T.U./lb_m
J	Conversion factor 778 ft lb_f = 1 B.T.U.
j	Volumetric flux m^3/m^2 sec or ft^3/ft^2 sec
j^*	Stability parameter (equations 3·108)
j_T	Colburn factor
K_g	Gas phase mass transfer coefficient
k	Thermal conductivity, B.T.U./hr ft °F, W/cm °C
k', k''	Thermal conductivity
k_∞	Thermal conductivity at T_∞
k_{App}	Thermal conductivity of HeII, see equation (3.168)
L	Characteristic length, ft
M	Mass kg or lb
M_0, M_s	Stability modulus, equations (3.24, 3.26)
N_s	Biot modulus, equation (3.27)
N_T	Number of transfer units
Nu	Nusselt number
p	Pressure, p.s.i.a., mmHg
p_0	Perimeter, ft
p	Vapour pressure
p_c	Critical pressure, p.s.i.a.
p_s''	Volumetric heat generation, B.T.U./hr ft^3
Pr	Prandtl number
q	Heat flux, W/cm²
(q/A)	Heat flux, B.T.U./hr ft^2
$(q/A)_{max}$	Maximum nucleate boiling heat flux
$(q/A)_{min}$	Minimum film boiling heat flux
$(q/A)_K$	Kapitza heat flux
Q_b	See equation (3.163)
R	Residual
R	Gas constant, $(lb_f/lb_m)\ (ft/°R)$
Re	Reynolds Number
Ra	Rayleigh number, $GrPr$
St	Stanton number
S	Entropy
s	Wall thickness of solid

T, T', T''	Temperature °R, °K
T_m	Temperature at maximum value of c_p, °R
T_{sat}	Saturation temperature, °R, °K
ΔT_{sat}	$T_w - T_{sat}$ °R, °K
ΔT_{lm}	Logarithmic mean temperature difference
T_w	Wall temperature, °R, °K
T_c	Critical temperature °R, °K
t	Time, hours, seconds
U	Overall heat transfer coefficient
u, u', u''	Velocity, ft/sec
v_{fg}	Volume for vaporization, ft^3/lb$_m$
V	Velocity, ft/sec
V_{av}	Average velocity
w	Mass flow rate, lb$_m$/hr, kg/hr
x	Coordinate, quality or vapour mass fraction of a two-phase flow
x^*	x/L
Δx	Grid size
X	Interface location
X_{tt}	Martinelli parameter equation 3.81
Y	Vapour volume void fraction
y	Co-ordinate
y^*	y/L
z	Co-ordinate; vertical distance, m or ft
z^*	z/L

Subscripts

b	Bulk
c	Cold
f	Film
g	Bulk gas or vapour phase
h	Hot
i	Inlet or interface
l	Liquid
o	Outlet
p	Packing (Regenerator)
v	Vapour
w	Wall
∞	Ambient
1	Primary heat transfer surface
2	Secondary heat transfer surface

Greek Symbols

α', α''	Thermal diffusivity, ft^2/hr
α	Absorptance
β	Expansion coefficient, °R^{-1}
δ	Disturbance parameter or film thickness (condensation)
$\Delta(t)$	Stratified layer thickness, ft
ε	Emittance
η	$z/\Delta(t)$, thermal efficiency of fin or regenerator

Λ	Reduced length of regenerator
μ	Viscosity, $lb_m/hr\ ft$
v	Kinematic viscosity, ft^2/hr
π	Reduced period of regenerator
ρ	Density lb_m/ft^3, g/cm^3
σ	Surface tension, lb_f/ft, dyn/cm
$\tau_{o,v}$	Shear stress in laminar sublayer; at condensate film surface lb_f/in^2
Φ_1	See equation 3.66 and Fig. 3.19
Φ_2	See equation 3.67
ϕ_1	Correction factor, Table 3.II
ϕ_2	Correction factor, equations 3.51 and 3.52
ϕ	Penetration factor, see equation 3.146

XII. References

Arpaci, V. S. (1966). "Conduction Heat Transfer", Addison–Wesley.

Arthur D. Little, Inc. (1958). "Atmospheric Heat Transfer to Vertical Tanks Filled with Liquid Oxygen." Special Report No. 50, to AFBMD, WDSOT.

Astruc, J. M., Perroud, P., Lacaze, A. and Weil, L. (1967). In "Advances in Cryogenic Engineering", Vol. 12, pp. 387–395. Plenum Press.

Baily, B. and Churgay, J. R. (1962). "Pressurant Gas Behavior in Pressurized Discharge Apparatus." Memo to J. A. Clark, Heat Transfer Laboratory, Department of Mechanical Engineering, University of Michigan.

Baker, O. (1954). *Oil and Gas J.* **53**, 184.

Balzhiser, R. E., *et al.* (1961). Final Report—Phase I, ASD Technical Report 61-594.

Banchero, J. T., Barker, G. E. and Boll, R. H. (1951). Project M834, Engineering Research Institute, University of Michigan.

Barakat, H. Z. and Clark, J. A. (1966). *J. Heat Transfer.* **88**, 421.

Barker, J. J. (1965). *Ind. Eng. Chem.* **57**, 43.

Barnett, D. O., Winstead, T. W. and McReynolds, L. S. (1965). In "International Advances in Cryogenic Engineering", Vol. 10, p. 314.

Barron, R. F. and Han, L. S. (1965). *J. Heat Transfer* **87**, 499–506.

Bartlit, J. R. and Williamson, Jr., K. D. (1966). In "Advances in Cryogenic Engineering", Vol. 5, pp. 561–568.

Benser, W. A. and Graham, R. W. (1962). "Hydrogen Convective Cooling of Rocket Nozzles". ASME Paper No. 62-AV-22.

Berenson, P. J. (1964). *Heat Transfer* **83**, 351–359.

Bergles, A. E. and Rohsenow, W. M. 1964 *J. Heat Transfer*, **86**, 365–372.

Bewilogua, L., Knoner, R. and Wolf, G. (1966). *Cryogenics* **6**, 36–40.

Borishanskii, V. M. (1956). *Zhurn. Tek. Fiz.* **26**, 452; Translated in *Soviet Physics–Technical Physics*, **1**, 438.

Boyko, L. D. and Kruzhilin, G. N. (1967). *Int. J. Heat and Mass Transfer*, **10**, 361.

Breen, B. P. and Westwater, J. W. (1962). *Chem. Engr. Progr.* **58**, 67–72.

Brentari, E. G., Giarratano, P. J. and Smith, R. V. (1965). National Bureau Standards Tech. Note No. 317, Boulder, Colorado.

Brentari, E. G. and Smith, R. V. (1965). "International Advances in Cryogenic Engineering", Plenum Press, New York.

Bretherton, A. (1970). Ph.D. Thesis. Bradford University.

Bromley, L. A. (1950). *Chem. Eng. Prog.* **46**, 221–227.

Burke, J. C. and Rawdon, A. H. (1965). ASME Paper 65-HT-37.

Caken, R. P., Gilcrest, A. S. and Zierman, C. A. (1964). *Advances in Cryogenic Engineering*, Vol. 9, pp. 457–463, Plenum Press, New York.

Carpenter, F. G. and Colburn, A. P. (1951). Proceedings of General Discussion of Heat Transfer, I.Mech.E. and A.S.M.E. pp. 20–26.

Carslaw, H. S. and Jaeger, J. C. (1959). "Conduction of Heat in Solids". Second Edition, Oxford University Press.

Chang, Y. P. (1959). *J. Heat Transfer* **81**, 1–12.

Chang, Y. P. (1963). *J. Heat Transfer* **85**, 89–100.

Chang, Y. P. and Snyder, N. W. (1960). *Chem. Eng. Symp. Series*, **56**, 25–38.

Chato, J. C. (1962). *Ashrae J.* **4**, 52.

Chen, J. C. (1963). ASME Paper No. 63-HT-34, ASME-AIChE Heat Transfer Conference, Boston.

Chen, M. M. (1961). *Trans. A.S.M.E. J. Heat Transfer Series C* **83**, 48.

Chenoweth, J. M. and Martin, M. W. (1955). *Petro. Refiner* **34**, 151.

Chrysler Corporation (1967). Technical Report HSM-R421-67, Contract NAS 8-4016, Schedule II, Fluid Mechanics and Thermodynamics Research Section, Space Division, Huntsville.

Clark, J. A. (1969). *In* "Advances in Heat Transfer" (Irvine, T. F. and Hartnett, J. P., Eds.), Vol. 5, pp. 325–517. Academic Press, London and New York.

Clark, J. A. "The Transport of Heat and Mass". In preparation. John Wiley and Sons, New York.

Clark, J. A. and Merte, H. (1963). Proceedings (Commission V), International Congress of Refrigeration, Munich.

Clark, J. A., Merte, H. and Barakat, H. Z. (1967). Proceedings Semi-International Symposium, Japanese Society of Mechanical Engineers, Tokyo.

Clark, J. A. and Rohsenow, W. M. (1954). *Trans. ASME*, pp. 553–563.

Class, C. R., DeHaan, J. R., Piccone, M. and Cost, R. B. (1960). *Adv. Cryo. Engr.* **5**, 254–261.

Clement, B. W. and Frederking, T. H. K. (1966). Proceedings International Institute of Refrigeration, Commission I. Boulder.

Clement, B. W. and Frederking, T. H. K. (1966). Proceedings 3rd International Heat Transfer Conference, Vol. 1, pp. 299–305.

Coeling, K. J. and Merte, H. M. (1971 "Incipent and Nucleate Boiling of Liquid Hydrogen". *Trans. A.S.M.E., J. Eng. for Ind.*

Cowley, C. W., Timson, W. J. and Sawdye, J. A. (1962). *Ind. Eng. Chem. Proc. Design Level* **1**, 81–84.

Core, T. C., Harkee, J. F., Misra, B. and Sato, K. (1959). "Heat Transfer Studies." WADD-60-239.

Cunningham, T. M. and Young, R. L. (1963). "Advances in Cryogenic Engineering,"Vol. 8, pp. 85–93. Plenum Press.

Desmond, L. G. and Sams, E. W. (1950). NACA, Research Memorandum, ESO H23.

Diehl, J. E. (1957). *Pet Refiner* **36**, 147.

Domingos, J. D. (1969). *Int. J. Heat Mass Transfer* **12**, 537.

Ellerbrook, H. H., Livingood, J. N. B. and Straight, D. M. (1962). NASA SP-20.

Fenster, S. K., Van Wylen, G. J. and Clark, J. A. (1960). *In* "Advances in Cryogenic Engineering", Vol. 5, p. 226. Plenum Press. New York.

Fleming, R. B. (1966). Paper R-8, Cryogenic Engineering Conference. Boulder.

Fleming, R. B. (1967). *In* "Advances in Cryogenic Engineering", Vol. 12, pp. 352–362. Plenum Press. New York.

Flynn, T. M., Draper, J. W. and Roos, J. H. (1962). "Advances in Cryogenic Engineering", Vol. 7, pp. 539–545. Plenum Press, New York.

Forslund, R. P., Rohsenow, W. M. (1968). *ASME J. Heat Transfer* **90**, 399.

Forster, H. K. and Greif, R. (1959). *J. Heat Transfer* **81**, 43–53.

Forster, H. K. and Zuber, N. (1955). *A.I.Ch.E. J.* **1**, 532–535.

Frederking, T. H. K. (1961). *Forschung* **27**, 17–62.

Frederking, T. H. K. and Clark, J. A. (1963). "Advances in Cryogenic Engineering, Vol. 8, pp. 501–506. Plenum Press, New York.

Frederking, T. H. K. and Schweikle, J. D. (1965). *In* "Low Temperature Physics-LT9" (Part A). Plenum Press, New York.

Fulk, M. M. and Reynolds, M. M. (1957). *In* "American Institute of Physics Handbook" (Gray, D. E., ed.). McGraw-Hill, New York.

Gardner, K. A. (1945). *Trans. A.S.M.E.* **67**, 621.

Gebhardt, B. (1961). "Heat Transfer." McGraw–Hill, New York.

George C. Marshall Space Flight Center (1961). "Cryogenic Subcooling by Helium Injection," Memo M-S+M-PE No. 327.

Giarratano, P. J. and Smith, R. V. (1966). *In* "Advances in Cryogenic Engineering, Vol. 11, p. 492.

Gilmour, C. H. (1958). *Chem. Eng. Prog.* **54**, 77–79.

Gluck, D. F. and Kline, J. F. (1962). *In* "Advances In Cryogenic Engineering," Vol. 7, p. 219. Plenum Press, New York.

Graham, R. W., Hendricks, R. C., Hsu, Y. Y. and Friedman, R. (1961). *In* "Advances in Cryogenic Engineering," Vol. 6, pp. 517–524. Plenum Press, New York. See also NASA TN D-765, 1961, by the same title and written by the same authors.

Grimison, E. D. (1937). *Trans. A.S.M.E.* **59**, 583.

Grimison, E. D. (1938). *Trans. A.S.M.E.* **60**, 381.

Handley, D. and Heggs, P. J. (1969). *Int. J. Heat Mass Transfer* **12**, 549.

Hargis, J. H. and Stokes, H. A. (1965). *Proceedings Conference on Propellant Tank Pressurization and Stratification*, Vol. I. George C. Marshall Space Flight Center, NASA, Huntsville.

Hausen, H. (1950). "Wärmeübertragung in Gegenstrom, Gleichstrom und Kreuzstrom." Springer Verlag, Berlin.

Hendricks, R. C., Graham, R. W., Hsu, Y. Y. and Medeiros, A. A. (1962). *J. Am. Rocket Soc.* February, pp. 244–253.

Hess, H. L. and Kunz, H. R. (1965). *J. Heat Transfer*, pp. 41–49.

Holdredge, R. M. and McFadden, P. W. (1966). *In* "Advances in Cryogenic Engineering", Vol. 11, pp. 507–516.

Holten, D. C. (1961). *In* "Advances in Cryogenic Engineering", Vol. 6, p. 499.

Hsu, Y. Y. and Westwater, J. W. (1958). *A.I.Ch.E. J.* **4**, 58–62. and (1960). *CEP Symp. Ser.* **56**, 15–24.

Hynek, S. J., Rohsenow, W. M. and Bergles, A. E. (1969). M.I.T. Dept. of Mech. Eng. EPL Report No. 70586-63.

Illiffe, C. E. (1948). *Proc. Inst. Mech. Engr.* **159**, 363.

Irey, R. K., McFadden, P. W. and Madsen, R. A. (1965). *In* "International Advances in Cryogenic Engineering", Vol. 10, pp. 361–371.

Kahl, W. H. (1962) Personal Communication to J. A. Clark from Linde Co., Tonawanda, New York.

Kao, S. (1965). *Trans. A.S.M.E.* **87**, 202.

Kapitza, P. L. (1941). *J. Phys. (U.S.S.R.)* **4**, 181.

Kays, W. M., Jain, R. K. and Saberwhal, S. (1968). *Int. J. Heat. Mass Transfer* **11**, 772.

Kays, W. M. and London, A. L. (1964). "Compact Heat Exchangers." McGraw–Hill, New York.

Kays, W. M. and London, A. L. (1964). "Compact Heat Exchangers," McGraw–Hill, New York.

Khalatnikov, I. M. (1952). *J. Exp. and Th. Phys.* (*U.S.S.R.*) **22**, 687.

Knudsen, J. G. and Katz, D. L. (1959). "Fluid Dynamic and Heat Transfer." McGraw–Hill, New York.

Kreith, F. (1965) "Principals of Heat Transfer." International Textbook Co., New York.

Kroeger, P. G. (1967). *In* "Advances in Cryogenic Engineering", Vol. 12, pp. 340–351. Plenum Press, New York.

Kutateladze, S. S. (1951). *Izv. Akad. Nauk, U.S.S.R., OTD. Teck. Navk* **4**, 529.

Kutateladze, S. S. (1959). "Heat Transfer in Condensation and Boiling." Translation Series AEC-tr-3770.

Labountzov, D. A. (1960). Teploenergetika **7**, 76–80.

Lambertson, T. J. (1958). *Trans. A.S.M.E.* **80**, 586.

Lane, C. T. (1962). "Superfluid Physics." McGraw–Hill, New York.

Larsen, P. S., Clark, J. A., Randolph, W. O. and Vaniman, J. L. (1968). *In* "Advances in Cryogenic Engineering", Vol. 8, pp. 507–520. Plenum Press, New York.

Laverty, W. F. and Rohsenow, W. M. (1965). ASME Paper 65 WA/HT-26.

Lemieux, G. P. and Leonard, A. C. (1968). *In* "Advances in Cryogenic Engineering", Vol. 13.

Lenfestey, A. G. (1961). Progress in Cryogenics **3**, 23.

Levy, S. (1959). *J. Heat Transfer* **81**, 37–42.

Liebenberg, D. H. and Edeskuty, F. J. (1965). *In* "International Advances in Cryogenic Engineering", Vol. 10, p. 284. Plenum Press, New York.

Lienhard, J. H. and Wong, P. T. Y. (1964). *J. Heat Transfer* **86**, 220–227.

Linde Co. (1962). "Superinsulation Applied to Space Vehicles." Engineering Development Laboratory, Cryogenic Products Department, Tonawanda, New York.

Lockhart, R. W., Martinelli, R. C. (1949). *Chem. Eng. Prog.* **45**, 39.

Löf, G. O. G., Hawley, R. W. (1948). *Ind. Eng. Chem.* **40**, 1061.

Loper, J. L. and Heatherly, E. R. (1955). Tech. Note No. G-002, Structures and Mechanics Laboratory, ABMA, Huntsville.

Lyon, D. N. (1964). *Int. J. Heat and Mass Transfer* **7**, 1097.

Lyon, D. N. (1965). "International Advances in Cryogenic Engineering." Vol. 70, pp. 371–380. Plenum Press, New York.

Lyon, D. N., Kosky, P. G. and Harnon, B. N. (1964). *In* "Advances in Cryogenic Engineering", Vol. 9, pp. 77–87.

Lytle, F. W. and Stoner, J. T. (1965). *Science* **148**, 1721–1723.

McAdams, W. H. (1954). "Heat Transmission." McGraw–Hill, New York.

McConnell, D. G. (1966). "Advances in Cryogenic Engineering," Vol. 11, pp. 328–338. Plenum Press, New York.

McDonald, R. (1967). Ph.D. Thesis, Bradford University.

Madsen, R. A. and McFadden, P. W. (1968). *In* "Advances in Cryogenic Engineering", Vol. 13.

Marto, P. J., Moulson, J. A. and Maynard, M. D. (1968). *J. Heat Transfer* **90**, 437–445.

9

Martinelli, R. C. and Nelson, D. B. (1948). *Trans. ASME,* **70,** 695–702.

Matheson, C. C. (1966). *Cryogenics* February, 1–10.

Meck, R. M. G. (1962). N.E.L. Rep. No. 54.

Merte, H. and Clark, J. A. *J. Heat Transfer* **83,** 233–243.

Merte, H. and Clark, J. A. (1962). *In* "Advances in Cryogenic Engineering", Vol. 7.

Merte, H. and Clark, J. A. (1964). *J. Heat Transfer* **86,** 351–360.

Merte, H. and Lewis, E. W. (1967). "Boiling of Liquid Nitrogen in Reduced Gravity Fields with Subcooling." Heat Transfer Laboratory, Department of Mechanical Engineering, University of Michigan, Ann Arbor.

Michenko, N. (1960). *Teploenergetika* **7,** 17–21.

Moissis, R. and Berenson, P. J. (1963). *J. Heat Transfer* **85,** 221–229.

Moore, B. C. (1962). "Effect of Gas Condensate on Cryopumps." American Vacuum Society, Los Angeles.

National Bureau of Standards (1961). Cryogenic Engineering Laboratory ABMA Project Order 1657-00-60, National Bureau of Standards Project 81411.

Noyes, R. C. (1963). *J. Heat Transfer* **85,** 125–132; See also discussion by Clark, J. A. p. 129.

Nusselt, W. (1916). *Z. Ver. deut. Ing.* **60,** 541.

Nusselt, W. (1927). *Z. Ver deut. Ing.* **71,** 85.

Nusselt, W. (1930). *Tech. Mech. Thermo-Dynam.,* Berlin **1,**

Randolph, W. O. and Vaniman, J. L. (1961). George C. Marshall Space Flight Center, MTP-S and M-P-61-19.

Rohsenow, W. M. (1952). *Trans. ASME,* **74,** 969-975.

Rohsenow, W. M. (1953). "Heat Transfer." University of Michigan Press, Ann Arbor.

Rohsenow, W. M. and Choi, H. Y. (1961). "Heat Mass and Momentum Transfer." Prentice Hall, New Jersey.

Rohsenow, W. M. and Griffith, P. (1956). *Chem. Eng. Progress Symposium, Series 18,* **52,** 47–49.

Rohsenow, W. M., Webber, J. H. and Ling, A. T. (1956). *Trans ASME,* **78,** 1637.

Richards, R. H., Steward, W. G. and Jacobs, R. B. (1961). National Bureau of Standards, NBS TN 122, Boulder Laboratories.

Rinderer, L. and Haenseler, F. (1959). *Helv. Phy. Acta* **32,** 322.

Rios, P. A. (1967). M.Sc. dissertation, Massachusetts Institute of Technology.

Rivers, W. J. and McFadden, P. W. (1966). *J. Heat Transfer* **88,** 343–351.

Roubeau, P. (1960). *Prog. Refr. Sci. Tech.* **11,** 49–53.

Ruccia, F. E. and Hinckley, R. B. (1967). *In* "Advances in Cryogenic Engineering", Vol. 12, pp. 300–308. Plenum Press, New York.

Platt, G. K., Nein, M. E., Vaniman, J. L. and Wood, C. C. (1963). Paper 687A, SAE-ASNE, National Aeronautical Meeting and Production Engineering Forum.

Pope, D. H., Killian, W. R. and Corbett, R. J. (1960). *In* "Advances in Cryogenic Engineering," Vol. 5, pp. 441–450. Plenum Press, New York.

Schalkwijk, W. F. (1959). *Trans. ASME,* Paper No. 58, A135.

Schneider, P. J. (1963). "Temperature Response Charts." John Wiley and Sons, New York.

Sciance, C. T., Colver, C. P. and Sliepcevich, C. M. (1967). *In* "Advances in Cryogenic Engineering", Vol. 12, pp. 395–409.

Scott, R. B. (1959). "Cryogenic Engineering." D. Van Nostrand, New York.

Scott, R. B., Denton, W. H. and Nicholls, C. M. (1964). "Technology and Uses of Liquid Hydrogen." Pergamon Press, Oxford.

Seader, J. D., Miller, W. S. and Kalvinskas, L. A. (1964). Final Report, Rockdyne Division, North American Aviation, Inc., NASA contract No. NAS 8-5337. George C. Marshall Space Flight Center, Report R-5598.

Soliman, M., Schuster, J. R. and Berenson, P. J. (1968). *J. Heat Transfer* **90,** 267.

Swalley, F. E., Ward, W. D. and Toole, L. E. (1966). Proceedings, Conference on Long-term Cryo-propellant Storage in Space." George C. Marshall Space Flight Center, NASA, Huntsville.

Szetela, E. J. (1962). *J. Am. Rocket Soc.* August, pp. 1289–1293.

Taylor, M. F. and Kirchgessner, T. A. (1959). NASA TN D-133.

Thompson, W. R. and Geery, E. L. (1963). *In* "Advances in Cryogenic Engineering," Vol. 7, pp. 391–401. Plenum Press, New York.

Tien, C. L. and Cravalho, E. G. (1967). Paper No. 306 A.I.Ch.E. Symposium, "Advances in Cryogenic Heat Transfer". A.I.Ch.E. National Meeting, New York.

Tipler, W. (1947). *Shell Tech. Rep.* ICT/14.

Tong, L. S. (1965). "Boiling Heat Transfer and Two-Phase Flow." John Wiley and Sons, New York.

Vance, R. W. (1964). "Cryogenic Technology." John Wiley and Sons, New York.

Van Gundy, D. A. and Uglum, J. R. (1962). *In* "Advances in Cryogenic Engineering", (Timmerhaus, K. D. ed.). Vol. 7.

Wallis, G. B. (1969). "One Dimensional Two Phase Flow." McGraw–Hill, New York.

Walters, H. H. (1961). *In* "Advances in Cryogenic Engineering", Vol. 6, p. 509.

Weil, L. (1951). Proceedings 8th International Congress in Refrigeration, London.

Weil, L. and Lacaze, A. (1950). *Acad. Sci. Tome* **230,** 186–188.

Weil, L. and Lacaze, A. (1951). *J. Phys. Rad.* **12,** 890.

Westwater, J. W. (1956). "Advances in Chemical Engineering", pp. 1–76. Academic Press, London and New York.

Wiederecht, D. A. and Sonnemann, G. (1960). ASME Paper 60-WA-82.

Willmott, A. J. (1962). BISRA Rep. IM/INT/6/62.

Wolf, H. and McCarthy, J. R. (1960). Abstract 100, A.I.Ch.E. Annual Meeting, Washington.

Wright, C. C. (1960). *In* "Advances in Cryogenic Engineering", Vol. 5, pp. 244–254. Plenum Press, New York.

Wright, R. D. and Colver, C. P. (1967). A.I.Ch.E. Report 24, 10th National Heat Transfer Conference, Philadelphia.

Yang, W. J. (1963). A.I.Ch.E. Preprint No. 48, Sixth National Heat Transfer Conference, Boston.

Yang, W. J., Larsen, P. S. and Clark, J. A. (1965). *J. Eng. for Industry,* (*ASME*) **87,** 413–419.

Zivi, S. M. (1964). *ASME J. Heat Transfer* **86,** 247.

Zuber, N. (1961). *Trans. ASME* **80,** 711–720.

Zuber, N. (1961). *J. Heat Transfer* **83,** 357.

Zuber, N. and Fried, E. (1961). American Rocket Society, Propellants, Combustion and Liquid Rockets Conference.

Zuber, N. and Tribus, M. (1958). "Further Remarks on the Stability of Boiling." UCLA Report 58-5.

Zuber, N., Tribus, M. and Westwater, J. W. (1959). Proceedings 2nd International Conference on Heat Transfer, Boulder.

Chapter 4

Insulation

W. Molnar

B.O.C.—Airco Cryogenic Plant Ltd., London

I. Introduction

The aim of this chapter is to present the topic of insulation from the practical viewpoint, enabling the designer to choose and size insulation systems intelligently. Many papers have dealt in detail with specific topics or have concentrated on the theory of the subject. The treatment given here will cover the complete field, including only sufficient theory to help the user understand the reasons underlying a choice of insulant.

Because there is available such a variety of insulants, with effective thermal conductivities ranging from 4×10^{-2} to below 3×10^{-5} W/m °K, the user no longer has to ask "can I achieve the desired heat inleak?" but "which insulant will best fit my requirements?" His aim is therefore to choose the system which fulfils the essential technical requirements at the lowest total cost.

A. PRACTICAL FACTORS TO BE CONSIDERED IN CHOICE OF INSULATION SYSTEM

The factors that have to be taken into account in choosing an insulating system are:

(i) The liquid being stored or conveyed, as this governs the level of heat inleak that can be tolerated. For example, in the case of liquid helium, because of its low latent heat of vaporization and the high cost of refrigeration involved in producing it, minimal heat inleak is of prime importance; whereas for liquid oxygen or nitrogen loss rate is generally of less importance than cost of equipment or mechanical reliability.

(ii) The type of use for which the unit is intended. If it is to be mobile there may be limitations on weight and size as well as choice of supports for the inner vessel.

(iii) The size of the unit and its complexity governs the type of insulation required; whether vacuum insulation is necessary or indeed practical.

(iv) The environment in which the unit is to be used. This must be considered in order to choose a suitably weather and corrosion resistant outer covering to protect the insulant from damage by water or other agents.

(v) Whether the usage is intermittent or continuous. With intermittent use cooling down loss rather than ultimate loss rate is the important factor.

(vi) Inflammability hazards—hydrogen, methane, etc., present a fire hazard. Because considerable care must be taken to prevent ignition of these liquids, the addition of inflammable materials presents little extra hazard. Oxygen on the other hand presents a special hazard since many materials not inflammable in air burn with explosive violence when the oxygen concentration of the permeating gas exceeds 25%. This means that for oxygen service all insulants or constructional materials containing organic constituents must be avoided.

Liquid nitrogen can also present a hazard in situations where air can penetrate the insulation and come into contact with a surface cooled by this cryogenic fluid. Partial condensation of the air will give a liquid containing about 40% of oxygen. Therefore organic materials should not be used as insulants for liquid nitrogen service unless a nitrogen gas purge of the interspace can be guaranteed.

A list of the most commonly available insulating systems and their properties is given in Section III. Before discussing these in detail the mechanism of heat flow will be briefly reviewed.

II. Review of Principles Governing Heat Flow

There have been many attempts to calculate heat flow in insulating media from first principles (Verschoor and Grebler, 1952; Johnson and Hollweger, 1969; Isenberg, 1963; Luikov *et al.*, 1968), but these have only met with moderate success. The basic principles are simple but the interaction of the several mechanisms and the effects of compression, moisture, etc., make the mathematical formulations of heat flow extremely complex. When one notes what extreme care has to be taken to get reproducibility in measuring the thermal conductivity of an insulating medium, it is not surprising that it is almost impossible to predict accurately from first principles the performance of a given insulation system.

A. MODES OF HEAT TRANSPORT

Heat is transported by two basic mechanisms (1) conduction and (2) radiation, and to obtain the lowest heat flow (the aim of all insulation) the *sum* of the contribution of both must be a minimum. This is a most important statement, which cannot be over-emphasized. It is no use cutting out radiation if at the same time conduction is allowed to increase excessively. At one extreme, the use of solid copper will completely eliminate radiation transfer but will maximize conduction. At the other extreme the use of a hard vacuum will nearly eliminate conduction but will accentuate the effect of radiation.

Since the phenomena are interlinked they should normally be taken into account together in a step by step calculation procedure.

1. *Conduction*

Conduction heat transfer in all phases is considered in Chapter 3. Comment will therefore be restricted to particular factors which arise in either the solid or gaseous phase within insulants.

a. Solid phase conduction. In most cryogenic insulants the solid material is subdivided into small fibrous or granular particles. The presence of breaks in the solid phase complicates the heat flow analysis because they are sources of high resistance to heat flow. In fact these contact resistances generally exert such a dominating influence that the conductivity of the material itself has little bearing on the overall heat transport. The main factor affecting heat flow across the breaks is the effective contact area between particles (which is a function of particle size, hardness and compressive load), the composition and pressure of gas between the particles, and the temperature. Compressive loads can be either externally applied or due to the weight of material itself, and for best results these loads should be as small as possible. This is particularly important in vacuum insulation. The variation of conductivity with applied pressure takes the form shown by Glaser *et al.* (1965) and the Division of Linde U.C. (1963) in which the increase in conduction varies as the two thirds power of the applied load.

The gas pressure and temperature affect the degree of adsorption of gas on the particles. This adsorbed film acts as a liquid and bridges across the points of contact, thus significantly increasing the heat conduction. The increase is quite separate from the contribution the gas itself makes to heat transport.

Such adsorption phenomena are the most likely explanation of the results for expanded perlite, quoted in Tables 3.I and 3.II, which show an increase in heat flow with increasing bulk density when determined at atmospheric pressure; but a reverse effect when the pressure is below 10^{-2} Torr.

At atmospheric pressure the increase in thermal conductivity with bulk density is linear. The heat flow under vacuum conditions is an order of magnitude lower than at atmospheric pressure and actually decreases with increasing bulk density. These results show that the conductivity of the solid phase itself has little bearing on the heat flow. However, at atmospheric pressure the increase in conduction achieved with increase in bulk density must be due to the greater mass of solid matter present and because the thermal resistances at points of contact are decreased by the wetting due to the adsorbed films. Thus when selecting an insulant for use at atmospheric pressures the lighter weight material should be chosen but for vacuum conditions the denser materials may be advantageous.

b. *Gas phase conduction.* Heat that is transported in the gas phase can be subdivided into true conduction and convective heat transfer. Conduction is a micro-phenomenon which depends upon momentum transfer by collisions between the gas molecules and the containing surfaces and between adjacent gas molecules. Under certain conditions the gas molecules striking a hot bounding surface, picking up extra energy, may rebound directly on to a colder surface where this energy is released. Alternatively, the heat can be transferred by successive collision with other molecules of gas before reaching the cold wall. The factor that governs which mechanism operates is the mean free path of the molecules which depends on both the temperature and the pressure.

When the mean free path is less than the distance between the bounding surfaces, whether these are the vessel walls or the surfaces of the insulating medium, then the heat transport at constant temperature differences and size is practically independent of pressure, because, as the pressure increases the number of molecules transporting the heat increases, but so do the number of collisions required to transport the energy between the bounding surfaces. For pressures at which the mean free path is greater than the distance between the bounding surface the conductivity is a linear function of the gas pressure since the number of molecules is a direct function of pressure. These pressures are usually below 1 Torr for most insulating media. For a further detailed analysis of the heat flow in this region see Knudsen (1911) and Smoluchowski (1911). For most insulating media at atmospheric pressure gas phase conduction accounts for over 80% of the heat flow.

The effect of gas pressure on thermal conductivity for powders is shown in Fig. 4.1, where conductivity is plotted against the logarithm of the pressure.

The apparent change of slope in the low pressure region is due to the logarithmic scale. In fact below about 1 Torr the conductivity decreases almost linearly to zero pressure where there is a residual conductivity due to solid phase conduction and radiation. The magnitude of these intercepts varies considerably with differing materials. For a reflecting layer insulant the residual is about 3×10^{-5} W/m °K whereas for a powdered insulant such as Santocel A the residual conductivity is $1 \cdot 6 \times 10^{-3}$ W/m °K.

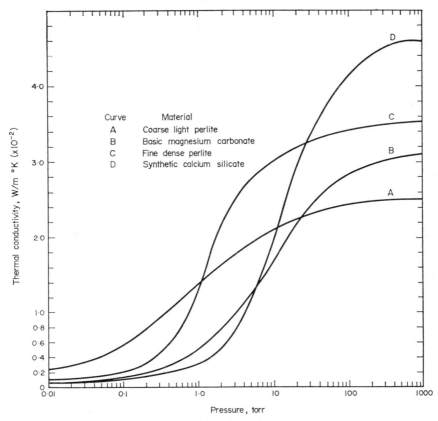

Curve	Material
A	Coarse light perlite
B	Basic magnesium carbonate
C	Fine dense perlite
D	Synthetic calcium silicate

Fig. 4.1. Thermal conductivity of powders versus gas pressure.

The absolute increase of conductivity with gas pressure is not the important criterion but rather the relative increase. Thus for a reflecting layer insulant a conductivity increase of say 5×10^{-5} W/m °K is serious but the same absolute increase for Santocel in vacuum would hardly be significant. The level of vacuum above which the different insulants begin to lose their maximum effectiveness rises from 10^{-4} Torr for conventional high polish high vacuum types, to 10^{-3} Torr for reflecting layer or reflecting powder types to 10^{-2} Torr for powdered insulants.

When working near ambient temperature another less effective method of reducing gas conduction, which may be useful in certain circumstances, is to use high molecular weight gases, such as the halogenated hydrocarbons, in the interspace. This will reduce heat flow by a factor of about 2. This factor accounts for the relatively good insulation performance of closed-cell foamed plastics, such as the polyurethanes, whose pores contain one of the Freon gases.

Convection is a macro-phenomenon which relies on the migration of large masses of gas. These migrations arise when there are density differences within a gas due to the existence of temperature gradients (except when it is vertical with the hotter gas on top). This mechanism is the main factor necessitating the presence of a solid matrix in insulating systems. The efficiency of the insulant is largely attributable to its efficiency in reducing convection currents. It is much more important at cryogenic temperatures than at ambient temperatures and above because the density of air is increased by a factor of 3 when the temperature drops from 300 to 100°K, whereas an equal temperature change from 300 up to 500°K only reduces the density to three-fifths of its value. For air the density difference between 300 and about 100°K is equivalent to a pressure difference of 1 mm water gauge/m height of cold surface.

Such pressure heads can give rise to enormous convection currents. The main factors that affect convection are the gas density and pressure, the upper and lower temperature, the size and geometry of the insulated space and the gas permeability of the insulant.

The mathematics of this situation are complex. Studies made by Martin and Haselden (1963) indicated that an insulant section where the thickness equals height gives the maximum convection effect. Katan (1952) in earlier work found that there was no change in conductivity of a mineral wool with change of height in the apparatus when the ratio of height to thickness varied from 2 to 18. However, it is probable that the thickness of insulation used by Katan was too small for the convection effect to have been significant.

Recent work by Baker and Haselden (1970) has led to simple equations for predicting the increased conductivity of a cube of insulation due to internal convection currents, and an approximate method for applying the same equations to other geometries. In practical situations the actual convection effects will depend greatly on the uniformity of the insulant and particularly on the presence of voids at the bounding walls and roof, or around buried pipes. The larger the thickness of the insulation the greater these effects will be.

The easiest way of stopping convection is to remove the gas in the interspace. The pressure at which convection becomes negligible even for very open pore insulants is much higher than that required to reduce conduction, varying between 10 and 50 Torr depending on the type of insulant.

The author's experiences using vacuum-powder insulant where convection is non-existent have given good agreement between calculations based on laboratory tests and the actual loss rate on full scale storage tanks, the differences being less than 5% even though there was a twelve-fold increase in insulation thickness.

Fine powders such as Basic Magnesium Carbonate or certain grades of perlite have very low gas permeabilities, and therefore bulk convection effects within them can be ignored except for the largest installations.

(c) *Radiation*. Radiation is an electro magnetic phenomenon, and is defined by the Stefan–Boltzmann law (Chapter 3). The contribution of radiation to heat flow becomes significant at cryogenic temperatures whenever vacuum is used.

The heat flow by radiation between the surfaces of two concentric bodies, which have no intervening matter between them is expressed by the equation

$$Q = \frac{\sigma \times A \times (T_2^4 - T_1^4)}{1/\varepsilon_1 + A_1/A2[(1/\varepsilon_2) - 1)]} \tag{4.1}$$

where Q is the heat flow in watts; σ is the Stefan–Boltzmann constant; A_1 and A_2 are the surface areas of inner and outer shells respectively in m^2; T_1 and T_2 are the temperatures of the inner and outer shells in °K. ε_1 and ε_2 are the emissivities of the inner and outer shells.

The emissivity ε is a non-dimensional parameter defined as the ratio of the amount of energy a surface actually emits to that which it would emit were it a perfect black body at the same temperature. For systems of fixed size and temperature limits, using a high-vacuum, high-polish system, the value of ε dominates the residual heat flow and must be as low as practicable.

Calculations show that less heat is transmitted through a vessel having 2 cm thickness of a good grade of perlite at atmospheric pressure than through a clear glass Dewar of the same size where the emissivity of the glass is approximately 0.86. These calculations have been confirmed in laboratory experiments and show that high vacuum alone without a good reflecting surface facing the vacuum space is quite inadequate.

To achieve a low emissivity the surfaces facing the vacuum must consist of pure metals free of all contamination especially grease, or metallic oxides. The metals in order of ascending emissivities are silver, copper, gold and aluminium, values for which are given in Table 4.IV, page 210. To obtain even lower values of heat flow by radiation additional radiation barriers must be interposed between the hot and cold surfaces. Assuming freely floating radiation shields, with surfaces all of the same emissivity, then the heat flow by radiation across the interspace is given by the equation:

$$Q = \frac{\sigma \times A_1 \times (T_2^4 - T_1^4)}{(n+1)[(2/\varepsilon) - 1]} \tag{4.2}$$

where n is the number of shields.

Such a system is impossible to achieve and a compromise solution is to use thin-metallic foils of aluminium as the radiation barriers interleaved with a material of low thermal conductivity.

By utilizing these techniques heat flows of about two to three times those obtainable using free floating shields have been achieved. Because a large number of reflecting layers can be packed in a small space this system has yielded lower values of overall conductivity than have been achieved by any other means.

With powders which are semi-transparent to infra-red the attenuation of radiation is very complex. Not only is radiation reduced by reflection at the surfaces but it is also reduced by adsorption and refraction. The attenuation of radiation under these conditions follows Lambert's law, which is expressed by the equation:

$$I = I_0 e^{-kx} \qquad\qquad (4.3)$$

where I and I_0 are the intensities of the emerging and incident radiation, k is the adsorption coefficient and x is the thickness.

Thus in the presence of absorption the attenuation is an exponential function of thickness whereas with multiple reflection surfaces the attenuation is a linear function of thickness. The true relationship for most powders lies somewhere between the two; the exception being highly opaque powders such as carbon black or the metallic powders. The latter type of powder is a very efficient radiation attenuator but suffers from high solid-phase conduction. However, by mixing selected powders, one being a good radiation attenuator and the other having high solid phase resistance, a very low overall conductivity can be achieved. Such a mixture is silica aerogel mixed with a metallic powder. The heat flow can be reduced by a factor as great as 10 as compared with the pure aerogel.

The characteristics of the various cryogenic insulation systems may therefore be summarized as follows:

When insulants are filled with a gas at atmospheric pressure the gas-phase conduction is usually the prime contributor either by true conduction or by convection. For very finely divided materials, well packed, convection is usually insignificant. For insulants operating under high-vacuum conditions solid-phase conduction and radiation are the main contributors to heat flow, and the important fact to know is the fraction of the total for which radiation and conduction are responsible. It is relatively easy to separate these contributions in laboratory experiments. The amount of radiation varies as the fourth power of the absolute temperature whilst conduction is virtually a linear function of temperature. The method to use is to determine the heat flow Q to the test vessel at various temperature levels and then to plot $Q/(T_2 - T_1)$ against $(T_1 + T_2)(T_1^2 + T_2^2)$. The slope of the curve gives the radiation contribution and the intercept the conduction. Although the values so determined are only approximate they are sufficiently accurate to separate the components and to decide which factor must be tackled to improve the performance of the insulant.

III. List of Insulating Materials

A. PROPERTIES OF INSULANTS

In this section the commonly used insulants are listed and are sub-divided into separate categories for easy reference. From the tables the most commonly used insulants are selected for detailed discussions.

TABLE 4.I. Properties of Insulants at Atmospheric Pressure

Insulant	Bulk density kg/m³ $\times 10^{-3}$	a Effective thermal conductivity W/mK°	Particle or fibre diameter mm	Oxygen compatibility	Notes
Powdered Insulants					
Expanded Perlite	50	$2\cdot6\times10^{-2}$	$0\cdot1$–$1\cdot0$	Yes	
Expanded Perlite	210	$4\cdot4\times10^{-2}$	$<0\cdot1$	Yes	
Basic Magnesium Carbonate	130	$3\cdot1\times10^{-2}$	$<10^{-2}$	Yes	
Silica Flour	220	$2\cdot1\times10^{-2}$	$<10^{-4}$	Yes	
Silica Aerogel	80	$1\cdot9\times10^{-2}$	<1	Yes	Agglomerates— ultimate particle size $2\cdot5\times10^{-5}$ mm
Vermiculite	120	$5\cdot2\times10^{-2}$	1–4	Yes	
Calcium Silicate (Synthetic)	360	$4\cdot6\times10^{-2}$	$<10^{-4}$	Yes	
Diatomaceous Earth	60	$2\cdot2\times10^{-2}$	$<10^{-1}$	Yes	
Asbestos	30	$4\cdot7\times10^{-2}$	$<10^{-1}$	Yes	
Fibrous Insulants					
Fibreglass Wool	110	$2\cdot5\times10^{-2}$	10^{-3}	Yes	These contain oil lubricants—must
Mineral Wool	130 to 320	$2\cdot9$ to $4\cdot3\times10^{-2}$	10^{-2} to 10^{-1}	Yes	be below $0\cdot2\%$ w/w for oxygen use.
Rock Wool	160	$3\cdot5\times10^{-2}$	10^{-2} to 10^{-1}	Yes	Mineral and Rock Wools can have
Cellulose Acetate Fibre	43	$2\cdot1\times10^{-2}$	2×10^{-2}	No	large quantities of non-fibrous matter which is not desirable.
Preformed Insulants					
Foam Glass	170	$5\cdot2\times10^{-2}$		Yes	
Calcium Silicate–Asbestos	330	$6\cdot9\times10^{-2}$		Yes	
Resin Bonded Fibreglass	30	$3\cdot2\times10^{-2}$		Yes	If resin content is
Asbestos Board	140	$4\cdot1\times10^{-2}$		Yes	below 4%.
Balsa Wood	110	$4\cdot9\times10^{-2}$		No	
Expanded Polystyrene	15	$2\cdot4\times10^{-2}$		No	
Polyurethane Foam	49	$2\cdot5\times10^{-2}$		No	
Foamed PVC Rigid	60	$2\cdot5\times10^{-2}$		No	
Phenolic Syntactic Foam	310	$5\cdot7\times10^{-2}$		No	
Urea Formaldehyde Foam	50	$3\cdot0\times10^{-2}$		No	
Cork Board	110	$3\cdot2\times10^{-2}$		No	

a All conductivities are for the temperature limits 303 and 90°K.

TABLE 4.II. Properties of Vacuum Insulants

Insulant	Bulk density kg/m³	[a] Effective thermal conductivity W/m °K	Particle or fibre diameter mm	Notes
Powdered Insulants				
Perlite fine	180	0.95×10^{-3}	<0.1	All pressures in
Perlite coarse	64	1.9×10^{-3}	$0.1–1.0$	insulant below 10^{-3} Torr
Silica Aerogel	80	1.6×10^{-3}	$<10^{-4}$	Agglomerates up to 2 mm
Calcium Silicate Synthetic	210	0.59×10^{-3}	$<10^{-4}$	
Calcium Silicate Synthetic	140	0.83×10^{-3}	$<10^{-4}$	
Silica Flour	110	1.6×10^{-3}	$<10^{-4}$	
Silica Flour	200	0.93×10^{-3}	$<10^{-4}$	
Basic Magnesium Carbonate	130	0.71×10^{-3}	$<10^{-3}$	
Kieselguhr	240	1.6×10^{-3}	$<10^{-2}$	
Fused Alumina	2,000	1.8×10^{-3}	<1	
Micro Sepiolite	260	1.6×10^{-3}	$<10^{-2}$	
Ppted Calcium Carbonate	380	2.0×10^{-3}	$<10^{-3}$	
Lampblack	200	1.2×10^{-3}	$<10^{-3}$	
Titania Aerogel	160	0.62×10^{-3}	$<10^{-2}$	All prepared by
Iron Oxide Titania Aerogel	180	0.55×10^{-3}	$<10^{-2}$	Author chemically
Iron Oxide Titania Aerogel	140	0.73×10^{-3}	$<10^{-2}$	
Iron Oxide Titania Aerogel	370	0.63×10^{-3}	$<10^{-2}$	
Fibrous Insulants				
Fibreglass random	50	1.7×10^{-3}		
Fibreglass ordered fibres	240	0.56×10^{-3}		
Mineral Wool				
Reflecting Powder Insulants				
50 : 50 w/w Copper– Santocel	180	0.33×10^{-3}		
43 : 57 w/w Aluminium– Santocel	160	0.35×10^{-3}		Author's work
40 : 60 w/w Aluminium Santocel	160	0.35×10^{-3}		
30 : 70 w/w Copper–Iron Oxide Titania Aerogel	200	0.21×10^{-3}		Author's work
Coprecipitated Copper Iron Oxide Titania	170	0.42×10^{-3}		Author's work
Silica–Carbon	80	0.48×10^{-3}		

[a] All conductivities are for the temperature limits 303 and 77.4°K.

The insulants presented in Table 4.II by no means cover all the available materials but give a useful cross section.

It is obvious from the above table that under vacuum the thermal conductivities are at least one-tenth the values for the same insulant at atmospheric pressure. The other notable feature is that as the bulk density of powder increases the conductivity decreases, whereas with the same powders at atmospheric pressure the conductivity increases. The reason for the decrease under vacuum is that the reduction in radiation transfer due to the increased mass is greater than the increase in solid phase conduction. At atmospheric pressure radiation is swamped by gas phase conduction and adsorption increases the value of the solid phase conduction.

TABLE 4.III. Properties of Reflecting Layer Insulants

Insulant	Layer density layers/cm	[a] Effective thermal conductivity W/m °K	Fibre diameter mm	
0·006 mm Aluminium Foil +0·15 mm Fibreglass paper	20	$3·7 \times 10^{-5}$	$<10^{-3}$	
Rayon Net 2 mm mesh+ Aluminium Foil 0·006 mm	10	$7·8 \times 10^{-5}$	$<2 \times 10^{-2}$	
Nylon Net 2 mm mesh+ Aluminium Foil 0·006 mm	11	$3·4 \times 10^{-5}$	$<2 \times 10^{-2}$	
NRC.2–Crinkled Aluminized Terylene Film 0·006 mm	33	$5·2 \times 10^{-5}$		Author's value
NRC.2–Crinkled Aluminized Terylene Film 0·006 mm	35	$4·2 \times 10^{-5}$		Hnilik (1959)
Dimplar alternating dimpled+smooth aluminized Terylene film	3–12	$4·2 \times 10^{-5}$		Kierulff Electronics Inc. (1966)

[a] All the conductivity values are for temperature limits 303 and 77·4°K.

With regard to the reflecting layer insulants, the values listed in Table 4.III are for laboratory determined thermal conductivities where the material has been packed under ideal conditions with virtually no self-applied loads. Large industrial units do not have such low values of conductivity. The loss rates are always higher than predicted from laboratory values by a factor as high as three, the magnitude of this factor depending upon the size and complexity of the vessel.

The variation of thermal conductivity with temperature for these insulants has been found to fit an equation of the form

$$K = a + b\sqrt{(T_1 T_2)} + c(T_1 + T_2)(T_1^2 + T_2^2) \qquad (4.4)$$

Where K is the effective thermal conductivity, a, b and c are constants, and T_1 and T_2 are the upper and lower boundary temperatures, °K.

Such an equation using appropriate values of the constants fitted laboratory measurements with a precision of $\pm 3\%$ when the upper temperature varied from 200° to 350°K and the lower temperature varied from 77° to 90°K. When some results over the temperature range 4° to 830°K reported by Riede and Wang (1959) were analyzed using equation 4.4, the mean deviation between calculated and observed conductivities was $\pm 7\%$, the largest percentage deviations being for values from 70° to 90°K where the absolute values are extremely low.

This equation predicts the observed phenomenon that the absolute value of the heat inleak to a vessel decreases when it contains liquid hydrogen or helium compared to its heat inleak when it contains liquid nitrogen even though the temperature difference has increased from 220° to 277° and 293° respectively. Although the form of the equation has no apparent theoretical basis, it is simple to apply and only needs determinations at three adequately spaced temperature levels to predict the behaviour over the whole temperature range.

As an aid to high vacuum high polish techniques the emissivities of selected commonly used materials are listed in Table 4.IV. The values are taken from the compilation by Reynolds et al. (1963) which gives a more complete list.

TABLE 4.IV. Low Temperature Emissivities of Clean Metal Surfaces

Material	Emissivity
Aluminium electropolished	0·03 (300°K)
Aluminium electropolished	0·018 (76°K)
Aluminium Commercial sheet	0·05 (76°K)
Brass polished	0·03 (76°K)
Copper commercial polish	0·03 (76°K)
Copper electrolytic polish	0·018 (76°K)
Carefully prepared surface of pure copper	0·008 (90°K)
Gold	0·01 (76°K)
Stainless steel	0·048 (76°K)
Silver	0·022 (300°K)
Silver	0·008 (76°K)
Tin 0·001 in foil	0·013 (76°K)

B. ASSESSMENT OF SELECTED INSULANTS

From Tables 4.I, 4.II and 4.III, insulants of particular interest are discussed in detail giving both their advantages and disadvantages.

Advantages	*Disadvantages*

1. *Expanded Perlite*

Low thermal conductivity.	Friable.
Low density.	Breaks down easily during handling
Cheap.	with increase in bulk density.
Ready availability.	When crushed it flows with difficulty.
Dimensional stability when in static	Dusty, a throat irritant.
installation.	Abrasive.
Non-inflammable.	Cannot be used to insulate moving
Not toxic.	parts.
Not hygroscopic.	When maintenance on plant is required
Easy to install.	all insulant must be removed from space.
Flows readily when new and is easily	Can only be transferred a limited num-
conveyed pneumatically.	ber of times before needing replacement.
Can be recovered even after soaking in	Non-elastic in compression.
water.	Needs protecting from atmosphere
Does not cake in use.	during use.
Can be produced on site saving transport	
and packaging costs.	
Useful in vacuum insulation; it is easy	
to install, in a fabricated vessel.	
Easy to evacuate when used in vacuum	
insulation.	

Typical Trade Names: Brelite, Cecaperl, Perlox, Pulvinsul, Ryolex.

2. *Fibreglass*

Low thermal conductivity.	Expensive.
Fine fibres comparatively easy to handle.	Needs careful even packing to obtain
No "shot" (non-fibrous large particles).	best performance, therefore expensive to
Dimensionally stable.	install.
Elastic in compression, it allows for	Difficult to install in vacuum insulated
contraction and expansion.	vessels.
Not hygroscopic, can be recovered after	Needs protecting from atmosphere
soaking in water.	during use.
Readily available.	

3. *Mineral or Rock Wool*

Cheap.	Because of high packing density needed
Easy to obtain.	it is more costly on a volume basis than
Dimensionally stable.	expanded perlite.
Can be elastically deformed slightly to	Can contain high shot content.
take up contraction and expansion.	Unpleasant to handle.
Not hygroscopic.	Requires expensive labour to install.

10

3. *Mineral or Rock Wool* (*contd.*)

Advantages (*contd.*)	*Disadvantages* (*contd.*)
Free from fine dust, can be used to insulate moving machinery. When maintenance needed only section near affected parts needs removing.	Careful control over supplier needed to obtain best quality as it can contain hydrolysable sulphide which can cause corrosion and it normally has organic agents to prevent dustiness. There must be a careful specification for the amount of the organic matter in the wool and must not be more than 0·2%. Needs protecting from atmosphere during use.

Typical Trade Names: Banroc, Eldorite, Fluffex, Insulwool, Rocksil, Stillite.

4. *Silica Aerogel*

Lightweight low thermal conductivity. Extremely free flowing, dimensionally stable when dry, does not shrink when larger particles are broken down. Slightly elastically compressible. After initial compaction it will not settle due to vibration. Not hygroscopic. Useful as the base for reflecting powders. Easy to evacuate in vacuum insulation.	Very expensive, the price is approx. ten times that of expanded perlite. When soaked in water it forms silica gel on drying and increases in bulk density by a factor of 10 and then is useless as an insulant. Needs protecting from atmosphere.

Trade Names: Aerosil, Carb-o-sil, Santocel.

5. *Basic Magnesium Carbonate*

Relatively cheap and readily available. Good insulating value. Dimensionally stable. Very good performance in vacuum insulation.	Slightly hygroscopic. At atmospheric pressure it cakes easily and when wet sets almost like concrete when it is difficult to regenerate. Normally contains large quantities of moisture making it difficult to evacuate when used as a vacuum insulant. Dissolves in acidic media. Needs protection from atmosphere.

6. *Vermiculite*

Cheap. Lightweight. Readily available. Dimensionally stable. Not hygroscopic. Not toxic, strong structure does not break down on handling.	High thermal conductivity which under the cryogenic temperature gradients is 3 to 5 times that of expanded perlite due to the large convection currents set up within the interstices and the high solid phase conduction.

Advantages (contd.)	*Disadvantages (contd.)*

7. Foamglass

Strong.	Expensive.
Readily available.	Needs skilled labour to install properly.
Closed pores, not permeable to water.	Needs external sealing to prevent water
Can be used as a support for the inner	ingress between the blocks.
tank.	Rather high conductivity.
Dimensionally stable.	
Not hygroscopic.	
Not inflammable.	

8. Foamed Plastics

Lightweight.	Relatively expensive to install.
Low thermal conductivity.	Needs skilled labour to install properly.
Cheap.	Allowance must be made for contraction
Readily available in many grades.	and expansion of plastic which is very
Strong, can be used as support for inner	great.
vessel.	Upper limit of temperature as low as
Available as closed pore insulants	80°C for expanded polystyrene, others
which are not permeable to water.	are slightly higher.
Certain grades can be foamed *in situ*,	Inflammable.
viz. the Polyurethanes making installa-	Although non-inflammable grades are
tion easy.	made these refer only to their burning
	in air.
	A slight oxygen enrichment will make
	all these inflammable.
	For example, a rigid foamed PVC which
	will not burn in air burns with the
	ferocity of cordite in an oxygen atmos-
	phere.
	These plastics are thus prohibited for
	liquid oxygen, and may be used with
	liquid nitrogen only if a nitrogen purge
	is introduced alongside the inner wall to
	prevent enriched air soaking the insu-
	lant.

Examples: Expanded ebonite, expanded rigid and flexible polyvinyl chloride, expanded polystyrene, and expanded rigid and flexible polyurethane.

9. Fibreglass Paper–Aluminium Foil Combination

Very low thermal conductivity.	Needs high vacuum for best perform-
Easily applied to cylindrical shapes.	ance.
Thermal conductivity does not increase	Limited numbers of suppliers for special
with compression as rapidly as other	paper needed.
forms of layered insulants.	Paper is relatively weak and is easily
Can be baked to high temperatures,	torn,

9. *Fibreglass Paper–Aluminium Foil Combination* (*contd.*)

Advantages (*contd.*)	*Disadvantages* (*contd.*)
about 250°C, to outgas insulant and vessel ensuring a good vacuum. Not inflammable in oxygen.	Will not resist appreciable vibration. Great expertise needed to apply properly. Difficult to fit round projections, without increasing heat flow.

10. *Synthetic Fibre Net–Aluminium Foil Combination*

Very low thermal conductivity. Organic fibre easily handled and very strong. Dimensionally stable. Will resist vibration. Easy to apply to cylindrical and simple shapes. Can be heated adequately to outgas. The organic nets give as good performance as the fibreglass but uses fewer reflecting layers. Easily obtained from a large number of suppliers.	Needs high vacuum for best performance. Difficult to fit round projections in interspace without increasing heat flow. Upper temperature limit of 150–200°C. Requires great care in application to obtain best performance. Inflammable. Restricted to nitrogen, argon and non-oxidizing liquids.

11. *Crinkled Aluminumized Terylene*

Only single layer being handled. Very low thermal conductivity. Can be fitted round obstructions easily. Strong. Easy to handle. Low lateral heat flow. Easily pumped down because it has already been outgassed for initial aluminizing.	Needs handling twice to apply, once to crinkle, once to apply. Does not resist compression. Of the superinsulants this material has the greatest increase of conductivity with compression. Needs high vacuum for best performance. Inflammable. Restricted to nitrogen, argon and non-oxidizing materials. Non-inflammable material is available e.g. aluminized P.T.F.E. but this is very expensive.

IV. Vessel Supports

In an insulated vessel the inner container must be kept apart from the outer casing. This may be achieved by the use of a load-bearing material, which may be the insulant itself, or a material introduced solely for this purpose. The problem of supports has become more acute in the case of high performance insulants not only because the heat inleak has been reduced but simultaneously the thickness of the insulant space in which to fit the supports has been considerably reduced. The main requirement of the supports is to be sufficiently strong to withstand all the stresses to which the tank is likely to be subjected and at the same time to contribute little to the heat inleak.

The two requirements are conflicting and a compromise has to be reached. In order to assess different materials a figure of merit has been established which is the numerical value of the maximum usable strength divided by the thermal conductivity, the larger the figure the better it is as a support material.

There are two basic types of support; those in which the active member is under tension and those in which the member is in compression.

The ultimate tensile or compressive strength is only a rough guide, the strength factor that has to be considered is the yield stress since there must be no permanent progressive stretch in the support materials. For most metals the design stress is approximately two-thirds of the yield-point but for plastics the maximum allowable stress is much less than this ratio because of creep.

The factors which affect heat flow are the cross-sectional area (fixed by the design stresses and loads), the thermal conductivity and the length of the supports. Usually it is desirable that the heat-leak contributed by the supports should account for less than 20% of the total, but this is a variable fraction depending on circumstances. The main questions to ask are "how much does the heat leak via the supports cost" and "how reliable are the supports"? Reliability is in fact the first consideration and only after this has been established should the cost aspect be analyzed.

Of the two types the tension supports are the easiest to design. The shape of the cross section, the support length, and the rigidity of the material are all vital factors to be considered in the mechanical design of compression supports, whilst for tension supports only the cross-sectional area is significant. Another advantage is that in general, both for metals and plastics, the tensile strengths are higher than the compressive strengths.

To take full advantage of tension supports they must be as long as possible and several ingenious methods of increasing their lengths have been proposed. Two examples are shown in Figs. 4.2 (a) and (b).

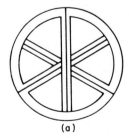
(a)

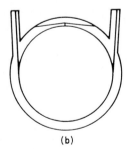
(b)

Fig. 4.2.

In the case of compression supports account must be taken of the bending moment as well as compression due to the difficulty of aligning the supports with the stresses. These supports are useful in vessels where space is limited since they are usually quite short. An interesting variation of the compression supports is the so-called "stacked washer" technique as proposed by Mikesell and Scott (1956) and further developed by Probert (1967). These utilize the high contact resistance in vacuum conditions to reduce heat flow. By stacking thin foils of a hard metal one can reduce the heat inleak to as little as 2% of

TABLE 4.V. Properties of Tension Support Materials

Material	Ultimate tensile strength mega N/m^2	Yield stress mega N/m^2	Thermal conductivity $W/m\,°K$	Figure of merit
Metals				
Copper annealed		82·7	475	0·104
Brass	412		100	1·65
Aluminium 2024		379	81·3	2·80
Aluminium 7075	606	482	86·5	3·35
Stainless steel (annealed)		251	10·2	14·8
Titanium	825	586	15·8	22·1
35 N_i–50 F_e–14 Cr Alloy	702		12·6	22·3
K Monel	up to 1,380	689	17·1	24·2
Hastelloy B		448	9·35	28·7
Stainless steel drawn wire		1,030	8·8	70·3
Titanium Alloy (4% Al–4% Mn)		1,000	5·87	102
Inorganics				
Fibreglass	1,720		0·92	750
68% Fibreglass 32% Resin	378		0·36	420
Organics				
Nylon high tenacity	606	138	0·245	247
Terylene (high tenacity Polyester)	730	138	0·15	487
Fortisan 36	1,070		0·20	535
Araldite 6060	830		0·17	490

Notes: The figure of merit is calculated as follows:

$$\left(\frac{\text{Design stress in mega } N/m^2}{\text{Thermal conductivity in } W/m\,°K}\right).$$

The Design stress is taken as 60% of yield stress or 40% of ultimate tensile strength for metals and inorganics, and 10% of ultimate tensile strength for organic materials.

Sources for thermal conductivity figures are R. L. Powell (1963) and "Modern Encylopedia of Plastics" (1959).

that of the parent metal. The technique is not straightforward since factors such as the cleanliness of the material and the weather conditions during the preparation greatly affect the heat flow, which can vary by as much as 100% even for the same assembly. From a study of Probert's development (1967) it is believed that this technique will be best suited to large vessels where heavy loads are being supported because of the bulky framework needed to hold the washers in place.

A list of likely support materials is included in tables 4.V and 4.VI. In table 4.V copper is included to show how wide the range is in merit rating for materials, the range varying almost 10,000-fold between copper, the worst material, and fibreglass, potentially the best material. No figures have been included for stacked washer techniques because these are so variable.

TABLE 4.VI. Properties of Compressive Support Materials

Material	Compressive strength mega N/m^2	Thermal conductivity W/m °K	Figure of merit
Inorganic mica-glass mixture	240	3·76	25·5
Polycarbonates	76	0·19	40·0
Glass		1·0	
Silica	1,130	1·4	320
Chemical stoneware	550	1·21	180
Alumina fused	2,890	48·0	24·1
Titanium carbide	3,440	17·1	80·6
30% TIC–70% Al$_2$O$_3$	1,900	1·04	730

$$\text{Note: Figure of merit} = \frac{\text{Design stress}}{\text{Thermal conductivity}}.$$

Design stress is 40% of compressive strength for inorganics and 10% of compressive strength for organics.

From Table 4.V the best materials would appear to be fibreglass alone or bonded with resin and other organic fibres. Fibreglass on its own has many problems, mainly those of end fixtures to secure the fibreglass to the inner and outer vessel, even tensioning of all strands and chafing because of the high friction coefficient of glass on glass. Resin-bonded fibreglass, although it has a somewhat lower strength, is easier to attach. The organic fibres have the drawback of stretching under load and some creep. Their potential figure of merit is so high that one can use lower stresses and still good performance, hence they are worth considering. End fixing is a problem but less so than with the plain fibreglass.

The compression materials in Table 4.VI show good potential, having performance indices as high as the organic fibres, but the best materials suffer because they are very brittle and hard. Unless extreme care is exercised in their use it is easy to get all the load concentrated at one point which will cause premature failure. The failures in these cases are catastrophic because if only one of the supports shatters the others just fall out of their places and the inner vessel is then completely unsupported. With well designed tensile members the supports tend to stretch or deform before complete failure and the remainder still offer some support to the inner.

V. Vacuum Techniques

It will be noted from Tables 4.I, 4.II and 4.III that vacuum insulation offers large reduction in heat inleak, the result being that it has become a major technique in liquefied gas storage vessels for sizes up to 100 tonnes. Apart from mechanical consideration the basic requirement in vacuum insulated vessels is to obtain and maintain the pressure in the insulating space below the level which gives the desired heat inleak. For powders the pressure is 10^{-2} Torr and for reflecting layer types 10^{-4} Torr.

There would be few problems if vessels could be made completely vacuum-tight because once pumped down nothing more would have to be done. However, no vessel is leak tight and the best that can be achieved is to reduce the leak rate to a low enough value to maintain the pressure rise within the allowable limit. The leak rate can be specified in many ways and the most logical one is the absolute rate of gas flow into the interspace. A convenient unit is the lusec which is the amount of gas required to raise the pressure in a 1 l volume at a rate of 1 micron Hg (10^{-3} Torr)/sec. The gas rate this represents is 4·72 ml at atmospheric pressure/hr.

Normally it is desirable to aim for a minimum life of two years (twice the normal guarantee period) before servicing is necessary and this means that the maximum leak rate for the following types of vessels having a 10 l interspace are:

(i) For a high vacuum high polish with no adsorbent getter where the total pressure rise is 1 micron it is $1·37 \times 10^{-7}$ lusec.
(ii) For a superinsulated vessel with no adsorbent getter where the total pressure rise is 10 microns it is $1·37 \times 10^{-6}$ lusec.
(iii) For a powder insulated vessel with an adsorbent getter where the total pressure rise is 100 microns it is $1·37 \times 10^{-5}$ lusec.

These are all very low leak rates needing specialized equipment to detect. They refer to the sum total of all the leaks in the vessel which could be made up of a single or many small leaks. Because of the difficulty in obtaining vessels with such a degree of leak tightness (except in the case of all glass Dewar vessels) the use of adsorbent materials in the interspace, cooled by the fluid being stored, is very beneficial. Dewar (1905, 1906) was the first to use charcoal to achieve the low pressures he desired. The use of such adsorption getters greatly increases the allowable leak rate and makes manufacturing and testing much easier. As little as 10 g of charcoal in a 10 l interspace will allow the leak to increase to $2 \cdot 42 \times 10^{-4}$ lusec for case (i) above; to $2 \cdot 9 \times 10^{-3}$ lusec for case (ii) and to $7 \cdot 3 \times 10^{-1}$ lusec for case (iii). These figures show how effective such materials are.

A. PUMPING AND MEASURING DEVICES

The range of pumps and measuring devices is extensive and only a brief survey will be given.

1. Pumping Devices

Both oil immersed rotary piston pumps and diffusion pumps are used. The former are for pressures down to 10^{-2} Torr and with these gas ballast is an essential feature when condensible vapours are present in the interspace. Diffusion pumps work below 10^{-1} Torr and act as first stage boosters to raise the pressure high enough for the rotary pumps to operate.

With the combined system pressures as low as 10^{-10} Torr can be achieved, although 10^{-5} Torr is sufficient for practically all insulation.

The rotary pump sizes vary from very small laboratory units with a swept volume of 30 l/min to large industrial units with a capacity of over 3,000 l/min. There are diffusion pump sizes to match the rotary sizes and reliable performance figures are given in the manufacturers' catalogues.

Other pumping devices having high pumping speeds but limited capacity are the adsorption types cooled in liquid nitrogen, or the liquid hydrogen or helium cooled cryopanels on which other gases are frozen.

2. Measuring Devices

The main types of pressure measuring device are as listed below.

a. The McLeod gauge. By far the most reliable where non-condensible gases are being measured. The type of gas, providing it cannot condense at room temperature, does not affect the calibration. The range of measurements is from 10 to 1×10^{-5} Torr; small portable versions are available.

b. The Pirani/thermocouple gauges. Portable and easy to use but can give false readings in the presence of draughts. Their response also varies with the gas in the interspace, being a function of the thermal conductivity. They are useful in giving an approximate indication of the pressure and are adequate for measuring the pressure in the insulation space. The range over which they operate is between 1 and 10^{-3} Torr.

c. The Ionization gauges. Useful below 10^{-2} Torr. They are portable but, like the Pirani, can give false reading due to variations in gas compositions.

10*

In some cases it is difficult to initiate ionization and virtual zero readings are recorded when the pressures are in fact quite high.

3. Leak Detection Methods

This is a crucial subject, and several methods of leak detection are listed below.

The soap bubble test. By use of a positive pressure in the interspace, this method is relatively crude and is used as an initial test to find the very large leaks. Special clear bubbling solutions are available which gives sensitivities down to 10^{-3} lusec. The method cannot be used for cold vessels.

The pressure rise method. This procedure is certain but time consuming, weeks being necessary for low leak rates. Outgassing can interfere with method by indicating a leak when there is none.

Halogen leak detection. This method uses Freon as test gas and is fairly rapid, reasonably sensitive, convenient, but must be used in a solvent free atmosphere (no trichlorethylene). The detector is easily poisoned by excess halogen.

Infra-red detection. This method uses nitrous oxide and has about the same sensitivity as the halogen leak detection and solvents do not interfere with the tests. The equipment is bulky and fairly slow in response.

Double Pirani leak detection. Using water vapour or hydrogen, this method has about the same sensitivity as the preceding methods. It can be interfered with by other contaminants.

The Hydrogen Palladium using hydrogen. This method has increased sensitivity because the selective nature of the Palladium diffusion cell reduces interference from other gases. It will detect leaks down to 10^{-4} lusec.

The Mass Spectrometer. This is the most sensitive and versatile method and uses hydrogen or preferably helium. It has been developed using helium gas as the tracer and its sensitivity is governed by the helium in the atmosphere giving a variable background. When coupled direct to the vacuum space the sensitivity is at least 10^{-7} lusec and when used with the sniffer probe attachment for locating the leaks the sensitivity is 10^{-4} lusec. The equipment is the most expensive but because of its reliability it is almost universally used in leak detection.

B. PRACTICAL PROBLEMS ASSOCIATED WITH EVACUATION

After the vessel has passed its leak tightness test, it should be connected to the evacuation system using the largest appropriate size of pipeline and joints.

A particularly acute problem is the outgassing of the vessel walls and insulant material. The most common contaminant is water and it is essential that the bulk of this be removed before evacuation is attempted. To illustrate this point consider $1 m^3$ of insulation containing $100 kg$ of a fine perlite with 0.9% adsorbed moisture (it can contain up to 5% of moisture if left exposed to the atmosphere). This water as vapour at 0.1 Torr, a typical pumping pressure, would occupy 9,000 m² and it would take a rotary pump of 3,000 l/min capacity 50 hr to remove it. A smaller pump (50 l/min capacity) would evacuate the same volume of perfectly dry insulant down to a pressure of 0·05 Torr in 6 to 18 hr, which illustrates clearly the importance of dryness in vacuum insulation.

A way to speed up pumping where moisture is present is to use a liquid nitrogen trap in the evacuating line. The governing rate is then not the size of the pumps but the conductance of the lines and insulant itself. Heating the vessel also helps because it increases the vapour pressures of the contaminants allowing them to be removed at the highest possible pressures. In practical terms pumping times of 1 day are to be aimed at as longer times tie up valuable equipment and space.

1. *Vacuum Powder Insulation*

Fine powders are difficult to pump down due to high flow resistance within the powder itself, and further resistance within the filter which must be employed. These difficulties can be overcome by using a number of large area filters connected to the vacuum space at a number of strategic points. The size and distribution of the filters is dependent upon the type of vessel being used and a little care in the initial choice will pay dividends in reducing pumping times.

Outgassing from the powder presents some problems but these are not too significant. Hydrogen outgassing from the metal presents no problem because the amounts involved do not raise the pressure by more than 10^{-2} Torr.

2. *Reflecting Layer Types*

Included in this section are the high polish high vacuum types. Here the problems are not so much gas flow through porous media, but subsequent outgassing. For liquid oxygen and nitrogen and higher boiling mixtures the hydrogen outgassing is a particularly acute problem since the normal adsorbent getters will not adsorb it at the prevailing temperatures, 77–110°K. The hydrogen problem has become particularly acute with the advent of stainless steel and aluminium construction.

The only satisfactory method of removing hydrogen is prolonged baking at temperatures in excess of 400°C. Days are required to outgass satisfactorily at 400°C and weeks would be needed at 200°C. Another approach is to insert an easily reducible metal oxide in the interspace after the initial evacuation. If it is introduced before evacuation the organic vapours will decompose the oxide and it loses its effectiveness. The technique involves the use of ingenious devices such as the one used by Union Carbide (British Patent 921,273) which consists of palladium oxide sealed in a glass ampoule placed in a side pocket on the vessel. After evacuation the ampoule is crushed by deforming the pocket releasing the oxide. Obviously other means can be used depending on the ingenuity of the designer.

One additional problem particular to layered insulants is the removal of gas from between the layers of insulant and is acute where the number of layers exceeds about 20. Down to about 10^{-2} Torr there is no problem but below this pressure where molecular conductance is the rate controlling factor, the problem becomes quite serious for large vessels, where path lengths of 1 m are quite common. Attempts to increase the rate of pumping by perforating the foil are successful but have to be applied with extreme care since every perforation forms a window with an effective area several times larger than the perforations. Another technique is to apply heat to the insulant to increase the mobility of the gas molecules.

3. Use of Getters

Getters serve two main functions:

(a) to prolong the service life of a vacuum insulated tank;
(b) to act as a pump so that the pressure to be reached during the initial evacuation is orders of magnitude higher than the ultimate required.

The increased pressures that can be tolerated are easily calculable. For instance, with a reflecting layer insulant the desirable pressure is below 10^{-4} Torr and the maximum allowable pressure is 10^{-3} Torr. Without a getter the former pressure would have to be achieved during the initial evacuation yet with only 10 g of charcoal in a 10 l interspace the initial pressure need only be 0·04 Torr because the pressure when cold will be well below the 10^{-4} Torr needed. The only effect of the initial high pressure will be to reduce the service life between overhauls by about 5% based on an adsorptive capacity at 10^{-3} Torr and $90°K$ of approximately 1 ml of air/g of charcoal.

Practically any porous material will act as a getter but many have capacities too low to be of practical interest. A list of possible materials has been published by Beher (1956). The most useful of these materials are charcoal, silica gel and molecular sieve 5A.

In use they are placed in good thermal contact with the inner vessel so that the getter is cooled to near the temperature of the contents. The presence of moisture on adsorbent getters reduce their capacity for air at low temperatures. For example, 5% by weight of water on silica gel reduces its adsorptive capacity for air at $-183°C$ by 50%. Most adsorbents are hygroscopic, therefore there is a problem to keep them dry once they have been regenerated and placed in position.

Other getters, useful mainly with layered insulants and plain vacuum vessels are the chemically active metals which are freshly deposited after evacuation. Deposition of the film is effected by electrically fusing a wire of the material in the interspace.

Getters when used properly considerably reduce the failure rate on vacuum insulated vessels and because they can cost so little to apply are a cheap form of insurance. A point to be stressed with them is that they are not a replacement for good constructional techniques but are merely a safeguard against accidental small undetected leaks.

VI. Practical Applications

The types of units are subdivided into the following categories:

A. SMALL PORTABLE VESSELS

This range covers the Dewar vessels up to about 500 l capacity which are used in laboratories or small workshops and are illustrated in Fig. 4.3. The various types of fabrication include:
1. All glass construction.
2. All metal construction for the storage and transport of liquid oxygen, nitrogen and argon.
3. Special dewars for storage of helium and hydrogen.
4. Special types such as cryostats for holding devices used at low temperatures, e.g. infra-red detectors, masers and superconducting magnets.

1. *All Glass Dewars*

Glass dewars are normally evacuated to about 10^{-5} Torr and the majority are silver plated to reduce heat inleak. The most common types are of open-ended cylindrical shape, and common sizes vary from 4 to 30 cm internal diameter and up to 2 m long.

For use at liquid nitrogen or oxygen temperatures very low loss-rate is usually of secondary importance compared to convenience and cheapness. For experiment using liquid helium and hydrogen double dewars, with the inner dewar immersed in a liquid nitrogen bath, are in common use. They are, however, being replaced by metal cryostats.

2. *Metal Dewars*

Historically the first metal dewars were soldered copper spheres insulated by the high-vacuum high-polish technique. These units are still being manufactured and the common size range is from 3 to 100 l liquid capacity. They are being superseded by cheaper and more efficient all-welded stainless steel or aluminium dewars. The vessel used most commonly in the U.K. is the 25 l size which, with its protective case, weighs about 25 kg empty, and has a loss rate of about 6 to 8% per day with liquid nitrogen. Some recently constructed all-welded dewars have either aluminium or stainless steel inner vessels and mild steel outer vessels. They have a reflecting layer insulant in the vacuum space. The 25 l size weighs between 10 and 15 kg empty and has a loss rate of $1\frac{1}{2}$ and $4\frac{1}{2}\%$ per day. The necks are shorter and wider than those of the older copper construction, and it is the size of the neck which mainly affects the loss rate. With this size of dewar the loss rate is generally of less importance than the convenience of having a wide neck and the reduced cost.

Fig. 4.3. A selection of small portable vacuum insulated, cryogenic vessels.

a. Soaker vessels.
b. 250 l liquid helium Dewar.
c. Liquid nitrogen container.
d. 50 l liquid helium dewar.
e. 35 l biological storage vessel.
f. 25 l superinsulated spherical vessel.
g. Twin-necked horizontal Dewar.
h. 250 l vessel for biological specimens.

Specialized dewars are required for storing and transporting frozen biological specimens, particularly bull semen. For this purpose vessels of all stainless steel construction have been developed. They all have wide necks for easy access, and have a reflecting layer insulant. Two common sizes for field use are the 10 and 35 l sizes with necks ranging from 5 to 11 cm in diameter. The loss rate is from 3% downwards for these vessels. Larger vessels of 250 l capacity upwards are used for bulk storage of specimens and these have neck sizes up to 50 cm in diameter.

Other specialized vessels are the vertical cylindrical multi-purpose containers of approximately 150 l liquid capacity, from which either liquid or dry gas at pressure of up to 15×10^5 N/M^2 (15 atm) can be withdrawn. They are all insulated with a reflecting layer insulant and the loss rate is approximately 1% per day. Although they are mobile they need special trolleys for easy handling.

Larger vessels are cylindrical in shape and are mounted on castors for mobility. The loss rates from these vary according to size. A 500 l size with twin necks has a loss rate of about 1% per day.

3. *Liquid Helium and Hydrogen Storage*

Traditionally liquid helium and hydrogen have been stored and transported in liquid nitrogen shielded copper dewar vessels. The sizes of these vessels ranged from 3 to 100 l with the 17 l size being the most popular. For this size the loss rate from liquid hydrogen is between 0·2 and 0·5% per day and with helium between 1 and 2% per day. A charged vessel weighs 60 kg. The nitrogen loss rate is about 10 to 15% per day of its capacity and if the level is allowed to fall appreciably the loss rate on helium increases rapidly— to over 200% of contents per day if all the nitrogen disappears.

New developments in helium storage, recovering the refrigeration of the cold escaping vapour, have made liquid nitrogen shielding unnecessary. By making use of a combination of gas-cooled shields and reflecting layer insulants, loss rates comparable to the nitrogen shielded dewars have been obtained. The sizes of the mobile tanks are from 25 to 500 l sizes. The most popular being the 100 l size, and this loses about 1 to 2% per day, yet only weighs 110 kg, with a full charge. The equivalent in copper construction weighs at least 200 kg. Air transportable vessels are also available in the 250 l size and upwards.

4. *Miscellaneous*

In this section is a small selection of the varied types of dewars and containers built for special purposes.

Of interest to the laboratory user who wants a small temporary storage container for liquid nitrogen is the type built from foamed plastics such as the expanded polystyrene flowerpots. The loss rate from these is only slightly higher than with a clear glass dewar yet they are more robust and a lot cheaper. These can be custom made to any size as described by Vedernikov (1969).

The metallic open-ended dewars (the so-called soaker vessels) made from stainless steel are gradually replacing the all-glass construction mainly because they are so robust. They have reflecting layer insulation and the loss rates are only slightly higher than the silvered glass dewars. The main heat inleak is by radiation down the neck opening, and plastic plugs have been used to cut this down. The sizes of such vessels range from about 8 cm to over 30 cm internal diameter and lengths from 16 cm to over 150 cm.

Large liquid helium cryostats with wide necks of between 15 and 50 cm diameter have been built for use with superconductivity devices. Some use liquid nitrogen shielding to reduce loss rates to acceptable values, but gas shielded helium cryostats have been built where the loss rates are down to 200 ml of liquid/hr.

This range covers vessels which are factory fabricated and either installed on site as a complete unit or used as road tankers for transporting liquefied gases and are illustrated in Fig. 4.4. Capacities range up to 100 tonnes of liquid.

Fig. 4.4. Large transportable cryogenic vessels, illustrated for the duty of transporting and storage of LNG. Note the insulated transfer hose.

1. *Storage Vessels*

The newer vessels of this class have vacuum insulation and range up to 100 tonnes liquid capacity. Their function is to act as liquid reservoirs to provide continuity of supply and cut out the need for frequent deliveries by road or rail tankers.

These vessels supply two types of user. One is primarily interested in the liquid for its refrigerating capacity, such as in shrink fitting of axles in railway wheels or in the freezing of foodstuff. The other requires the gas, but in quantities where supply in cylinders is inconvenient.

Both types are vacuum powder insulated and have stainless steel inner with mild steel outer shells. The main differences between them are in the thickness of the inner shell and the external pipework. The former have thinner walled inner vessels because they work at low pressure of up to 2 kg/cm² gauge whereas the latter operate at up to 15 kg/cm². They range in size from about 500 l liquid capacity to over 40,000 l capacity.

The approximate performance of two typical examples is as follows:

(i) Liquid capacity   4,000 l
     Outer diameter   2 m
     Empty weight   3·5–4 tonnes
     Loss rate   0·3 to 0·6% per day
(ii) Liquid capacity   20,000 l
     Outer diameter   2·5 m
     Empty weight   14 tonnes
     Loss rate   0·1 to 0·3% per day

2. *Road and Rail Tankers*

With road and rail tankers the main problems are the overall weight of the unit in relation to the liquid being carried, and the provision of sufficient strength to withstand buffeting. Road tankers were traditionally copper or brass spheres insulated with a powder at atmospheric pressure. To save weight modern tankers are of all aluminium construction and have vacuum powder insulation.

The special conditions prevailing in the U.K. do not require tankers with loss rates much below 2% per day because few journeys last longer than 8 hours. Convenience in use and layout of pipework, which minimizes transfer losses, are equally important considerations. However, in the United States, Australia, etc., low loss rate is more important.

The size of modern road tankers varies between 5,000 and 20,000 l liquid capacity, a typical example being one which holds 15 tonnes of liquid oxygen (approximately 13,000 l) and weighs 28 tonnes overall. These have vacuum powder insulation.

For the transport of liquid methane in the U.K., tankers of about 20,000 l liquid capacity insulated with foamed plastics are being used.

Rail tankers have a capacity of about 50,000 l, which means they hold approximately 50 tonnes of liquid oxygen and have an overall loaded weight of 100 tonnes. They are made with stainless steel inner and mild steel outer shells and are insulated with vacuum powder. Their loss rates are below 0·6% per day.

C. LARGE STATIC INSTALLATIONS

These are large site-fabricated units for the bulk storage of the liquefied gases with capacities of 4,000 tonnes of liquid or more. Included in this category are the "cold-boxes" which insulate the cold parts of plant for the production of liquid oxygen, nitrogen, methane, etc.

The large oxygen and nitrogen storage tanks are all above-ground and consist of either an aluminium or stainless steel inner shell with a mild steel outer one which are insulated with expanded perlite. Because of the trouble with some of the earlier smaller cold boxes for air separation plants due to frost heave, the bases of all large static units are now mounted on concrete pillars to allow air circulation underneath.

The large size of such tanks poses problems due to the shrinkage of the inner vessel during cooldown which is such as to leave a potential gap of between 1 and 3 cm if the powder did not move. After several cooling and warming cycles the powder will compact so much that it will cause damage to the inner vessel. Various devices, such as the use of resilient blankets of compressed fibreglass attached to the inner vessel, have been proposed so as to take up the contraction of the inner shell. A typical example of this type of vessel is one holding 3,000 tonnes of liquid oxygen. It has an outer diameter of 18 m and its loss rate is about 0.1% per day.

Two points need particular attention. One is the need to achieve leak-tightness of the outer vessel to prevent moisture ingress, and the other, applying only to liquid nitrogen tanks, is to provide a nitrogen purge in the insulant space. Similar considerations apply to liquid hydrogen storage tanks where only helium or hydrogen in the interspace is permissible.

It is worth noting that cryogenic tanks on rockets are not insulated except for a thin layer of polyurethane foam on tanks containing liquid hydrogen. This layer is to prevent solid air freezing on to the surface and causing a large increase of weight.

For the storage of methane the insulant space needs purging either with nitrogen or with methane itself to prevent explosive mixtures forming. In the latter case an interesting development by the Chicago Bridge and Iron Company (1968) is their suspended roof insulation where the inner vessel has an open top, and a flat tray carrying the top insulant is suspended from the outer vessel. Thus the evaporated gas freely vents the whole insulation space.

The fact that in recent years the quantity of methane being handled as liquid at single sites has increased enormously and that it will shortly be much greater than either liquid nitrogen or oxygen has led to the search for cheaper reservoirs than those built by the conventional techniques described above. One development has been to use inground storage tanks where the container walls as well as the insulation is the frozen soil itself thus eliminating the need for separate inner and outer container.

It was first proposed by Conch International Methane Ltd. (1960) and briefly the method consists of the following; selection of a suitable soil structure preferably marshy or silty soil, freezing a ring of the required hole size and depth, excavating the central unfrozen area, covering the hole with an insulated roof so that if forms a gas tight seal with the frozen ring and filling it with liquid. The concept is simple but to carry it out in practice there are many practical problems to solve one of which is the roof support ring which has to be flexible and strong enough to cope with frost heave in the initial stages of freezing and then later shrinkage as the tank is cooled to its final working temperature. The techniques to overcome this and other problems met with during the construction of the inground storage tanks at Canvey Island are described by Ward and Hildrew (1968) and Ward and Egan (1969).

The first storage tank in the United States was a small 10 m diameter hole built at Lake Charles, Louisiana in 1961. This was successful but since then some full-scale holes in the ground have been failures. The one at Hopkington, Massachusetts for the Tennessee Gas Pipeline Company (approx. 50 m diameter and 53 m deep) could only be filled to one-sixth of its capacity before the boil-off rate became as large as the liquefaction capacity. A similar tank for Transcontinental Gas Pipelines (1968) at Jersey Meadows had a high boil-off rate when only half full. These have since been abandoned and above ground tanks have been built. The only completely successful in-ground liquid methane storage tanks are at Arzew in Algeria.

There are three main drawbacks with this simple technique of which the prime one is that it is vital to have exactly the right soil structure and this may not be available where the storage is needed. Another is that the loss rates will be at least twice that from conventional tanks and will mean that, where the main usage of liquid is for peak shaving, the size of the liquefier will have to be 20% or more greater than if conventional tanks were used.

The third one is the enormous cooling down load. At least 10 m thickness of soil with a density of 1,500 kg/m^3 or more has to be cooled down whereas with conventional vessels only the inner metal of about 10 mm thickness plus 1 m of insulant having a density of 50 to 100 kg/m^3 have to be cooled. This large cooling load means that it takes ten to twenty years to reach full equilibrium which is to be compared with two weeks to one month for powder insulated vessels.

The development reported by Duffy and Shoupe (1966) could remove most of the drawbacks and make this technique practicable. They use a flexible liner in the hole which can be combined with conventional insulation to give the advantage of cheaper hole in the ground construction costs without the above-mentioned drawbacks.

1. Cold Boxes for Air Separation and other Cryogenic Plants

Cold boxes present many special problems due to the complex nature of the pipework (see Fig. 4.5), the fact that there may be moveable parts requiring a separate insulant, and the need to provide access points for the maintenance of items such as valves.

The basic construction of cold boxes is the same as the large static tanks, i.e. they are mounted on stilts or have underground heating to prevent frost heave.

For ease of maintenance the cold boxes are compartmented according to function. The heat exchangers and regenerators are separated from the distillation columns and associated equipment.

Fig. 4.5. Cold box for large tonnage oxygen plant partially complete to show maze of pipework contained therein. Note the size of the man at the top of the casing. (By permission of BOC-Airco Cryogenics Plant Ltd.)

In the U.K. the larger plants are insulated with powder at atmospheric pressure except for the components with moving parts which are insulated with mineral wool. In the United States both types are also used.

In the case of expanded perlite insulation the initial capital cost is low and maintenance can be kept low by having all welded construction of pipework and a high standard of workmanship. With mineral wool, if the trouble spot is known, then only the section affected has to be opened up and a small quantity of insulant removed. However, with large plant units it is not always easy to identify the source of the leak.

The large size of modern plants means that volumes of insulation in excess of 2,000 m³ (over 40 tonnes in weight) may have to be handled each time any work needs to be done on the cold parts of the plant.

Two interesting developments are the hollow chamber, "the tea cosy", technique of Linde (Weishaupt, 1962) in Germany and the vacuum insulated shell used by the Linde Division of Union Carbide (1966) on their cold boxes.

The former technique involves leaving the cold vessels and pipework free of insulant by building a framework around them covered with a retaining medium such as wire mesh or loose fitting plates. A gap is left between this retaining medium and the outer leak tight shell. The gap alone is filled with insulant. This means that maintenance is easy since only one access door need be provided into the inner compartment, however there is likely to be increased heat inleak compared with conventionally insulated boxes and inevitably there is increased heat exchange between the warm and colder components within the cold box due to convection.

The vacuum insulated shell allows much smaller quantities of insulant to be used but it poses many formidable design problems if the capital cost is not to become excessive.

Common to all types of insulation (except vacuum types) is the need to prevent moisture ingress. This is accomplished by means of a purge of dry gas, usually the waste nitrogen from the process plant, which keeps the insulation at a pressure slightly above atmospheric.

2. Pipelines and Valves

For the transfer of liquid hydrogen and helium vacuum insulated pipelines must be used. For other liquefied gases the choice between vacuum and normal insulation depends on many factors such as the length, duty and intermittentcy of the flow.

For liquid oxygen and nitrogen, the traditional transfer lines consisted of copper tubing surrounded by a galvanized mild steel ducting sealed with bitumen or other water-proofing which was either filled with powder or mineral wool, the thickness of insulation being about 15 to 20 cm. It is difficult to seal the outer casing against the weather, and the ingress of moisture soon increases the loss rate to double or treble the predicted figures.

More recently developed vacuum insulated lines use reflecting layer insulant or plain vacuum.

The loss rate from lines of 5 cm bore having atmospheric insulation varies from 10 to 60 W/m, the value depending upon the insulation thickness and its condition. The same size of pipeline with powder filled vacuum insulation has a leak of between 3 and 7 W/m, whereas those insulated with a reflecting layer type have heat inleaks of below 2 W/m, the exact values again depending upon the thickness of insulation.

For any duty there is an economic balance to be struck between the cost of the pipeline and the savings accrued from the reduced liquid loss on transfer. The optimum design will minimize the total operating cost and a method of working out this total is described by Tantam and Robb (1969) which takes into account all of the important factors.

Because vacuum insulated lines rely critically on maintaining a good vacuum, most performance problems are due to vacuum failures. The tendency is to use factory fabricated units which are joined together on site. This has presented problems in designing low heat-inleak end fixtures which are mechanically robust.

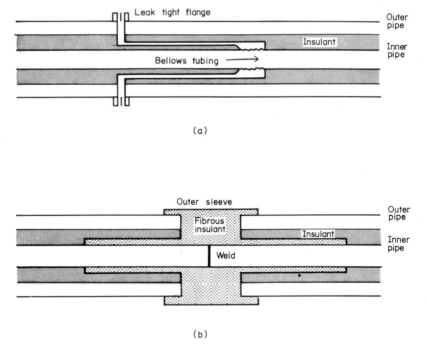

Fig. 4.6. Designs of joints for vacuum insulated pipelines. (a) Clamped bayonet type. (b) Double re-entry type.

Two basic types have been evolved and are shown in Fig. 4.6.

The type (a) has the lowest heat inleak since the non-vacuum insulated gap is the smallest, but it is also the trickiest to assemble since precise alignment is necessary during fixing and any subsequent movement can cause vacuum failure. This becomes increasingly difficult as the length of the standard units exceeds about 6 m.

Most of the articles on vacuum insulated pipelines are by authors associated with a particular manufacturer, e.g. Sergeant (1966). It is thus difficult to obtain an unbiased assessment. An article by Value Engineering (1969) has reviewed some of the problems a potential buyer faces and these are mainly due to lack of experience in cryogenics so that he cannot specify exactly his requirement.

A new development in the use of coaxial corrugated lines now known as the Andrews type (Cryogenic Technology 1968) which overcome most of the pitfalls of the rigid lines, viz. the need to know the exact dimensions of the pipe run and having to provide rigid elbows etc. for these lines. For non-branched lines between two widely spaced points such a line offers significant advantage in cost since only two end fittings are required as against the dozen or more that may be used with rigid lines. When first introduced the lines were made of copper but, because outgassing in service gave trouble, stainless steel has replaced the copper and these are claimed to be satisfactory. Two drawbacks with these lines are, the leak rate is a fixed value for a given size which cannot be altered as in the case of rigid lines, and the pressure drop is higher than for smooth bore pipes which mean that larger sizes need to be ordered to do the same duty. Because of the former drawback such lines will probably be limited to nitrogen, oxygen and argon use.

The construction of cryogenic valves involves the use of extended spindles, with both the stem extension and the sleeve made from stainless steel. Thus the gland stays warm enough to prevent moisture freezing inside it and causing the spindle to seize. Another precaution is that they should normally be mounted pointing upwards, otherwise liquid will flow down the sleeve and freeze the gland. Apart from these precautions and the need to use stainless steel or bronze the valves are no different from those used in the chemical industry. For oxygen use special PTFE glands have been developed to remove a combustion hazard. Bernstein (1969) and Cryogenic Technology (1969) have reviewed valves for cryogenic use and have detailed the source of trouble as well as recommendations and standards to avert these.

For liquid oxygen or nitrogen duty the valves range from a nominal $\frac{1}{2}$ cm to 30 cm bore and for normal duty these are simply lagged with mineral wool covered by a protective layer. These are even used with vacuum insulated lines because the cost of the vacuum jacketed valves does not justify the saving in liquid.

For liquid helium and hydrogen vacuum jacketed valves are essential and special valves ranging in size from the 1 mm bore valves used in small helium liquefiers to large 20 cm bore valves for space use have been developed. Initially the whole of the valve body including the extended casing were made as a separate vacuum unit. The newest technique is for manufacturers to fabricate the jackets onto the valves and to incorporate this into the vacuum insulated pipeline during the manufacturing stage so that the valve becomes an integral part of the line. This results not only in lower cost valves but also lower heat inleak because two sets of end closures are disposed of.

Acknowledgements

I wish to thank the directors of BOC–Airco Cryogenic Plant Limited for permission to publish this paper and Mr. Carne for his advice.

VII. References

Baker, C. K. and Haselden, G. G. (1970). Paper presented at London meeting of Comm. II and III at the International Institute of Refrigeration.

Beher, J. T. (1956). Proceedings of the 1956 Cryogenic Engineering Conference, p. 182.

Black, I. A. and Glaser, P. E. (1965). *Adv. in Cryogenic Eng.* **11,** 26.

Bernstein, B. A. (1969). *Cryogenics and Ind. Gases* p. 24.

Chicago Bridge and Iron Co. (1968). *Chem. Eng.* **75,** 73.

Cockett, A. H. and Molnar, W. (1960). *Cryogenics* **1,** 21.

Conch International Methane Ltd. (1960). British Patent 894, 762.

Dewar, J. (1905). *Proc. Roy. Inst.* **18,** 177.

Dewar, J. (1906). *Proc. Roy. Inst.* **18,** 433.

Duffy and Shoupel (1967). *Chem. Eng. Prog.* **63,** 55.

Huilicka, M. P. (1959). *Adv. in Cryogenic Eng.* **5,** 199.

Isenberg, L. (1963). National Symposium on Materials for Space Vehicle Use III, Article 10. Seattle.

Johnson, C. L. and Hollweger, D. J. (1969). *U.S. Govt. Res. Rep.* AD-60, 7891, **39,** 72.

Katan, L. L. (1952). Internal B.O.C. Communication.

Kierulff Electronic Inc. (1966). Private Communication.

Knudsen, M. (1911). *Ann. Phys. Leipzig* **35,** 983.

Luikov, A. V. et al. (1968). *Int. J. Heat and Mass Transfer* **11,** 117.

Linde Div. Union Carbide (1966). *Cryogenic Eng. News* **1,** 26; 44.

Martin, G. and Haselden, G. G. (1963). *Proc. XIth Int. Congress of Refrig., Munich* **2,** 241.

Mikesell, R. P. and Scott, R. B. (1956). *J. Res. Nat. Bur. Stand.* **57,** 371.

Modern Encyclopedia of Plastics. (1959). Vol. **36,** No. 14. McGraw-Hill Inc.

Powell, R. L. (1963). "American Institute of Physics Handbook." McGraw-Hill Publishing Co., New York.

Probert, S. D. (1967). H.M.S.O. publication. TRG. Rept. no. 1455 (R/X), 31.

Reynolds, M. M. (1963). "American Inst. of Physics. Handbook", 2nd edn. McGraw-Hill Publishing Co., New York.

Riede, P. M. and Wang, D. I.-J. (1959). *Adv. in Cryogenic Eng.* **5,** 199.

Sergeant, E. A. (1966). *Modern Refrig.* **69,** 893.

Tantam, D. H. and Robb, J. (1968). Proceedings of the Second International Cryogenic Engineering Conference, Brighton.

Transcontinental Gas Pipelines (1968). *Cryogenic Eng. News* **3,** 8.

Union Carbide (1963). British Patent 942,371.

Union Carbide (1962). British Patent 921,273.

Value Engineering Inc. (1969). *Cryogenic Tech.* **5,** 258.

Vedernikov, M. V. *et al.* (1969). *Cryogenics* **9,** 386.

Verschoor, J. D. and Grebler, P. (1952). *Trans. Amer. Soc. Mech. Eng.* **74,** 961.

Von Smoluchowski, H. (1911). *Ann. Phys. Leipzig* **35,** 984.

Ward, J. A. and Hildrew, R. H. (1968). *Proc. 1st. L.N.G. Conf., Chicago,* **7.**

Ward, J. A. and Egan, P. C. (1969). Proceedings of the Conference on L.N.G., London.

Weischaupt, J. (1962). *Linde Berichte* No. 13. May, 3.

Chapter 5

Fluid Dynamics

R. V. SMITH

Cryogenics Division, Institute of Basic Standards,
National Bureau of Standards, Boulder, Colorado, U.S.A.

I. Introduction

The purpose of this chapter is to introduce information on fluid dynamics as it is related to cryogenic systems. It is intended primarily to serve the design and operational engineer. In this context, research will be treated in terms of its usefulness to the design and operational situation rather than its relevance to the general understanding of the hydrodynamic phenomena.

Cryogenic fluids for this treatment will refer to fluids in the temperature range at or below the normal boiling temperature of methane. Much of the development of fluid mechanics has been related to the behaviour of conventional fluids or those which are in very common use such as air and water. The treatment of cryogenic fluids is somewhat different from that of conventional fluids. This difference stems from two primary sources. The first is a special situation created by operating at temperatures far below ambient which makes the use of special equipment and procedures necessary. The second difference arises from some unique properties of cryogenic fluids. The pressure, volume and temperature behaviour varies substantially from that of a conventional fluid particularly in the cases of hydrogen and helium. Liquid helium is markedly compressible and fluid flow studies should recognize this point. Transport properties of the cryogens also exhibit a unique behaviour. For example, at the normal point, the viscosity of water is about 100 times that of helium while the density is about 8 times the helium value. Additionally, the viscosity of low-pressure, liquid helium is gas-like in character. Finally, with respect to the properties of cryogenic fluids, the dividing lines for phase transitions are compressed into narrower pressure and temperature ranges. The entire saturation range for helium extends over only about 3 K° in temperature and slightly more than 2 atm in pressure. Therefore, with cryogenic fluids one is often forced to deal with mixed phases and phase changes. Further, these situations cannot be avoided by substitution of other fluids because, in general, there are none.

II. Pressure Drop in Simple Systems

The generalized form of the conservation equations for flow are attributed to Navier–Stokes. A full development of these expressions may be found in Bird *et al.* (1960), Shapiro (1953) or Rohsenow and Choi (1961). In a Cartesian frame of reference, the equation expressing the momentum conservation for the x (flow) direction with constant viscosity and for an incompressible fluid is

$$\frac{\partial}{\partial t}(\rho u) = -\frac{\partial}{\partial x}(\rho u^2) - \frac{\partial}{\partial y}(\rho uv) - \frac{\partial}{\partial z}(\rho uw) - \frac{\partial p}{\partial x}$$
$$+ \mu\left(\frac{\partial^2 u}{\partial x^2} + \frac{\partial^2 u}{\partial y^2} + \frac{\partial^2 u}{\partial z^2}\right) + F_b \quad (5.1)$$

The density is represented by ρ, u, v, and w are the velocities of fluid particles in the x, y, and z directions respectively, μ is the Newtonian viscosity, and F_b is a body force. The pressure is denoted by p and the time by t.

There is no general solution for these equations. Therefore only numerical solutions are available for this form of expression. Alternately, assumptions can be made which substantially simplify the equations which limit their application to simple systems with specific flow processes. These simplified equations are quite useful because they describe many flow situations with sufficient accuracy for design studies. Modification of these equations to include some flow cases, which are excluded by the simplifying assumptions in the following development, will be treated later in the section on geometry effects.

If the flow is one dimensional and steady then,

$$w = v = 0,$$

$$\partial/\partial z = 0,$$

$$\partial/\partial t = 0.$$

Body forces, F_b, represent forces per unit volume in the co-ordinate direction such as a weight force of a mass in a gravitational field. These forces are very often negligible, for example, when the fluid completely fills a horizontal conduit. Then,

$$0 = -\rho u \frac{du}{dx} - \frac{dp}{nx} + \mu \frac{d^2 u}{dy^2} \tag{5.2}$$

The preceding treatment involves local derivatives. Subsequent equations will use velocity terms obtained by integration over the velocity profile. Therefore, the velocities become bulk fluid velocities. The last term in equation 5.2 is associated with shear resulting from forces at the solid surfaces. These surfaces can be conduit walls or submerged bodies. They are basically the same type of forces but conventionally the empirical expressions which are used to represent the shear terms are expressed differently according to the geometry of the system. For conduit walls the geometry variable is expressed in terms of wall length (Δx) and for submerged objects the projected area A_{DR} is represented by the area of the solid on a plane perpendicular to the fluid approach velocity.

For the wall shear, $\mu(d^2 u/dy^2)$ may be expressed as

$$-2f \frac{\rho u^2}{D_h} \tag{5.3}$$

where f is the Fanning friction factor which may be defined as

$$f = \frac{\tau}{\rho u^2/2} \tag{5.4}$$

The shear stress is τ and the hydraulic diameter, D_h, is defined as the diameter of a circular conduit and, for a non circular section, four times the ratio of the cross-sectional area to the wetted perimeter. Experimentally determined values of f are presented as a function of Reynolds number (Re) in Fig. 5.1.

For submerged bodies, $\mu(d^2u/dy^2)$ is commonly expressed as

$$-C_{DR}\frac{\rho u^2 A_{DR}}{2} \tag{5.5}$$

where C_{DR} is referred to as the drag coefficient. Since the length (Δx) is not a variable term, each submerged geometry must have a separate C_{DR} value. These drag coefficients, f, and C_{DR}, are empirically determined for both

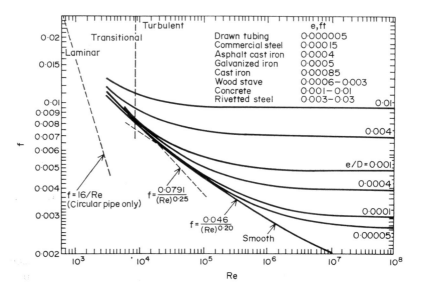

Fig. 5.1. Fanning friction factor for clean commercial steel and wrought iron pipe. From Moody (1944) with permission from ASME.

laminar and turbulent flow and values are shown in Figs. 5.1 and 5.2. It should be noted that the drag force, associated with C_{DR}, is composed of the form drag and the surface drag. The form drag is a result of the pressure difference across the body set up by a separation of the fluid stream from the body in the downstream portion.

Therefore, the working momentum equation is

$$dp = -\rho u\, du - zf\frac{\rho u^2}{D_h}\, dx - C_{DR}\frac{\rho \bar{u}^2 A_{DR}}{2} \tag{5.6}$$

For incompressible fluid flow (ρ = constant), (5.6) integrates to

$$p_2 - p_1 = \frac{\rho(u_2^2 - u_1^2)}{2} - zf\frac{\rho\bar{u}^2}{D_h}(x_2 - x_1) - C_{DR}\frac{\rho\bar{u}^2 A_{DR}}{2} \tag{5.7}$$

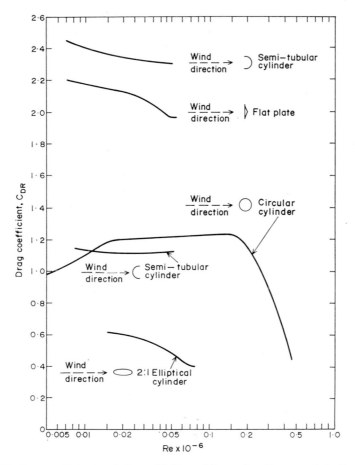

Fig. 5.2. Drag-coefficient variation with Reynolds number for several bodies having infinite aspect ratio. From Dwinnel (1949) after Lindsey (1938).

where $\bar{u}$ indicates the effective velocity for the flow length this velocity is defined as

$$\bar{u} = \frac{\displaystyle\int_{x_1}^{x_2} u\,dx}{x_2 - x_1}$$

If there are no submerged bodies and the flow cross-sectional area is constant, then,

$$p_2 - p_1 = -2f \frac{\rho u^2}{D_h}(x_2 - x_1) \tag{5.8}$$

Mass conservation for one-dimensional steady state, is

$$\frac{dm}{dt} = \rho A u \tag{5.9}$$

or with logarithmic differentiation

$$\frac{du}{u} + \frac{d\rho}{\rho} + \frac{dA}{A} = 0. \tag{5.9a}$$

Energy conservation for one-dimensional, adiabatic, horizontal, steady-state flow becomes

$$dh = d\left(\frac{u^2}{z}\right)$$

where h is the specific enthalpy, expressed in appropriate units. Also,

$$h_0 = h + \frac{u^2}{z}$$

where h_0 is the stagnation enthalpy.

Example 5.1

A smooth tube, 30 m in length, is to carry liquid oxygen at a rate of 60 kg/min. The oxygen is at a pressure of 2 atm and 76°K. The design criteria of the systems requires that the pressure drop cannot exceed 35×10^3 N/m², yet the tube size must remain as small as possible. Find the optimal tube inside diameter. The density of the oxygen is $1·16 \times 10^3$ kg/m³.

Solution

The solution will be iterative since the value of the friction factor must be obtained from Fig. 5.1. If the Reynold's number range were known, one of the empirical equations shown in Fig. 5.1 might be used. First we will try an inside tube diameter of 1 cm.

From equation 5.9

$$u = \frac{dm/dt}{\rho A}$$

$$= \frac{1 \text{ kg/s}}{1·16 \times 10^3 \text{ kg/m}^3 \times \pi/4(1 \times 10^{-2})^2 \text{ m}^2}$$

$$= 11 \text{ m/sec}$$

$$Re = \frac{uD\rho}{\mu}$$

$$= \frac{11 \text{ m/s} \times 1 \times 10^{-2} \text{ m} \times 1\cdot16 \times 10^3 \text{ kg/m}^3}{2\cdot9 \times 10^{-4} \text{ kg/m sec}}$$

$$= 4\cdot4 \times 10^5$$

From Fig. 5.1

$$f = 3\cdot4 \times 10^{-3}.$$

From equation 5.8

$$p_1 - p_2 = \frac{2f\rho u^2(x_2 - x_1)}{D_h}$$

$$= \frac{2 \times 3\cdot4 \times 10^{-3} \times 1\cdot16 \times 10^3 \text{ kg/m}^3 \times \overline{11}^2 \text{ m}^2/\text{sec}^2 \times 3 \times 10^1 \text{ m}}{1 \times 10^{-2} \text{ m}}$$

$$= 2\cdot6 \times 10^6 \text{ N/m}^2$$

This pressure drop is too large to satisfy the design requirements. We will next try a larger diameter of 2·5 cm.

$$u = \frac{1 \text{ kg/sec}}{1\cdot16 \times 10^3 \text{ kg/m}^3 \times \pi/4(2\cdot5 \times 10^{-2})^2 \text{ m}^2}$$

$$= 1\cdot76 \text{ m/sec}$$

$$Re = \frac{uD\rho}{\mu}$$

$$= \frac{1\cdot76 \text{ m/s} \times 2\cdot5 \times 10^{-2} \text{ m} \times 1\cdot16 \times 10^3 \text{ kg/m}^3}{2\cdot9 \times 10^{-4} \text{ kg/m sec}}$$

$$= 1\cdot76 \times 10^5$$

From Fig. 5.1,

$$f = 4 \times 10^{-3}$$

$$p_1 - p_2 = \frac{2 \times 4 \times 10^{-3} \times 1\cdot16 \times 10^3 \times \overline{1\cdot76}^2 \text{ m}^2/\text{sec}^2 \times 3 \times 10^1 \text{ m}}{2\cdot5 \times 10^{-2} \text{ m}}$$

$$= 3\cdot4 \times 10^4 \text{ N/m}^2$$

This tube size satisfies the design criteria.

III. Compressible Flow

The difficulties which arise in analysing compressible fluid flow compared to incompressible fluid flow may be found in an examination of equation 5.6. Although equation 5.6 was derived for incompressible fluid flow it may be used with a variable density for compressible flow studies. With the density variable, the integration of that expression will be substantially more complex, and will require information regarding the p, V, and T behaviour for the fluid.

11

Physically, there are two very significant characteristics of compressible fluid flow. One is that as the pressure is decreased along the flow passage the volume of the fluid is increased. Thus, if the flow cross-sectional area is constant the velocity of the fluid must be increased, or conversely if the velocity is to remain constant the flow cross-sectional area must be increased. A part of this behaviour is covered in a later section on Fanno flow. A second significant factor in compressible fluid flow has to do with Mach number considerations and is discussed in the Section A following.

All fluids are compressible to some degree, however, one may usually distinguish between compressible fluids and incompressible ones by noting the values of the compressibility terms which are the coefficient of volume expansion

$$\beta = \frac{1}{V}\left(\frac{\partial V}{\partial T}\right)_p \tag{5.10}$$

and the isothermal compressibility

$$\kappa = \frac{1}{\rho}\left(\frac{\partial \rho}{\partial p}\right)_T \tag{5.11}$$

One can see that the compressibility expressions are functionally dependent on the variation of the density or specific volume of a fluid as the pressure or temperature changes. Generally, we consider that liquids have very low compressibility values and that the values shown in equations 5.10 and 5.11 are only significant for gases. Air, for example, is about 20,000 times more compressible than water. This incompressible assumption is not necessarily valid for the cryogenic liquids, however, particularly for hydrogen and helium. Liquids with low boiling points have weak molecular bonds. Thus, low energy levels are required to condense the substance from a gas to a liquid. Therefore, even in a liquid state, some compressibility of the liquid is experienced and consequently cryogenic liquids in general must be considered as potentially compressible.

The treatment in this section only serves to outline the major points of compressible fluid flow. More comprehensive treatments of this subject may be found in Shapiro (1953) and Chambel and Jennings (1958). The treatment here is intended to give the reader an awareness of the subject in order that he may determine whether or not he will have to consider the complexities of compressible fluid flow.

A. MACH NUMBER CONSIDERATIONS

The Mach number is defined as the fluid velocity divided by the velocity of sound in that fluid or

$$M = \frac{u}{a}$$

and

$$a = \left(\frac{\partial p}{\partial \rho}\right)_s^{\frac{1}{2}}$$ (5.12)

Then, for an ideal gas

$$a = (\gamma RT)^{\frac{1}{2}}$$

One can see from this expression that, as previously stated, the sonic velocities in an essentially incompressible fluid such as water are very high (about 1,500 M/sec) and the Mach number for all practical flows would be extremely low. When the Mach number is very much less than Mach 1 consideration of the Mach number is negligible and a simple calculation will often suffice to show that further consideration of this factor is not important. Conversely as the Mach number approaches one it must be considered in the flow equations. The following discussion will consider one specific case.

For isentropic flow with variable area we may develop the following expression, see Smith (1968).

$$du = \frac{-u\,dA}{A\left[1 - \dfrac{u^2}{\partial p/\partial \rho}\right]}$$ (5.13)

We can immediately see the significance of the Mach number and the sonic velocity from the bracketed term in the denominator of equation 5.13. As the Mach number approaches one the denominator becomes zero and the expression is undefined. In summary, the consequences of equation 5.13 are as follows.

1. For Ideal, Convergent and Divergent Flow

Mach 1 signals the point at which the section must change from the convergent to divergent condition, with subsonic flow in the former section and supersonic flow in the latter.

2. Non-Ideal Flow

If the flow is not ideal in that the pressures and flow cross-sectional areas do not match the ideal flow conditions, then shock waves will develop in the flow. These shock waves are characterized by pressure discontinuities, substantial irreversibilities and large drag forces.

3. For a Converging Section Alone

If the discharge pressure is below the critical pressure indicating the Mach 1 condition (discussed later), the flow is said to be choked or critical flow at the point of discharge. This means that further reduction in the downstream pressure will not increase the velocity of the flow at the critical point or increase the mass rate of flow in the system. This condition has special significance because in systems in which the flow is controlled by downstream pressures, this control is completely lost once critical flow has been achieved.

The critical pressure indicating critical or choking flow for a given flow inlet can be computed for ideal conditions. The expression is usually given for a critical pressure ratio, that of the critical pressure over the stagnation pressure for the isentropic flow of an ideal gas. That expression is

$$\frac{p_c}{p_0} = \left(\frac{2}{\gamma+1}\right)^{\gamma/(\gamma-1)} \tag{5.14}$$

here γ is the ratio for specific heats. The stagnation pressure is that pressure which would be achieved if the flow velocity were isentropically reduced to zero. Thus,

$$p_0 = p + \frac{\rho u^2}{2} \tag{5.15}$$

is valid for Mach numbers very much lower than one. For higher Mach numbers the stagnation pressure may be expressed as

$$p_0 = p\left(1 + \frac{\gamma-1}{2}M^2\right)^{\gamma/(\gamma-1)} \tag{5.16}$$

for an ideal gas.

B. FANNO FLOW

Fanno flow is a special case of compressible fluid flow which approximates many actual flow cases and, most fortunately, permits a closed form solution. Fanno flow is defined as one dimensional, steady, adiabatic, horizontal flow. The basic equation for Fanno flow is (5.6), omitting the body force term.

An important characteristic of Fanno flow is that one cannot cross between supersonic and subsonic flow within the conduit. Therefore, in the extreme case of flow acceleration (or deceleration from a supersonic entrance) the fluid will be at Mach 1 at the conduit exit. Therefore, it is conventional in the final Fanno flow equation to integrate between the limits of conduit entrance to the maximum conduit length, real or imaginary, which would provide the Mach 1 flow condition. The expression is

$$\int_{x=0}^{x=L_{max}} 4f\frac{dx}{D_h} = \int_M^1 \frac{2(1-M^2)}{\gamma M^3\{[\frac{1}{2}(\gamma-1)]M^2+1\}}\, dM \tag{5.20}$$

Thus, it may be seen that a general formulation involving the friction factor, the length to the critical or choking condition and the diameter of the flow section may be expressed as equal to an integral in which the variables are the Mach number and fluid properties. Solutions to this expression are usually given in tables for the case of an ideal gas with a specific heat ratio of 1·4. These may be found in the references cited earlier in this section.

Further insight into Fanno flow may be found in examinations of Figs. 5.3 and 5.4. In Fig. 5.3, ratios of variable flow properties over a starred value of the same property are shown. The starred values in the denominator designate the condition for that variable at the point of Mach 1 flow. Thus, the ratio of p/p^* has a value of 1 at the Mach 1 condition. Figure 5.4 shows an enthalpy plot versus entropy for Fanno flow. It may be seen that, from either subsonic or supersonic inlet, the flow proceeds to a maximum entropy condition which is identical with the Mach 1 flow condition.

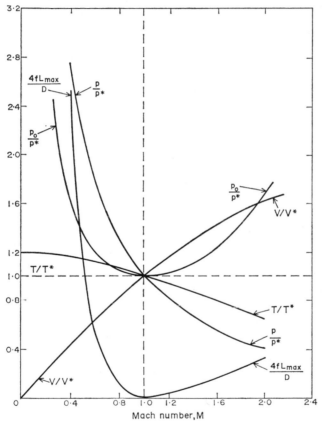

Fig. 5.3. Functions for constant-area flow with friction, assuming $Y = 1.4$. From Cambel and Jennings (1958) with permission from McGraw-Hill.

C. CRYOGENIC FLUIDS

For some of the preceding cases discussed there are no specific cryogenic fluid flow data. In this section we present the cryogenic fluid data available indicating modifications and special considerations where required. These are all for the single-phase flow case.

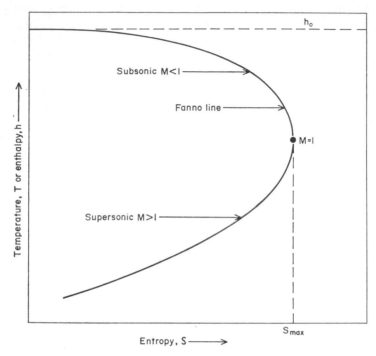

Fig. 5.4. *T–S* or *H–S* for Fanno flow illustrating Fanno line. From Cambel and Jennings (1958) with permission from McGraw-Hill.

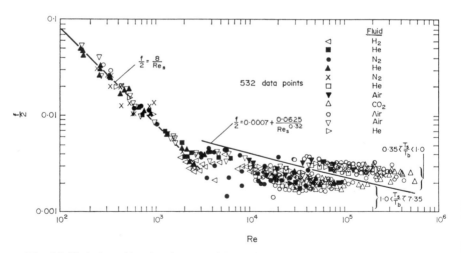

Fig. 5.5. Variation of local and average friction factors with modified Reynolds number. Density for coefficient evaluated at bulk temperature; viscosity in Reynolds number evaluated at surface temperature. From Taylor (1967) with permission of Pergamon Press.

Taylor (1967) discusses a friction factor correlation for the special case where the surface temperature, T_w, is very much higher than the bulk temperature, T_b. This is often the case with cryogenic fluids because the ratio of T_w/T_b can be very high without exceeding the metallurgical limit for plastic behaviour. Taylor's data were primarily for gases at higher temperatures, however the fluids tested such as hydrogen, helium and nitrogen are often associated with cryogenic temperatures. In Fig. 5.5, the nitrogen data reported by Perkins and Worsøe–Schmidt (1964) was at cryogenic temperatures. Taylor showed that the relationship between the friction factor and the Reynold's number as shown in Fig. 5.1 is valid providing a modified Reynold's number is used,

$$Re = \left(\frac{\rho D u}{\mu_w}\right)\frac{T_b}{T_w}. \tag{5.21}$$

The modification adds the bulk-to-wall temperature ratio and evaluates the viscosity at the wall conditions. These data indicate that fluids often used at cryogenic temperatures behave much the same as other fluids at higher temperatures and nitrogen at lower temperatures shows no special behavioural characteristics.

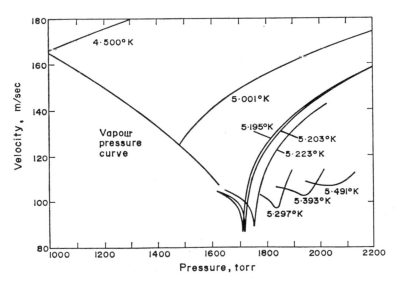

Fig. 5.6. Overall view of the sound velocity of He^4 as a function of pressure near the critical point. From Hendricks et al. (1969) after Williamson and Chase (1968).

Regarding the Mach 1 condition, Fig. 5.6 shows the sonic velocity of liquid helium along the vapour pressure curve is quite low compared to more conventional liquids, in fact about one tenth that of water. As one approaches the critical condition this sonic velocity becomes very low indeed. Thus, the

Mach 1 condition might be expected to be found at relatively low flow velocities in this region. Experimental data showing choking or shock have not been reported for liquid helium although there have been some unpublished inferences of this with liquid hydrogen. Near the thermodynamic critical point one would expect non-equilibrium conditions which would make the critical velocity higher than that indicated by sonic velocity. Figure 5.6 indicates that the designer should be wary of critical flow with cryogenic liquids or with any fluids near the thermodynamic critical condition.

IV. Geometry Effects

The previous treatment in this chapter has dealt with simple systems, for example, straight tubes of constant cross section with no entrance or other geometry effects. This section will primarily be concerned with the variations of the previous equations which can be employed to suit geometries other than the simple cases.

A. CURVES, VALVES AND FITTINGS

In simple systems the flow is quite complex but still well organized with respect to the flow through other geometries. For example, secondary flow, in which the radial components are significant on a macroscopic level appears in curved tubes. This secondary flow introduces additional irreversibilities which are evidenced as additional pressure losses. Flow through valves and fittings is even more complex because separation, in which the flow streamlines separate from the conduit surface, often exists. Therefore, empiricism is increasingly prevalent as one proceeds through this section from curved tubes to valves and fittings.

Curved Tubing

Two classical works related to conventional fluids are reported by Ito (1959) and Dean (1928). The work of Ito has proven successful in describing pressure drops. He suggests that the ratio of the friction factor for a curved tube to that for a straight tube may be expressed as

$$\frac{f_{\text{curved}}}{f_{\text{straight}}} = \left[Re\left(\frac{r}{r_{\text{curve}}}\right)^2 \right]^{0.05} \tag{5.22}$$

the radial ratio, r/r_{curve} is the inside radius of the tube divided by the radius of the tube curvature. Pressure drop data for the flow of cryogenic fluids in curved tubes have not been reported. Taylor (1968) used the hydrogen work of Miller (1966) and of Thompson and Geery (1960) and showed that the Nusselt number ratio $Nu_{\text{curved}}/Nu_{\text{straight}}$ can be correlated with the Ito parameter shown in (5.22). Then, by the analogy between heat and momentum transport, one would expect the friction factor ratio would also correlate for hydrogen.

TABLE 5.I. Representative Equivalent Length in Pipe Diameters (L/D) of Various Valves and Fittings. (From Crane Technical Paper 410 (1969). Courtesy Crane Co.)

	Description of product		Equivalent length in pipe diameters
Globe Valves	Conventional	With no obstruction in flat, bevel, or plug type seat	Fully open	340
		With wing or pin guided disc	Fully open	450
Angle Valves	Conventional	With no obstruction in flat, bevel, or plug type seat	Fully open	145
		With wing or pin guided disc		
Gate Valves	Conventional Wedge Disc, Double Disc, or Plug Disc		Fully open	200
			Fully open	13
			Three-quarters open	35
			One-half open	160
			One-quarter open	900
	Conduit Pipe Line (8-in. and larger)		Fully open	3 [a]
Butterfly Valves			Fully open	40
Fittings	90 Degree Standard Elbow			30
	45 Degree Standard Elbow			16
	90 Degree Long Radius Elbow			20
	90 Degree Street Elbow			50
	45 Degree Street Elbow			26
	Square Corner Elbow			57
	Standard Tee	With flow through run		20
		With flow through branch		60
	Close Pattern Return Bend			50

[a] Exact equivalent length is equal to the length between flange faces or welding ends.

11*

Valves and Fittings

As mentioned previously, flow through these devices is quite complex and is usually expressed empirically. The pressure loss through a valve or fitting is usually expressed in terms of an equivalent length of straight tubing. Thus, in order to calculate the pressure drop through a valve or fitting one simply substitutes the equivalent length into equation 5.8 using the best estimate for the friction factor related to the surface roughness of the fitting or valve employed. A thorough documentation of these equivalent lengths has been published by the Crane Company (1969) and some of the more important values are shown in Table 5.I.

B. ENTRANCE EFFECTS IN STRAIGHT TUBES

In the entrance of a smooth, straight tube the pressure-drop is influenced by both the development of the velocity profile and by effects of abrupt entrances such as sharp edges. First, the profile development can be treated analytically to a reasonable degree of satisfaction. The classical solution to the entrance effects may be attributed to Langhaar (1942). Both his work and that of Deissler (1955) showed that entrance effects persist only about 10 diameters downstream from the entrance point. The experimental data expressed as a friction factor has been found to reach a constant value before the velocity profile is fully developed.

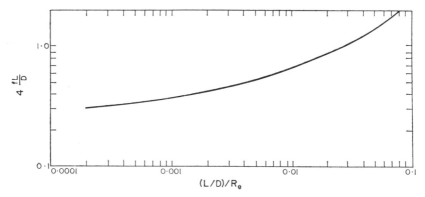

Fig. 5.7. Average friction factor for laminar flow in entrance region of circular tubes. From Langhaar (1942) with permission from ASME.

For abrupt entrances there is an additional pressure drop at the entrance or disturbance point. The pressure at either side of this disturbance is fundamentally similar to that of a sharp edged orifice to be considered later. The influence of the developing profile may be expressed as a variable friction factor as shown in Figs. 5.7 and 5.8.

C. TUBE BUNDLES AND FINNED TUBING

The basic concepts of flow through tube bundles and fins are the same as for pipe flow but are expressed mathematically in a slightly different form. For example, the friction factor defined as

$$f = \frac{\tau}{\rho u^2/2}$$

gives a total wall force of

$$F_w = \tau A_w$$

and a pressure drop of

$$\Delta p = \frac{F_w}{A}$$

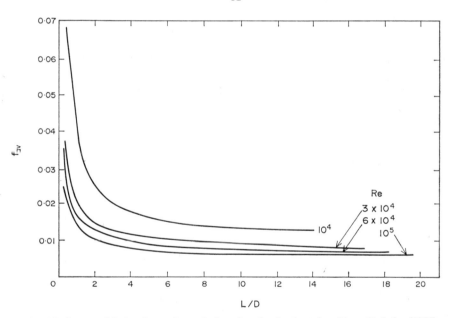

Fig. 5.8. Average friction factor for turbulent flow in circular tubes. From Deissler (1955).

Form drag, if significant, is empirically added to the force and shear terms. Then, combining the above equations produces the expression

$$\Delta p = \frac{f\,(\rho u^2/2)A_w}{A} \tag{5.24}$$

where A_w is the total area over which the fluid flows (wets) and A is the minimum cross sectional area for the flow.

In the case of tube bundles there is an entrance and exit effect of contraction and expansion resulting in a pressure drop in one end and a pressure

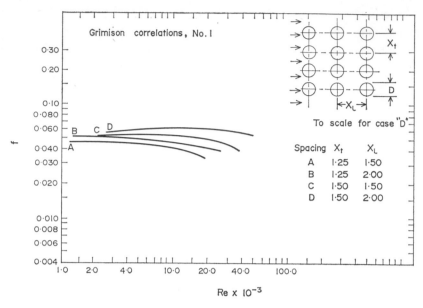

Fig. 5.9. Flow normal to in-line tube banks. From Kays and London (1958) (Figs. 5.9-5.22. with permission from McGraw-Hill.)

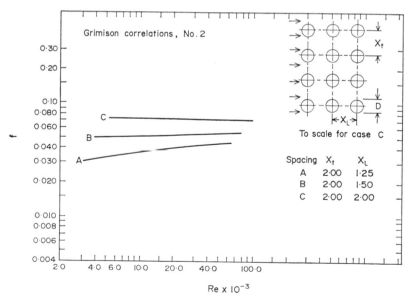

Fig. 5.10. Flow normal to in-line tube banks. From Kays and London (1958).

rise in the other so that there is only a small pressure change resulting from this combination. A complete treatment of these entrance and exit pressure drop effects may be found in Kays and London (1958). In the following treatment the pressure drop will be calculated using an overall friction factor without consideration of the entrance and exit effects. Values for this friction factor are shown in Figs. 5.9 and 5.10. The Reynold's number used in these expressions should be based on the minimum free flow area.

Finned tubing is often employed because it gives an efficient increase of the heat transfer surface and reduces the space and material requirements for the heat exchanger. The pressure drop for such tubing, however, is substantially increased. The results of extensive studies by Kays and London (1958) on various types of finned arrangements are shown in Figs. 5.11 to 5.22. It should be noted here that the Reynold's number is computed using the hydraulic diameter for the finned tubing. Although these data are for specific fluid and geometric conditions, the variables are dimensionless and the figures can be used for similar geometries and flow situations.

D. PACKED BEDS

A packed bed consists of a bed of spheres, cylinders, granules or other shapes through which the fluid flows. It provides a very effective means of transferring heat. They differ from tube bundles and fins in that the fluid is assumed not to channel in flowing through a packed bed. Because this process is exceedingly complex, empirical expressions are used to estimate the pressure drop. In a recent review of pressure drops in packed beds, Yen (1967) evaluated three generalized approaches. From these he selected the method of Brownell (1950) as the most accurate and reliable. This method was selected by applying the results from three predictive systems to the reported data. The data included experiments with a variety of packed beds and with the fluids air, a mixture of CO_2 and nitrogen, and oil and water. The standard deviation for the Brownell approach was in the vicinity of 25%. The Brownell method follows the conventional frictional pressure drop approach. Its special characteristics are that it employs a modified Reynold's number

$$Re_{BR} = ReF_{Re}$$

Values for F_{Re} and for the modified friction factor may be determined from Fig. 5.23(a)–(c). Using these values the pressure drop over a packed bed may be determined from

$$\Delta p = \frac{f_{BR}G^2L}{2\rho D_p} \tag{5.25}$$

where $D°$ is an equivalent diameter defined in Fig. 5.23(a). Equation 5.25 is a modified version of equation 5.6 used for a straight tube.

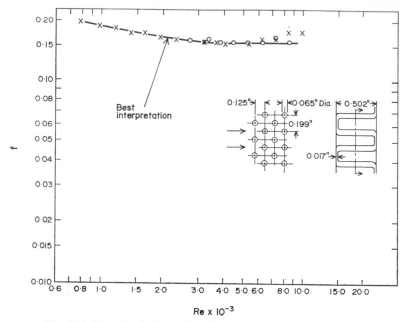

Fig. 5.11. Flow in pin-fin surface. From Kays and London (1958). $4r_h = 0.0186$ ft. Heat transfer area/volume = 140 ft²/ft³. Fin area/total area = 0·704. *Note.* Minimum free-flow area is in spaces transverse to flow.

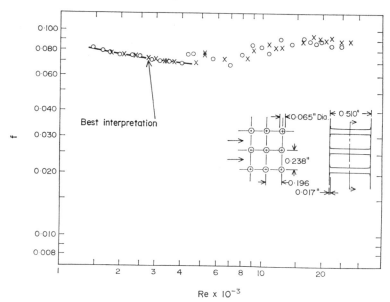

Fig. 5.12. Flow in pin-fin plate-fin surface. From Kays and London (1958). $4r_h = 0.0297$ ft. Heat transfer area/volume = 96·2 ft²/ft³. Fin area/total area = 0·546.

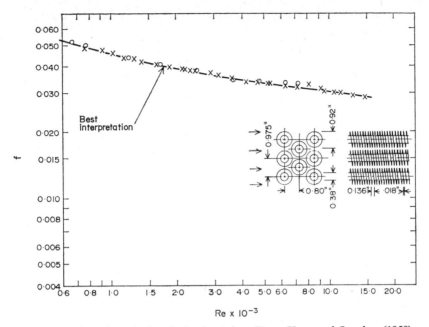

Fig. 5.13. Flow through finned circular tubes. From Kays and London (1958). $4r_h = 0.0154$ ft.   Free flow area/frontal area $= 0.538$.   Heat transfer area/total volume $= 140$ ft²/ft³.   Fin area/total area $= 0.892$.

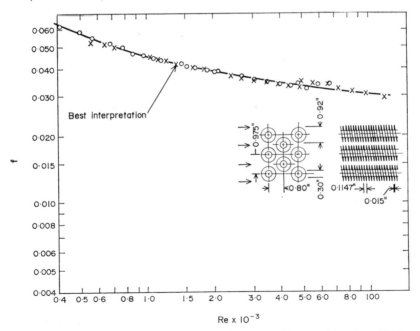

Fig. 5.14. Flow through finned circular tubes. From Kays and London (1958). $4r_h = 0.01288$ ft.   Free-flow   area/frontal   area $= 0.524$.   Heat   transfer   area/total volume $= 163$ ft²/ft³.   Fin area/total area $= 0.910$.

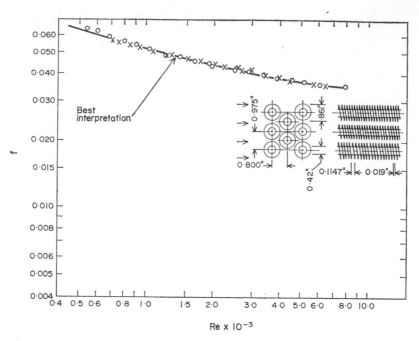

Fig. 5.15. Flow through finned circular tubes. From Kays and London (1958). $4r_h = 0.01452$ ft. Free-flow area/frontal area $= 0.494$. Heat transfer area/total volume $= 136$ ft^2/ft^3. Fin area/total area $= 0.876$.

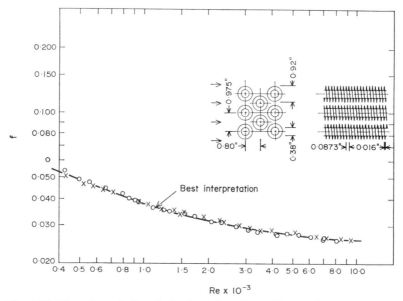

Fig. 5.16. Flow through finned circular tubes. From Kays and London (1958). $4r_h = 0.00976$ ft. Free-flow area/frontal area $= 0.510$. Heat transfer area/total volume $= 209$ ft^2/ft^3. Fin area/total area $= 0.931$.

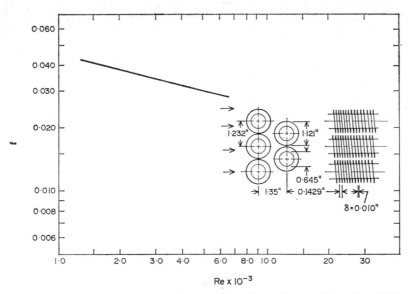

Fig. 5.17. Flow through finned circular tubes. From Kays and London (1958). $4r_h = 0.0219$ ft.   Free-flow area/frontal area $= 0.449$.   Heat transfer area/total volume $= 82$ ft²/ft³.   Fin area/total area $= 0.830$.

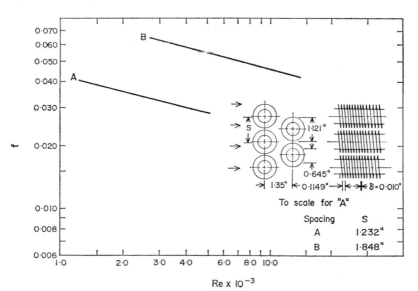

Fig. 5.18. Flow through finned circular tubes. From Kays and London (1958). Fin pitch $= 8.7$ per in.   Fin area/total area $= 0.862$.

	A	B
Flow passage hydraulic diameter, $4r_h$	0·01797	0·0383 ft
Free-flow area/frontal area	0·443	0·628
Heat transfer area/total volume	98·7	65·7 ft²/ft³

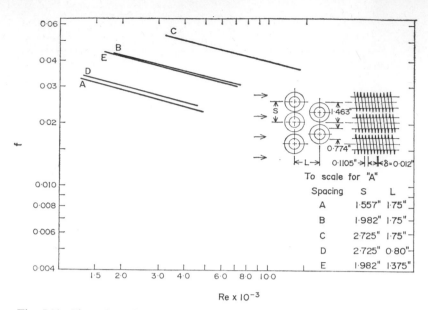

Fig. 5.19. Flow through finned circular tubes. From Kays and London (1958). Tube outside diameter = 0·774 in. Fin pitch = 9·05 per in. Fin thickness = 0·012 in. Fin area/total area = 0·835. For D, in which the minimum free-flow area is in the diagonals.

	A	B	C	D	E
Flow passage hydraulic diameter, $4r_h$	0·01681	0·02685	0·0445	0·01587	0·02108 ft
Free-flow area/frontal area	0·455	0·572	0·688	0·537	0·572
Heat transfer area/total volume	108	85·1	61·9	135	108 ft²/ft³

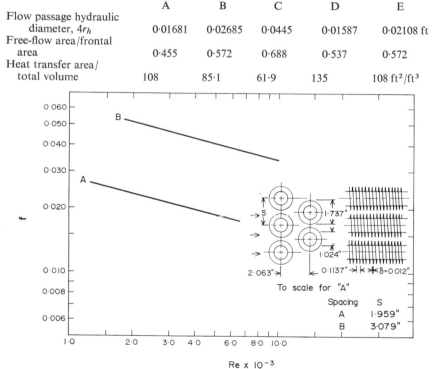

Fig. 5.20. Flow through finned circular tubes. From Kays and London (1958). Fin area/total area = 0·825.

	A	B
Flow passage hydraulic diameter, $4r_h$	0·01927	0·0443 ft
Free-flow/frontal area	0·439	0·643
Heat transfer area/total volume	91·2	58·1 ft²/ft³

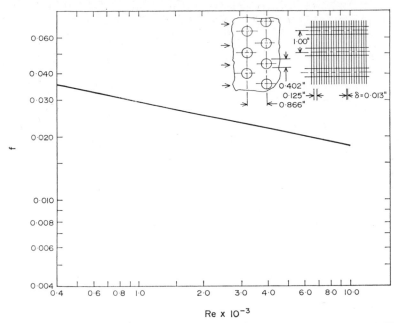

Fig. 5.21. Flow through finned circular tubes. From Kays and London (1958).
$4r_h = 0.01192$ ft. Free-flow area/frontal area, 0·534. Heat transfer area/total volume,
179 ft²/ft³. Fin area/total area, 0·839.

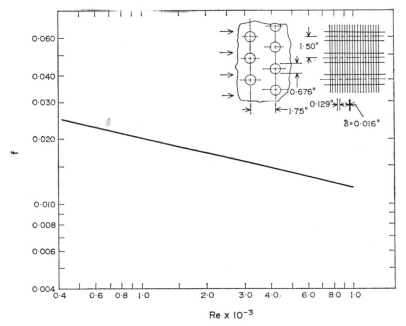

Fig. 5.22. Flow through finned circular tubes. From Kays and London (1958).
$4r_h = 0.01268$ ft. Free-flow area/frontal area = 0·497. Heat transfer area/total volume =
157 ft²/ft³. Fin area/total area = 0·905.

E. ANNULI

The form of the expressions used to determine the friction factor and pressure drop for an annuli is similar to that used for flow inside straight tubes. For the annular tube, the Reynold's number is

$$Re = \frac{[2(r_2 - r_1)]u\rho}{\mu}, \tag{5.26}$$

where $2(r_2 - r_1)$ is the hydraulic diameter. Then, one may use the conventional smooth tube friction factor and arrive at the following expression for pressure drop.

$$\Delta p = \frac{-fu^2\rho\Delta L}{(r_2 - r_1)}. \tag{5.27}$$

Some authors, for example Davis (1943) and others, have included an additional term which is a function of the radius ratio for the pressure drop in an annulus.

It should be noted, however, that this is an overall friction factor which cannot be related to the shear stress in the same way as it is for circular tubes. Therefore, this concept cannot be used in analogies for heat transfer calculations. Rothfus *et al.* (1955) have shown that one can define separate friction factors for the inner and outer walls. These would be more helpful if a detailed study of the shear stress is required—for example, in the case of analogies between momentum and energy transport.

F. VENTURI, NOZZLES, AND ORIFICES

Each of these constrictions produces a pressure drop in the flow and may be used to measure the flow rate. Each has specific advantages and disadvantages. The pressure losses are shown in Fig. 5.24. Details of the standard construction and arrangement of these meters in a flow situation may be found in the ASME publication "Fluid Meters and Their Theory and Applications" (1959). Orifices and venturi tubes have been used with reasonable success in cryogenic flow rate measurements. From equation 5.7 and 5.9, for a horizontal installation neglecting frictional terms a general volumetric flow rate equation is usually expressed as

$$Q = \frac{dm/dt}{\rho} = A_{\min} C \sqrt{\left(\frac{2\Delta p}{\rho}\right)}. \tag{5.28}$$

In equation 5.28 the empirical coefficient, C, is a function of the $(A_{\min}/A_1)$ ratio.

1. Venturi Meter

As shown previously in the A.S.M.E. reference (1959), the venturi meter is an efficient and accurate means of measuring flow rate. It consists of a convergent and divergent section. Pressure measurements to indicate the flow rate are made at the entrance and the throat. The flow losses are low but sometimes small pressure differentials are difficult to measure accurately for low flow rates. Its installation influences a substantial portion of a flow system.

In equation 5.28, the discharge coefficient, C, accounts for frictional and contractional losses. It is a function of Reynold's number and its value may be determined from Fig. 5.29. In cryogenic systems the liquid is often very near saturation conditions and pressure reductions at the throat may become sufficiently large to nucleate the vapour phase. This is usually referred to as cavitation and a cavitation study with cryogenic liquids has been reported by Edmonds and Hord (1969).

Purcell et al. (1960) has shown that a venturi may be used as a satisfactory measuring device for liquid hydrogen flow. Figure 5.25 shows the discharge coefficient. Close (1969) estimates that, with cryogenic fluids, orifices and venturi meters have an error of $\pm 3\%$ and a repeatability of 1%.

2. Nozzles

Nozzles might be described as the upstream portion of a venturi from the entrance to the throat section. In general, however, nozzle converging contours are somewhat different from those in a venturi. The nozzle has the advantage that it is shorter but the disadvantage that the net flow losses are substantially greater. For the same flow quantities, pressure differentials are generally larger for the entrance to throat in a nozzle than those in a venturi affording more sensitivity in the measurements. The equation for the flow rate through a nozzle is the same as for a venturi, equation 5.28, where C, becomes the nozzle discharge coefficient. Values for the nozzle coefficients are shown in Fig. 5.26.

3. Orifice Meter

A thin annular flat plate installed in the flow stream is by far the simplest installation for measuring the flow of fluids. It may be easily placed between flanged sections. These installation procedures are detailed in "Fluid Meters, Their Theory and Application" (1959). The orifice meter has the disadvantage of relatively high pressure losses.

Again, the basic meter equation 5.28 may be used for an orifice meter provided one obtains the proper discharge coefficient. Figure 5.27 shows the flow coefficient for an orifice installation with flange taps. The flow coefficients for orifices are substantially lower than that for nozzles and venturi tubes because of the inefficiency and irreversibility of the process. The coefficients also have a weak dependence on the diameter ratio as shown in Fig. 5.27.

Richards *et al.* (1958), have studied the use of orifices to measure the flow of hydrogen and nitrogen. They were concerned with the applicability of water and other conventional fluid discharge coefficients for use with cryogenic liquids. The major concern was whether or not vaporization would be a substantial factor since sub-cooling at the inlet is usually very slight for cryogenic fluids and pressures below the saturation pressure may exist in the flow through the orifice. The authors found very little evidence that vaporization occurred as long as single phase flow existed upstream of the orifice. It was found that the orifice performance was not influenced when static pressures measured in the meter were as much as 10 in. of mercury below the vapour pressure for the liquid. The basic conclusion of the report is that orifices may be used to measure the flow of liquid hydrogen and nitrogen with approximately the same degree of confidence as that for water flow.

Brennan (1964) studied the flow of liquid hydrogen and nitrogen through very small orifices which were 0·0145 and 0·0184 in. diameter. These orifices discharged into a pressure field below the triple point of the fluid. It was thought that these very low downstream pressures might produce phase changes in the fluid which would influence the flow metering accomplished by pressure measurements. It is found, however, that the orifice discharge coefficients were not affected by phase changes so that the usual discharge coefficients could be used for this special case.

4. *Compressible Flow Through Venturi, Nozzles and Orifices*

For compressible subsonic flow the meter equation must be modified to account for changes in the expansion of the fluid. Consideration of these factors produces the empirical equation

$$Q = Y M_v C A \left(\frac{2(p_1 - p_2)}{\rho} \right)^{\frac{1}{2}} \tag{5.29}$$

where Y is the expansion factor determined by Figs. 5.28 and 5.29 and

$$M_v = (1 - \beta^4)^{-\frac{1}{2}} \tag{5.29a}$$

and

$$\beta = \frac{D_{\text{throat}}}{D_{\text{pipe}}}$$

C is the discharge coefficient determined by Fig. 5.30.

G. NON-CIRCULAR CONDUITS

For frictional pressure drop, the simplest approach is to use the hydraulic diameter of the non-circular conduit to determine the Reynold's number and subsequently the friction factor from Fig. 5.1. Then, one can use equation 5.7, omitting the body forces and acceleration terms to evaluate the frictional pressure drop. Nikuradse (1930) and Schiller (1923) reported that this system, shown in Fig. 5.31, is satisfactory for many design cases. No pressure drop or friction factor data for these kinds of sections using cryogenic fluids are available. This procedure is essentially empirical and has even less theoretical foundation than the use of the friction factor system for circular conduits.

There has been substantial work, particularly with triangular sections, to place these studies on a more theoretical basis. Among these are Deissler and Taylor (1958), Eckert and Irvine (1960), and Carlson and Irvine (1961). The more recent data are shown in Fig. 5.32 and the reader is referred to these works if he needs a specific set of data and more basic information than that provided by the simple hydraulic diameter system.

The previous discussion has dealt with fully developed flow. Entrance effects have also been investigated in non-circular conduits. It has been found that the entrance effects or the equivalent length to develop a steady velocity profile is much longer than in circular tubes. Eckert and Irvine (1957) have shown that this entrance length varies with Reynold's number and have reported a value of entrance effect length over a hydraulic diameter of 20 for a Reynold's number of 140. This value increased to a ratio equal to 140 for a Reynold's number of 4,950. Irvine and Eckert (1958) suggested the use of an equation for the entrance pressure drop as

$$\Delta p_{entrance} = \frac{\rho u^2}{2} \left(\frac{f \, \Delta L}{2r_h} + K_{IE} \right) \tag{5.30}$$

The value for K_{IE} in this equation has been empirically and theoretically determined to be approximately 2·5 for laminar flow. This value is within $\pm 20\%$ agreement with experimental data for isosceles triangles and rectangular cross-sectional ducts. For turbulent flow the authors report that the entrance pressure drop is usually only moderately larger than $\rho u^2/2$.

V. Two-Phase Flow

Since two-phase flow is a very broad and complex subject the presentation will be deliberately restricted to a minimum survey of background material coupled with present recommended equations and procedures for the particular two-phase flow situations most likely to occur in design studies. Two-phase flow studies are quite important to cryogenic systems because the liquids are almost always very near the saturation temperatures and two-phase flow sometimes cannot be avoided in the flow processes. Also, as in other fluid systems, the two-phase process is sometimes desirable or necessary for the optimal operation of a system. Two-phase flow is almost sure to occur in systems during cool-down.

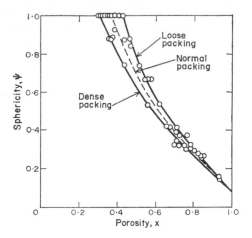

Fig. 5.23(a). Chart for estimating sphericity from porosity X.

D = conduit diameter
Dp = equivalent diameter (diameter of sphere having same volume as particle)
f_{mod} = modified friction factor ($f \times F_f$)
Re_{mod} = modified Reynold number (Re/F_{Re})
X = porosity of bed (volume of voids/volume of bed)
ψ = particle sphericity (area of sphere having same volume as particle/area of particle)

Pressure drop calculation

(1) Plot porosity X on Fig. 5.23a knowing packing, to find sphericity ψ.
(2) Plot X, ψ on Fig. 5.23b find Reynolds number function F_{Re}.
(3) Plot X, ψ on Fig. 5.23c to find friction factor function F_f.
(4) Calculate modified Reynolds number:

$$Re_{BR} = \frac{D_p G F_{Re}}{\mu}$$

(5) On standard friction factor chart (Fig. 5.1) plot Re_{mod} and find friction factor f.
(6) Calculate modified friction factor:

$$f_{BR} = f \times F_f$$

(7) Calculate pressure drop across bed:

$$\Delta P = \frac{f_{BR} G^2 L}{2 \rho D_p}$$

A. FLOW PATTERNS

A number of studies have attempted to define distinct flow patterns and predict transitions from one flow pattern to another. Perhaps the classical work in this field is that of Baker (1954) who proposed the plot shown in Fig. 5.33 to generally describe two-phase flow patterns for many fluids. Bronson *et al.* (1962) has reported some agreement with the Baker plot using cryogenic fluids.

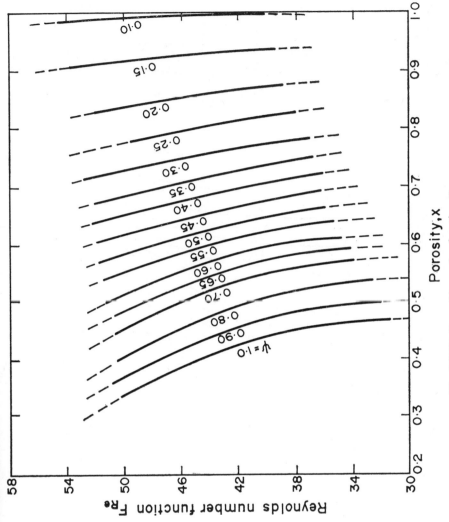

Fig. 5.23(b). Reynolds number function F_{Re}.

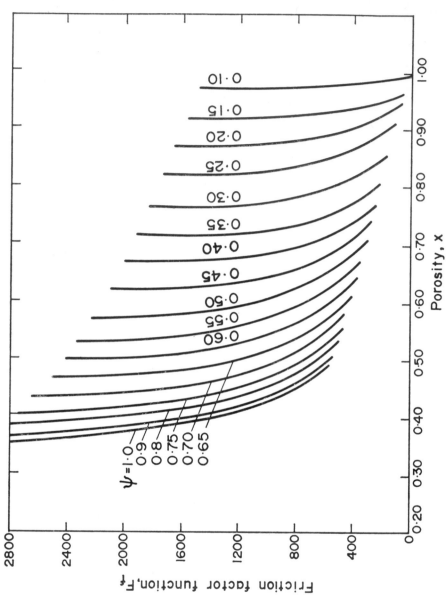

Fig. 5.23(c). Friction factor function F_f. All from Brownell (1950) with permission from AIChE.

Almost all these studies agree that a knowledge of flow pattern is important in making an analysis of the flow or in presenting experimental data. Unfortunately, however, the development of a system to predict flow patterns or understanding of the proper use of flow pattern information in analytical

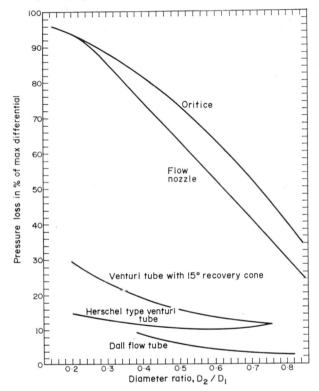

Fig. 5.24. Overall pressure loss through several primary elements. From ASME Fluid Meters (1959) with permission.

work has not progressed to the point where designers of cryogenic systems can make optimal use of such information. Therefore, in this chapter, flow pattern information will be used only as background material. The correlations and techniques presented will not, in general, be dependent upon flow pattern information.

B. TWO-PHASE PRESSURE DROP

1. Simple Systems

The pressure drop in two-phase flow will be considered initially for systems involving simple geometry, and conventional fluids. This will be followed by a section devoted to cryogenic fluids and finally, a section discussing more complex geometries.

Equations 5.6 and 5.7 show the momentum and frictional pressure drop terms. In two-phase flow the compressibility of the gas component often produces a momentum pressure drop which is much greater than the frictional pressure drop. Therefore, for many applications the calculation of the

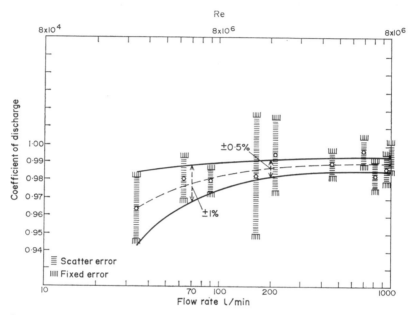

Fig. 5.25. Discharge coefficients for venturi tube carrying liquid hydrogen. From Purcell *et al.* (1960).

momentum pressure drop may be sufficient to describe the total pressure drop. This was the case for hydrogen flowing in heated tubes as reported by Hendricks *et al.* (1961). The simplest model for two-phase flow calculations is one which assumes a homogeneous mixture and thermal equilibrium. This means that the liquid and gas are uniformly mixed throughout the fluid and that thermal equilibrium is maintained at all times. The thermal equilibrium assumption allows a calculation of quality based only on the pressure change in the fluid. Hendricks *et al.* (1961) were able to use this model to calculate the momentum pressure drop in their system and the calculated results were within the range of experimental error. Therefore, for this case it was not necessary to consider the frictional pressure drop or was it necessary to consider velocity differences in the gaseous and liquid phases or the possibility of non-equilibrium between the phases.

When two-phase frictional pressure drop must be calculated one normally turns to the work of Martinelli and his co-workers reported in Lockhart and Martinelli (1949) and Martinelli and Nelson (1948). Dukler *et al.* (1964) in an exhaustive review study of the frictional pressure drop have shown that

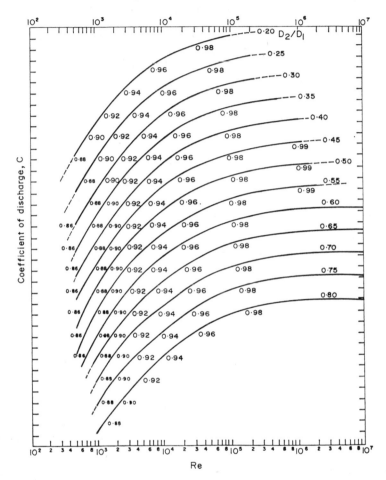

Fig. 5.26. Discharge coefficients of ASME long-radius flow nozzles with pressure taps at $1D$ and $\frac{1}{2}D$ or $\frac{3}{2}D$. (For nozzles in 2 in. pipe and larger) with permission from ASME Fluid Meters (1959).

the Martinelli correlation is still one of the most effective. Furthermore, almost all data for cryogenic fluids have been analyzed by use of the Martinelli approach.

The Martinelli method proposes that the pressure gradient experienced by either of the phases flowing alone may be used to determine the two-phase pressure gradient. Specifically, they develop a correlating term which is the ratio of the pressure gradient for the gas flowing alone to that of the liquid phase flowing alone,

$$\chi_{tt} = \left[\frac{(\Delta p/\Delta x)_g}{(\Delta p/\Delta x)_f}\right]^{\frac{1}{2}} \qquad (5.31)$$

The subscript, tt, indicates turbulent flow for both phases which is the case for most cryogenic flow applications. For small pressure increments indicating

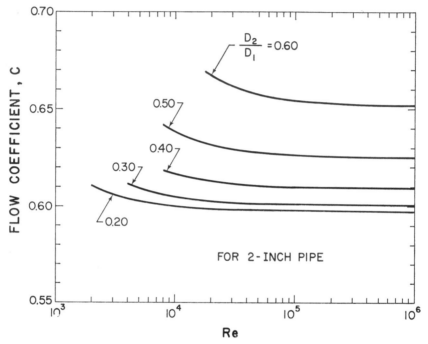

Fig. 5.27. Flow coefficient C for thin-plate orifices. (Flange taps-pressure taps 1 in. upstream and 1 in. downstream from flange.) Taken from ASME Fluid Meters (1959) with permission.

small property changes, χ_{tt} may be expressed in terms of the fluid properties and quality as

$$\chi_{tt} = \left(\frac{v_f}{v_g}\right)^{0.571}\left(\frac{\mu_f}{\mu_g}\right)^{0.143}\left(\frac{1}{x_q}-1\right) \qquad (5.32)$$

This correlating term may be then expressed as a function of pressure and quality for a single component because the properties must fall along the saturation line. Figure 5.34 shows χ_{tt} as a function of pressure and quality for hydrogen, nitrogen, and oxygen.

This Martinelli parameter, χ_{tt}, is then empirically related to the ratio of the two-phase pressure gradient to the pressure gradient of the total flow if it were in the liquid phase alone as,

$$\Phi_l = \left(\frac{(\Delta p/\Delta x)_{TP}}{(\Delta p/\Delta x)_{f,\text{tot}}}\right)_{\text{friction}} = f(\chi_{tt}) \qquad (5.33)$$

This relationship does not require a knowledge of the flow pattern, the degree of equilibrium or the gas and liquid velocity. The correlating curve is shown in Fig. 5.35 which was obtained from data for air and water. This correlation

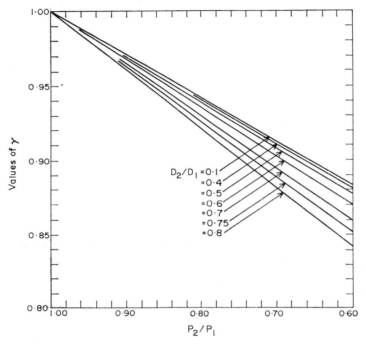

Fig. 5.28. Expansion factor for square-edged orifices with $y = 1\cdot4$. With permission from ASME Fluid Meters (1959).

has also proved effective for single component flow for a wide range of fluids both in adiabatic and non-adiabatic flow.

It should be emphasized that the Martinelli curve shown is for fluids far removed from the thermodynamic critical point. To account for the necessary changes as the thermodynamic critical is approached Martinelli and Nelson (1948) have proposed a series of curves for water. A similar set of curves have been prepared by Rogers (1964) and Rogers and Tietjen (1969) for hydrogen and nitrogen which are shown in Fig. 5.36. They assume that the Martinelli curve (Fig. 5.35) is universally applicable for all fluids at a pressure of 1 atm.

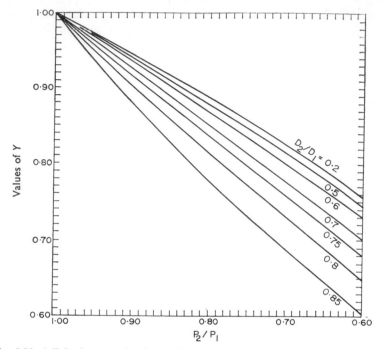

Fig. 5.29. Adiabatic expansion factor for use with venturi tubes and flow nozzles with $\gamma = 1.4$. With permission from ASME Fluid Meters (1959).

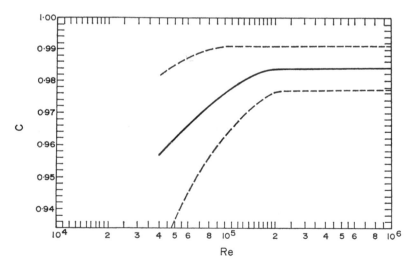

Fig. 5.30. Discharge coefficients for venturi tubes as a function of pipe Reynolds number. (Applicable to diameter ratios from 0.25–0.75 in pipes 2-in. and larger. The tolerance limits are shown by the dotted lines.) With permission from ASME Fluid Meters (1959).

It is sometimes desirable to use different phase velocities in calculating the momentum pressure drop. For this, one may use the void-fraction curves shown in Fig. 5.35 identified as R_f and R_g. For this case the momentum pressure drop becomes

$$\Delta p_{TP}_{\text{mom}} = G^2\left[\frac{x_q^2 v_g}{R_g}+\frac{(1-x)^2 v_f}{R_f}\right]_1 - \left[\frac{x_q^2 v_g}{R_g}+\frac{(1-x)^2 v_f}{R_f}\right]_2 \tag{5.34}$$

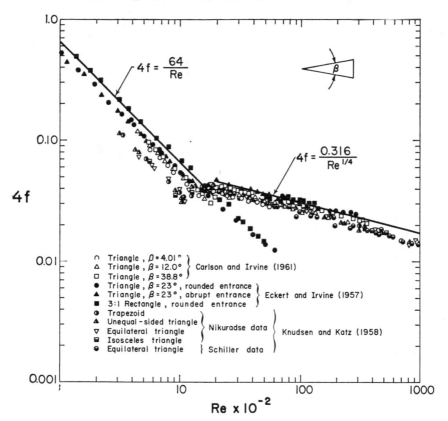

Fig. 5.31. Plot of friction-factor data for conduits of non-circular cross section.

2. Two-phase Pressure Drop for Cryogenic Fluids

Cryogenic fluids have different properties than the fluids for which the Martinelli and other frictional pressure drop correlation were developed. For two-phase flow, the major difference is shown in the density ratio column of Table 5.II. For this reason some investigators have chosen to use the halide refrigerants (trade name Freon) as a means of simulating nitrogen and oxygen.

12

Timmerhaus and Sugden (1965) and Sugden *et al.* (1967), showed that some refrigerant two-phase, friction factor data could be plotted as a function of Reynold's number with the quality as a parameter. This method was simple and straightforward but the accuracy and range of the data were insufficient to verify the reliability of the method. Hatch and Jacobs (1962) found that the Martinelli correlation generally predicted the halide, Refrigerant-11 and hydrogen pressure drop data as shown in Figs. 5.37 and 5.38.

TABLE 5.II. Typical Gas–Liquid Density Ratios of Cryogenic Fluids and Other Systems at the Normal Boiling Point

Fluid	ρ_f/ρ_g	Boiling temperature °K
Water–Steam	1,599·59	373
Water–Air	819·74	
Refrigerant–11	256·41	297
Refrigerant–12	231·71	244
Oxygen	254·58	90·4
Argon	241·71	87·4
Nitrogen	175·86	78·5
Hydrogen	52·74	20·3
Helium	7·59	4·2

Lapin and Bauer (1967) performed some experiments to determine the two-phase flow frictional pressure drop for nitrogen and methane in straight circular tubes. Their data are shown compared to the Martinelli and a similar, Chenoweth–Martin, correlation in Figs. 5.39, 5.40, and 5.41. Better agreement might have been achieved if the pressure correction had been used as suggested by Martinelli and Nelson (1948) and by Rogers (1964).

Turney and Cox (1967) studied pressure drops in a high-pressure hydrogen flow facility simulating a nuclear rocket nozzle. Their data were taken as the system was cooled. Therefore, gaseous hydrogen first entered the nozzle followed by a two-phase fluid entrance somewhat later and finally there was a liquid phase at the entrance. They compared their experimental data with the Martinelli correlating curve and found reasonably good agreement as shown in Fig. 5.42. In this work a correction was added to account for the curvature of the tubes flowing along the nozzle contour. However, the pressure corrections shown in Fig. 5.36 were not made. The authors suggested that the deviation between experimental and predictive data may be explained by non-equilibrium flow producing incorrectly calculated qualities.

Forced convection pressure drops were reported for the flow of two-phase helium by de la Harpe *et al.* (1968). They tried several correlations and reported that the homogeneous model best predicted their experimental pressure drop data. They found, however, that the Martinelli parameter, χ_{tt}, was useful in correlating their heat transfer data.

Bartlit (1964) has developed a method which provides an effective quality for calculating pressure drop providing the final exit quality is known and the entrance quality is zero. This system avoids the tedious step-wise calculation associated with the use of the Martinelli method. His work, shown in

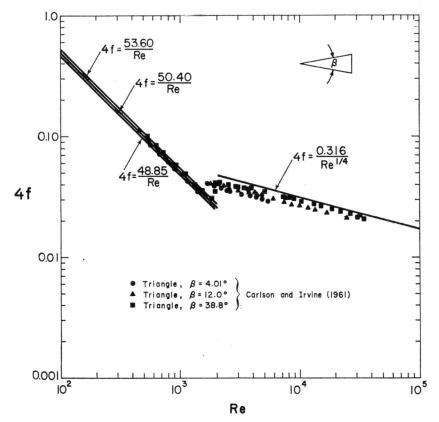

Fig. 5.32. Experimental data of Carlson and Irvine (1961). Lines are laminar solutions of Yen (1957).

Figs. 5.43 and 5.44 is specifically for hydrogen but the system is applicable for any single component fluid.

3. Geometry Effects

For contractions and expansions the momentum pressure drop probably represents the total pressure drop to a reasonable degree of accuracy.

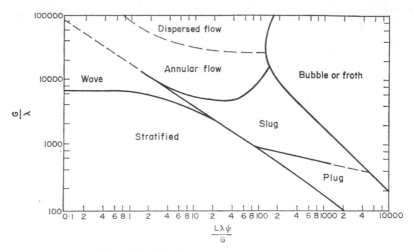

Fig. 5.33. (*a*) Plot of two-phase flow regimes.

$$\lambda = \left[\left(\frac{\rho_g}{0 \cdot 075}\right)\left(\frac{\rho_l}{62 \cdot 3}\right)\right]^{\frac{1}{2}}, \quad \psi = \left(\frac{73}{\nu}\right)\left[\mu_l\left(\frac{62 \cdot 3}{\rho_l}\right)\right]^{\frac{1}{2}}$$

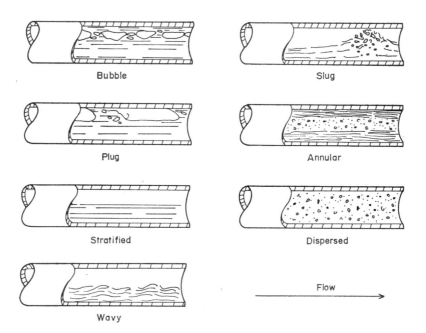

(*b*) Sketches of flow patterns as suggested in Fig. 33*a*. From Baker (1948) with permission from Petroleum Publishing Co.

Expansions would tend to produce a separated-phase, flow-model and momentum pressure drop using that model was developed by Romie *et al.* (1960).

$$p_2 - p_1 = \rho_f u^2 \left[1 - \left(\frac{A_1}{A_2} \right)^2 \right] \left[\frac{x_q^2 \rho_f}{\alpha \rho_g} + \frac{(1 - x_q)^2}{1 - \alpha} \right] \tag{5.36}$$

Fig. 5.34(*a*). Martinelli–Nelson correlating term χ_{tt} as a function of quality and pressure for hydrogen. From Smith (1963a).

This expression was tested by Lottes (1961) who found it to be the best of several in predicting experimental data obtained using water over a range of pressures.

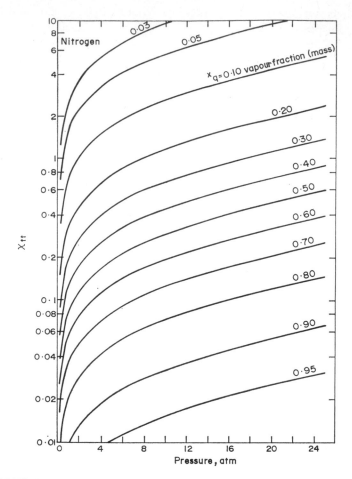

Fig. 5.34(*b*). Martinelli–Nelson correlating term χ_{tt} as a function of quality and pressure for nitrogen. From Smith (1963a).

For contractions, the tendency would be toward a homogeneous model as exemplified by Mendler (1963) and Geiger (1964) have proposed the use of this model. Geiger's expression is

$$p_2 - p_1 = \left(1 + \frac{x_q v_{fg}}{v_f}\right)^{-1} \frac{\rho_f u^2}{2}\left(1 - \left(\frac{A_1}{A_2}\right)^2 + K_G\right) \tag{5.37}$$

where

$$K_G = \left(\frac{1}{C_c} - 1\right)^2$$

C_c given by Weisbach (1855) is

A_1/A_2	0	0·2	0·4	0·6	0·8	1·0
C_c	0·617	0·632	0·658	0·712	0·813	1·00

Murdock (1962) studied the flow through orifices with steam–water, air–water, natural gas–water, natural gas–salt water, and natural gas–

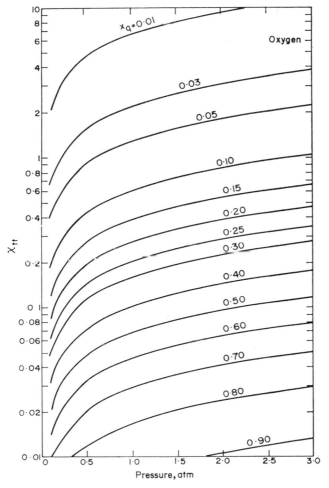

Fig. 5.34(c). Martinelli–Nelson correlating term χ_{tt} as a function of quality and pressure for oxygen. From Smith (1963a).

distillate combinations. He developed an expression for the flow rate through orifices considering single-phase pressure drop terms. The agreement with the experimental data was remarkably good.

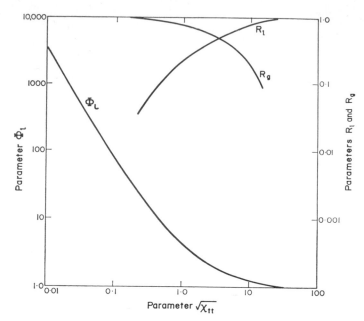

Fig. 5.35. Two-phase pressure-drop function Φ_l and volume fractions R_l, R_g. From Martinelli and Nelson (1948) with permission from ASME.

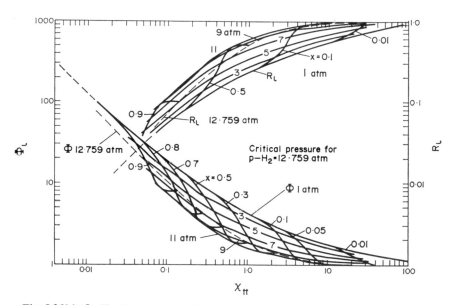

Fig. 5.36(a). Φ_l–X_{tt}–P–χ_q surface at lower left and R_l–χ_{tt}–P–χ_q surface at upper right for hydrogen. From Rogers (1964) with permission from Plenum Press.

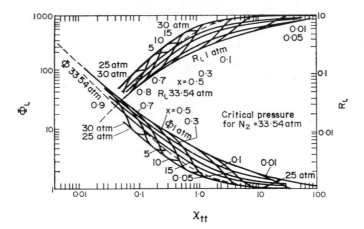

Fig. 5.36(b). $\Phi_l-X_{tt}-P-\chi_q$ surface lower left and $R_l-\chi_{tt}-P-\chi_q$ surface upper right for nitrogen. From Rogers and Tietjen (1969) with permission from AIChE.

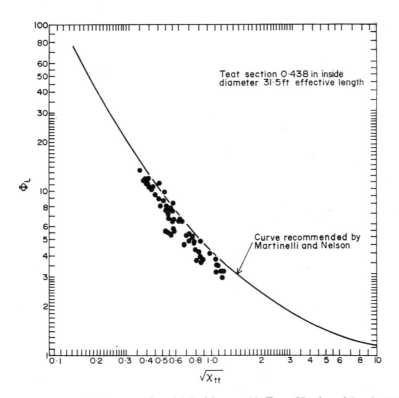

Fig. 5.37. Tests with heat transfer with Refrigerant 11. From Hatch and Jacobs (1962).

12*

C. COOL-DOWN

Cool-down in this section is defined as the process which occurs when a cold liquid is introduced into a system producing gaseous and two-phase flow until finally all liquid flow prevails, and the system is cooled to essentially the temperature of the cryogenic liquid. This process is transient and very complex and substantially involves the heat transfer process. In this section we will deal with the cool-down of relative long paths where the fluid flow is a primary factor.

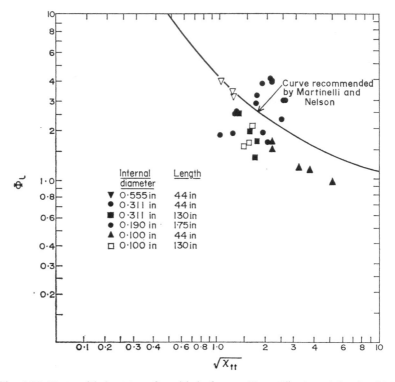

Fig. 5.38. Tests with heat transfer with hydrogen. From Hatch and Jacobs (1962).

Two specific subjects will be treated in this section; that of cool-down time (the period from the introduction of the cryogenic fluid until the system is completely cooled) and secondly, oscillations during cool-down of both pressure and flow rate. For flow sections with a length-over-diameter ratio of a thousand or more a physical and analytical model may be obtained by study of experimental data. The cooling of each station over a relatively short period of time, compared to the total cool-down time, indicates the flow past the station is first a gas, probably relatively warm, secondly, a two-phase section which is short and followed closely by all-liquid flow. Bronson *et al.*

(1962) data shown in Fig. 5.46 also observed this "cold front" type of cool-down. Oscillations with large amplitudes have been observed in most cool-down studies. Steward *et al.* (1969) showed that oscillation pressures may be several times the fluid driving pressure. They also found that reversed flow can occur in the upstream portion of the flow line caused by a development of a peak pressure in the midpoint of the line higher than the driving pressure of the dewar supplying the cool-down liquid.

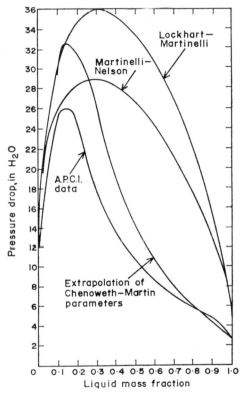

Fig. 5.39. Pressure drop characteristics in two-phase methane stream at 400 p.s.i.a., $-144.5°F$, 1.42×10^5 lb. m/hr ft². From Lapin and Bauer (1967) with permission from Plenum Press. (APCI indicates data of Air Products and Chemicals, Inc.)

1. *Cool-down Time*

Four studies, Burke *et al.* (1960), Bronson *et al.* (1962), Drake *et al.* (1961) and Steward *et al.* (1969), have analytically and experimentally considered cool-down time. They have used a model which was previously described involving a "cold front" consisting of a short two-phase section or essentially a liquid-vapour interface proceeding along the line until this front reaches the exit at which time the cool-down time is completed. Thus, this model is essentially a gas column followed by a liquid column. Except for Steward *et al.* (1969), the authors have assumed a flow restriction at the exit of the section. It follows that the escaping gas flow through that restriction, usually at sonic velocity, regulates the flow rate during the cool-down process. Good results were reports by Burke *et al.* (1962) and Drake *et al.* (1961) in comparing predicted and experimentally measured cool-down times. Bronson *et al.* (1962) also showed good results. These studies are limited to cases where there is substantial restriction at the exit of the line to be cooled. Steward (1968) and Steward *et al.* (1969) used essentially the same model but did not add the requirement of a restriction at the exit. These results are illustrated in Fig. 5.45 and must be regarded as quite satisfactory for longer insulated lines. Steward *et al.* (1969) have provided a simplified computational procedure for cool-down time which is shown in Fig. 5.46.

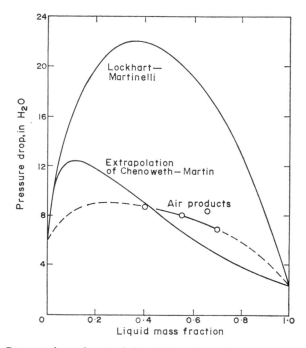

Fig. 5.40. Pressure drop characteristics of two-phase nitrogen stream at 400 p.s.i.a., −240°F, 1.71×10^5 lb. m/hr ft². From Lapin and Bauer (1967) with permission from Plenum Press.

2. *Cool-down Oscillations*

Cool-down oscillations are transient and persist only for a portion of the cool-down time in contrast to fluid oscillations discussed later which are essentially steady state in their behaviour. Steward (1965) and Steward *et al.*

(1969) have made extensive experimental and analytical studies of these oscillations. In the first study, a cold-front type model was used, discussed previously, and in the second study a model of homogeneous segments was used. These models do not differ as much as one might suspect because a model consisting of a series of homogeneous segments becomes a cold-front model when there are only a few segments in the two-phase region which have essentially liquid segments upstream and gas segments downstream. Both models were reasonably successful in predicting the experimental cool-down oscillations in amplitude and frequency using liquid nitrogen, as shown in Fig. 5.47.

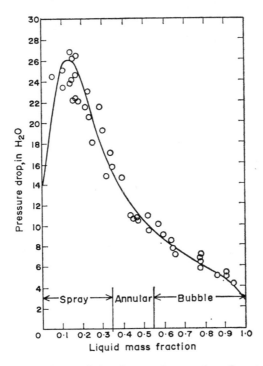

Fig. 5.41. Pressure drop characteristics of a two-phase methane flow stream at 400 p.s.i.a., 1.4×10^5 lb. m/hr ft². From Lapin and Bauer (1967) with permission from Plenum Press.

D. FLUID OSCILLATIONS

Fluid oscillations, primarily associated with the density differences in the fluid, often play a major role in two-phase flow behaviour. These conditions also occur when the fluid is near the thermodynamic critical point and similar oscillation behaviour is found. The frequency of these oscillations may vary from $\frac{1}{20}$ to 10,000 Hz and the amplitude may be substantial (about

one-third of the system static pressure). These oscillations can produce
system failures, particularly if the flow is controlled by a pressure or temperature signal. Structural failures may also be caused by high amplitude pressure
oscillations.

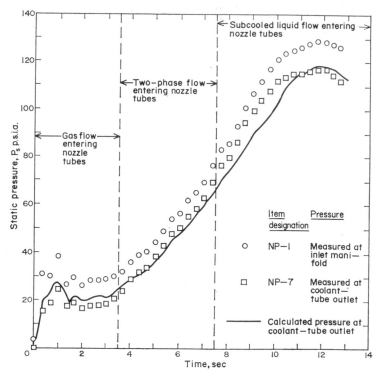

Fig. 5.42. Comparison of measured and computed P_s (static pressure at nozzle outlet) at
RN–2 nozzle coolant tube outlet as a function of time for liquid hydrogen test. The experimental pressures are higher than those predicted, indicating actual pressure drop was smaller
than that predicted. From Turney and Cox (1967).

   In an extensive review of fluid oscillations, Zuber (1966) reported eighteen
studies involving severe oscillations in supercritical systems. Certain rules
may be established in determining whether or not oscillations are likely to
be present and what geometric and flow situations are likely to increase or
decrease the oscillating characteristics. Zuber (1966) suggested the map shown
in Fig. 5.48 to indicate the likelihood of oscillating behaviour.
   Flow and process parameters which influence oscillations are:
1. A high heat input will tend to increase the flow instabilities because it
   enhances the density changes.
2. Increased pressures for single component fluids tend to reduce the oscillation amplitudes because the density difference between the liquid and gas

and hence the mixture density changes become smaller. As the reduced pressure (p/p_{crit}) is increased in the two-phase region to 1 and greater, these oscillations tend to persist.

3. Increased resistance at the exit of an oscillating system decreases the stability of the system and an increased resistance at the entrance to the

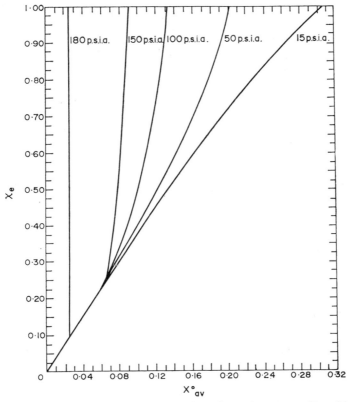

Fig. 5.43. Average exit quality (x_e) as a function of x_{avg} (average quality with constant wall temperature) and pressure. The plot is for hydrogen flowing isothermally in a pipe. From Bartlit (1964).

(1) Knowing the inlet pressure and exit quality, use Fig. 5.43 to find average quality with constant wall temperature, Set $X_{avg} = X_{avg}^\circ$.

(2) If wall temperature is not constant but ΔT_w is known, find an X_{avg}°, then use Fig. 5.44 to calculate X_{avg}.

(3) Calculate Lockart–Martinelli parameter:

$$X_{tt} = \left(\frac{1-X_{avg}}{X_{avg}}\right)^{0.9}\left(\frac{\mu_g}{\mu_f}\right)^{0.1}\left(\frac{\rho_f}{\rho_g}\right)^{0.5}$$

(4) Fig. 5.35 gives the two-phase friction factor Φ_l.

(5) Knowing Φ_l and the liquid-phase-only pressure drop, determine the two-phase pressure drop:

$$\left(\frac{\Delta P}{\Delta \chi}\right)_{TP} = \Phi_l^2 \left(\frac{\Delta P}{\Delta \chi}\right)_f.$$

Note: For Figs. 5.43 and 5.44 $X = X_g =$ quality.

system increases its stability. This is because the resistance at the inlet to the system will tend to produce a high driving pressure and clear the variable-density section. Conversely, a resistance at the exit will produce a high back pressure which will tend to hold the variable density fluid in the section.

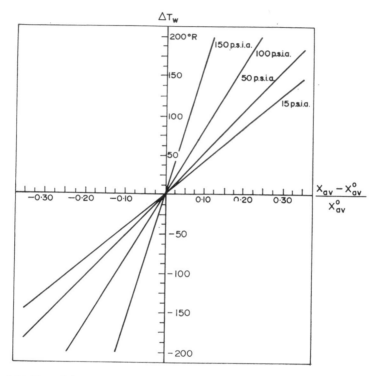

Fig. 5.44. Plot of ΔT_w (pipe wall inlet temperature minus outlet temperature). X_{avg} is the average total quality used for pressure drop. From Bartlit (1964) with permission from Los Alamos Scientific Laboratory. See (1)–(5) Fig. 5.43

These fluid oscillations may be divided into two general types, omitting the gas-like oscillations above 300 Hz. They are the thermal acoustic oscillations (5 to 300 Hz) and the system oscillations ($\frac{1}{20}$ to 5 Hz). The thermal acoustic oscillations tend to behave more as a wave phenomena in the fluid where the systems oscillations involve motions in the entire fluid system.

1. *Thermal Acoustic Oscillations*

Thurston (1966) and Thurston *et al.* (1967) have separated these oscillations into two types. The first type, a Helmholtz oscillation, is generated in a system with a cavity connected to a large plenum with a substantial restriction between the two. The natural frequency of this system may be expressed by the equation

$$W_{nat} = \frac{A}{2\pi}\left(\frac{\gamma p}{mv_g}\right)^{\frac{1}{2}}$$

(5.38)

This Helmholtz resonance has not been reported in many systems and may be presumed to occur only when the system fits the analytical model rather closely.

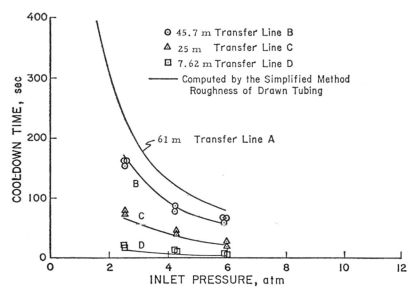

Fig. 5.45. Cool-down time for subcooled liquid nitrogen showing the effect of length. From Steward (1968).

The second type of thermal acoustic oscillation is a sonic wave resonance and is associated with the velocity of sound and the length of fluid in the flow loop. The natural frequency of this system may be expressed by

$$f_{nat} = \frac{a}{2L}$$

(5.39)

The period is the time required for the acoustic wave to traverse one-half the fluid length. The frequency of these oscillations has been found experimentally to be a weak function of the heating rate and has been shown to change appropriately with fluid length thereby substantiating equation 5.39. Thurston et al. (1967) have also proposed empirical relationships to determine the onset of the oscillations. Their criteria for stability are

$$S_r \leqslant 0.0045(S_v)^{0.5}$$

where

$$S_r = \frac{q}{G h^*_{fg}}$$ (5.40)

and

$$S_v = \frac{v^*_{fg}}{v_f}$$

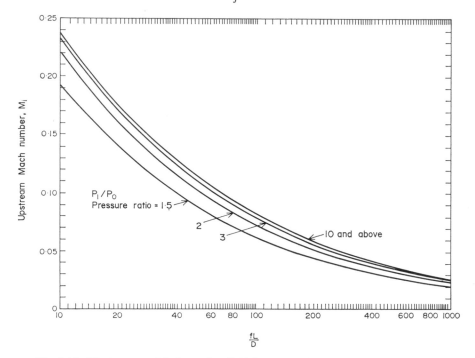

Fig. 5.46a. Upstream gas Mach number (M_1) for Fanno flow. L is total pipe length, D is diameter. From Steward *et al.* (1967).

a_s = sonic velocity
Δh_f = liquid sub-cooling enthalpy
$\Delta h_t = h_g - h_{f_{sat}}$
ρ_{gi} = density of upstream warm gas.

An iterative procedure must be followed to find f, from which the cool-down time can be determined.

(1) Using an initial estimate of f, calculate fL/D.
(2) Knowing the pressure ratio P_1/P_0, use Fig. 5.46a to find an initial Mach number M_i.
(3) Calculate a first estimate of the Reynolds number:

$$\therefore Re = \frac{\rho_g D M_i u_s}{\mu_g}$$

(4) On Fig. 5.1 find new estimate of friction factor.
(5) Repeat the process until f has been determined with sufficient accuracy.
(6) Using Fig. 5.46b, $ta/\beta L$ and then t, the cool-down time, can be found.

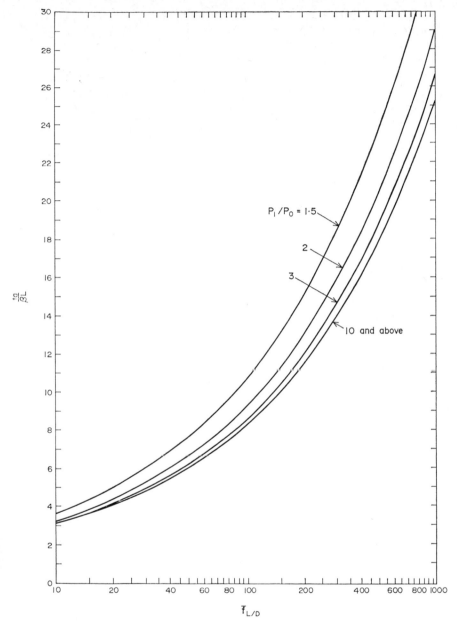

Fig. 5.46*b*. Cool-down time parameters, *t* is cool-down time, *B* is the enthalpy change parameter

$$B = 1 + \frac{\rho_w A_w \Delta h_w}{\rho_{gi} A_f (\Delta h_f + \Delta h_\tau)}$$

From Steward *et al.* (1967).

The terms denoted by $(\overset{*}{_{fg}})$ are the phase change terms for the two-phase case and pseudo two-phase behaviour in the near critical region. The symbol S_r is for the Stermann or boiling parameter. Other authors have identified this threshold by applying stability criteria to the fluid and employing the general conservation equations. These will be described later in this section.

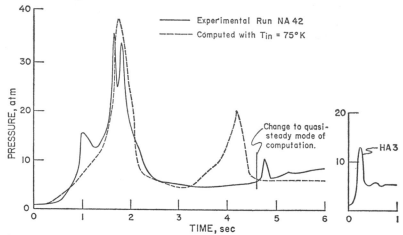

Fig. 5.47. Subcooled liquid nitrogen maximum pressure surge 6·1 m from reservoir along a 61 m transfer line. (Corresponding maximum hydrogen surge at right for comparison.) $P_{in} = 6\cdot1$ atm. From Steward (1968).

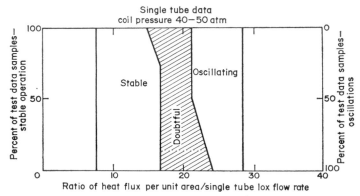

Fig. 5.48. Percentage of heat exchanger test data samples showing steady operation versus heat flux per unit area per unit mass flow rate. Taken from Zuber (1966).

2. Systems Oscillations

These large amplitude oscillations, often found in two-phase and near-critical flow, are by far the most serious of the two types and can prevent the operation of a flow system.

Analyses are based on a boiling-like phenomenon with large density changes in both the two-phase and near-critical region. The basic approach is to supplement the time dependant, one-dimensional conservation equations of continuity, energy, and momentum with an equation of state which exhibits a discontinuity at the saturated liquid line or at the equivalent point in the near critical region.

Two general methods have been used to predict the onset of oscillations. The first attributed to Bouré, and shown in Craya and Bouré (1966), might be called the density effect model. This method has been used and extended by a number of authors among them Harden and Boggs (1964), Walker and Harden (1964), and Harden and Walker (1967). Walker and Harden (1964) found six independent variables which would influence the inception point. The two geometric parameters and the friction factor, could be considered essentially constant for a given system. The remaining three variables are fluid velocity, gravitational force, and sub-cooling. Walker and Harden (1964) were able to compute a threshold surface involving the three remaining variables and to predict instabilities in a Refrigerant-114 natural convection loop with good reliability.

The second criteria used is the Nyquist-type test to determine the stability of the system. This method is widely used for electrical systems and can be employed in fluid systems by utilizing analogous behaviour. For example, the resistance and capacitance of a fluid system components may be modelled electrically. Examples of such modelling may be found in the work of Zuber (1966) Friedly et al. (1968) and Wallis and Heasley (1961). These methods have produced reasonably good success in predicting the onset of systems oscillations. This would indicate that the proposed mechanism is qualitatively correct, however, the lack of agreement and uncertainties involved in these calculations point up the complexity of the phenomena.

Hendricks et al. (1965) found some oscillating characteristics in the heat transfer study of cryogenic hydrogen flowing upward vertically with static pressures from 800 to 2,500 p.s.i.a. In the region nearest the critical point they found both vertical and lateral oscillations. The vertical oscillations did not appear to influence the heat transfer data and were at a lower frequency than the lateral oscillations (about 1,000 Hz). The lateral oscillations were observed to have a frequency of 4,500 to 8,000 Hz and substantially influenced the heat transfer process. It should serve as both a surprise and warning to find these oscillations occurring at approximately four times the critical pressure.

E. CRITICAL (CHOKING) TWO-PHASE FLOW

In single-phase flow of a compressible fluid a critical or choking condition will be reached if the pressure gradient is sufficiently large, for example, when flowing through a converging nozzle or a constant-area conduit. In both cases the choking or critical condition is reached at the exit plane.

Thus, with the upstream pressure unchanged, the critical condition is defined as the maximum flow rate or the flow rate which is not increased by further reductions in downstream pressure for either single-phase or for two-phase flow. Equation 5.13 is again generally applicable with the expression undefined for the critical condition.

Although the phenomena are not completely understood the predominate characteristics of critical two-phase flow have been identified as below.

1. The effective, critical velocities based on a mixture density are significantly lower than the critical velocities for either the liquid or gas phase alone. For the mixture, an analytical model which assumes homogeneous and thermal equilibrium conditions underpredicts the critical mass flow rate. This latter lack of agreement may be attributed, primarily, to differences in the phase velocities and nonequilibrium conditions. The axial pressure profile just upstream of the critical condition is very steep in both single and two-phase flow, and its shape is reasonably well described (although quantatively wrong) by this model.

2. Over a wide range of experimental conditions, geometry changes have a secondary effect on critical flow. This is also the case for critical, single-phase flow. This has been demonstrated by comparisons of reported data from various systems as shown by Smith (1968, 1963). The exception to this rule has been found in flow with nozzles with high entrance angles and with orifices. Here, the geometry effects are significant and have been reported by Edmonds and Smith (1965) Hoopes (1957) and Chisholm and Watson (1965).

3. Experimental data show that different fluids have the same general critical two-phase flow behaviour.

4. Similarities between critical two-phase and single-phase flow are found in the observation of shock waves reported by Muir and Eichorn (1963), Eddington (1967) and by Smith (1968). Data of Smith (1968) and Henry (1968) show evidence that critical flow pressure ratios, (p_c/p_0), are not substantially different from those with flow of the gas phase alone.

Some analytical models retain the thermal equilibrium assumption of the homogeneous, thermal equilibrium model but introduce values different from one for critical slip velocities. Among these are Cruver and Moulton (1966), Moody (1965) and Zivi (1964). For the critical slip velocity ratio they develop

$$\left(\frac{u_g}{u_f}\right) = \left(\frac{\rho_f}{\rho_g}\right)^{\frac{1}{3}} \tag{5.41}$$

Fauske (1961) suggested a slightly different relationship

$$\left(\frac{u_g}{u_f}\right) = \left(\frac{\rho_f}{\rho_g}\right)^{\frac{1}{2}} \tag{5.42}$$

The use of these relationships provides a reasonably good prediction for critical mass flow rates but the critical velocity ratios are in disagreement with experimentally determined ratios reported by Fauske (1965) and Klingbiel (1964).

Turning to models which consider non-equilibrium flow, Cruver and Moulton (1967) have shown that if the lower, experimentally-determined, slip velocities of Fauske (1965) and Klingbiel (1964) are considered along with some non-equilibrium assumptions, one can predict critical two-phase flow as well as with a thermal equilibrium model using equations 5.41 or 5.42. Henry (1968) has shown that low quality flow may be predicted quite well assuming essentially no critical slip velocity and by use of a non-equilibrium model.

Another approach, the vapour choking model, assumes the two-phase flow will be choked when the gas phase reaches the choking condition. Ryley (1965) has shown that this model describes steam–water flow if certain geometrical distributions of the droplets are assumed. Smith (1968) has shown that this model successfully describes air–water flow in a venturi both at the point of choking and for stations upstream of the choking. As in the case of Ryley, however, certain geometrical features in the flow have to be assumed and Smith described them by use of an empirical term.

The special property behaviour of cryogenic fluids could make their critical two-phase flow somewhat different from that of the more conventional fluid. Examples of these property deviations are low surface tension, relatively low liquid-to-gas density ratios, and possibly the compressible behaviour of the liquid. Smith (1963) has calculated the probable upper and lower critical flow rates for a number of cryogenic fluids. The lower limit was assumed to be the homogeneous thermal equilibrium model results. The upper limit employed a combination of models first assuming frozen flow (no phase change) for very low qualities, secondly, an empirical correlation in the middle quality regions and thirdly, for the highest quality region the vapour-choking model using slip velocities as indicated by the Martinelli and Nelson (1968) relationships. For a general guide the homogeneous model flow solutions are shown in Figs. 5.49, 5.50 and 5.51. The experimental data of Brennan *et al.* (1968) indicates that these limits generally encompass the experimental data.

Bonnet (1967) studied critical, two-phase flow of oxygen and nitrogen through orifices. For analytical comparison, he used the Moody (1965) equations modified to provide variable slip ratios. He reported best agreement for critical mass flow rates when the slip ratio was approximately equal to 1. These computed flow rates do not appear to be in agreement with those reported by Smith (1963b) for the homogeneous, thermal equilibrium case. Bonnet's experimental critical pressure was measured some distance from the point of critical flow. This does not allow a direct comparison with the experimental data of other investigators.

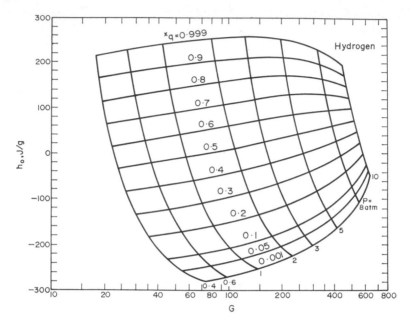

Fig. 5.49. Mass-limiting flow predictions at the point of choking for helium; homogeneous, thermal equilibrium model. From Smith (1963a).

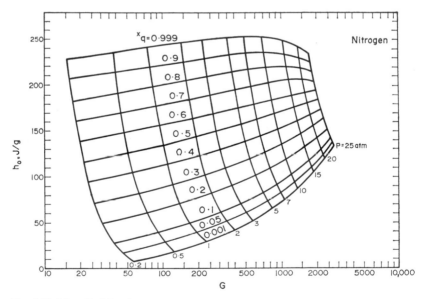

Fig. 5.50. Mass-limiting flow predictions at the point of choking for nitrogen; homogeneous, thermal equilibrium model. From Smith (1963a).

Bissell *et al.* (1969) considered two-phase flow in liquid hydrogen pumps employed in rocket engines. They found it necessary to consider the possibility of choking conditions at the inlet because its presence would substantially influence the pump performance. Different model assumptions predicted substantially different flow performance at the pump inlet.

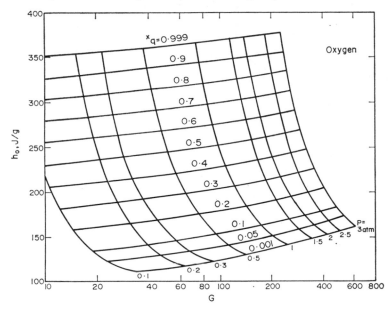

Fig. 5.51. Mass-limiting flow predictions at the point of choking for oxygen; homogenous, thermal equilibrium model. From Smith (1963a)

VI. Super-fluid Helium

The thermodynamic states of helium are shown in Fig. 5.52. We immediately notice the absence of a triple point together with the large pressures (about 25 atm) required to produce solid helium. Also there is a discontinuity in the specific heat curve at a temperature of 2·172°K, indicating a second order transition at lambda-point. The lambda-transition, still not completely understood, is easily recognized in practice since all visible boiling suddenly ceases and the liquid becomes very quiesent. Here, energy is transported so effectively that bubbles cannot develop. This efficient transport is sometimes called super-conductivity in the fluid, resulting from internal convection which will be discussed in more detail later. The liquid above the lambda-point behaves as a normal, viscous liquid and is called HeI. The behaviour of helium below the lambda-point however, cannot be described by the classical laws of fluid mechanics and instead fits a two-fluid model and is termed HeII.

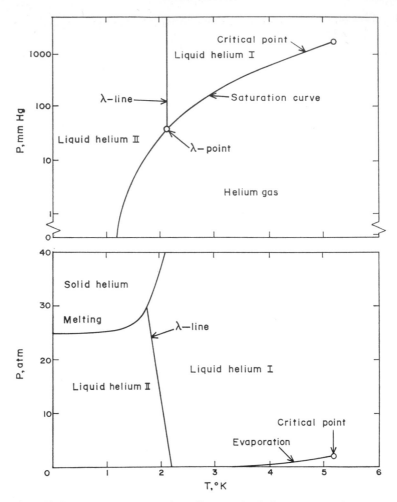

Fig. 5.52. Pressure-temperature phase diagrams for helium. From Smith (1969).

Although there are relatively few data with respect to flow and heat transfer characteristics of HeI, there are indications that it could be expected to exhibit a behaviour similar to that of a conventional fluid. HeII however, behaves quite differently and requires special treatment both in applications and analysis.

A. TWO-FLUID MODEL

The explanation of the special behaviour of HeII comes from the two-fluid model which suggests that the fluid may consist of superfluid and normal components. It should be emphasized at the outset that this model is an abstract concept and no single atom may be considered to be either superfluid or normal fluid. The model is an analytical concept which is successful in describing much of the HeII flow and heat transfer behaviour. First, the normal and superfluid components are primarily related by densities as;

$$\rho = \rho_N + \rho_{SUP} \tag{5.43}$$

where N and SUP subscripts represent the normal and superfluid, respectively. When a heat source is introduced in a bath of HeII, the model assumes that the components act in counter flow as shown in Fig. 5.53. There is no net mass flow for this case so

$$\rho_{SUP} u_{SUP} = -\rho_N u_N$$

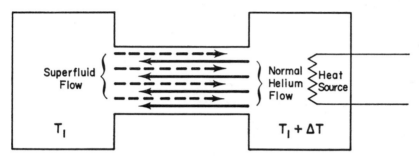

Fig. 5.53. Sketch of heat conduction by counterflow of the normal and superfluid components of helium.

The hydrodynamic equations of motion for the two-fluid model are similar to the Navier–Stokes equations. They are;

$$\rho \frac{du_N}{dt} = \frac{-\rho_N}{\rho}\nabla p - \rho_{SUP}S\nabla T + \mu\nabla^2 u_N + F_{SN} \tag{5.44}$$

$$\rho_{SUP} \frac{du_{SUP}}{dt} = \frac{-\rho_{SUP}}{\rho}\nabla p + \rho_{SUP}S\nabla T - F_{SN} \tag{5.45}$$

The term F_{SN}, called mutual friction, represents the inter-action between the normal and superfluid components. The onset of this interaction occurs at a critical velocity.

B. UNIQUE BEHAVIOUR

Much of the super-fluid behaviour may be summarized in noting two important characteristics of the super-fluid component. It behaves as a fluid with no viscosity and with zero entropy. This explains the efficient counter flow of the normal and super-fluid components, producing its high effective thermal conductivity (as much as 800 times that of copper at room temperatures) and also suggests the enormous advantage of HeII in a heat transfer situation. This high conductance of thermal energy is limited however,

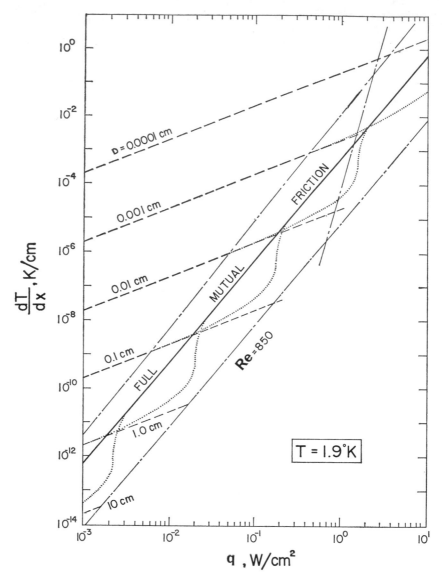

Fig. 5.54. Thermal gradient as a function of thermal flux at 1·9°K. From Arp (1969).

because vortices or turbulent-like behaviour develop in the super-fluid component or in the normal fluid component or in both. The onset of this condition is analogous to a critical Reynolds number in a conventional fluid flow indicating the onset of turbulence in that fluid. These limiting or critical velocities are a function of the temperature gradient in the direction of the heat flow and various geometries and fluid properties and history. In a recent review Arp (1969) has summarized the heat transfer behaviour of HeII and its behaviour and limiting conditions are shown in Figs. 5.54 and 5.55.

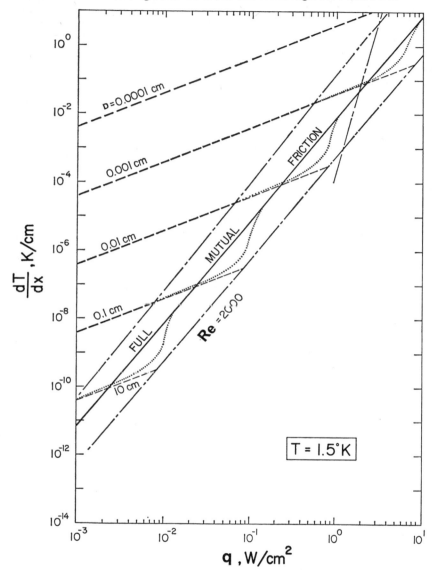

Fig. 5.55. Thermal gradient as a function of thermal flux at 1·5°K. From Arp (1969).

Another unique characteristic of HeII is its behaviour in a very closely packed media such as the narrow passage packed with jeweller's rouge or the employment of vycor glass. This narrow passage prevents the flow of an ordinary fluid which would include the normal component of HeII. The superfluid, however, can pass through this media and this flow is called a superleak. The grain size of the powder is likely to be less than 1 micron and the passage dimension substantially less than that.

A rather spectacular example of the superleak is shown in a sketch in Fig. 5.56. If energy is supplied downstream of the HeII reservoir, either at the superleak packing or downstream of it, the super-fluid portion will flow toward the warmer energy source leaving behind the normal fluid. A compensating backflow of normal fluid is prevented because it cannot flow through the superleak passage. Thus, there is a one-directional fountain or pumping effect produced by the employment of a superleak and a heat source at or downstream of the superleak plug. Lyneis *et al.* (1968) have shown an analysis which will indicate the column of helium which one might theoretically obtain from such a system. They report this column height could be in excess of 30 m.

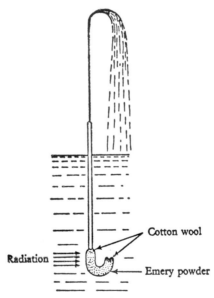

Radiation

Cotton wool

Emery powder

Fig. 5.56. The liquid helium fountain.

Van Alpen *et al.* (1969) experimentally determined a critical flow rate for the superfluid helium through such a media. This flow rate was found to be essential independent of the pressure head and proportional to the density. They also found this critical velocity to be independent of temperature below the lambda point. The critical velocities for this superfluid flow may be substantial and are reported to be as high as 20 to 30 cm/sec. The evidence thus far shows the critical velocity to be inversely proportional to the effective flow-passage diameter.

Film flow is another unique characteristic of HeII. In this, the helium flows as a film on its container walls. When there is a differential gravitational potential between the contained liquid and another body or film of helium it will flow first up the containing wall then downward on the other wall to the lower energy level. Thus, if a beaker of HeII is placed in a HeII bath, there will be film flow on the beaker walls until the liquid levels in the beaker and in the bath are equal.

Therefore, HeII exhibits a rather strange siphon effect without the siphon conduit required by other fluids. The explanation is in the superfluid or negligible viscosity of HeII flow. The film on the walls could be expected from any wetting fluid on walls in equilibrium with its saturated vapour. The film flow is possible because a net driving force exists (gravitational potential) and because there is no resistance in superfluid flow. This flowing film is limited both in thickness and in velocity. Therefore, the maximum rate of transfer is governed by the minimum periphery dimension of the wall over which it flows.

VII. Nomenclature

A	Area, cross-sectional
A_{DR}	Projected area for use with drag coefficient
a	Velocity of sound
Bo	Boiling number
C	Discharge coefficient
C_c	Contraction coefficient
C_{DR}	Drag coefficient
c_p	Specific heat capacity at constant pressure
c_v	Specific heat capacity at constant volume
D	Diameter
F	Force
F_a	Factor to account for thermal expansion of orifice
f	Fanning friction factor
F_{Re}, F_f	Modification functions (Brownell, 1950)
G	Mass flow rate per unit area
g	Gravitational acceleration
h	Enthalpy
h_{WTF}	Effective differential head produced by two-phase orifice flow
k	Coefficient in entrance-effect equation
K	Flow coefficient
K_G	Coefficient (5.37)
L	Specified length
m	Mass
m	Mass flow rate
M	Mach number
M_v	Velocity of approach factor
p	Pressure

q	Rate of heat transfer
Q	Volume flow rate
R_g, R_l	Volume fraction of a single phase
R	Gas constant
r	Radius
Re	Reynold's number
S	Entropy
S_r	Stermann parameter (5.40)
S_v	Specific volume number (5.40)
T	Temperature
t	Time
u	Axial velocity (x direction)
V	Volume
v	Specific volume (with subscript) or Velocity in y-direction
w	Velocity in z-direction
x	Axial length
X_P	Porosity
x_q	Quality, weight percent of gas phase
Y	Expansion factor

Greek letters

α	Void fraction
β	(1) Coefficient of volumetric expansion
	(2) Ratio of throat diameter to pipe diameter (5.29)
	(3) Enthalpy change parameter from Steward *et al.* (1967)
γ	Specific heat ratio, c_p/c_v
ε	Pipe interior roughness
κ	Coefficient of isothermal expansion
μ	Coefficient of viscosity
ρ	Density
τ	Shearing stress
Φ_l	Martinelli two-phase friction factor
χ	Martinelli correlation parameter (5.31, 5.32)
ψ	Sphericity
ω	Frequency

Subscripts

1,2	Successive stations in flow direction
b	Bulk or body condition
c	Choked or critical flow
BR	Brownell (1950) modification (5.25)
d	Submerged object
e	Exit
f	Liquid phase
fg	Vaporization condition
g	Gas phase
h	Hydraulic radius or diameter
IE	Irvine–Eckert term for flow coefficient

i	(1) Interface, inside wall
	(2) Upstream condition
n	Normal fluid
max	Maximum
min	Minimum
mom	Momentum
nat	Natural
O	Stagnation condition
P	Constant pressure process
SUP	Superfluid
S	Constant entropy process
SN	Normal-superfluid interaction
T	Constant temperature process
TP	Two-phase
tot	Total
v	Constant volume process
w	Wall condition

Superscripts

°	Condition of constant wall temperature
*	(1) Equivalent condition
	(2) Mach 1 condition.

VIII. References

Allen, J. F. and Jones, H. (1938). *Nature* **141**, 243.

ASME (1959). "Fluid Meters, their Theory and Application", 5th ed. ASME, New York.

Arp, V. (1969). *Cryogenics*. To be published.

Baker, O. (1954). *Oil and Gas J.* **53**, 185–190.

Bartlit, J. R. (1964). Los Alamos Scientific Laboratory Report LA-3177-MS. Los Alamos, New Mexico.

Bissel, W. R., Wong, G. S. and Winstead, T. W. (1969). Presented at AIAA 5th Propulsion Joint Specialist Conference. Colorado Springs, Colorado.

Bonnett, F. W. (1967). "Advances in Cryogenic Engineering", Vol. 12, pp. 427–437. Plenum Press, New York.

Brennan, J. A. (1964). "Advances in Cryogenic Engineering", Vol. 9, pp. 292–303. Plenum Press, New York.

Brennan, J. A., Edmonds, D. K. and Smith, R. V. (1968). NBS TN 359.

Brennan, J. A., Edmonds, D. K., and Smith R. V. (1969). "Advances in Cryogenic Engineering", Vol. 14, pp. 294–298. Plenum Press, New York.

Bronson, J. C., Edeskuty, F. J., Fretwell, J. H., Hammel, E. F., Keller, W. E., Meier, K. L., Schuch, A. F. and Willis, W. L. (1962). "Advances in Cryogenic Engineering", Vol. 7, pp. 198–205. Plenum Press, New York.

Brownell, L. E., Dombrowski, H. S. and Dickey, C. A. (1950). *Chem. Eng. Progr.* **46**, 415–422.

Burke, J. C., Byrnes, W. R., Post, A. H. and Ruccia, F. E. (1960). "Advances in Cryogenic Engineering", Vol. 4, pp. 378–394. Plenum Press, New York.

Cambel, A. B. and Jennings, B. M. (1958). "Gas Dynamics." McGraw–Hill, New York.

13

Carlson, L. W. and Irvine, T. F., Jr. (1961). *ASME J. Heat Transfer* **83**, 441.

Chenoweth, J. M. and Martin, M. W. (1955). *Petrol Refin.* **34**, 151.

Chisholm, D. and Watson, G. G. (1964). Symposium on Two-Phase Flow, Vol. 2. University of Exeter, Devon, England.

Clark, J. A. (1968). *In* "Advances in Heat Transfer", Vol. 5. Academic Press, London and New York.

Close, D. L. (1969). Cryogenics and Industrial Gases, Vol. 4, no. 8, pp. 19–23.

Crane Co. Technical Paper 410 (1969). Flow of Fluids through Valves, Fittings, and Pipe. Chicago, Ill.

Craya, A. and Bouré, J. (1966). *Acad. Sci. Paris Compt. Rend., Ser. A* **263**, 477–480.

Cruver, J. E. and Moulton, R. W. (1966). A.I.Ch.E., Fundamentals of Fluid Mechanics Symposium, Detroit.

Cruver, J. E. and Moulton, R. W. (1967). *A.I.Ch.E. J.* **13**, 52.

Davis, E. S. (1943). *Trans. ASME* **65**, 755.

Dean, W. R. (1928). *London, Edinburgh, and Dublin Phil. Mag. and J. Sci.* **5**, 673–695.

Deissler, R. G. (1955). *Trans. ASME* **77**, 7; also NACA TN 3016.

Deissler, R. G. and Taylor, M. F. (1958). NACA TN 4384.

de la Harpe, A., Lehongre, S., Mollard, J. and Johannes, C. (1968). "Advances in Cryogenic Engineering", Vol. 14, pp. 170–177. Plenum Press, New York.

Drake, E. M., Ruccia, F. E. and Ruder, J. M. (1961). "Advances in Cryogenic Engineering", Vol. 6, pp. 323–333. Plenum Press, New York.

Dukler, A. E., Wicks (III), M. and Cleveland, R. G. (1964). *A.I.Ch.E. J.* **10**, 38–43; 44–51.

Dwinnel, J. H. (1949). "Principles of Aerodynamics", 1st ed. McGraw–Hill, New York.

Eckert, E. R. G. and Irvine, Jr., T. F. (1957). Proc. Fifth Midwestern Conference on Fluid Mechanics, University of Michigan Press.

Eckert, E. R. G. and Irvine, Jr., T. F. (1960). *ASME J. Heat Transfer*, **82**, 125–137.

Eddington, R. B. (1967). Cal. Inst. of Tech. Jet Propulsion Lab., Tech. Report 32-1096.

Edmonds, D. K. and Hord, J. (1969). "Advances in Cryogenic Engineering", Vol. 14, pp. 274–281. Plenum Press, New York.

Edmonds, D. K. and Smith, R. V. (1965). Proc. Symposium on Two-Phase Flow, Vol. 1. University of Exeter.

Fauske, H. K. (1961). Proc. 1961 Heat Transfer and Fluid Mech. Inst., p. 79. Stanford University Press, Stanford.

Fauske, H. K. (1965). Proc. Symp. on Two-Phase Flow, Vol. 1. University of Exeter.

Friedly, J. C., Maganaro, J. L. and Kroeger, P. G. (1969). "Advances in Cryogenic Engineering", Vol. 14, pp. 258–270. Plenum Press, New York.

Geiger, G. E. (1964). Ph.D. Thesis. University of Pittsburgh.

Harden, D. G. and Boggs, J. H. (1964). Proc. 1964 Heat Transfer and Fluid Mechanics Institute, pp. 38–50. Stanford University Press.

Harden, D. G. and Walker, B. J. (1967). ASME Paper 67-WA/HT-23.

Hatch, M. R. and Jacobs, R. B. (1962). *A.I.Ch.E. J.* **8**, 18–25.

Hendricks, R. C., Graham, R. W., Hsu, Y. Y. and Friedman, R. (1961). NASA TN D-765.

Hendricks, R. C., Simoneau, R. J. and Friedman, R. (1965). NASA TN D-2977.

Hendricks, R. C., Simoneau, R. J. and Smith, R. V. (1969). NASA TMX-52612.

Henry, R. E. (1968). Argonne National Laboratory, ANL-7430.

Hoopes, J. W. (1957). *A.I.Ch.E. J.* **3**, 268–275.

Irvine, Jr., T. F. and Eckert, E. R. G. (1958). *J. Appl. Mech.* **25**, 288.

Itō, H. (1959). *ASME J. Basic Eng.* **81**, 123–134.

Kays, W. M. and London, A. L. (1958). "Compact Heat Exchangers." McGraw–Hill Book Company, New York.

Klingbiel, W. J. (1964). Ph.D. Thesis, Univ. of Washington, Seattle.

Knudsen, J. G. and Katz, D. L. (1958). "Fluid Dynamics and Heat Transfer." McGraw–Hill Book Company, New York.

Koshar, M. M. and Hoerning, J. (1960). An internal report of the Martin Company, Denver.

Langhaar, H. L. (1942). *J. Appl. Mech.* **9**, A55.

Lapin, A. and Bauer, E. (1967). "Advances in Cryogenic Engineering", Vol. 12, pp. 409–419. Plenum Press, New York.

Leonhard, K. E. and McMordie, R. K. (1961). "Advances in Cryogenic Engineering", Vol. 6, pp. 481–498. Plenum Press, New York.

Levy, S. (1963). *ASME J. Heat Transfer* **85**, 137–152.

Lindsey, W. F. (1938). NACA TR 619.

Lockhart, R. W. and Martinelli, R. C. (1949). *Chem. Eng. Prog.* **45**, 39–48.

Lottes, P. A. (1961). *Nucl. Sci. Eng.* **9**, 26–31.

Lyneis, C. M., McAshan, M. S. and Schwettman, H. A. (1968). Proceedings of the 1968 Brookhaven Summer Study on Superconductivity Devices and Accelerators, BNL 50115 (C-55), Part 3.

Macinko, J. (1960). M. S. Thesis, University of Colorado, Boulder, Colorado.

Martinelli, R. C. and Nelson, D. B. (1948). *Trans. ASME* **70**, 695.

Mendler, O. J. (1963). Ph.D. Thesis, University of Pittsburgh.

Miller, W. S. (1966). AIAA Second Propulsion Joint Specialist Conference, Colorado Springs, Colo., Paper #66-580.

Moody, F. J. (1965). *ASME J. Heat Transfer* **87**, 134.

Moody, L. F. (1944). *Trans. ASME* **66**, 671.

Muir, J. F. and Eichorn, R. (1963). Proc. Heat Transfer and Fluid Mech. Inst. Stanford University Press, Stanford.

Murdock, J. W. (1962). *ASME J. Basic Eng.* **84**, 419–433.

Nikurdase, J. (1930). *Ingen. Arch.* **1**, 306–332.

Perkins, H. C. and Worsøe-Schmidt, P. (1964). Stanford Univ., TR SU 247-7.

Pierre, B. (1964a) *ASHRAE J.* **6**, 58–65.

Pierre, B. (1964b) *ASHRAE J.* **6**, 73–77.

Purcell, J. R., Schmidt, A. F. and Jacobs, R. B. (1960). "Advances in Cryogenic Engineering", Vol. 5, pp. 282–288. Plenum Press, New York.

Richards, R. J., Jacobs, R. B. and Pestalozzi, W. J. (1958). "Advances in Cryogenic Engineering", Vol. 4, pp. 272–285. Plenum Press, New York.

Rogers, J. D. (1964). "Advances in Cryogenic Engineering", Vol. 9, pp. 311–315. Plenum Press, New York.

Rogers, J. D. and Tietjen, G. (1969). *A.I.Ch.E.* **15**, 144–146.

Rohsenow, W. M. and Choi, H. Y. (1961). "Heat, Mass, and Momentum Transfer." Prentice–Hall, Inc., Englewood Cliffs, New Jersey.

Romie, F. E., Brovarney, S. W. and Giedt, W. H. (1960). *ASME J. Heat Transfer* **82**, 387–388.

Rothfus, R. R., Monrad, C. C., Sikchi, K. G. and Heideger, W. J. (1955). *Ind. Eng. Chem.* **47**, 913.

Ryley, D. J. (1965). Proc. Symposium on Two-Phase Flow, Vol. 2. University of Exeter.

Schiller, L. (1923). *Zeit. Ang. Math. and Mech.* **3**, 2.

Shapiro, A. H. (1954). "The Dynamics and Thermodynamics of Compressible Fluid Flow." Ronald Press Co., New York.

Smith, R. V. (1963a). NBS TN 179.

Smith, R. V. (1963b). "Advances in Cryogenic Engineering", Vol. 8, pp. 563–573. Plenum Press, New York.

Smith, R. V. (1968). Ph.D. Thesis, University of Oxford, England.

Smith, R. V. (1969). *Cryogenics* **9**, 11–19.

Steward, W. G. (1965). "Advances in Cryogenic Engineering", Vol. 10, pp. 313–322. Plenum Press, New York.

Steward, W. G. (1968). Ph.D. Thesis, Colorado State University, Fort Collins, Colorado.

Steward, W. G., Smith, R. V. and Brennan, J. A. (1968). *Developments in Mechanics* **4**, 1513–1525.

Steward, W. G., Smith, R. V. and Brennan, J. A. (1969). Preprint of paper to be published in "Advances in Cryogenic Engineering", Vol. 15 (1970).

Sugden, R. P., Timmerhaus, K. D. and Edmonds, D. K. (1967). "Advances in Cryogenic Engineering", Vol. 12, pp. 420–426. Plenum Press, New York.

Taylor, M. F. (1967). *Int. J. Heat Mass Transfer* **10**, 1123–1128.

Taylor, M. F. (1968). *J. Spacecraft Rockets* **5**, 1353–1355.

Thompson, W. R. and Geery, E. L. (1960). Aerojet-General Corp., Report 1842 (AFFTC-TR-61-52, DDC AD-263 465).

Thurston, R. S. (1966). Ph.D. Thesis, University of New Mexico.

Thurston, R. S. (1965). "Advances in Cryogenic Engineering", Vol. 10a, pp. 305–312. Plenum Press, New York.

Thurston, R. S., Rogers, J. D. and Skoglund, V. J. (1967). "Advances in Cryogenic Engineering", Vol. 12, pp. 438–451. Plenum Press, New York.

Timmerhaus, K. D. and Sugden, R. P. (1965). "Advances in Cryogenic Engineering", Vol. 10, pp. 367–374. Plenum Press, New York.

Turney, G. E. and Cox, E. (1967). NASA TN D-3931.

Van Alphen, W. M., De Bruyn Ouboter, R., Olijhoek, J. F. and Taconis, K. W. (1969). *Physica* **40**, 490–496.

Walker, B. J. and Harden, D. G. (1964). ASME Paper 64-WA/HT-37.

Wallis, G. B. and Heasley, J. H. (1961). *ASME J. Heat Transfer* **83**, 363–369.

Weisbach, J. (1855). "Die Experimental Hydraulik." J. S. Engelhart, Frieberg.

Williamson, R. C. and Chase, C. E. (1968). *Phys. Rev.* **176**, 285–294.

Yen, I. K. (1967). *Chem. Eng.* **74**, 173–176.

Yen, J. T. (1957). WADC TR 57-224, Wright-Patterson Air Force Base, Ohio.

Zivi, S. M. (1964). *ASME J. Heat Transfer* **86**, 247.

Zuber, N. (1966). Final Report NAS 8-11422, NASA Marshall Space Flight Center, Huntsville.

Chapter 6

Materials of Construction and Techniques of Fabrication

Chapter 6

Materials of Construction and Techniques of Fabrication

D. A. WIGLEY

Department of Mechanical Engineering,
University of Southampton, Southampton, England

P. HALFORD

Petrocarbon Developments Limited, Manchester, England

I. Introduction

There are a large number of factors which must be considered when selecting a material for use in a piece of cryogenic equipment, and the final choice often involves a compromise between conflicting requirements, e.g. that it should be strong, have a high or low thermal or electrical conductivity, be weldable and, almost invariably, be the most economic material available. To make such decisions the cryogenic engineer should ideally have a sound knowledge of the physical and mechanical properties of a wide range of materials and possess a clear understanding of their relative advantages.

It is obvious that such expertise is not to be gained by merely reading a chapter in this or any other textbook. Rather, the aim of this section is to create an awareness of some of the more important properties of materials at low temperatures and to assist in the interpretation of the large amount of data now available. Of particular importance is an appreciation of the effects that the presence of impurities or physical defects can have on certain physical and mechanical properties. For example, the electrical and thermal conductivities, toughness, strength and ductility of most materials are highly dependent on their physical and chemical purities, whereas their specific heats, expansion coefficients and elastic moduli are virtually unaffected. As the structural integrity of a piece of equipment depends primarily on its strength and toughness it is essential that the relevant values of these quantities are obtained for the grade, thickness, condition etc. of material under consideration and if such information is not available it must be determined by the appropriate tests. Because of the large range of materials used in the construction of cryogenic equipment and the impossibility of including design data on all of them, very little data is in fact included in this chapter. Instead, the emphasis is placed on presenting the underlying principles which determine the mechanical and physical properties of materials.

Discussion of the physical properties of materials has been reduced to a minimum as these topics are covered quite well in most textbooks on solid state physics and in a number of more specialist works on low temperature physics (Mendelssohn, 1960; Rosenberg, 1963; White, 1968). There are also texts available which cover certain topics such as specific heat (Gopal, 1966) and electrical conductivity (Meaden, 1965) in great detail. Furthermore, a number of excellent data compilations now exist, many of which originate from the Cryogenic Engineering Laboratories of the National Bureau of Standards. These include monographs on the specific heat and enthalpy (Corruccini and Gniewek, 1960) thermal expansion (Corruccini and Gniewek, 1961) and thermal conductivity (Powell and Blanpied, 1954) of both pure substances and technical solids. Further information on these three properties is contained in Phase I of a compendium edited by Johnson (1961).

It is somewhat more difficult to get an overall picture of the mechanical properties of materials at low temperatures. There are many good sources of data, foremost among which are the "Cryogenic Materials Data Handbook"

(Schwartzberg, 1964) and further monographs from the N.B.S. (McClintock and Gibbons, 1960; Warren and Reed, 1963; Reed and Mikesell, 1967) while the annual volumes of "Advances in Cryogenic Engineering" (Timmerhaus, 1955 onwards) are a rich source of information and comment on all facets of cryogenic engineering. The published proceedings of other conferences often provide valuable references, the cryogenic properties of polymers being particularly well covered in a recent publication with the same title (Serafini, 1968 and in Mark, 1966). Certain aspects of the subject are thoroughly dealt with in specialist texts, review articles and original papers, but in general the information is rather fragmented and scattered. A substantial text on "The Mechanical Properties of Materials at Low Temperatures", by one of the authors (Wigley, 1971) is in course of publication. The present chapter is in part a summary of this work.

The justification of a study of the properties of engineering materials must be found ultimately in the application of this knowledge to the construction of real equipment. As indicated earlier, this usually involves the optimization of a number of different factors, some of which are more stringent than others. The prime characteristic of most equipment designed for use at low temperatures is that it should be tough, i.e. that it should not be liable to failure in a brittle manner. The traditional method of meeting this requirement is to design according to some recognized code such as ASME VIII (1965) or BS 1500 (1958) but this can, under certain circumstances, lead to the inefficient and uneconomic use of materials as these codes do not take into account the improvements in mechanical properties which are shown by many materials at low temperatures.

It is concluded that the essence of good cryogenic design lies in achieving the most efficient and economic solution which is compatible with the overriding need for adequate toughness. The cost and availability of materials and the ease with which they may be fabricated and joined are just some of the factors to be considered in striking the final balance and these topics are discussed in the last section of this chapter.

II. The Physical Properties of Metals and Non-Metals

A. SPECIFIC HEAT

The Debye function has been found to give a good approximation to the specific heat at constant volume of many pure, isotropic, crystalline solids, i.e. $C_v = aD(\theta_D/T)$ where θ_D is the Debye characteristic temperature of the substance. Values of $aD(\theta/T)$ have been tabulated (e.g. Gopal, 1966) and if θ_D is known, the required specific heat may be calculated. To a first approximation θ_D may be assumed to be a constant, although in fact it varies slightly, particularly at temperatures below $\sim\theta_D/5$. Values of θ_D quoted in the literature are usually either the limiting value as T tends to absolute zero, θ_{D_0}, or the value which gives the best fit to the experimental data in the range $\theta_D/2$ to θ_D.

In metals there is also a contribution from the free electrons,

$$C = \gamma T,$$

but in most metals γ is so small that the electronic contribution is negligible in comparison to that of the lattice except at very low temperatures ($T \leqslant \theta_D/50$). Even in the transition metals which have relatively large values of γ, the electronic specific heat is only significant below $\sim 20°K$.

Phase, magnetic and order-disorder transformations can also give rise to specific heat contributions which may be quite large over a limited temperature range and this "excess" specific heat can occasionally be put to good use. It has, for example, been suggested that the magnetic specific heat of neodynium could be utilized in the construction of more efficient regenerators for very low temperature cooling cycles.

It is, however, the lattice specific heat which dominates at most temperatures and which determines the form of the observed specific heat. At this stage it should be noted that the Debye theory gives C_v, the specific heat at constant volume, whereas it is C_p, the specific heat at constant pressure, which is measured experimentally. They are related by the formula:

$$C_p - C_v = AC_v^2 T$$

where A is a constant. At temperatures in the range $\theta_D/2$ to θ_D, $C_p - C_v$ is of the order of a few per cent while at lower temperatures it becomes very small and may be neglected. In fact for many substances the difference between C_p and C_v is less than the errors in the experimental data and so may be ignored.

The measured specific heats of aluminium, copper, iron and lead are plotted as a function of temperature in Fig. 6.1. At high temperatures the specific heats tend towards their classical value of $3R$/g mole (the logarithmic scales used for the axes tend to exaggerate the effect), while at temperatures below about $\theta_D/20$ $C_p \propto T^3$ as shown by the line of slope 3 on the log–log scales. The departure from linearity below $\sim 4°K$ is due to the electronic contribution to the specific heat.

An important point illustrated in Fig. 6.1 is that the rapid decrease in C_p commences at $T \approx \theta_D$ and thus a material with a low θ_D still has a high specific heat at low temperatures. This can be readily seen by comparing the curves for, say, lead and aluminium. In the temperature range 20°–4°K, the specific heat of lead is almost an order of magnitude greater than that of aluminium. This can be of considerable practical importance as many solders used at low temperatures contain high proportions of lead and other metals with low characteristic temperatures. Over-zealous use of the soldering iron can thus result in a situation in which the joint has a higher heat capacity than the components being joined! A further point worth emphasizing is that

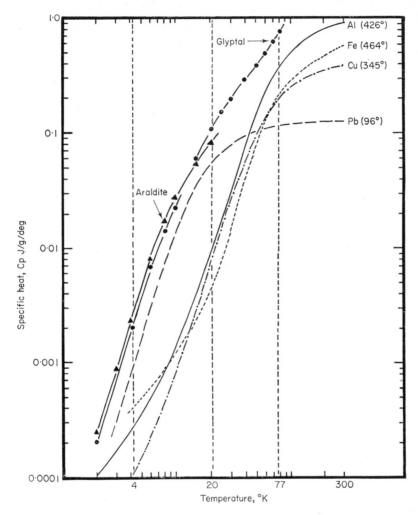

Fig. 6.1. The specific heat of some common materials between 2° and 300°K. Debye temperatures are given for the metals. (Replotted from data in Corruccini and Gniewek (1960).)

most references give the specific heat per unit mass whereas in situations where a component of given dimensions could be made from a range of different materials it is the specific heat per unit volume which is the more relevant parameter.

13*

As long as the relevant Debye θ, and perhaps the electronic γ of a pure crystalline solid are known it is easy to calculate its specific heat at any required temperature. Unfortunately, most technologically significant materials are alloys or compounds and calculation of their specific heats poses greater problems. At room temperature the Kopp–Joule rule states that the specific heat of many compounds and alloys can be obtained by a linear combination of the specific heats of their constituent elements, but this rule breaks down at lower temperatures. There appears to be no really satisfactory substitute for experimental data on the specific heat of alloys at low temperatures, but in its absence there is little alternative but to extrapolate from the data relating to the constituents. In this context it is worth noting that the specific heats of the f.c.c. 18/8 stainless steels at low temperatures are closer to those calculated for f.c.c. γ iron than for those measured on b.c.c. α iron, an illustration of the importance of lattice structure in determining the specific heat.

One final point which should be made is that the physical condition of a material does not seem to have much effect on its specific heat. Cold working has been found to lower the heat capacity slightly and size effects come into consideration if the material is in a finely divided form. For most practical purposes, however, these effects are small enough to be neglected.

As the Debye theory is only valid for crystalline solids it does not hold for a number of important materials including polymers, resins, elastomers and glasses. In general their specific heats do not level off to a constant value but rise continuously to the melting point. In addition, most polymers and elastomers go through "glass" transitions which are accompanied by specific heat anomalies which can be very sharp. The most important glass transition is the one below which the material becomes very rigid and brittle and this point will be reconsidered in the section dealing with mechanical properties. In Fig. 6.1 it can be seen that the specific heats of common resins and varnishes can be considerably larger than those of metals at low temperatures and hence their over-lavish use is to be avoided. It is difficult to calculate the heat capacities of these non-Debye solids and measured values are much to be preferred where available.

Finally it may be noted that for many purposes, such as heat balance calculations, it is not C_p, the specific heat, which is used directly but the Enthalpy, $H = \int C_p \, dT$. Accordingly some of the most useful references (Corruccini and Gniewek, 1960) also tabulate values of the Enthalpy relative to some reference temperature which is usually absolute zero, viz.

$$H - H_0 = \int_0^T C_p \, dT$$

In the absence of these data it is possible, with some loss of accuracy, to use the internal energy $U = \int C_v \, dT$ instead of the enthalpy. For a Debye-type solid values of $(u - u_0)/T$ have been tabulated as a function of θ_D/T and hence internal energies may be calculated if θ_D is known.

B. THERMAL EXPANSION

It can be shown that the linear expansion coefficient α of a crystalline solid is related to its specific heat at constant volume by the Grüneissen expression

$$\alpha = \tfrac{1}{3}\gamma\chi\frac{C_v}{v}$$

where χ is the isothermal compressibility, C_v/v the specific heat per unit volume and γ is the Grüneissen constant whose value ranges from about 1·5 for b.c.c. lattices to 2·3 for f.c.c. structures. In practice γ may be taken as a constant although variations similar to those shown by the Debye θ occur at low temperatures.

As the compressibility is only a weak function of temperature, the Grüneissen expression indicates that α is proportional to C_v or, remembering that C_v and C_p are very similar at low temperatures, the expansion co-efficient α at any temperature is proportional to its specific heat measured at this temperature. Thus as the temperature is reduced α decreases slowly until, at about $T = \theta_D$, it drops rapidly to an almost negligible value at temperatures below about $\theta_D/5$.

This close relationship between expansion coefficient and specific heat can sometimes be used to obtain approximate values for α at low temperatures if data is available for higher temperatures such as room temperature. If we assume

$$\frac{\alpha(T)}{\alpha(293)} \approx \frac{C_v(T)}{C_v(293)}$$

then $\alpha(T)$ may be calculated if the specific heat is known. If measured values of C_p are available at both $T°$ and 293°K (or any other suitable temperature) the calculation is simple. Alternatively if the characteristic temperature of the material is known, low temperature values of C_v may be obtained from the Debye function. Should measured values of expansion coefficient be available at temperatures where specific heat data is absent, the relationship may be used in reverse. Care must, however, be taken in interpreting the results obtained at very low temperatures because of the possible existence of anomalies in the expansion coefficients.

In many references, e.g., Corruccini and Gneiwek (1961) expansion data are presented in terms of the total linear contraction between a temperature T and some reference temperature (often 293°K) relative to the length at this temperature, i.e. $(L_{293} - L_T)/L_{293}$. This quantity is shown in Fig. 6.2 for a number of materials commonly used at low temperatures. It can be seen that many metals have virtually finished contracting once the temperature has dropped to between 80° and 50°K (about $\theta_D/5$ for materials with θ_D between 200 and 400°K). Thus the bulk of the contraction occurs on cooling from room to liquid nitrogen temperatures ($(L_{293}-L_{77})/L_{293}$ is usually 20 to

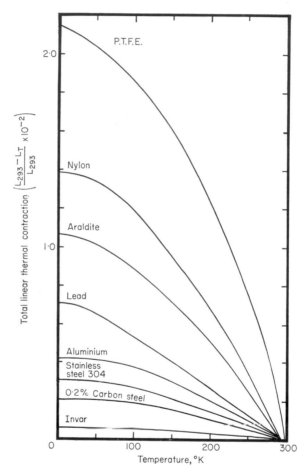

Fig. 6.2. The total linear thermal contraction of some common materials in the range 0°–300°K. (Replotted from data in Corruccini and Gniewek (1961).)

40×10^{-4}) and very little on subsequent cooling to liquid helium ($(L_{77}-L_4)/L_{293}$ is a few $\times 10^{-4}$).

There are a few practical points which might be added at this stage. Firstly, there are a few applications for which the absolute value of the contraction is required, one example being the calculation of the dimensions of a resonant cavity or wave guide which has to operate at low temperatures. On a much larger scale, one often wishes to minimize the total contraction in long runs of cryogenic transfer line and Invar (Nilo 36) is frequently used for this purpose as its contraction of about 5 cm/100 m run is considerably less than that of aluminium or stainless steels (between 20 and 40 cm/100 m run).

In most cases, however, it is the differential contraction between two or more components which is important and most difficulties can be attributed to one of two basic causes:

1. One Material, Different Temperatures

This is shown schematically in Fig. 6.3a. Coaxial vessels are joined along their rims and the space between them is thermally insulated. When the inner vessel is filled with cryogenic fluid it contracts relative to the warm outer vessel and imposes large, often fatal, stresses on the joints. Such problems can be avoided by intelligent design which may require the inclusion of expansion members.

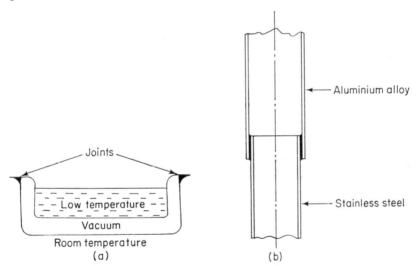

Fig. 6.3. Two typical examples of design errors which can lead to failure due to differential contraction between dissimilar metals.

2. Different Materials Cooled to the Same Temperature

One common design fault illustrated in Fig. 6.3b involves two coaxial tubes made of dissimilar metals and again joined. If the whole assembly is cooled down one metal will contract more than the other and put one tube in tension and the other in compression. If, for example, one tube is made of an aluminium alloy and the other of stainless steel, a differential contraction of approximately 0·1 % would be set up by cooling from 300 to 77°K. Assuming Young's modulus of aluminium to be 10^6 p.s.i. the tensile stress existing in the aluminium tube would be 10^3 p.s.i., which exceeds the yield stress of pure aluminium and some of its alloys. Repeated temperature cycling could lead to joint failure.

Differential expansion can be particularly severe if one component is made from certain non-crystalline solids, especially plastics. It can be seen from Fig. 6.2 that the total linear contractions of Araldite, nylon and PTFE are very much larger than those of most metals. This is particularly unfortunate in the case of PTFE which, as we shall see later, is almost the only polymeric material which retains some ductility at very low temperatures. Its good bearing qualities make it suitable for use as a valve seat or liner if the differential contraction problem can be solved. The usual method of doing this is to load the PTFE with graphite, glass or some other material having a low expansion coefficient so that the contraction of the resultant composite matches that of the metal casing. In a similar manner glass fibre reinforced epoxy or polyester composites have expansion coefficients which lie between those of their components and which can be varied by altering the volume fractions of the constituents. It should also be noted that the total linear contraction of 1·2% shown in Fig. 6.2 for Araldite is for one particular combination of resin and hardener and that values down to 0·4% can be obtained for other unfilled resins. Furthermore, loaded resins and adhesives can be produced which have even lower contractions.

Finally, it should be noted that many materials have expansion coefficients which are highly directional. Non cubic crystals expand anisotropically along different crystal axes and even polycrystalline specimens of these materials can be quite anisotropic. Forming operations such as rolling and drawing can produce a preferred orientation in many metals and in virtually all plastics based on long chain molecules, while most fibre reinforced composites contract least in the direction parallel to the fibre axes. It is, thus, obvious that considerable care is necessary in interpreting data on these materials.

C. ELECTRICAL CONDUCTIVITY

The electrical resistance of a metal is due to the scattering of electron waves by two basic mechanisms: thermal vibrations of the lattice and static defects such as impurity atoms. Mattheissen's rule states that these two mechanisms may be considered separately and that their contributions to the total resistivity are additive, viz.

$$\rho = \rho_0 + \rho_i(T)$$

where ρ_0 is the temperature independent "residual" resistivity and ρ_i the "ideal" resistivity which is a function of the temperature. The variation of ρ_i with temperature is given by the Grüneissen–Block relationship which contains a further characteristic temperature, θ_R, which, for many metals, has a value similar to that of the Debye characteristic temperature, θ_D. The ideal resistivity of a metal with a low θ_R increases more rapidly with rise in T than that of a metal with a high θ_R. At high temperatures, $(T \geqslant \theta_R/2)$ the Grüneissen–Block relationship simplifies to $\rho_i \propto T$, while at low temperatures $(T \leqslant \theta_R/10)\rho_i \propto T^n$, where n is ideally 5 but can be as low as 2. Thus the ideal resistivity of a metal drops off very rapidly at the lowest temperatures reaching, for many metals, values of about 10^{-8} ohm cm or less at 20°K.

At 4°K ρ_i has dropped to a negligible level and the measured resistivity is equal to the residual resistivity ρ_0. The value of ρ_0 is very sensitive to the presence of physical and chemical impurities and very low values are possible in pure, well annealed metals. In fact, measurements of ρ_0, or more often the resistance ratio $\rho_{293}/\rho_{4°K}$, give one of the most accurate methods of determining the purity of a metal.

The attainment of resistivities of the order 10^{-9} to 10^{-10} ohm cm at, and below, 20°K in copper and aluminium has led to speculation on the possibility of low loss transmission of electric power at these temperatures. There are however a few factors, such as magneto resistance and, for a.c. transmission, skin effect and eddy current losses, which reduce the attractiveness of this proposition. For example, Kohler's rule shows that the magneto resistivity $\Delta\rho$ in a field H, for a metal of resistivity ρ is given by

$$\frac{\Delta\rho}{\rho} = f_n\left(\frac{H}{\rho}\right)$$

and hence pure metals with low values of ρ can have relatively large magneto resistivities.

In certain metals, the electrical resistivity vanishes completely when they are cooled below a temperature known as their transition temperature, and this phenomenon, super-conductivity, is described in detail in Chapter 9.

D. THERMAL CONDUCTIVITY

1. *Metals*

There are two basic mechanisms by which heat is conducted through a metal: by the conduction electrons and by quantized lattice vibrations (phonons). In pure metals the electronic conductivity is dominant and the phonon conductivity may be neglected. The electrons are again scattered by two mechanisms, lattice vibrations and static defects. The thermal analogue of Mattheissen's rule can be written as

$$\frac{1}{K_e} = w = w_i + w_r$$

where w_i, the "ideal" thermal resistance due to the scattering of conduction electrons by lattice vibrations, has been shown to be proportional to T^2 and

w_r, the resistance due to impurity scattering, is proportional to $1/T$. Thus

$$\frac{1}{K_e} = \alpha T^2 + \frac{\beta}{T}$$

At high temperatures the conductivity of pure metals is controlled by lattice scattering and as the temperature is reduced this scattering decreases and the thermal conductivity rises. This can be seen in Fig. 6.4 for copper and

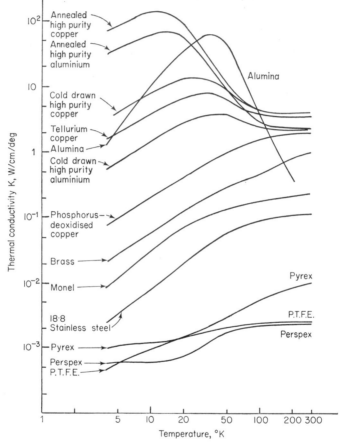

Fig. 6.4. The thermal conductivity of a number of materials in the range 4–300°K. (Replotted from data in White (1968), and Powell and Blanpied (1954).)

aluminium. In a perfectly pure metal the conductivity would rise indefinitely but in practice the scattering due to impurities and defects present dominates at very low temperatures and the conductivity passes through a maximum. The purer the metal the higher the value of this conductivity maximum and the lower the temperature at which it occurs (as may be seen from a comparison of the different curves for copper).

This has some rather important practical consequences as the conductivity at 20°K of commercially available copper ranges from about 25 W/cm/deg for very pure annealed material, through 6 W/cm/deg for free cutting tellurium copper to only 0·4 W/cm/deg for phosphorus-deoxidized copper, the variety used for most copper tubing and some plate and sheet. It is, therefore, important to know which grade of material is being used if an accurate value of its low temperature thermal conductivity is required. Direct measurement is, of course, the ideal, if rather inconvenient solution to this problem, but White and Woods (1968) have shown that it is possible to estimate the thermal conductivities of many metals from measurements of their electrical resistivities.

If the same scattering mechanism is responsible for both the electrical resistivity, ρ, and the thermal resistivity, w, the Wiedemann–Franz law says that $\rho/wT = L$, the Lorenz ratio which has the value $2·45 \times 10^{-8}$ W ohm deg^{-2}. Such a situation exists at very low temperatures (4·2°K) where both conductivities are determined by impurity scattering and, with a slightly lower degree of accuracy, at high temperatures ($T \geqslant \theta_R$) where lattice scattering is dominant. The relationship does not hold at intermediate temperatures but semi-empirical interpolation formulae have been developed to cover this temperature range. For many purposes, however, the thermal conductivity at 4·2°K calculated from $K_{4·2} = (L \times 4·2°)/\rho_{4·2}$ is all that is required to select the best conductor from a range of available metals. Selection on the basis of chemical analysis would be far less satisfactory as some impurities are far more effective than others, their scattering efficiency depending on the electronic band structures of host and impurity atoms. Furthermore the metallurgical structure of the alloy is also important as impurities in solid solution in the lattice have a much greater effect on the conductivity than those which are precipitated at grain boundaries. Lastly, it is difficult to estimate the degree of cold work which a metal has received and, as can be seen from Fig. 6.4, this can have a strong influence on the conductivity at low temperatures.

The very low conductivities shown by 18/8 stainless steel, monel and some other alloys in the temperature range below ambient are of great technological significance as such materials are often required for the construction of heat breaks in cryogenic equipment. Their low conductivities arise from a combination of high alloy concentration, cold work and other effects such as small grain size which reduce the electronic conductivity so much that it becomes of the same order as the lattice conductivity. Thus in complex alloys heat is conducted by both electrons and lattice vibrations, the Wiedemann–Franz ratio is no longer valid and it is not possible to obtain accurate values for thermal conductivity from measurements of the electrical resistivity.

2. *Non-metals*

In non-metals there are no free electrons present to conduct heat, but their absence also means that the phonons are not scattered by electrons and high thermal conductivities can occur in pure dielectric crystals. Although these materials have a very restricted practical application, the curve for alumina is included in Fig. 6.4 to illustrate the basic mechanism of phonon conductivity. The conductivity at high temperatures is limited by the mutual scattering of phonons but as the temperature decreases this process becomes less efficient and the conductivity rises. In a perfect, infinite crystal the conductivity would rise without limit but in a real material phonon scattering occurs at the crystal boundaries and this causes the conductivity to fall off at low temperatures. In polycrystalline materials scattering occurs at the grain boundaries and as the crystallite size is reduced the conductivity maximum decreases in magnitude and is displaced to higher temperatures.

In less pure materials, phonon scattering by impurity atoms, lattice defects and even the different isotropic species, all tend to lower and flatten the thermal conductivity curve. The limiting case is reached in Pyrex and the silicate glasses which have no long range crystalline order and whose conductivities are effectively determined by scattering from the "boundaries" of the small regions of local order. Their thermal conductivities are small and become steadily lower as the temperature decreases.

Perspex, PTFE and many other plastics behave in some respects like organic glasses at low temperatures. In most cases their structures are amorphous and their conductivities are very low as may be seen from Fig. 6.4. These values should, however, only be taken as a rough guide since variations in density, percentage crystallinity and degree of molecular orientation can alter their thermal conductivities by about an order of magnitude.

Glass fibre reinforced plastic and other composites are another group of materials which have very low thermal conductivities. Not only do their constituents have intrinsically low conductivities but the large number of fibre–matrix interfaces further reduces the total conductivity. Boundary and contact resistances are of particular importance in considering the thermal conductivity of solids at low temperatures. Very low effective conductivities can be obtained from stacks of thin discs of materials such as stainless steel because of the large contact resistances between successive discs. Furthermore this configuration has the added advantage of being able to support considerable compressive loads.

This brings us to one final, general point. There are many applications in which constructional materials are required to combine good mechanical strengths with either very bad or very good thermal conductivities. The combination of high strength with low thermal conductivity is relatively easy to achieve. In metals, the processes such as alloying, precipitation hardening, cold working and reduction in grain size which are used to increase the strength also decreases the thermal conductivity. Similarly, treatments which strengthen glasses and plastics usually decrease their conductivities. It is, however, the glass-fibre reinforced plastics which are undoubtedly the most promising materials on a strength/conductivity basis. In Table 6.I σ_y/K, the ratio of yield strength at 300°K to the average thermal conductivity in the range 300°K to 20°K, is shown for a number of constructional materials both in absolute terms and relative to that of stainless steel. Once the present porosity and jointing problems have been overcome, these materials should prove increasingly useful for the construction of cryogenic equipment.

TABLE 6.I. Strength–Thermal Conductivity Ratios for Various Materials

Material	$\sigma_y{}^a$ $(10^3$ p.s.i.)	K^b mW/cm/°K	σ_y/K	(σ_y/K)Rel
Stainless steel, type 304 annealed	35	109	0·32	1
Stainless steel type 304, cold drawn to 210,000 p.s.i.	150	104	1·44	4·5
Titanium alloy, 5 Al–2·5 Sn	130	45	2·89	9·0
PTFE	2	3	0·66	1·9
Nylon	10	2·8	3·6	11·2
Epoxy–glass–fibre laminate	75	3·5	21·4	67

[a] σ_y is the yield stress at 300°K.
[b] K is the average thermal conductivity between 20°K and 300°K.

The combination of high mechanical strength with good thermal and electrical conductivity is a much more difficult requirement to meet. Some increase in the strength of copper without too serious a decrease in its conductivity can be achieved by light cold rolling or certain precipitation and dispersion hardening treatments. A more satisfactory approach is, however, probably that currently used for large superconducting magnets where a superconductor–copper–stainless steel composite is formed in which the copper provides the required high conductivity and the strength comes from the stainless steel.

III. The Effect of Temperature on the Yield and Plastic Deformation of Pure Metals and Alloys

The elastic moduli of metals are relatively insensitive to changes in temperature. A useful rule of thumb states that they increase by about 0·03%/°K decrease in temperature and hence moduli at 4°K are about 10% higher than at room temperature. In most metals elastic behaviour is terminated by the onset of plastic deformation at strains of about 0·1% and established theories of solids explain this deformation in terms of the movement and interaction of dislocations. These processes are influenced by a number of factors including the deformation temperature, the thermal and mechanical histories of the material, its purity and, above all, its crystal structure.

Pure metals and alloys with face centred cubic structures are particularly suitable for use at low temperatures as many of their mechanical properties improve as the temperature falls. In contrast, body centred cubic metals are of more limited use as most of them undergo a transition from ductile to brittle failure at low temperature. Steels are the most important metals of this type and their transition temperatures depend on a number of factors including alloy content, grain size, strain rate and the severity of any notches or other stress raisers. The hexagonal close packed metals have properties which are in many ways intermediate between those of f.c.c. and b.c.c. metals. Many of them undergo a ductile–brittle transition below room temperature; but there are others, the most important of which is titanium, which can remain ductile down to 4°K if they have a low enough concentration of interstitial impurities.

A. FACE CENTRED CUBIC METALS AND ALLOYS

1. *Copper Alloys*

Most alloys used at cryogenic temperatures are based on copper, aluminium, nickel or the austenitic form of iron. Copper and its alloys were among the first metals to be used in the construction of cryogenic equipment and they provide a convenient illustration of the effect of temperature on the deformation of f.c.c. metals. Figure 6.5(*a*) shows a series of stress–strain curves for O.F.H.C. copper stressed at 300, 195, 76 and 20°K; the following points may be noted:

(i) Yield, which starts gradually at a stress of about 10,000 p.s.i., is virtually unaffected by changes in deformation temperature.

(ii) The strain hardening rate (slope of stress–strain curve) increases as the temperature is lowered, the metal is thus able to deform further before necking commences and hence:

(iii) The uniform elongation increases at low temperatures. Furthermore, as a result of both the increased rate and extent of strain hardening,

(iv) The ultimate tensile stress (U.T.S.) increases significantly as the temperature falls.

(v) Failure is extremely ductile at all temperatures and is preceded by a large reduction in area (an indication of which is the large decrease in stress between the U.T.S. and fracture).

A common method of summarizing such results is that used in Fig. 6.5(*b*) where the yield and tensile stresses, the percentage elongation and reduction in area are plotted as a function of temperature. The trends shown by these parameters are typical of most pure f.c.c. metals and of many of their solution hardened alloys, viz. they have small changes in yield stress coupled with significant increases in both tensile strength and percentage elongation at low temperatures. The reduction in area tends to decrease as the temperature is lowered but the manner of its variation varies from metal to metal.

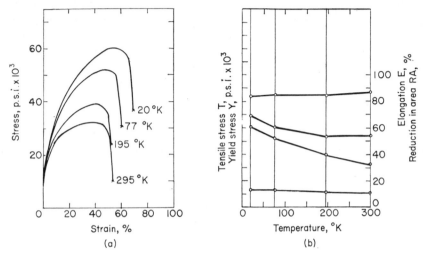

Fig. 6.5. (a) Stress-strain curves for oxygen-free high conductivity copper tested at 20, 77, 195 and 295°K.
(b) The same data replotted to show the temperature dependence of the yield and tensile stresses, the elongation and the reduction in area before fracture. (Data from Warren and Read, 1963.)

The main disadvantages of pure f.c.c. metals lies in their low yield strength in the annealed condition and although significant strengthening can often be obtained by cold working, it is more convenient to specify one of their alloys. The most familiar copper based alloys are the alpha brasses, a series of copper–zinc solid solutions whose strength increases with zinc concentration. The two phase α–β brasses formed at zinc concentrations above 39wt.% are even stronger but are rarely used for cryogenic purposes as their ductilities are much lower. Free cutting brasses, which contain a few per cent of insoluble lead, are liable to be porous especially if hard soldered and thus their use should be avoided where possible, particularly in the manufacture of thin walled components.

2. Aluminium Alloys

Aluminium and its alloys are widely used in the construction of cryogenic equipment as they have a number of favourable characteristics including high strength/weight ratios, good availability and ease of fabrication. They may be divided into two groups on the basis of their principle strengthening mechanisms; those which combine solid solution hardening with cold work and those which can be heat treated to produce precipitates.

The main solution hardened alloys are those with manganese (3000 series) and magnesium (5000 series) as the principle alloying additives. The aluminium–manganese alloys have only moderate yield strengths in the annealed condition and are widely used for tubes and other fabricated components. They are easily welded and, as they contain little or no copper or magnesium alloying additives, they can be brazed in molten salt baths using an aluminium silicon brazing alloy. For this reason Al–Mn alloys are widely used in the construction of tube plate heat exchangers.

The stronger aluminium–magnesium alloys are also weldable if the proper filler alloys and welding techniques are used. Furthermore, they require no post-weld heat treatments. In cold worked alloys the strength of the heat affected zone is reduced to that of the annealed condition but this can be quite high for certain alloys, e.g. 42,000 p.s.i. tensile strength for type 5456. This alloy in particular is widely used for the construction of large storage tanks and transporters as well as heat exchangers and other items of cryogenic plant.

The main heat-treatable aluminium alloys are those with copper (2000 series), magnesium and silicon (6000 series) and zinc (7000 series) as the principle alloying additions. In the optimum aged condition they have higher yield and tensile strengths than the non heat-treatable alloys but this extra strength is obtained at the expense of reduced ductility and toughness. There is also an increased risk of notch sensitivity at low temperatures. The aluminium–magnesium–silicon alloys have the lowest strength of the heat-treatable alloys and are also the most weldable. Type 6061, in particular, is widely used and is available in most product forms. In the as-welded condition these alloys have lower strengths than the aluminium–magnesium alloys but if post-weld heat treatment or artificial ageing is possible, much of the lost strength can be regained.

The high strength aluminium–copper series alloys are less widely used at cryogenic temperatures as welding is less straightforward. Nevertheless, these problems can be overcome and alloy 2014 has been successfully used for the cryogenic fuel tanks of Saturn rockets, an application in which its high strength/weight ratio is particularly advantageous. Most of the very high strength aluminium–zinc alloys are not usually considered to be weldable. They also suffer from a severe loss of notch toughness below 77°K and are thus rarely used at cryogenic temperatures.

3. *Nickel Based Alloys*

These have a very high resistance to corrosion and are used widely in the petrochemical industry. Inconel, Monel and German Silver have the added attraction of possessing very low thermal conductivities and they are commonly used in cryogenic equipment where heat influxes have to be minimized. All three alloys are available in the form of thin walled tubes, although the use of German silver is decreasing as it is less reliable and more corrodable than the alloys with higher nickel contents. The nickel superalloys, Hastalloy, Inconel X and K Monel, were originally developed for their excellent high temperature properties but their toughness and very high strength/weight ratios also make them suitable for certain cryogenic applications, such as expansion turbines and fasteners in cryogenically fuelled rockets.

4. *Ferrous Alloys*

The high temperature austenitic form of iron can be retained in a stable state at room temperature and below by the addition of a high concentration of certain alloying elements, the most important of which is nickel. If insufficient nickel is added the steel is only partially stabilized and it is liable to transform to martensite either spontaneously on cooling or during plastic deformation at low temperatures. The prime purpose of the 36% nickel present in Invar is to produce its very low coefficient of expansion but it also has the secondary effect of stabilizing the f.c.c. phase and so making the alloy suitable for use at low temperatures. It is widely employed in the construction of transfer lines and other pipework where its low coefficient of expansion can reduce the number of expansion joints that would otherwise be required. Some troubles were initially experienced in welding Invar but these have now been overcome by the use of a modified filler metal.

The austenitic stainless steels are, however, the main iron based alloys used at low temperatures. Very few of the many available grades are produced specifically for cryogenic applications, the majority having been developed for various combinations of corrosion resistance, strength and weldability. They have a wide range of compositions, and in consequence their stability with respect to martensite transformation varies considerably. In general high nickel, manganese and carbon contents tend to stabilize the austenite, while chromium and other b.c.c. elements have the opposite effect.

Martensite, the transformation product, is hard, brittle and ferromagnetic. Its strength can be advantageous if high strength–weight ratios are required and this property is exploited in "cryogenic stretch forming", a process in which martensite is deliberately produced by a controlled amount of plastic deformation at 77°K. Its presence can however also lead to excessive notch brittleness at low temperatures and grades known to suffer from this fault are to be avoided for components likely to suffer impact loads in the presence of stress raisers. A further disadvantage is the large dimensional change which accompanies martensitic transformations, an obvious drawback in components requiring close tolerances such as turbine shafts. If it is impossible to specify a non-transforming grade of stainless steel, this problem may be circumvented by prolonged soaking at 77°K prior to final machining.

It is, however, its ferromagnetism which is often the most undesirable feature of martensite as many types of cryogenic equipment have to operate in a magnetic field. For these applications, e.g. bubble chamber bodies and superconducting magnet formers, non transforming grades have to be specified. Special high nickel-content alloys such as Kromarc-55 and modified CK20 have been developed recently for bubble chamber castings but these alloys are not, as yet, readily available in the wrought form. If very high strengths are required the precipitation hardenable A-286 alloy (25% Ni, 15% Cr and 2·1% Ti) can be used but it too has a limited availability.

The most widely available non-transforming stainless steel is the AISI type 310 which contains ∼20% Ni and 25% Cr. There are also a number of British and Continental stainless steels which have chemical compositions similar to that of type 310, but care must be taken if these "equivalents" are specified. In particular it is important to check that their specifications come within the prescribed limits for the AISI type 310. The main disadvantage of type 310 is its relatively low yield strength at room temperature (ca. 30,000 p.s.i.). However, if some loss of ductility can be tolerated, considerable strengthening can be achieved by cold working.

Alloys with lower nickel contents such as 316, 321 and 347 do not transform spontaneously on cooling to 77°K and below, but some transformation does occur if they are deformed at these temperatures. It is, however, the widely available 18/8 stainless steels, and in particular type 304, which have the most variable transformation characteristics. All grades will transform during deformation at 77°K and below and in addition some transform spontaneously on cooling to these temperatures. Careful work by Gunter and Read (1962) has shown that this is mainly due to variations in their carbon and nitrogen contents and, as these elements are strong austenite stabilizers, alloys with low carbon and nitrogen contents transform more readily than those with higher concentrations. There is also a further advantage to be gained from having a high nitrogen content in this type of steel; that is, a significant increase in yield (or proof) stress. The recently developed "hi-proof" grades of 304, 316 and 347 stainless steel contain 0·2% nitrogen and have proof and tensile stresses about 15,000 p.s.i. higher than those of the normal grades.

The austenitic stainless steels are generally considered to be readily weldable but "weld decay" can occur in some types when they are heated within the range 400–800°C due to the intergranular precipitation of very thin films of chromium carbides, an effect known as "sensitization". These films are formed by excess carbon combining with chromium and so depleting the layer of metal adjacent to the grain boundary that its corrosion resistance is seriously impaired. Furthermore the impact strength of sensitized stainless steels such as type 304 decreases rapidly below ca. 100°K due to the embrittling effect of the carbide film. Sensitization may be avoided by reducing the carbon content to 0·03 % (e.g. 304 L) but this in turn increases the probability of martensitic transformation at low temperatures. Alternatively, carbide forming elements such as titanium and niobium can be added to the alloy to combine with the excess carbon and prevent chromium depletion. Thick sections of niobium stabilized stainless steels are difficult to weld because of hot cracking, but sections less than 0·75 in. are satisfactory. On the other hand, titanium stabilized steels occasionally suffer from a defect known as "titanium streaking" in which titanium stringers can cause leak paths though the material especially if it is in the form of thin-walled tubes. For filler rods niobium stabilization is preferred as titanium readily oxidizes during arc welding.

Finally, if high chromium-content austenitic or ferritic steels are heated to between 600 and 950°C for long periods, an ordered iron-chromium structure called sigma phase is formed. This can cause severe embrittlement at room temperature and procedures leading to its formation should be avoided. Further discussion of this topic is, however, outside the scope of this text.

B. HEXAGONAL CLOSE-PACKED METALS AND ALLOYS

The h.c.p. metals show a large variation in their mechanical properties at low temperatures, e.g. zinc undergoes a ductile–brittle transformation while cadmium retains its high ductility at 4°K and below. For most engineering purposes, however, a combination of strength and ductility is required and this characteristic is found to co-exist only in magnesium, zirconium, beryllium and, in particular, titanium. Very high strength/weight ratios are obtainable from titanium alloys at low temperatures and special Extra Low Interstitial grades have been developed for cryogenic use. As very small concentrations of interstitial impurities play a critical role in determining the mechanical properties of h.c.p. metals at low temperatures, it is relevant to consider briefly the effect of impurity atoms on the three main crystal lattices.

In f.c.c., b.c.c. and h.c.p. metals edge dislocations interact relatively weakly with the symmetrical lattice distortions created by substitutional impurity atoms, and impurity concentrations of the order of many percent are needed to achieve significant strengthening. Interstitial impurities in f.c.c. metals also produce symmetrical distortions in the lattice and interact weakly with edge dislocations. In contrast, asymmetrical distortions are formed by interstitial impurities in h.c.p. and b.c.c. lattices and these interact strongly with both edge and screw dislocations. Very small concentrations of the common interstitial impurities, hydrogen, carbon, nitrogen and oxygen can pin dislocations, prevent plastic deformation and render b.c.c. and h.c.p.

metals liable to cleavage failure at low temperatures. Thus the concentration of interstitial impurities must be minimized if h.c.p. and b.c.c. metals are to remain ductile at low temperatures. As fabricated the E.L.I. grades of titanium have sufficiently low interstitial concentrations to prevent low temperature embrittlement and care must be taken to ensure that they are not contaminated during any subsequent fabrication and welding operations.

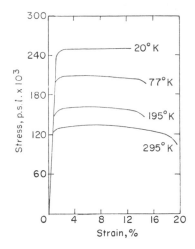

Fig. 6.6. Stress-strain curves for a stabilized titanium alloy, Ti–5Al–2·5 Sn tested at 20, 77, 195 and 295°K. (Data from Warren and Read, 1963.)

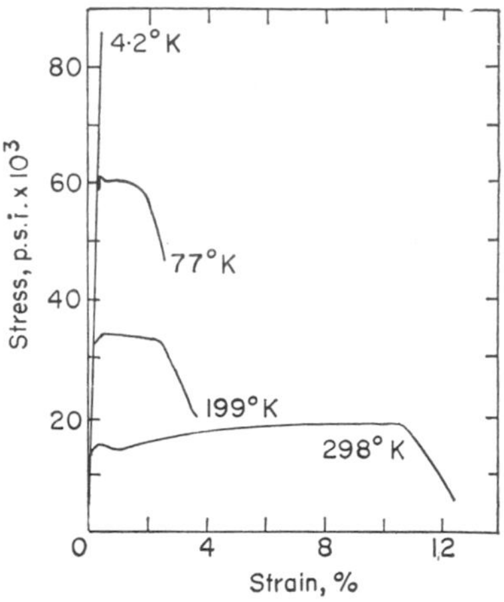

Fig. 6.7. Stress-strain curves for commercial purity Armco iron (after Smith and Rutherford, 1957.)

The only titanium alloys considered suitable for use down to 20°K and below are the α stabilized types, typical of which is the 5 Al–2·5 Sn alloy shown in Fig. 6.6. It may be seen that both yield and tensile stresses increase rapidly as the temperature decreases and that the strain hardening rate is low at all temperatures. The ductility decreases slightly at low temperatures but it is still adequate at 20°K. The very high strengths (and strength/weight ratios) achieved at 20°K have been exploited in the construction of liquid helium pressure vessels for Apollo spacecraft. Titanium alloys are produced in two other forms: the b.c.c. β-stabilized alloys which are rarely used below room temperature and the duplex α–β alloys which are considered safe to use down to 77°K but which lack adequate toughness at lower temperatures.

C. BODY CENTRED CUBIC METALS AND ALLOYS

1. *Pure Metals*

Although β brass and the alkali metals remain ductile down to 4°K, the majority of b.c.c. metals become brittle at low temperatures and in consequence their applicability is severely restricted. The effect of decreasing temperature on the mechanical properties of b.c.c. metals is typified by the behaviour of commercially pure iron shown in Fig. 6.7. At room temperature, sharp yield is followed by moderate strain hardening and failure occurs in a ductile manner. As the temperature is reduced there is a large increase in yield stress but the metal loses its ability to work harden and the ductility decreases, until at 4°K cleavage failure occurs. This is in direct contrast to the behaviour of f.c.c. metals in which the yield stress is almost temperature independent and the ductility increases as the temperature drops. The h.c.p. metals combine some of each of these characteristics in that they show a large increase in yield stress at low temperatures without severe loss of ductility.

2. Low Alloy Steels

From a practical point of view plain carbon and low alloy steels are the most important class of materials which show a ductile–brittle transition. In Fig. 6.8(a) a series of stress strain curves is shown for a plain carbon steel

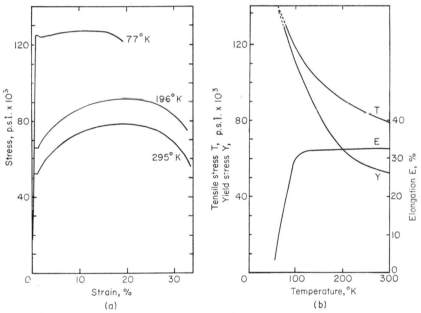

Fig. 6.8. (a) Stress-strain curves for a 0·2% plain carbon steel normalized at 880°C and tested at 77, 195 and 295°K.

(b) The effect of temperature on the yield and tensile stresses and percentage elongation of the same steel.

tested at and below room temperature, while in Fig. 6.8(b) the resultant values of yield strength, tensile strength and elongation are shown as a function of temperature. It can be seen that below about 200°K the yield strength increases more rapidly than the tensile strength and that the two curves coalesce at about 50°K. The elongation also falls off to zero over this temperature range, and below 50°K the steel is unquestionably brittle.

As the loading conditions imposed during tensile tests on smooth specimens are less severe than those liable to be found in service, impact tests are more frequently used to investigate the transition behaviour of metals. In Fig. 6.9 the tensile elongation and impact energy are shown for the same material as a function of temperature, and it can be seen that the toughness transition given by impact testing occurs at much higher temperatures than the ductility

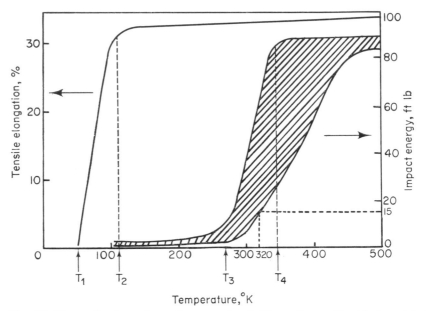

Fig. 6.9. The tensile elongation and energy absorbed in a Charpy V notch impact test shown as a function of temperature for the same 0·2% carbon steel as used for the tests of Fig. 6.8.

transition obtained from tensile tests. One can say that at temperatures above T_4 the material is ductile under all normally encountered strain rates, while below T_1 it is brittle. Between T_2 and T_3 the material will be ductile if tested at low strain rates under uniaxial stress conditions but brittle if subjected to triaxial stresses at high loading rates.

We shall return shortly to the subject of suitable design criteria for selecting metals resistant to brittle failure, but at this stage we will assume that the metal must absorb a minimum of 15 ft/lb. (2·1 kg/m) in a standard Charpy V notch impact test. Thus the steel shown in Fig. 6.9 would meet this criterion at temperatures above 320°K, taking the lower bound values for impact energy. The aim of much development work has been the economic realization of alloy steels which have low transition temperatures and there is now a range of steels available for use in various temperature ranges below ambient. Ordinary plain carbon steels start to become brittle at or just below room temperature, the actual transition temperature depending on a number of variables including carbon content, grain size and the previous thermal and mechanical histories of the metal. Carbon–manganese steels can be used down to about 223°K (− 50°C) due mainly to the intense grain refining action of the manganese, but for lower temperatures one of the nickel bearing steels must be specified.

3. *Nickel Steels*

The effect of nickel content on the Charpy K impact energy is shown in Fig. 6.10 for a range of steels. In practice only $2\frac{1}{4}$, $3\frac{1}{2}$ and 9% nickel steels are readily available as these three grades have been found sufficient to cover most common cryogenic needs. The $2\frac{1}{4}$ and $3\frac{1}{2}$% nickel steels may be used

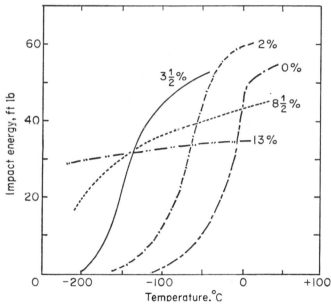

Fig. 6.10. The Effect of variation in nickel content on the tough-brittle transition temperature as measured by Charpy K impact tests. (International Nickel Co. data.)

down to 213°K (−60°C) and 173°K (−100°C) respectively, but many designers prefer $3\frac{1}{2}$% nickel steel at the higher temperatures because of the additional safety margins conferred by its lower transition temperature. The $3\frac{1}{2}$% nickel steel is readily welded with electrodes of the same composition and may be used without post-weld stress relief.

Nine per cent nickel steel, now widely used in the construction of large storage vessels for liquid nitrogen and liquid methane, is the only ferritic material accepted for service down to $-200°C$. Its low cost and high proof strength make it an attractive material especially if the vessel can be designed to yield point codes such as B.S. 1515. Even with the less advantageous UTS codes such as ASME 8, it compares favourably with most aluminium alloys and stainless steels. The material is easily welded if the correct filler metals are employed but the common nickel–chromium fillers produce welds which have strengths lower than those of the parent metal. Higher strength welds are obtainable using the complex austenitic-type electrodes such as Inconel, but the expansion coefficients of these fillers do not match those of the parent metal as well as the nickel–chromium compositions and high stresses can be set up in the welds during thermal cycling. The high cost of the correct electrodes detracts from the favourable economics offered by the parent metal, and this is a restriction on the more widespread application of 9% nickel steel.

IV. Fracture in Metals

Most engineering design criteria are based on the assumption that the fracture strength of a material is always greater than its yield strength and at least equal to its ultimate tensile strength. Safety factors derived mainly from feedback from previous experience are then used to keep the applied stresses well below the assumed fracture strength. Structures based on such criteria can, under certain circumstances, be unsafe as they offer no protection against unstable fractures occurring at stresses below the general yield stress of the material. Such failures can occur, for example, in constructional steels at or just below room temperature, and it is usually found that they initiate at notches, welds or other discontinuities which trigger off local yielding, even though the applied stress is below the general yield stress. Furthermore, once initiated, cracks can become self propagating and spread through a structure at speeds approaching that of sound. Under such circumstances design criteria must ensure reliability against brittle fracture either by avoiding the *initiation* of such cracks or by preventing their *propagation* in a low energy mode.

The only really satisfactory method of assessing the suitability of a material for service in a particular application is to evaluate full scale components under the most embrittling conditions which could ever be met in practice. Such tests are, however, expensive and time consuming and more convenient methods of evaluation, such as the Charpy V notch impact test, are required for materials selection. If the criteria obtained from the small scale tests can be satisfactorily correlated with those determined from full scale tests, the simpler tests may often be used in place of the full scale tests. Such correlations can, however, only be established rigorously for specific materials and are *not universally valid*, e.g. although it has been found that some grades of mild steel may be used at temperatures where a minimum of 15 ft lb. of energy is absorbed in a Charpy V notch test, it cannot be assumed that a 15 ft lb. Charpy energy criterion will be valid for other materials.

The mode of construction can also play a significant role in determining the liability of a structure to brittle fracture. Welds in particular can influence the fracture behaviour in two main ways. Firstly, cracks can be initiated at or near imperfect welds due to the residual stresses or microcracks produced during cooling. Secondly, a welded structure is continuous and a crack can thus propagate further than in a discontinuous structure. It was, for example, a combination of both factors which led to the large number of catastrophic failures in the all-welded Liberty ships: most fractures were initiated at bad welds or other stress raisers and propagated all round the hull instead of stopping at the edge of the plate containing the flaw.

Any satisfactory approach to designing with materials susceptible to unstable fracture must recognize that cracks or other stress raisers are liable to develop or intensify during service and that the stability of the structure will depend on the relationship between the flaw size and the applied stresses. This is the basis of the Fracture Toughness (or Fracture Mechanics) approach to design which is gradually becoming accepted as the most efficient and reliable basis for ensuring the safety of structures made from materials prone to unstable fracture.

It is not possible to give in this section more than the barest outlines of the many factors which determine the fracture behaviour of materials. A very comprehensive account of the fracture of structural materials is given in a book by Tetelman and McEvily (1967), while the metallurgical aspects of the topic are covered in many textbooks and review articles.

A. BASIC MECHANISMS OF DUCTILE AND BRITTLE FAILURE

On an atomic scale fracture occurs when atomic bonds are broken, and fresh crack surface is created. If the concentrated *tensile* forces at a crack tip exceed the cohesive stress of the material, bonds perpendicular to the fracture plane are ruptured and *cleavage* failure occurs. *Shear* failure is caused by the rupture of these same bonds by forces applied *parallel* to the fracture plane. In crystalline materials both cleavage and slip (shear deformation) take place preferentially on certain crystallographic planes, the cleavage and slip planes respectively, and the relative ease with which these two processes operate depends principally on the type of crystal structure concerned but also on the testing temperature, strain rate, applied stress system, purity of the material and a number of other factors.

In many crystalline solids slip takes place at stresses far below the theoretical shear stress of the crystal because dislocations can move at these lower stress levels. If it is always possible for such plastic deformation to occur and relieve potential stress concentrations, stress levels can never exceed the cleavage strength of the crystal and cleavage failure cannot occur. This is the situation which exists in pure f.c.c. metals at all temperatures. Their shear strengths are always lower than their cleavage strengths and they always fail in a ductile manner. In b.c.c. and h.c.p. metals, however, it is possible for dislocations to be pinned and slip to be inhibited. High tensile stresses can then be built up at the tips of microcracks and cleavage failure is possible. Cleavage and shear are thus alternative modes of failure in these metals. If slip is completely inhibited cleavage failure occurs before general yield, e.g. in b.c.c. metals at very low temperatures. Often, however, yield and plastic deformation commence because the shear strength is initially lower than the cleavage strength and cleavage failure does not occur until strain hardening has raised the shear strength above the cleavage strength.

Although a small amount of plastic deformation is possible in some non-metallic crystalline solids, their failure mode (especially at low temperatures) is usually by cleavage. Amorphous solids are usually brittle and fail by cleavage because dislocation slip is only possible in crystalline structures and thus amorphous materials lack the mechanisms by which stress concentrations may be relieved. As we shall see in Section V, polymeric materials can deform plastically at temperatures above their glass transition, but at lower temperatures they too fail by cleavage.

On a larger scale cleavage and shear failures may be distinguished by the modes in which they propagate and by the appearance of the fracture surfaces they produce. In polycrystalline materials the grains have a range of orientations with respect to the tensile axis and some grains are more favourably orientated for cleavage than others. Thus although on a macroscopic level a cleavage crack characteristically propagates perpendicular to the axis of maximum tensile stress and has a "flat" appearance, fracture on a microscopic scale follows the cleavage planes of individual grains. In very brittle materials the cleavage crack can propagate continuously from grain to grain, but in less brittle materials such as mild steel, fracture is discontinuous; cleavage occurring in the more favourably orientated grains with subsequent linkage by tearing. Brittle fracture can also propagate along grain boundaries in an *intergranular* mode as opposed to the more usual transgranular cleavage. Such intergranular fractures occur when impurity atoms segregate out at grain boundaries, and this is the only type of brittle failure possible in f.c.c. metals. Finally, the most distinctive characteristic of a cleavage fracture surface is its bright crystalline appearance (Fig. 6.11(d)), a result of the high reflectivity of freshly cleaved grains and their random orientations.

In contrast, the surface of a ductile fracture has a characteristic dull, fibrous appearance caused by the extensive plastic deformation which takes place during failure. In very pure metals slip becomes concentrated on one or two major shear planes to produce "slant" and "knife edge" fractures respectively. In most engineering materials, however, ductile fracture is intimately associated with the presence of inclusions and voids. Cracks are nucleated at the interfaces between the second phase particles and the matrix

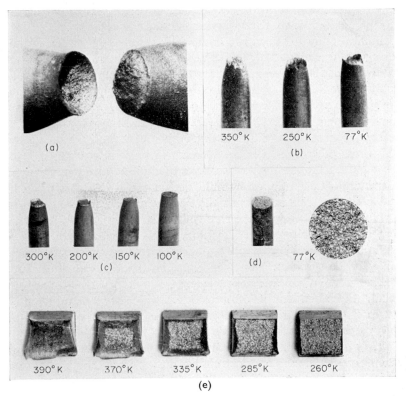

Fig. 6.11. (a) A perfect cup and cone fracture of mild steel at ambient temperature.

(b) The change in profile of three annealed copper tensile specimens fractured at 350, 250 and 77°K.

(c) The ductile-brittle transition in plain carbon steel as shown by the change in form of the fracture halves. Specimens tested at 300, 200, 150 and 100°K.

(d) Two views of a similar specimen which failed by cleavage at 77°K. Note the bright, crystalline appearance of the fracture surface.

(e) The tough-brittle transition in the same steel as shown by the change in the proportion of the fracture surface which shows cleavage failure. Tests carried out at 390, 370, 335, 285 and 260°K.

and these grow to form voids which subsequently coalesce by intense localized deformation along planes of maximum resolved shear stress. On a macroscopic scale the direction of crack propagation is determined by a combination of the applied tensile stresses and the resultant triaxial stress system set up by plastic constraints and the presence of included second phase particles.

14

Two distinct modes of propagation can be identified, *normal* and *shear* rupture. Normal rupture initiates at the centre of a bar or plate where the triaxial tensile stresses are at a maximum and propagates outwards on a plane normal to the applied tensile force. On a microscopic scale, however, the crack advances by void coalescence on alternate planes inclined at about 45° to the tensile axis to give a jagged fracture surface. It is quite unusual for normal rupture to propagate completely through the material and the final stages of fracture usually take place by shear rupture on planes inclined at 45° to the tensile axis to give a characteristic "shear lip". This sequence of normal rupture followed by shear lip formation is responsible for the characteristic "cup and cone" appearance of a ductile failure in a round tensile specimen shown in Fig. 6.11(*a*). Such fractures can be very coarse and jagged as in the copper specimens shown in Fig. 6.11(*b*). (The other feature illustrated by this series of specimens is the increase in uniform elongation prior to necking and fracture at low temperatures.) On the other hand, the cup and cone fracture obtained by testing mild steel at 150°K (Fig. 6.11(*c*)) exhibits only fine scale markings on both the normal and shear rupture surfaces. Figure 6.11(*c*) also illustrates the change in fracture appearance as mild steel undergoes its ductile to brittle transition. As the temperature decreases the shear lip becomes less and less pronounced becoming completely absent in the fully brittle specimen shown in Fig. 6.11(*d*) which also illustrates the crystalline nature of the fracture surface.

Finally there are a number of occasions in which more than one mode of fracture can operate during the failure of a specimen or structure and under these circumstances the fracture is described as "mixed". The mode in operation at any instant is that which requires the smallest strain at the tip of the advancing crack and this factor can change during propagation. For example, fracture initiates by plastic deformation in the notch root of a mild steel Charpy V specimen tested at room temperature (Fig. 6.11(*e*)). Once initiated, fracture spreads by cleavage across much of the section of the specimen but shear lips are developed at the sides and bottom during final separation of the fracture halves. In general, normal rupture and cleavage occur at the centre of thick structures where triaxial stresses are greatest and fracture takes place under *plane* strain conditions. In contrast shear lips and 45° slant fractures occur under *plane stress* loading in very thin sheets and at the extremities of thick plates and bars. The stress state thus plays an important role in determining the mode of fracture and this topic will be reconsidered later.

B. CRACKS, NOTCHES AND FLAWS: FRACTURE TOUGHNESS

As indicated earlier, the presence of a small defect such as a microcrack, casting flaw or surface scratch can have a very deleterious effect on the integrity of structures made from certain notch sensitive materials. Any void in a material acts as a stress raiser and the degree of *elastic* stress concentration which it causes depends largely on its form and sharpness. Very large elastic stress concentrations can occur at the tips of atomically sharp microcracks and thus permit the stresses of the order of E/10 needed to break atomic bonds to be generated even though the applied stresses are many orders of magnitude lower.

Notches can also act as a constraint to *plastic* deformation. If a symmetrical notch is placed in a tensile specimen and a load is applied, the material beneath the root of the notch will be in a state of triaxial tension and plastic deformation will be constrained by the unyielded material above and below the notch. Under these circumstances the *nominal yield stress* (tensile yield load–notched sectional area) is almost three times larger than the uniaxial

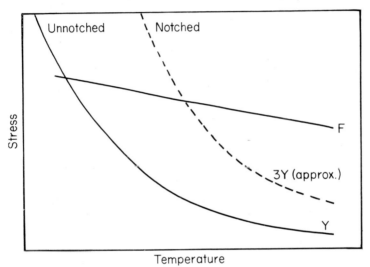

Fig. 6.12. The Ludwik–Davidenkow–Orowan interpretation of notch brittleness in metals having a highly temperature dependent yield stress.

tensile yield stress. Orowan has shown that this constraint to plastic deformation will encourage the transition from ductile to brittle fracture in those metals susceptible to such a transition. Plastic flow and brittle fracture are assumed to be independent processes each with its own characteristic tensile stresses Y and F, respectively. The flow stress Y increases rapidly with decrease in temperature while the fracture stress F is assumed to be relatively temperature insensitive as shown in Fig. 6.12. Then at high temperatures where $F > 3Y$ the material is simply ductile; at low temperatures where $F < Y$ it is simply brittle. At intermediate temperatures where $Y < F < 3Y$ the material is notch brittle, i.e. ductile in plain tensile tests but brittle in tests on notched tensile specimens.

The most important influence exerted by a notch or crack lies in the relationship between the strength of a material or structure and the *size* of largest flaw which it contains. Many years ago Griffith showed that a cleavage crack in a brittle material such as glass will propagate spontaneously at constant applied stress once the crack reaches a certain critical length. This critical length is determined by the criterion that the elastic strain energy released by a small extension of the crack equals the surface energy gained by such an extension. The resultant relationship between applied stress, σ, and critical crack length, a, is given by the relationship

$$\sigma \simeq \sqrt{\frac{E\gamma}{a}}$$

where E is Young's modulus and γ is the true surface energy of the material.

Metals are never completely brittle and a certain amount of work must be done to produce plastic deformation at the tip of a moving crack. Orowan has shown that the Griffith criterion for unstable fracture can be modified to include non-ideally brittle materials if the energy required to produce plastic deformation, p, is added to the true surface energy, γ,

$$\sigma \simeq \sqrt{\left(\frac{E(\gamma+p)}{a}\right)}$$

The larger the plastic work term, p, the more difficult it is for unstable fractures to occur and the greater the critical crack length, a. In many metals, $p \simeq 10^2$–$10^3\gamma$ and hence critical crack lengths are much larger than in very brittle materials like glass which typically have values of a of the order of microns under conditions where a for metals like mild steel would be of the order of millimetres.

A similar analysis by Irwin based on fracture mechanics gives the result that unstable fracture occurs at a stress of σ_F when a parameter, G, defined as the *crack extension force* reaches a critical value G_c. The relation between fracture stress and flaw size is then given by:

$$\sigma_F \simeq \sqrt{\frac{EG_c}{a}}$$

This has the same form as the Griffith–Orowan equation with G_c replacing the $(\gamma+p)$ term.

In Irwin's notation G_c may be written

$$G_c = \frac{K_c^2}{E}$$

where K_c is called the *fracture toughness* of the material. By substituting in the previous equation and rearranging we are left with the relationship

$$K_c = \sigma\sqrt{a} \text{ (units : stress}\sqrt{\text{length)}}$$

The fracture toughness of a material is thus a measurement of its ability to resist the unstable propagation of a crack; the greater its toughness, the larger the crack that a material can tolerate when loaded to any given stress.

For efficient design good fracture toughness is usually required in combination with a high yield stress, but this is generally difficult to achieve as in most alloys an increase in yield stress leads to a decrease in fracture toughness. Very high strength materials ($\sigma_y < E/150$) fail at all temperatures in a low energy mode either by rupture or cleavage. To use these materials safely, the operating stress must at all times be kept below the fracture stress determined by the size of the largest flaw which can remain undetected. Medium strength materials (whose room temperature yield strengths lie between $E/150$ and $E/300$) usually have a high fracture toughness at room temperature and above and fail in a high energy shear mode. However, the yield stresses of many of these alloys (certain grades of aluminium, titanium and stainless steel in particular) increase significantly as the temperature falls and there is a corresponding drop in their fracture toughness. Failure at low temperatures then occurs by a low energy tear mechanism and the materials are said to be *notch brittle*.

The failure mode of low strength materials ($\sigma_y < E/300$) depends on their liability to cleavage failure; f.c.c. metals cannot fail by cleavage and hence are always tough, b.c.c. metals fail at high temperatures by a high energy shear mechanism, at low temperatures by cleavage and at intermediate temperatures by a mixture of both mechanisms. Their toughness thus varies with temperature and some of the tests and criteria used to determine their suitability for use at a particular temperature will be discussed in Section IVc.

The fracture toughness of a material does not have a unique value as it is dependent on such factors as plate thickness, specimen geometry and loading conditions. For example as the thickness increases from zero the fracture toughness, K_c, rises rapidly to a maximum and then decreases slowly until it reaches a limiting value K_{Ic}, the plane strain fracture toughness. Furthermore, the fracture toughness of a material may be decreased during forming or welding operations and hence a welded structure may lack adequate toughness even if the parent metal is satisfactory.

Care must therefore be taken in applying the concepts of fracture toughness to design procedures. The approach can, however, offer a bonus as it is possible to apply the concept of fracture toughness to *whole structures* and not just to individual materials and components. Consider for example a pressure vessel required to operate at a certain stress, σ_{op}. The stress σ at which the vessel would fracture in an unstable mode if it contained a flaw of length a is given by:

$$\sigma = \frac{K_c}{\sqrt{\{\pi a + \frac{1}{2}[(K_c/\sigma_y)]^2\}}}$$

which is a somewhat refined form of the equation for σ_F but containing a plasticity term, $\frac{1}{2}(K_c/\sigma_y)^2$ (see Telelman and McEvily (1967), p. 65). This relationship is shown in Fig. 6.13. Failure would occur at the operating stress, σ_{op}, if a flaw had the critical length a_0. In order to ensure some margin of safety, pressure vessels are always "proof tested" to some stress σ_p which is higher than the operating stress by a factor laid down by the appropriate design code. If the vessel passes such a test it may be inferred from Fig. 6.13 that it cannot contain a flaw larger than a_p, that given by the critical flaw size curve for a stress of σ_p. This also establishes a flaw size safety factor for the vessel.

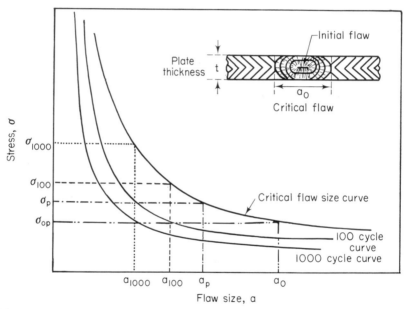

Fig. 6.13. Critical flaw size curves for both static and cyclic applied stresses. *Inset*: the relationship between initial flaw size a, critical flaw size a_0 and plate thickness t.

The analysis can be taken a stage further by considering the probability of a crack growing to the critical value, a_0, by fatigue, hydrogen embrittlement, stress corrosion or any other relevant mechanism. By cycling a number of specimens containing flaws of a given initial length to various levels of stress, the number of cycles required to enlarge the flaws to the critical length can be determined. This data can then be plotted to give cyclic life curves as shown for example in Fig. 6.14 for 100 and 1,000 cycles. Thus at the operating stress, σ_{op}, 100 cycles are required to enlarge a flaw initially of size a_{100} to the critical size a_0. However, if the vessel were proof tested at a stress of σ_{p100} one could be sure that no flaws of size greater than a_{100} existed in the vessel.

Hence it could be cycled 100 times to the operating pressure σ_{op} before a critically sized flaw could develop. If a life of 1000 cycles is required, proof testing at σ_{p1000} would provide the required degree of protection.

Finally, this analysis also forms the basis for design according to a "leak before break" criterion which is valid even in those cases where the crack characteristics and growth rate are unknown. If the operating stress is such that stable growth produces a crack which penetrates the full wall thickness of the vessel, t, before it reaches the critical flow size, a_0, then the vessel will leak before it breaks catastrophically. In cryogenic equipment such a leak should either depressurize the vessel or produce a "cold spot" which is easily noticed. The relationship between t and a_0 depends on crack geometry and therefore cannot be specified uniquely, nevertheless a critical crack size of at least twice the wall thickness would seem to be a minimum requirement.

C. FATIGUE, EMBRITTLEMENT AND CORROSION

As indicated above the three principal mechanisms by which cracks may form or intensify during service are fatigue, hydrogen embrittlement and corrosion, in particular stress corrosion. None of these is specifically a low temperature phenomenon, indeed the fatigue lives of many metals increase at low temperatures, while the rate at which most corrosion reactions occur drops very rapidly as the temperature is lowered. Rather, they increase the probability of unstable failure under service conditions which would normally be considered satisfactory. Their effect is due basically to one or more of the following effects; they lower the toughness of the metal, they provide a mechanism by which a crack may sharpen and thus increase the degree of elastic stress concentration, or they allow a subcritical crack to grow at stresses below the yield stress until it reaches the critical length required for unstable propagation.

Fatigue failure can occur in a metal subjected to cyclic or fluctuating stresses even though the maximum applied stress is below its yield and it has been estimated that about 80% of failures encountered in general engineering practice are caused by fatigue. A large amount of effort has been expended in attempting to understand the fundamental nature of this phenomenon and in developing methods by which the fatigue life of a component or structure may be predicted. These results are well documented and are discussed further in Tetelman and McEvily (1967) and many other textbooks and research papers.

Fatigue must be considered as a possible mode of failure in any piece of cryogenic equipment subjected to vibration, e.g. pumps and turbines, or to periodic changes in pressure such as reversing heat-exchanges or storage vessels. The predicted life of a component subjected to a given stress cycle is usually obtained from an S–N diagram (Stress vs. Number of cycles to

failure) similar to that shown in Fig. 6.14(*a*). This has one of two characteristic forms; for most ferrous materials the *S–N* curve tends to level off at
about 1 million cycles to give a *fatigue limit* at a stress of about $\frac{1}{2}$ UTS.
There is, however, no fatigue limit for many non-ferrous metals and the
allowable stress continues to drop as the number of cycles is increased, e.g.
for $N \sim 10^9$ cycles it is $\sim \frac{1}{3}$UTS. As we saw in an earlier section of this chapter,
the UTS of most metals increases as the temperature is lowered and it is thus
reasonable to expect that their endurance limit is higher at low temperatures.
This is generally true for most metals and is illustrated for aluminium in the
curves of Fig. 6.14(*a*).

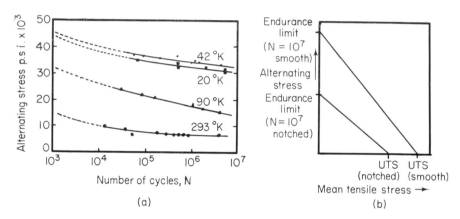

Fig. 6.14. (*a*) A series of *S–N* curves for aluminium specimens fatigue tested at 4·2, 20, 90
and 293°K (after McCammon and Rosenberg, 1957).
   (*b*) A Goodman diagram for fatigue under a combination of static and alternating applied
stresses for both notched and unnotched specimens.

In many applications it is found that the alternating stresses are superimposed on a mean tensile or compressive stress. It is therefore necessary to be
able to predict the life of components under the effect of these combined
stress systems and one way of doing this is to use a Goodman diagram as
shown in Fig. 6.14(*b*). For most metals the experimental values for smooth
specimens are found to lie quite close to the straight line joining the fatigue
to the ultimate tensile stress. For notch-sensitive metals, however, both tensile
strength and fatigue limit are lowered by an amount which depends on the
severity of the notch and thus the lower line in Fig. 6.14(*b*) defines the upper
boundary of the allowable combined stresses.

Most fatigue failures encountered in practice may be attributed directly to
the introduction of stress concentrators such as notches, bad welds, badly
radiused corners and other fabrication defects which act as nucleation sites
for fatigue cracks. Furthermore, even in the absence of such obvious defects,
the surface finish of a component has an important effect on its fatigue life,

smooth surfaces being more reliable than rough ones. Finally, the surface stress has a strong influence on the fatigue life of a component: a compressive stress is beneficial and such stresses are often introduced deliberately by processes such as shot peening to improve the fatigue life.

Thermal fatigue. This is a particular problem in items of cryogenic plant such as regenerators which are subject to periodic changes in temperature. Such cycles can generate thermal stresses if the temperature gradients are non-linear or if free expansion on contraction is restricted by external constraints. Large plastic strains can be developed in this way and fatigue failures can occur after a relatively small number of cycles. This is the so called "low cycle" or "high strain" type of fatigue where failure is due to accumulated persistent plastic deformation and it has been found that for many metals the number of cycles to failure, N, and the reversed plastic strain range, ε_p, are related by the Manson–Coffin law, $N^{\frac{1}{2}}\varepsilon_p = C$, a constant. As a rough rule of thumb $\pm 1\%$ strain causes failure in 1,000 cycles.

Corrosion fatigue. This is yet another hazard liable to cause failure in cryogenic equipment, especially air separation plants located by the sea or near chemical installations. Although the loss of metal by corrosion is negligible the endurance limit of many metals can be severely reduced by this mechanism. Aluminium alloys are particularly prone to corrosion fatigue unless treated with an impervious coating, while the endurance limit of plain carbon steel at 10^7 cycles is lowered by $\sim 50\%$ for fresh water and $\sim 75\%$ for salt water environments. Stainless steel is to be preferred if such conditions are likely to be encountered, but even this materials has been known to fail due to "shrouding" and crevice corrosion in brazed joints.

A related type of failure which is due to a combination of a mildly corrosive environment and residual internal stresses is known as *stress corrosion*. The stresses are normally introduced by plastic deformation during fabrication or more often by incorrect welding procedures. Such failures are common in wrought brass components, where the effect is known as season cracking, and are also found in welded aluminium and magnesium alloys and occasionally in stainless steels. It can be overcome or minimized by annealing the affected component to relieve the internal stresses.

Hydrogen embrittlement. This is another form of delayed fracture which can occur minutes or months after the application of a static or slightly increasing load. High strength steels are particularly susceptible to this type of failure, as are titanium, zirconium and their alloys. The effect is due basically to the presence of hydrogen in concentrations which exceed the equilibrium solubility limit of the metal and it is thus more serious in b.c.c. and h.c.p. metals which have low solubility limits. Hydrogen pick-up occurs typically in pickling or cathodic protection processes and during some types of welding process. Hydrogen is able to dissolve readily in the molten metal but is rejected from

14*

solution during subsequent cooling. The "excess" hydrogen then collects in pores or inclusions and builds up high internal pressures which can nucleate cleavage cracks in brittle materials or form blisters in more ductile ones. Hydrogen can be removed from a metal by vacuum baking at 350°C but it is better to prevent its initial pick-up where possible. 9% nickel steel, like other high tensile steels, is prone to this phenomenon.

D. DESIGNING AND TESTING FOR FRACTURE RESISTANCE

For some applications it is necessary to calculate the stress at which a material of a certain thickness will operate safely at a given temperature. Design stresses based on conventional proof or tensile stress criteria cannot give the optimum value for this stress as they take no account of the possible presence of flaws in the material and the only rigorous basis for calculation is that provided by fracture toughness, especially if the material has a high yield stress and low toughness. Reliable values of K_c or K_{Ic} must be obtained for the thickness and grade of material to be used in order to calculate the stress, σ_F, at which unstable fracture will be propagated from the largest flaw which can develop without detection. The operating stress, σ_{op}, must then be maintained below σ_F.

A less exact form of fracture toughness analysis is implicit in the everyday use of many low strength materials: they have such high values of K_c that σ_F always exceeds their yield stress unless very large, easily visible, flaws are present. There are, however, many situations, especially in cryogenic applications, in which the toughness of the material concerned decreases as the temperature is lowered and an assessment is required of the probability of catastrophic failure under the conditions expected in service. Alternatively, a more positive requirement would be to find the temperature at which a structure will function safely at a given level of applied stress. This can be determined as before if the relevant values of K_c are known for the temperature range concerned. As such data are often not available and K_c tests are rather laborious and expensive to carry out, less specific tests are usually used to determine the likelihood of catastrophic failure: the three most common tests being the Charpy V Notch Impact Test, the Drop Weight Tear Test and Notched Tensile Tests.

The Charpy test measures the energy absorbed by a notched bar of standard dimensions when fractured by a blow from a pendulum hammer, but this energy value has no direct application, i.e. it cannot be used in any equation to give the fracture stress of a material. The *change* in impact energy from a high to a low value over a range of testing temperatures does, however, indicate a decrease in toughness and a change in the mode of fracture from a high energy shear mode to either of the low energy modes, cleavage or low energy tear. Most work on Charpy tests has been done on mild steels in an effort to correlate Charpy data with that obtained from large scale tests and actual structures and in Fig. 6.15 the energy absorbed in a Charpy test is related to three of the recognized transition temperatures which describe the relative ease with which unstable fracture can be initiated or propagated. At very low temperatures cleavage fracture is readily initiated and propagated and the energy absorbed is low. The Nil–Ductility Temperature (N.D.T.) is defined as the highest temperature at which a small flaw (less than about $\frac{1}{4}''$

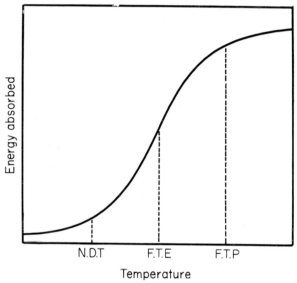

Fig. 6.15. A schematic representation of the relationship between the energy absorbed in a Charpy impact test and the Nil Ductility Transition (N.D.T.), the Fracture Transition Elastic (F.T.E.) and the Fracture Transition Plastic (F.T.P.) temperatures.

long) can initiate fracture at the yield stress of the material and above this temperature fracture is initiated by plastic deformation in the notch root. The relationship between the N.D.T. and impact energy is *not* constant from material to material. For example, for low strength carbon steels N.D.T. = 8 ft lb. absorbed energy, while in semi-killed steels and high strength quenched and tempered steels the values are approximately 25 and 50 ft lb., respectively. As the temperature increases above the N.D.T. crack initiation becomes more difficult and its propagation requires more energy as the ratio of shear fracture to cleavage fracture increases. Above the Fracture Transition Plastic (F.T.P.) crack propagation is totally by shear or fibrous rupture and the energy absorbed levels off to a more or less constant value.

The appearance of the fracture surface provides still further insight into the mechanisms involved in fracture and it is, in many respects, a more reliable guide to the probability of cleavage failure than is the impact energy.

Criteria based on the percentage of the fracture surface covered by cleavage failure usually give the most reliable correlation between Charpy impact tests and service failures and it has been found, for example, that if the surface of a Charpy bar fractured at a certain temperature is less than 70% crystalline, there is a high probability that cleavage fracture will not occur at this temperature if the applied stresses do not exceed half the yield stress. For some mild steels the impact energy corresponding to 70% crystallinity is 15 ft lb. and this has led to the erroneous belief that any material which can absorb 15 ft lb. of energy in a Charpy test cannot fail in a brittle manner. Such generalizations are not justified as in other steels 70% crystallinity corresponds to impact energies as high as 50 ft lb.; hence criteria based on fracture appearance are to be preferred to those based on impact energy. If a structure has to operate at stresses approaching the yield stress the safe operating temperature can be obtained from the more conservative 50% crystallinity criterion, while even higher safety margins can be obtained by working above the temperature at which no crystallinity appears on the fracture surface.

The Charpy test has two major limitations. Firstly, it is unsuitable for evaluating f.c.c. metals as these do not fail by cleavage and their susceptibility to low energy tear failure is not shown up by a marked change in impact energy or fracture appearance. The more sensitive Drop Weight Tear Test which uses much larger and thicker specimens is to be preferred for testing f.c.c. and h.c.p. metals. Secondly, impact tests cannot be carried out on thin sheets or plates less than 10 mm thick and these are best evaluated by Notched Tensile tests which favour the plane stress loading conditions conducive to unstable fracture in sheet materials.

In a typical test the net ultimate tensile stress of a symmetrically notched specimen is determined under the same testing conditions as that of a similar, but unnotched specimen. The Notch Sensitivity Ratio (N.S.R.) is then defined as the

$$\frac{\text{U.T.S. of the notched specimen}}{\text{U.T.S. of the unnotched specimen}}.$$

In tough ductile materials the N.S.R. can be greater than unity due to the plastic constraint effect discussed earlier, while in higher strength, low toughness materials the N.S.R. is often considerably less than unity. Although the N.S.R. cannot be used directly for design purposes it is generally considered that if the N.S.R. of a material is less than about 0·7 the material is excessively notch brittle. Recently there has been a move towards the adoption of a notched tensile/unnotched *yield* strength criterion (the N.Y.R.) as it is claimed that this is a more realistic parameter. If the N.Y.R. is less than unity fracture will propagate from the root of a notch before general yield occurs in the rest of the material, while the larger the N.Y.R. the lower the risk of service failure due to accidental scratching or other surface flaws.

V. Non-metals

Glass was one of the first materials to be employed in the construction of low temperature equipment and even today there are still many laboratories which use glass dewars for containing cryogenic fluids. In its bulk form, however, glass has a limited utility and it is now much more widely used in the divided form of glass fibre. Some polymeric materials have characteristics which make them attractive for low temperature applications, in particular their low thermal conductivities, low densities and relatively high strengths. Many types of material, including thermoplastics, elastomers, thermosetting resins, adhesives, films and foams are generally covered by the term "plastics" and they all share to some degree two basic disadvantages for cryogenic applications; they have high coefficients of expansion compared to those of metals and they become brittle at low temperatures. Both of these drawbacks are much reduced when plastics are reinforced with glass or carbon fibres and these composite materials have considerable potential for cryogenic applications.

A. POLYMERS

The mechanical properties of polymers differ considerably from those of metals and it is appropriate to consider them from a fundamental position. Polymers consist of long chains of molecules built up from relatively simple molecular units. If the plastic is thermosetting these chains become heavily cross linked during curing to give a rigid and rather brittle material. If on the other hand it is a thermoplastic the chains are held together by weak secondary bonds and it is relatively easy for the molecular chains to slide over each other under the influence of an applied stress. The more the chains are cross linked the more limited is this easy deformation and the greater the rigidity of the material. (Elastomers are basically lightly cross linked polymers which are able to undergo large amounts of reversible elastic deformation at room temperature.) At high temperatures, and this often includes room temperature, entire segments of the molecular chains are in thermal motion about their equilibrium positions and it is relatively easy for molecular sliding to occur during deformation. A considerable amount of yielding can then take place by the uncoiling and straightening of the chains until they are orientated parallel to the applied stress.

As the temperature is lowered, the thermal motion becomes less intense and the Van der Waals forces between adjacent chains become stronger, making yield by molecular straightening more difficult. This effect is most pronounced in a temperature range of about $10°$ known as the *glass transition temperature*, T_g, which is usually $\frac{1}{2}-\frac{2}{3}$ of the melting point. Below this transition the polymers are organic glasses and behave, in practically every respect, like ordinary inorganic glass. For example, they are almost completely brittle and show the same type of relation between crack size and tensile strength as found by Griffith for ordinary glass. As most of the significant polymeric materials melt at temperatures between 300 and 700°K, they undergo their glass transitions at temperatures above about 150°K. Thus virtually all polymers and elastomers are brittle at cryogenic temperatures, the only notable exception being PTFE which will be considered in more detail.

The increase in strength of the Van der Waals forces during the glass transition and the consequent reduction in separation of adjacent molecular chains are also responsible for the higher elastic moduli at low temperatures. For example, the tensile modulus of nylon increases by a factor of six between 300°K and 77°K while the modulus of PTFE at 20°K is over 20 times that at room temperature. Significant increases in strength are also shown by most polymers when cooled to low temperatures, the tensile strength of most plastics at 20°K being between 2 and 10 times that at room temperature. The actual values depend on a number of factors including the molecular weight, crystallinity and degree of molecular orientation of the material. Furthermore it should be noted that the higher the moisture content of materials such as nylon, the more brittle they are at low temperature.

It has already been pointed out in Section IIB that most polymers have a very large contraction on cooling and that this poses severe compatibility problems in using these materials in conjunction with metals at low temperatures. Their most serious drawback is, however, their small, or often non-existent, ductility at low temperatures. Most polymers are completely brittle even at 77°K and only PTFE shows a measurable elongation (about 1%) at 4°K. The exceptional properties of PTFE are believed to be due to its unique molecular structure in which crystallites are formed having a tight spiral formation of fluorine groups. These groups are unable to pack closely together and hence the Van der Waals forces cannot become strong enough to make the material go through a glass transition. Two unfortunate side effects of this molecular structure are the phenomenon of stress relaxation and cold flow under pressure, both of which are due to the ability of the long chains to slide over each other and rearrange themselves so as to relieve internal stresses. This tendency to cold flow reduces the usefulness of PTFE for applications such as valve seats and bearings unless it is possible to increase the load to compensate for stress relaxation. In order to overcome this difficulty, composites of PTFE and glass fibre have been developed which do not suffer from cold flow. The glass fibre improves the tensile and compressive properties of the material while the PTFE gives the composite sufficient ductility to accommodate plastically the strains developed. Furthermore, it has the additional advantages of reducing the contraction of the composite to a value more nearly compatible with that of most metals.

B. COMPOSITES

In the PTFE–glass fibre composite just considered, the matrix was slightly ductile; in most composites, however, both fibre and matrix are brittle at and below room temperature. Despite this fact epoxy, phenolic or polyester resins reinforced with carbon or glass fibres and cloths are able to absorb far more energy during fracture in an impact test than would be absorbed by an equivalent piece of bulk glass or resin. To understand this discrepancy it must be recalled that once fracture is nucleated in a homogeneous brittle material, it becomes self-sustaining and propagates very rapidly through the rest of the sample in a low energy mode. In contrast a crack nucleated in one fibre of a composite spreads rapidly through that fibre but slows down as it passes through the matrix. A separate crack has then to be nucleated in each successive fibre in order that the fracture can proceed. In this way the total energy absorbed during fracture is more nearly characteristic of the sum of the energies of the compound fibres than that of the bulk material. Thus although these composites are not ductile, their ability to fracture progressively is for many purposes an adequate substitute for ductility. Furthermore by varying the efficiency of the fibre–matrix interfacial bond the toughness and strength of the composite can be altered; increase in toughness being obtained as usual at the expense of tensile strength.

The toughness of Glass-Reinforced Plastic (GRP) laminates as measured by impact or notch tensile tests shows little significant change with temperature, while moduli in general, increase by $\sim 10\%$ on cooling from 300 to 4°K. There is, however, a marked increase in the tensile strengths of GRP laminates which increase by 40–100% between 300 and 77°K with little further increase on lowering the temperature to 4°K. Strengths of 75–100,000 p.s.i., typical of GRPs at 77°K, are by any standard quite high, but when one also realizes that these materials have low densities it will be appreciated that very high strength/weight ratios can be achieved. This is one of the reasons why such composites are being used in the construction of fuel tanks and structural members in rockets and other aerospace applications. Carbon Fibre-Reinforced Plastics (CFRPs) have considerably higher moduli than GRPs and comparable specific strengths at room temperature, but the tensile strengths of CFRPs only increase by about 10% on cooling from 300 to 77°K. They are thus likely to replace GRPs only in situations where their higher moduli and thermal and electrical conductivities are advantageous.

As a result of their disordered structures both plastics and glass fibres have inherently low thermal conductivities especially at cryogenic temperatures. When fabricated together the resultant composite has an even lower thermal conductivity. Combining this characteristic with the high specific strengths typical of GRPs and it can be appreciated that very high ratios of strength/thermal conductivity are possible. (See, for example, the ratio of σ_y/K given for GRP in Table 6.I.) The combination of high tensile and compressive strengths and very low thermal conductivity offered by GRPs thus makes them highly suitable for use as load bearing thermal insulators and they are finding applications in storage vessels and other situations which require low heat inleaks.

Glass reinforced plastics and other laminates have, however, a few serious drawbacks which limit their more widespread use. Their mechanical properties are highly directional, being much better parallel to the fibre axis than perpendicular to it. Two dimensional reinforcement can be obtained by cross plying or filament winding the fibres but efficient reinforcement in all three directions is not practicable. Joints between two or more composites or between a composite and a metal are difficult, but it is often possible to design around the problem. Furthermore, as the thermal contraction of a GRP composite lies between the low value of the glass fibre and the high value of the resin, its contraction can be matched quite successfully to that of most metals.

A less tractable problem is that of porosity, especially in thin filament-wound structures. It can be particularly severe if the composite is stressed beyond the point at which its modulus drops from its initial value to a lower secondary value, for this is the point at which micro-cracking occurs at the resin–fibre interface and the matrix can be seen to craze. Some control over this porosity has been achieved by electrodepositing a thin layer of nickel on, or bonding a thin sheet of aluminium to the composite surface, but attempts at bonding mylar and other thin plastic films have generally been less successful due to the differential contraction between film and composite. As carbon reinforced plastics have higher moduli than GRPs it can be expected that porosity will be less of a problem in CFRPs.

There are a number of situations which call for more specific data than that obtained from simple tensile, compressive or impact tests, in particular knowledge of the flexural strength and modulus are often needed and where cyclic loading occurs the fatigue life must be known. Such additional data often tips the balance more strongly in favour of one material than another. For example, from static tests epoxy resins appear most promising for use as matrices in composites for cryogenic use, followed by the polyesters and phenolics. However, when the tests are extended to include fatigue life, the epoxies and phenolics continue to show excellent properties while the polyesters fail much sooner.

C. ADHESIVES, FILMS AND FOAMS

The challenge provided by the development of large cryogenically-fuelled rockets has not only encouraged the development of new and improved structural materials but also of novel methods of construction and fabrication. In particular it has led to the development of structural adhesives which can be used at low temperatures without the serious problems of embrittlement and shrinkage which occur with most plastics. The key to these advances lies in the use of fillers whose main functions are to match the expansion coefficient of the adhesives as closely as possible to that of the adherent and also to redistribute thermal stresses throughout the adhesive instead of concentrating them at the adhesive–substrate interface. As a general rule the best results are obtained by applying the adhesives in a thin and uniform layer and this has the additional advantage of improving the thermal shock resistance of the bond, since otherwise the low thermal conductivity of most adhesives can lead to a temperature differential between the adhesive and a metal substrate.

The joint configuration and loading system can also have a strong influence on the strength at low temperature. For example in butt joints the main stresses induced by cooling are tensile and are largely offset by the increase in tensile strength of the adhesive at low temperatures. In contrast, the shear stresses induced in the plane of the adhesive in a lap joint are too large to be accommodated by the increase in shear strength at low temperatures and as a result the joint strength decreases.

The toughness of an adhesive is usually measured by its peel strength; a high peel strength indicating considerable resistance to further failure under conditions where the adhesive already contains voids, defects or cracks Of the two adhesives most commonly used for cryogenic applications (the epoxy–nylons which cure at about 100°C and the polyurethane pastes which cure at room temperature) the epoxy nylons have higher peel and shear strength at room temperature but the polyurethanes are superior at and below 77°K. Most structural adhesives are reinforced by a fabric called a "structure" or "carrier" which is usually a non-woven, woven or knitted structure of glass or polymer filaments. Non-woven, single filament glass fibre mat has generally been found the most efficient as the adhesive can easily penetrate the fabric and cause complete wetting of the filaments. The main action of the glass mat is to produce a uniform bond line, even out the stresses and reduce the contraction of the adhesive relative to the substrate. It also produces a more rigid bond which does not creep under stress at ambient temperatures.

Significant improvements in adhesion may be obtained by the use of silane primers which cause chemical reactions to occur between adhesive and adherent, in addition to the usual intermolecular forces. The silanes are ambifunctional coupling agents one end of which can be hydrolyzed to bond to an hydroxyl containing substrate, while the other end reacts with the resin to form a molecular bridge between resin and substrate. Silane additives not only produce a stronger and more reproducible bond, they also greatly improve its ageing properties especially in mildly corrosive environments.

Thin films and fibres of polymers such as Mylar (terylene) are of particular use in certain cryogenic applications such as the electrical insulation of wires, the thermal insulation of flexible cryogenic pipeline and the construction of expulsion bladders in rocket fuel tanks. In virtually every case the material is in a brittle condition and its high flexibility is due to the extreme thinness and its consequent ability to be bent around radii without exceeding the elastic limit of the material. Some problems are, however, associated with the relatively large thermal contractions shown by these highly orientated fibres and films and great care has to be taken in making leak proof joints in films especially if they have to withstand a number of cooling or stress cycles.

The final major use of plastics at low temperatures is in their expanded form, known commonly as "foams". They are cheap, convenient and relatively efficient insulating materials, their efficiency being superior to most mineral insulants but inferior to vacuum–powder and superinsulation, and they are thus used mainly for insulating storage tanks for the higher boiling point cryogens. Epoxy, polystyrene, ebonite and polyurethane are commonly used for producing foams but their thermal conductivity is more dependent on the type of gas used to blow the foam than on the type of plastic: if the blowing gas liquefies at cryogenic temperatures a partial vacuum is created and the thermal conductivity is reduced. The mechanical properties of most foamed plastics deteriorate at low temperatures due to the inherent brittleness of the plastic below its glass transition and also to the "internal notching" effect of the cell corners. If the foam is required to support more than the smallest loads it is best reinforced with a suitable honeycomb structure prior to blowing. Additional thermal data for foams is given in Chapter 4.

VI. Design and Materials Selection Criteria

There are a large number of factors which have to be considered when selecting materials for cryogenic applications and there is no unique formula for successful design. Each different job has its own priorities, e.g. weight minimization in aerospace applications—low heat capacity in transfer lines, high strength and low thermal conductivity in load bearing insulators. However, in most cases the exercise reduces to one of finding the most practicable and economic method of producing a finished item which is technically adequate for its intended duty.

The two most important technical requirements for cryogenic equipment are for adequate toughness under all possible operating conditions and for compatibility with its working environment. To a certain extent these can be considered as mandatory requirements, but this does not imply that safety margins shall be over-generous as this almost invariably leads to inefficient design and unnecessarily high costs. In assessing the cost of an item, it is necessary to consider not only the price of the raw material but also the additional expense of forming components, joining them together and finishing the job to the required standards such as cleanliness. Furthermore, the availability of suitable processing facilities and techniques, of the desired materials, or even of suitable low temperature design data for these materials, are all factors which have to be taken into consideration during the optimization process. As it is not possible to consider all these factors in detail attention will be concentrated on two particular aspects—the economic implications of designing pressure containing equipment to codes such as ASME VIII or B.S. 1515, and a review of the various techniques used in jointing materials for service at cryogenic temperatures.

A. TOUGHNESS, DESIGN CODES AND THEIR ECONOMIC IMPLICATIONS

In Section IV, the factors involved in determining the ability of a metal to resist brittle fracture were considered from a fundamental point of view and it was pointed out that the most satisfactory basis for efficient design is given by fracture mechanics. In this, the safe operating stress is determined by the size of the largest flaw present in a structure, which is considered safe if either the maximum sized defect which could escape detection by the available non-destructive testing techniques is smaller than the critical flaw size, or if this critically sized flaw is larger than that required to penetrate the full wall thickness of the vessel.

Such criteria have been successfully applied in designing cryogenic fuel tanks and other pressure vessels used in American rocket systems. As yet, however, they are not accepted as a basis for the design of commercial equipment which has to satisfy national insurance authorities such as the British Insurance Inspection Companies in the U.K. Such equipment has to be designed and manufactured in accordance with the codes of practice laid down by these authorities, and these in essence define the minimum temperature at which a given material may operate when subjected to a certain design stress. The effects of plate thickness, thermal stress relieving and other factors are taken into consideration and in many cases minimum impact energy values are specified in order to ensure that the material is not brittle at the design temperature. The minimum operating temperatures are usually at least 25°C above the Nil ductility temperature, but, as pointed out in Section IV the NDT does not correspond to the same impact energy value for every metal. A further point to note is that the permissible stress levels given by these codes are derived from the tensile properties *measured at room temperature* and thus no advantage can be gained from the large increases in yield and tensile stresses shown by most metals at cryogenic temperatures.

Of the relatively large number of design codes available to designers, only a small number specifically deal with the the design of vessels for operation at subzero temperatures. Of these the most commonly used are listed in Table 6.II.

TABLE 6.II. British, U.S.A. and German Pressure Vessel Design Codes

Standard	Materials	Temperature range	Factors considered	Comments
BS. 1500: 1958 Part 1, Appendix C	Carbon and low alloy steels BS. 1501–101 151 161 221 621 BS. 1510 LTO, 15, 30, 50, 100, 190	0 to −190°C	Special steels, impact testing, stress relief, plate thickness	BS. 1501–101 151 161 Not available with subzero impact tests
BS. 1500: 1965 Part 3, Aluminium	Aluminium alloyed with magnesium or manganese BS. 1470–1477	Below 0°C	Section 2 states that these alloys do not suffer from brittle fracture and are suitable for operation below 0°C. No special requirements are given	
BS. 1515: 1965 Part 1, Appendix C	Carbon steels BS. 1501–101 151, 161	0 to −30°C	Impact tests, stress relief, plate thickness	BS. 1501–101 151 161 Not available with subzero impact tests
	Carbon manganese Steel (semi-killed) BS. 1501, 211, (26,28) BS. 1501, 211, 32 (032)	0 to −50°C 0 to −50°C	Impact tests, stress relief, plate thickness	Available with impact tests to; −15°C (LT 15) 0°C (LT 0)
	BS. 1501, 211 (26, 28, 30) LT 0, LT 15 Carbon Manganese Steel (semi-killed) Niobium treated BS. 1501, 213 (28, 30, 32) LT 30 Carbon Manganese steel (semi-killed) BS. 1501, 221 (26, 28, 30, 32) LT 0, LT 15	0 to −50°C 0 to −15°C 0 to −50°C 0 to −30°C −5 to −40°C	Impact test, stress relief, plate thickness Impact test, stress relief, plate thickness Impact test, stress relief, plate thickness	Available with impact test to −30°C (LT 30) Impact tests to −15°C (LT 15) Impact tests to 0°C (LT 0)
	Carbon manganese steel (silicon-killed aluminium treated) BS. 1501, 224 (26, 28, 30, 32) LT 50	0 to −60°C 0 to −60°C	Impact test, stress relief	Available with impact tests to −50°C (LT 50)
	Carbon manganese steel BS. 1503–221	0 to −50°C	Impact test, stress relief	
	3½% Nickel steel BS. 1501–503 BS. 1503–503 BS. 1506–503	0 to −50°C 0 to −50°C 0 to −50°C	Impact test, stress relief	Available with impact tests to −50°C (LT 50)
	BS. 1510: 1958 Impact tested steels LT 0 LT 15 LT 30 LT 50 LT 100	0 to −50°C −5 to −20°C −20 to −35°C −35 to −50°C −50 to −60°C −100°C	Impact test Impact test, stress relief, plate thickness	

Table 6.II (*contd.*)

Standard	Materials	Temperature range	Factors considered	Comments
ASME VIII 1965 Unfired pressure vessels UCS 65, 66, 67	Carbon and low alloy steels with exception of materials to SA–7, SA–36, SA–113 or SA–283	Below −29°C	All vessels stress relieved. Charpy V impact test 15 ft lbs. at minimum temperatures	Steels most commonly used for vessels ASTM A–201B A–212B
ASME VIII 1965 Unfired pressure vessels U.N.F. 65	Aluminium and aluminium alloy, copper and copper alloys, nickel and nickel alloys	For wrought aluminium alloys to −254°C For copper, nickel and cast aluminium alloys to −198°C	No special requirements	For lower temperature impact test required
ASME VIII 1965 UHA–51	High alloy steels Types 304, 304L, and 347 (Stainless steels)	Below −254°C	Impact tests required	
	Other materials Ferritic chromium stainless steels. Austenitic stainless steels carbons 0·1%.	Below −198°C	Impact tests	
	Austenitic stainless steels whose analysis is outside AISI standard composition range.		Impact test required	
	Materials in cast form. Material in the form of deposited weld metals.	Below −20°C	Impact test required	
	Type 309, 310, 316, 309 (b), 310 (b)	Below −20°C	Impact tests required	
ASME VIII 1965 Unfired pressure vessels U.C.L.–27	Integrally clad plates	Below −20°C	Shall satisfy impact test requirements as UCS 66, 67, UHA–51 or UNF 65 for base metal	

		Duty Heavy	Lesser	Low	
A. D. Merkblätter Sheet W.10	Killed steels A.D. Data sheets W1, W2, W4, W5, W7, W8, W12, W13	√10	−50	−100	Safety factor used stress relief, permissible deformation, wall thickness, impact tests
	Non-ageing steels as: DIN 17135				Safety factor used, stress relief, permissible deformation wall thickness, impact test
	A. st. 35	−30	−80	−120	
	A. st. 41	−25	−75	−120	
	A. st. 45	−20	−70	−120	
	A. st. 52	−20	−70	−120	
	Cold ductile steel, as steel iron date sheets 680–60				Safety factor used, stress relief, wall thickness, impact test, permissible deformation
	TT, St, 35N.	−35	−105	−150	
	TT, St, 41N.	−30	−90	−150	
	TT, St. 45N.	−20	−80	−150	
	TT, St, 35V.	−80	−130	−200	
	TT, St, 41V.	−70	−120	−200	
	TT, St, 45V.	−55	−110	−200	
	25 Cr. Mo 4,	−60	−110	−273	wall thickness, impact test, permissible deformation
	14 Ni 6,	−110	−160	−273	
	12 Ni 19.	−160	−210	−273	
	X40 Mn Cr 22.	−200	−253	−273	Wall thickness, impact test, permissible deformation
	X12 Mn Cr 1811.	−195	−245	−273	
	X12 Cr Ni 189.	−253	−273	−273	
	X10 Cr Ni Ti. 1810	−253	−273	−273	

1. *A. D. Merkblätter W 10* (*Draft, December 1963*)

Section 3 of the Code deals with the intended duty of the vessel. Three kinds of duty are considered.

(a) *Pressure vessels for heavy duty.* Pressure vessels come under this heading when their materials are used to the limits permitted by the A.D. Data Sheets with regard to pressure and diameter and where they are constructed in accordance with the A. D. Data Sheets, especially A.D. Data Sheet H1. The safety factor used is equal to the usual safety factor. Stress relieving is mandatory for wall thicknesses over 30 mm.

(b) *Pressure vessels for lesser duties.* Under this heading come pressure vessels in which peak stresses are avoided by reliable design and manufacture. The safety factor used must be at least 4/3 the usual safety factor for wall thicknesses 5 mm and above. Stress relieving is only required for materials whose composition is outside A.D. Data Sheet H1.

For vessel thickness above 5 mm the vessel must be stress relieved.

Pressure vessels of wall thickness between 5 and 15 mm which have not been stress relieved are allowed if the safety factor is twice the normal value.

(c) *Pressure vessels for low duty.* Under this heading are considered pressure vessels, where the safety factor used to calculate the design stress is four times the usual safety factor. Stress relieving is conditional to material.

A footnote allows a vessel to be used at lower than normal temperatures where the lower temperatures result from a reduction in vapour pressure of the vessel contents provided that certain conditions listed are complied with.

2. *B.S. 1500: 1958*

B.S. 1500 : 1958 Appendix C, Table 20, gives the minimum operating temperatures for carbon steels, carbon manganese steels, chromium molybdenum steels, and impact tested steels (L.T. steels). It takes into account the effects of plate thickness, thermal stress relieving and impact testing. The temperature range covered is from 0 to $-60°C$ in the case of non-impacted tested steels and -5 to $-190°C$ in the case of impact tested (L.T. steels).

Table 21 of the specifications gives the maximum design stress to be employed when the pressure is due solely to vapour pressure of contents and where low temperatures are only coincident with low pressure. Under these conditions the special requirements with regard to materials, plate thickness, stress relieving and impact testing may be relaxed.

3. *B.S. 1515: 1965*

B.S. 1515 : 1965 Part 1, Appendix C, Table 24, gives the minimum operating temperature for carbon steels, carbon manganese (semi-killed) steel, carbon manganese (silicon killed), carbon manganese (silicon killed, aluminium treated), $3\frac{1}{2}$ nickel steel, chromium molybdenum steel, impact tested steel.

Table 25 is similar to Table 21 of B.S. 1500.

B.S. 1515 comments that "brittle" failures are not likely to occur except when both the following conditions occur simultaneously:

(i) The material exhibits very little notch ductility at the service temperature.

(ii) A tensile force which may be produced by the applied loads or residual stressing of a magnitude sufficient to cause plastic deformation is present at an existing crack or other severe notch. A brittle fracture will not propagate unless the general tensile membrane stress is sufficiently high to supply the necessary energy.

4. *ASME Boiler and Pressure Vessel Code—1965, Section VIII*

The ASME Boiler and Pressure Vessel Code—1965, Section VIII, Unfired Pressure Vessels, sub-section UCS–65, deals with low temperature operations and gives mandatory requirements for vessels operating below $-20°F$ ($-29°C$). The Code exempts from impact testing and stress relieving vessels where the operating pressures are small at the low operating temperatures. See UCS–66.

B. ECONOMICS

The following section specifically deals with the economic factors that affect the choice of a material for a particular duty.

1. *Design Code*

The use of codes where a design is expressed as a fraction of the yield stress such as B.S. 1515, in the U.K., and the A. B. Merkblätter in Germany, allow the wall thickness of pressure vessels to be reduced when compared with vessels designed in accordance with codes where the design stress is expressed as a fraction of the U.T.S. such as: B.S. 1500 and ASME VIII.

The saving in material costs are, however, sometimes offset to a certain extent by the more sophisticated design, fabrication and inspection techniques required.

2. *Strength Cost Factors*

For a given shell diameter the shell thickness for a given material is related to the operating pressure by the equation:

$$t = \frac{PR}{SE}$$

Where t is the thickness of the vessel shell:

R = Shell radius
L = Shell length
P = The design pressure
S = The design stress for the given material
E = The joint efficiency.

If C = the material cost of the shell for a particular duty:

c = Material cost per lb.
ρ = The material density lb./ft^3

$$C = \frac{c\rho LPR2\pi}{SE}$$

For any given vessel for a given duty the L, P, R and E are constant and the comparison between material a and material b is as follows:

$$\frac{C_a}{C_b} = \frac{c_a\rho_a S_b}{c_b\rho_b S_a}$$

$$= \frac{c_a\rho_a}{S_a} \bigg/ \frac{c_b\rho_b}{S_b}$$

Using the cost per ton, the ASME allowable design stress, and the density given in Table 6.VI a comparison can be made between the four most commonly used materials of construction, Aluminium Type 5083, Stainless Steel Type 304L, 9% Nickel Steel, Copper SB152; for comparison purposes carbon steel is also included. The results are presented in Table 6.III.

Table 6.III illustrates the effect of using different design codes. In one example, ASME VIII, the design stress is taken as quarter the UTS of the material, whilst in the case of B.S. 1515 the design stress is taken as two-thirds the yield stress. In (1969) 9% nickel steel only possessed marked cost advantages over aluminium 5083 when its high yield strength is used to advantage. In 1968 9% nickel steel had apparent cost advantages over 5083 aluminium even when vessels were designed to meet the requirements of ASME VIII. This comparison shows the close cost balance which may exist.

The cost comparison is based on the cost of material only, and it does not necessarily follow that the cost of the finished vessel would be in the same ratios as those given in the Table 6.III. For example, a comparison based on 1963 figures, for the field erection of a 40 ft diameter spherical container capacity is given in Table 6.IV.

It must be stressed that each individual case must be examined using up-to-date costing data to permit an accurate comparison to be made.

A number of vessels for incorporation in cryogenic plant, particularly air separation plant, operate at pressures below 50 p.s.i.g. and are under 5 ft diameter; under these conditions the material thickness is governed by

TABLE 6.III. Comparative Costs of Structures Metals for Low Temperature Pressure Vessels

	£ Cost per ton		Cost indices 1968		Cost indices 1969	
Material	1968	1969 Nov.	ASME VIII stress	BS. 1515 stress	ASME VIII stress	BS. 1515 stress
Aluminium NP8 (5083)	378	500	4·5	5·9	4·4 (4·4)	5·75
Stainless Steel 304 L	380	525	7·3	11·0	7·5 (6·0)	11·2
9% Nickel Steel	260	420	3·7	3·4	4·4 (4·85)	4·0
Copper Phosphorous Deoxidized SB. 152	610	857	33·0	47·0	35·5	51·0
Carbon Steel BS. 1501–161–26A	46	62	1·0	1·0	1	1·0

Notes

1. In the case of the ASME VIII Cost Indices, the design stresses used are based on values of UTS given in Table 6.III and in the case of the BS. 1515 Indices the design stress is taken as two-thirds of the yield stress values given in Table 6.VI.

2. Costs vary with quantity and form and are approximate only, the figures given for 1969 are for the month of November and include the value of the surcharge on all nickel containing alloys.

3. Comparative costs are for a given vessel diameter, and design pressure, the design temperature is taken as 20°C and there is no allowance for corrosion allowance.

4. No allowance has been made for the cost of welding which, owing to the high cost of welding electrodes for 9% nickel steel, would have an adverse effect upon the cost index for this metal.

5. The figures in brackets give an approximate comparison between the costs of the finished products but the actual result varies with size of vessel and service.

rigidity rather than strength consideration, and often the same plate thickness would be employed irrespective of the material of construction used. When these conditions apply, aluminium is an obvious choice. For example, a 2 ft diameter vessel designed to ASME VIII with a joint effluency of 0·85 would have the thickness shown in Table 6.V for design pressures of 50 p.s.i.g. and 700 p.s.i.g., respectively.

C. JOINTING METHODS

1. General

The leakage of process fluids from cryogenic equipment will usually give rise to a hazardous situation as well as representing a refrigeration loss which would affect the efficiency of the process. A serious leakage generally requires that the equipment be shut down and heavy financial losses can be incurred

TABLE 6.IV

Cost	5083 Aluminium Design pressure 15-p.s.i.a.	5083 Aluminium Design pressure 50-p.s.i.a.	9% Nickel Design pressure 15-p.s.i.a.	9% Nickel Design pressure 50-p.s.i.a.	Type 304 Stainless Design pressure 15-p.s.i.a.	Type 304 Stainless Design pressure 50-p.s.i.a.
	Material					
Material $	15,800	36,350	15,200	30,000	36,000	64,000
Labour and electrodes	9,720	17,050	11,060	19,520	8,900	11,820
Total $	25,520	53,400	26,260	49,520	44,900	75,820
Material cost % of total	62	70	58	60	80	85
Ratio labour/material	0·61	0·465	0·72	0·65	0·25	0·185

(Taken from "Low Temperature and Cryogenic Steels Material Manual 1968", U.S. Steel Corporation.)

TABLE 6.V. A Comparison of Pressure Vessel Wall Thicknesses (Taking Rigidity into Account)

Material	Design pressure 50 p.s.i.g. Calculated thickness in	Design pressure 50 p.s.i.g. Actual thickness in	Design pressure 700 p.s.i.g. Calculated thickness in	Design pressure 700 p.s.i.g. Actual thickness in
Aluminium 5083	0·095	0·125	1·25	1·25
Stainless steel 304 L	0·04	0·125	0·575	0·625

whilst the leakage is being repaired. In order to reduce the possibility of leakage occurring, equipment designers make the maximum use of butt welded joints in both fabricated equipment and pipework. Flanged and bolted, rivetted and soldered joints are reduced to the absolute minimum.

The jointing methods available for service at cryogenic temperatures are reviewed in Table 6.VII below, together with their fields of application.

Flanged and bolted joints are used where it is necessary for maintenance purposes to disconnect items of equipment such as valves from pipework or where removable closures are required in vessels, i.e. to recharge adsorbents or catalysts. Such joints are also used to connect together pipework made from dissimilar materials when welding is not possible.

Where valves are flanged into pipe lines, ASA series or B.S. 1560 series flanges are normally employed. Narrow faced 1/16 in. or 1/32 in. thick

TABLE 6.VI. Design Data for Common Constructural Materials used in Cryogenic Plant

Materials	Cost per ton 1969 Nov. £	Temperature °C	U.T.S. p.s.i. (min)	0·2% Yield stress p.s.i.	% Elongation in 4D	Impact strength 5 mm Charpy U notch ft lb.	Impact strength Charpy V notch ft lb.	ASME max allowable design stress p.s.i.	Design stress 2/3 yield stress p.s.i.	Density gm/cc	Density lb./ft³
Aluminium ASA 5083	500	38°C	39,000	16,000	23	15		9,750	10,600	2·7	166
		−196°C	—	—	33	15		9,750		—	—
Stainless steel type 304L annealed	525	20°C	70,000	25,000	66	75	140	17,500	16,600	7·84	500
		−196°C	—	—	36	81	80	17,500		—	—
Stainless steel type 304L high yield	525	20°C	85,200	42,600	—	—	—	17,500	28,400	7·84	500
Stainless steel type 321		20°C	90,000	38,000	60	—	110–125		25,200	7·84	500
		−196°C			43	—	110				
Copper annealed phosphoro-dioxidized SB152	857	20°C	30,000	10,000	48	57	43	6,700	6,600	8·84	560
		−196°C	—	—	57·5	75	50	6,700		—	—
9% Nickel steel normalized and tempered ASTM A 353 B	420	20°C	100,000	75,000	22	52	78	23,750 (as welded)	36,600 (as welded)	7·8	491
		−196°C			32	26·5	39	23,750			
Carbon steel B/501/161/26	62	20°C	58,000	—	28	—	—	14,500		7·8	495
Carbon steel ASTM A 201 B		20°C	60,000	32,000	28	—	—	15,000	21,200	7·8	495

TABLE 6.VII. Jointing Methods

Method	Materials	Application	Limitation
Bolting	Stainless steels, aluminium alloys, copper alloys, 9% nickel steel	Flanged joints when used with suitable gaskets	Unless special bolts and/or sleeves are used are subject to leakage if thermal cycling occurs
Rivetting	Copper and its alloys, aluminium alloys	Distillation trays	Only satisfactory for low pressure duties. To obtain leak tight joint, rivets have to be flooded with solder
Soldering (soft solder)	Copper and copper alloys	Distillation trays. Tube to tube sheet joints. Used as a sealant	Joint is brittle and has low tensile strength. Must be mechanically supported
Silver solder and brazing	Copper, copper alloys, stainless steel	Flanges to pipe-work. Column sections to column flanges	Generally limited to copper alloys. Expensive, requires skilled labour
Welding "Tungsten Inert Gas" (T.I.G.) Welding electric arc with protective argon–helium shroud, non-consumable tungsten electrode	Copper and alloys, stainless steels, aluminium alloys	Pipework, pressure vessels, exchangers	Usually used up to 3/16 in. thick aluminium, requiring skilled labour
"Metal Inert Gas" (M.I.G.) electric arc with protective shroud consumable electrode	Copper and alloys, stainless steel, aluminium alloys, 9% nickel steel	Pipework, pressure vessels, heat exchangers	Requires skilled labour
Friction welding	Aluminium alloys. Stainless steels. Aluminium to stainless steel	Transition pieces between stainless steel pipe and aluminium pipe	Limited by size of machinery available and shape
Dip brazing	Aluminium magnesium alloys	Extended surface heat exchangers. Automatic	Size, shape and material used limit application

C.A.F. (Compressed Asbestos Fibre) gaskets are employed, generally without the use of any jointing compound. If the two flanges used in a flanged joint are made from the same material it is necessary to use bolts which are also made from this material in order to avoid differential contraction effects.

2. *Flanged and Bolted Joints Between Dissimilar Metals*

Care has to be taken when flanges of dissimilar metals such as aluminium alloy and stainless steel are joined together. As may be seen from Fig. 6.2, the total linear contraction of aluminium alloy is greater than that of stainless steel. Thus, if an aluminium alloy bolt were tensioned enough to obtain the necessary gasket seating load at room temperature, there would be a distinct risk of its failure on subsequent cooling to low temperatures. Conversely, if a stainless steel bolt were used it would be necessary, in order to prevent leakage when at low temperatures, to apply a bolt load at ambient temperature which could transfer to the aluminium flange a stress in excess of its design value.

Various types of joints using extension sleeves have been developed to overcome this type of problem. Figure 6.16 represents a joint due to Usher (1963). He recommends that the bolt and sleeve be made of HE 15 aluminium alloy, the stainless steel flanges from Type AISI 310 stainless steel (25/20) and the aluminium flange from B.S. 1470 NB 5/6.

Usher reports that the most satisfactory results were obtained when the bolt tension was measured by using a micrometer to measure the bolt elongation. It was found that when a torque wrench was employed to indicate the bolt tension, there was a large variation in bolts loads even after the bolts had been tightened to the same torque setting.

AISA Type 310 (25/20) stainless steel is used in preference to other grades of austenitic stainless steel, as it is not subject to dimensional changes at liquid nitrogen temperatures, the effect of which is to reduce the tension in the

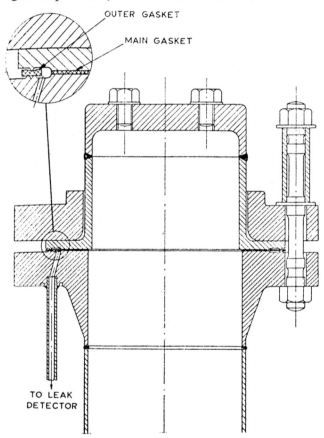

Fig. 6.16. A compensated sleeve joint which overcomes the problems of leakage when flanges of dissimilar metals are bolted together.

bolts. A compensated sleeve joint was developed by one of the authors is similar to the one illustrated in Fig. 6.16. This joint uses Monel sleeves, AISA Type 310 stainless steel bolts and washers, standards ASA series weld neck flanges, and

1/16 in. thick C.I.F. narrow faced gaskets. Monel metal has a lower coefficient of expansion than stainless steel, which in turn has a lower coefficient than aluminium. Thus the length of a Monel metal sleeve can be calculated so

Fig. 6.17. A cold box assembly in which transition pieces were used to connect stainless steel control values into aluminium alloy pipework.

that the tension in the bolt is independent of temperature. This type of joint has been successfully used on 18 in. diameter columns operating at 110 p.s.i.g. and $-180°C$ and on pipework and vessels operating at $-170°C$ with 650 p.s.i.g. hydrogen. Fig. 6.17 illustrates the use of this joint in connecting stainless steel control valves into aluminium alloy pipework.

A similar compensated sleeve joint which uses Nilo 36 alloy, a nickel iron alloy with a very low thermal contraction (similar to Invar), for the sleeves and 9% nickel bolts, has been described, and was successfully used on the L.N.G. lines at Canvey Island.

3. *Transition Joints between Stainless Steel and Aluminium*

A number of transition joints have been produced that enable leak free connections to be made between stainless pipe and aluminium pipe without the use of flanges or bolts. All of the couplings have one end stainless steel, that can be welded to a stainless steel pipe and one end aluminium, that can be welded to an aluminium pipe. The actual joint between the stainless steel and aluminium is produced in various ways.

Friction welding was developed by A. I. Chedekov in the U.S.S.R. during the late 1950s. In the U.K. the process has been developed by the British Welding Research Association. This type of joint is produced by rotating one of the two components to be jointed in contact with the second component, which is prevented from rotating. When sufficient frictional heat has been generated, pressure is applied to weld the components together. Such joints, if produced correctly, are free from porosity and inclusions.

Friction welded couplings between Type 6061 aluminium and Type 304 stainless steel pipe have been used extensively in the fabrication of cold box pipework. The joint has been widely tested and has been found to be resistant to thermal cycle stresses, e.g. one hundred temperature reversals between $-193°C$ and $50°C$ did not produce any detectable leakage, when examined using a mass spectrometer leak detector. Furthermore, when the joint was subjected to tensile testing it failed in the body of the aluminium and not at the joint line. To date the friction welded joints do not appear to be available in sizes above 6 in.

Care should be exercised when welding the free ends of the transition joints to the interconnecting pipework, as the joint can be destroyed by temperatures above $400°C$. It is usual during the welding operation to protect the joint from overheating by the use of a thermal shunt. Quality control methods have been developed using ultrasonic and helium leak detection techniques.

Brazed aluminium to stainless steel transition joints are produced by treating a stainless steel tube with a special flux, followed by coating with aluminium alloy and finally furnace brazing the stainless steel into an aluminium sleeve using an inert atmosphere. This type of joint has been used on hydrogen service in sizes up to 8 in. operating at pressures up to 650 p.s.i.g. and temperatures as low as --200°C. This joint is resistant to thermal cycling.

As with the friction welded transition pieces the joint can be destroyed if subjected to high temperatures during welding operations; it is important therefore, that thermal shunts are used to prevent heat flow from the welded zone to the joint.

The third method is named the "Al–Fin" transition joint. The stainless steel tube is specially prepared and then immersed in a bath of molten aluminium-silicon alloy, which forms an intermetallic layer of $Fe \cdot Al_3$ about 0·001–0·004 in. thick on the surface of the stainless steel. The stainless tube so treated is placed in a mould and aluminium cast round it. This type of joint has not been as extensively used as the friction welded and brazed type joint.

4. *Welding Aluminium Alloys*

Perhaps the most widely used materials for application in the cryogenic field are the non-heat treatable aluminium alloys which contain $3\frac{1}{2}$–5% magnesium (B.S. 1470 NP8, NP 5/6, ASA 5454, ASA 5083). These alloys are generally welded by either of two processes:

    (i) Tungsten Inert Gas Welding (T.I.G.)

    (ii) Metal Inert Gas Welding (M.I.G.)

Both processes use an electric arc and both prevent oxidation by providing a protective shroud of argon gas. In the case of the T.I.G. process the arc is struck between the non-consumable tungsten electrode and the work piece and the filler wire is fed in separately. In the case of the M.I.G. process the arc is struck between the filler wire and the work pieces, and the filler wire is fed forward through the torch.

The M.I.G. process is generally used for plate thicknesses above 3/16 in., whilst the T.I.G. process is used for thinner plates and pipework. The T.I.G. process offers complete control when welding in all positions.

In addition to the processes described above, the so called "high current technique" Hackman (1959) can be used for automatically welding aluminium in thicknesses greater than 10 mm. Using this technique, it is possible, using a single pass from each side of the joint, to make butt welds in thicknesses as great as 30 mm. Inherently high radiographic standards can be achieved.

It is reported that "pulsed arc welding" Boughton and Lucey (1965) has been used to automatically weld aluminium sheets as thin as 1·6 mm. Compared with the T.I.G. process, welding speeds are two to three times as high, the heat input is reduced and distortion is reduced. This process is useful for welding nozzles into thin walled (5 mm) vessels without burning or distorting the vessel wall.

In order to obtain good welds in aluminium that satisfy high standards of X-ray acceptance, the following factors are important:

(a) *Cleanliness*. When welding aluminium it is very important to obtain high standards of cleanliness. The edge preparation should be clean and smooth and free from inclusions, grease and moisture. Immediately prior to welding it is necessary to clean the joint faces with a stainless steel brush and if necessary wipe with a suitable solvent.

(b) *Operator skills*. If leak tight joints and a high standard of X-ray acceptance are to be obtained then welder qualifications are of supreme importance; there is no other single factor which more critically affects the progress and success of a job. Many welders, whose performance is first class on stainless steels cannot pass a welder qualification test when welding aluminium alloys.

5. Welding Stainless Steel

Type 304L 18/8 chromium nickel stainless steel is used extensively for cryogenic purposes and does not present any particular welding problems; it can be welded by the following methods:

(a) *Tungsten inert gas (T.I.G.)*. This method is suitable for the highest quality butt weld. A filler rod is used for thicknesses above 0·060 in.

(b) *Metal inert gas* (M.I.G.). This process is very fast and well suited to fully or semi-automatic welding.

(c) *Submerged arc welding*. This type of welding finds its greatest use in automatic welding processes and quality welds can be achieved by this method at high speed and low cost.

(d) *Shielded metal arc welding*. Shielded metal arc welding with flux coated electrodes can be used for the manual welding of stainless steel. A source of trouble with flux coated electrodes is their tendency to adsorb moisture from the atmosphere, which can give rise to a porous weld deposit. It is advisable therefore to store the electrode in a drying cabinet at about 175°F.

6. Welding 9% Nickel Steel (ASTM A353)

There has been a growth in the use of 9% nickel steel for cryogenic services during recent years. 9% nickel steel has been used for L.N.G. storage, both ship-board and land terminal, liquid oxygen and nitrogen static storage, liquid carbon dioxide and ammonia road and rail tankers.

The ASME Code Case 1308-4 (special ruling) (1962) allows the use of 9% nickel steel in the quenched and tempered or double normalized and tempered form in thicknesses up to 2 in. and temperatures as low as $-196°C/-320°F$ without thermal stress relief after welding. The Code Case 1308-4 requires a minimum Charpy V notch impact value of 25 ft lb. at $-320°F$, and allows a design stress of 25,000 p.s.i.g. for the parent metal and 23,700 p.s.i.g. for weldments.

15

The Italian Authorities (A.N.C.C.) issued a ruling in circular n 9010 of 12 April 1963 covering the acceptance of 9% nickel steel for pressure vessel construction.

Whilst a number of welding methods are available none are currently accorded a completely matching proof stress. The two basic welding processes have been successfully employed to weld 9% nickel steel are described below.
(a) *Flux coated electrodes.* The production of a 9% nickel steel weld electrode having acceptable notch toughness at low temperatures has not been possible. Two classes of electrode have been used:

    1. Inconel type electrodes, e.g. Inco weld A and Inconel 182, of which the latter can be used for positional welding.

      ("Nyloid 1"—Used for on site construction of Liquid Oxygen tanks).

    2. Stainless steel electrodes (according to R. Machine, 1966).

The French E.S.A.B. have successfully developed an 18% Cr–10% Ni–2% Mo. electrode designated OKSP 184, which was used during the construction of the L.N.G. plant at Arzew and is suitable for positional welding.
(b) *Gas shielded welding.* Both T.I.G. and M.I.G. welding have been used with an argon shield. Three types of electrodes can be used.

    1. Inconel 92

    2. Inconel 82 (does not suffer from age hardening tendency as Inconel 92)

    3. NC 8020 (an 80% nickel, 20% chromium electrode).

The mechanical properties of Inconel 92 and 82 and NC 8020 are given in Table 6.VIII along with Inconel 182, Inco-Weld A and NC 8020.

TABLE 6.VIII

Electrode	0·2% Proof stress p.s.i.	U.T.S. p.s.i.	Elongation %	Charpy V ft lb.
Inco–Weld A	52,400	92,500	45	55
Inconel 182	54,700	94,000	40	70
OKSP 184	71,000	98,500	40	40
Inconel 92/82	53,800	69,500–91,800	40	40
NC 8020	70,500–77,300	105,000	15/30	35/60

It is reported (Jordan and Heath 1964) that two versions of the M.I.G. process are used, the conventional and the fine wire process. The fine wire process employs a wire of 0·030 in. diameter and has advantages for positional welding.

To obtain good welds cleanliness is essential. Mill and flame cutting scale should be removed from the weld area by grinding, blasting or power brushing. Between weld passes flux should be removed.

VII. References

ASME VIII (1965). American Society for Mechanical Engineers Boiler and Pressure Vessel Code, Section VIII–Unfired Pressure Vessels.

British Standards Institution. (1958). Standard No. B.S. 1500.

Boughton P. H. and Lucey, J. A. (1965). April, *Brit. Weld. J.* **12,** 159–166.

Corruccini, R. J. and Gniewek, J. J. (1960). N.B.S. Monograph 21.

Corruccini, R. J. and Gniewek, J. J. (1961). N.B.S. Monograph 29.

Gopal, E. S. R. (1966). "Specific Heats at Low Temperatures." Plenum Press, New York.

Gunter, C. J. and Reed, R. P. (1962). *Trans. Quart. A.S.M.* **55,** 399–419.

Hackman, R. C. (1959). *Brit. Weld. J.,* pp. 676–684.

Hargreaves, R. and Tempton, D. H. (1961). *Proc. Inst. Mech. Eng.,* **175,** 941.

Johnson, V. J. (1960). U.S.A.F. Wadd Report 60–56. (Also published by Pergamon Press, Oxford.)

Jordan, D. E. and Heath, D. J. (1964). *Brit. Weld. J.* **11,** 2–5.

Machine, R. (1966). *In* "Welding and Metal Fabrication." July, 266–269.

Mark, H. F. (1966). "The Encyclopedia of Polymer Science and Technology", Vol. 4, pp. 415–449. John Wiley and Sons Inc., New York.

McCammon, R. D. and Rosenberg, H. M. (1957). *Proc. Roy. Soc.* **A242,** 203, 370, 379, 380.

Powell, R. L. and Blanpied, W. A. (1954). N.B.S. Circular 556.

Reed, R. P. and Mikesell, R. P. (1967). N.B.S. Monograph 101.

Rosenberg, H. M. (1963). "Low Temperature Solid State Physics." Clarendon Press, Oxford.

Schwartsberg, F. R. (1964). "Cryogenic Materials Data Handbook." Martin Co. Denver, Colorado. ML TDR-64-280 (AF 33 657 9161) (Supplement 1966).

Serafini, T. T. and Koenig, J. L. (1968). "The Cryogenic Properties of Polymers." Marcell Dekkar Inc., New York.

Tetelman, A. S. and McEvily, A. J. (1967). "Fracture of Structural Materials." J. Wiley and Sons Inc., New York.

Timmerhaus, K. D. (1955). "Advances in Cryogenic Engineering" (being the proceedings of the annual Cryogenic Engineering Conferences). Plenum Press, New York.

U.S. Steel (1968). "Low Temperature and Cryogenic Steels–Materials Manual." United States Steel, International Inc., New York and London.

Usher, J. W. C. (1963). Applied Mechanics Group, Institute of Mechanical Engineers, London.

Warren, K. A. and Reed, R. P. (1963). N.B.S. Monograph 63.

White, G. K. (1968). "Experimental Techniques in Low Temperature Physics." Clarendon Press, Oxford.

Wigley, D. A. (1971). "The Mechanical Properties of Materials at Low Temperatures". Plenum Press, New York.

Chapter 7

Adsorption

K. WILSON

Air Products Limited, New Malden, Surrey, England

I. Introduction

A. FIXED BED ADSORPTION

Adsorption has been defined as that operation which uses surface forces on solid bodies, called adsorbents, to achieve the concentration of volatile materials. Because adsorption takes place on the surface of a solid, practical adsorbents must have large surface area to volume ratios, and the cavities implied by this condition must be reasonably accessible. A wide range of adsorbents has been developed for industrial use in the separation of fluid mixtures. They are usually granular in form and are supported in beds of suitable geometry through which the mixture to be separated is passed.

The adsorber usually takes the form of a vertical cylinder with fluid flow parallel to the axis. To avoid the danger of particle attrition, the direction of the process flow will normally be downward.

In general, some of the components of the mixture are almost completely adsorbed, while the remainder pass out of the bed unchanged to form the effluent. The adsorbate from the feed is initially deposited in the region near the inlet to the bed. As the adsorbent in this region becomes saturated, the adsorption zone moves further along the bed. In front of this zone both feed and adsorbent are essentially free of adsorbate. Behind the zone the feed retains its initial composition and the adsorbent is in equilibrium with it. When the front of this zone reaches the end of the bed, the effluent composition begins to rise and eventually reaches the feed composition. The time at which adsorbate appears in the effluent is known as "breakthrough", and the concentration-time relationship for the effluent is known as the "breakthrough curve". It is usually desirable to stop the flow before breakthrough occurs. Semi-continuous operation can be achieved by switching the flow to an identical bed while the first one is "reactivated". The desorption involved in this process is achieved by raising the temperature of the adsorbent or by lowering the partial pressure of the adsorbate or by a combination of both.

The design of an adsorption system will involve consideration of the following:

(i) Adsorbent Properties
(ii) Fixed Bed Dynamics
(iii) Reactivation
(iv) Equipment Design

As with any other item of chemical plant, there are generally a number of different designs which are all technically acceptable solutions to the same problem. The designer's task is to find the solution that involves the lowest overall cost.

B. ADSORPTION IN CRYOGENIC PROCESSES

The cost of an adsorption unit will be found to vary significantly with the initial concentration of the impurity. (This contrasts with some other separation processes such as distillation.) Adsorption is usually most economical for the removal of impurity concentrations below about 1%.

In many cryogenic processes, it is necessary to remove from a feed gas small quantities of an impurity which would otherwise solidify at low temperatures. Some examples of this are shown in Table 7.I.

It should not be inferred that fixed bed adsorption is necessarily limited to the removal of trace impurities. Numerous examples have been given in the literature (Barry, 1960; Carter, 1961; Leavitt, 1962), of the separation by adsorption of feeds with much higher concentrations.

TABLE 7.I. Examples of Purification by Adsorption in Cryogenic Processes

Process	Stream	Impurity	Typical impurity concentration
Air separation	Air feed	H_2O/CO_2	0·5%/350 p.p.m.
Air separation	Crude oxygen	Hydrocarbons/C_2H_2	300 p.p.m./1 p.p.m.
Air separation	Pure oxygen	C_2H_2	1 p.p.m.
LNG	Natural gas feed	$H_2O/CO_2/H_2S$	
Hydrogen liquefaction	Hydrogen	O_2	1 p.p.m.

II. Physical Properties of Adsorbents

The main selection criteria for adsorbents used in cyclic purification or separation systems are capacity, selectivity, mechanical strength and chemical stability.

The adsorption capacity, measured as the ratio of adsorbate to adsorbent weight, will normally be 10 to 20% although the actual equilibrium capacity for a particular application may be much less than this. It is important not only because it determines the size and therefore the capital cost of the adsorber, but because it affects the operating costs arising from regeneration and the process pressure drop. High adsorbent capacity is the result of two properties: a pattern of molecular potential which provides sites for adsorbate molecules, and a large surface-area-to-volume ratio.

Moreover, the equilibrium capacity should change significantly with modest changes in pressure or temperature. This requirement follows from the fact that the cost of reactivation (heater, blower, power, reactivation gas) often represents a substantial proportion of the total process cost.

Selectivity, α, is defined as $\dfrac{X_{A,g} X_{B,a}}{X_{B,g} X_{A,a}}$ where $X_{A,g}$ is the mole fraction of component A in the gas phase, and so on. High selectivity is a desirable feature of an adsorbent. Normal adsorbents, such as charcoal and silica gel, are highly selective for components with widely differing boiling points. It is less easy to generalize about molecular sieve adsorbents, since for these the size and shape of the adsorbed molecules also play an important role in determining selectivity.

The adsorbent must be mechanically strong. When adsorbent pellets are packed in a vessel the combination of bed weight and flow pressure drop may exert large forces on the pellets at the bottom of the vessel. This consideration can place uneconomical limitations on bed height and pressure drop. Particularly large forces can be generated during pressure let-down when reactivation by pressure cycling is used.

Adsorbent manufacturers describe the strength of their product in various ways. Generally some form of grinding or shaking test is prescribed and the weight per cent of fines produced in a certain time is used as a measure of hardness. But since the grinding method, the time for grinding, and the definition of fines can vary from one manufacturer to another, the relative strengths of the various adsorbents are difficult to determine. In addition, it is not clear what the relationship is between this index of strength and the type of strength required to resist bed weight and pressure drop. As a consequence, it is sometimes necessary to resort to a crushing experiment which simulates bed weight and pressure drop, in order to determine whether or not a proposed design is acceptable.

The adsorbent must be able to withstand the repeated heating and cooling necessary for satisfactory reactivation. Not only must it be chemically stable during heating, but its coefficient of thermal expansion must be sufficiently similar to that of its container to avoid the generation of stresses high enough to cause particle fracture.

There are, in addition, certain secondary properties which must be borne in mind. The adsorbent particles should be of fairly uniform shape with no sharp edges or projections. Non-uniform size tends to increase bed pressure drop and smooth particles are somewhat less liable to cause problems of abrasion or dusting.

The mean size of an adsorbent pellet will affect both the bed pressure drop and the diffusional resistance of the pellet. In small adsorbent beds, pressure drop is usually not important, whereas the role of the diffusional resistance of the pellets may be quite significant. In large beds, the reverse of both these statements is frequently true. It is therefore common to increase pellet size as the adsorber itself increases in size. Fortunately, most adsorbents are available in a variety of sizes, so that this consideration seldom influences the choice of adsorbent type.

The adsorbent should have a low specific heat if it is to be reactivated by heating, particularly if one defines this specific heat as "heat required per unit of temperature rise per unit of mass of adsorbate adsorbed".

The physical properties of some commercial adsorbents are listed in Table 7.II.

III. Equilibrium Adsorption

Since capacity is a function of the two parameters, pressure and temperature, the characteristics of any particular adsorbent–adsorbate system may be represented by a curved surface in capacity–pressure–temperature space. Contours may be drawn on this surface to represent lines of constant capacity (isosteres), constant pressure (isobars) and constant temperature (isotherms). Figure 7.1 is a representation of a hypothetical equilibrium surface.

TABLE 7.II. Physical Properties of Some Adsorbents

Adsorbent	Activated carbon	Silica gel	Activated alumina	Molecular sieves
Surface area (m²/kg)	$1 \cdot 15$ to $1 \cdot 25 \times 10^6$	$0 \cdot 75$ to $0 \cdot 80 \times 10^6$	$0 \cdot 32$ to $0 \cdot 36 \times 10^6$	not relevant
Solid density (kg/m³)	$1 \cdot 76$ to $2 \cdot 08 \times 10^3$	$2 \cdot 2 \times 10^3$	$3 \cdot 0 \times 10^3$	$1 \cdot 55 \times 10^3$
Particle density (kg/m³)	$0 \cdot 6$ to $0 \cdot 9 \times 10^3$	$1 \cdot 2 \times 10^3$	$1 \cdot 12$ to $1 \cdot 2 \times 10^3$	$0 \cdot 91$ to $1 \cdot 15 \times 10^3$
Bulk density (kg/m³)	$0 \cdot 42$ to $0 \cdot 54 \times 10^3$	$0 \cdot 72 \times 10^3$	$0 \cdot 74$ to $0 \cdot 80 \times 10^3$	$0 \cdot 61$ to $0 \cdot 75 \times 10^3$
Average pore diameter (m)	20 to 100×10^{-10}	22×10^{-10}	22 and 44×10^{-10}	$3,4,5$ and $10 \cdot 3 \times 10^{-10}$
Maximum reactivation temperature (°K)	410	590	530	700
Specific heat (J/kg °K)	$0 \cdot 84 \times 10^3$	$0 \cdot 92 \times 10^3$	$1 \cdot 0 \times 10^3$	$1 \cdot 0 \times 10^3$

Isosteres may be used to determine the value of the heat of adsorption in a manner quite analogous to the determination of heat of vaporization from a vapour–pressure curve from the Clausius–Clapeyron relation. The log of the pressure is plotted against the inverse of the absolute temperature for points along an isostere. The slope of this curve gives the value of the heat of adsorption for a particular capacity.

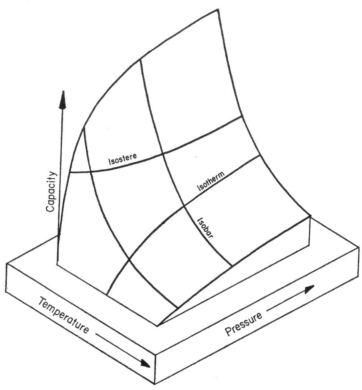

Fig. 7.1. An adsorption equilibrium surface.

If the heat of adsorption does not vary much with temperature, the curve will be approximately linear. Figure 7.2, from data presented by Brunauer (1945), shows a typical plot. It is sometimes possible to use this approach to interpolate or extrapolate known data to obtain predictions of capacity in the region of interest.

Isobars are occasionally measured in the course of experimental work or plotted for the purpose of checking the consistency of a family of isotherms. Aside from these circumstances, however, they are not generally useful.

Isotherms, on the other hand, are extremely useful, and most adsorption equilibrium data are both determined and presented in this form. Isotherms are uniquely significant in the design of fixed bed adsorbers, since these units generally operate at essentially constant temperature, while the partial pressure of the adsorbate and the quantity adsorbed are both changing.

The equilibrium relationship for an adsorbent–adsorbate system can be described by curves which represent experimental data, or by empirical equations which approximate these curves, or by equations derived from various adsorption theories. Design of process equipment is usually based on experimental curves. The various theoretical models are useful principally in extrapolating known data, or as aids in understanding and using observed phenomena.

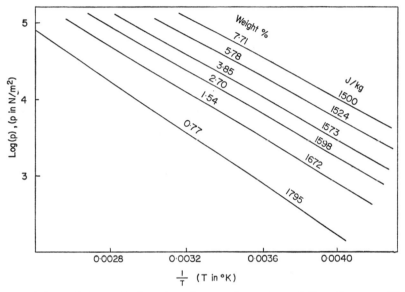

Fig. 7.2. Isosteric heats of adsorption of ammonia on charcoal. From Brunauer (1945).

A. ADSORPTION THEORIES

Two texts which serve to introduce the extensive theoretical background to this subject are those of Brunauer (1945) and Young and Crowell (1962). Only a few of the main theories will be mentioned here.

1. Langmuir Monolayer Theory

Langmuir (1918) proposed that the energy states of molecules adsorbed at preferred sites on the adsorbent are independent of the presence of other adsorbed molecules in the neighbourhood. The adsorbed molecules are assumed to be in dynamic equilibrium with those in the gas phase.

From the kinetic theory of gases, the number of molecules striking the surface per unit time is proportional to pressure, $n = kp$. If a fraction, θ, of the adsorbent sites are already filled, only $(1-\theta)n$ of these molecules have a chance of being adsorbed. If the fraction of these molecules adsorbed is f, then the rate of adsorption will be $(1-\theta)fn$.

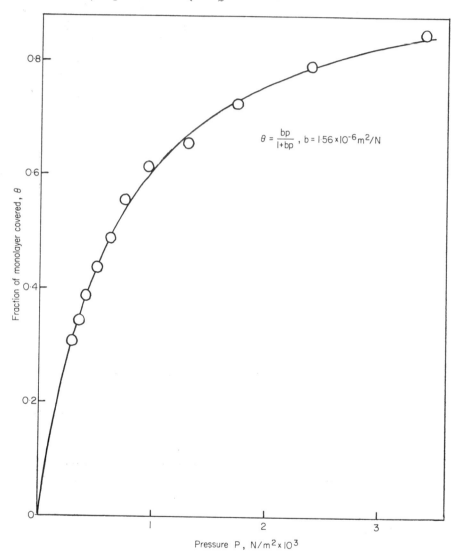

Fig. 7.3. Adsorption isotherm for nitrogen on mica at 90°K. From Langmuir (1918).

Assuming that the rate of desorption of molecules from a complete mono-layer is m, the desorption rate at partial coverage will be θm. For equilibrium,

$$\theta m = (1-\theta)fn$$

Combining these equations

$$\frac{fkp}{m} = \frac{\theta}{1-\theta}$$

Letting $fk/m = b$, and rearranging

$$\theta = \frac{bp}{1+bp}$$

This is the form of Langmuir's Isotherm. Figure 7.3, from Langmuir (1918), demonstrates the shape of such isotherms and their quality of fit to some experimental data. Other isotherm equations of the same general form have been derived by more sophisticated approaches (Young and Crowell, 1962).

For any particular system, the value of b and the relationship between θ and the weight fraction adsorbed, w, is best determined experimentally. If one assumes $w = \theta a$ the isotherm expression may be rewritten

$$\frac{p}{w} = \frac{p}{a} + \frac{1}{ab}$$

If experimental values of p/w are plotted against p, the slope of the line will be $1/a$ and the intercept $1/ab$. Figure 7.4, from Brunauer and Emmett (1937), illustrates this technique. The Langmuir Isotherm has been found to be particularly useful in fitting the data for a variety of adsorbents on charcoal.

2. Polanyi Potential Theory

This theory (Polanyi, 1914) does not lead directly to the prediction of adsorption isotherms but it relates isotherms of a given adsorbate at various temperatures, or the isotherms of different adsorbates on a particular adsorbent. Young and Crowell (1962) give an excellent review of its basis, modifications and present utility. Very briefly, a characteristic curve of adsorption potential, E, against volume of gas ϕ adsorbed at potentials greater than E, is calculated from one experimental isotherm. Since this curve is assumed independent of temperature, it can be used as the basis for the calculation of isotherms at other temperatures. Figure 7.5, which is based on Berényi's (1920) calculations from Titoff's (1910) data, is an example of a characteristic curve. The isotherms of Fig. 7.6 were calculated from this curve. The Potential Theory has proved useful with a wide range of adsorbates, and with adsorbents such as activated carbon, silica gel and activated alumina.

3. BET Multilayer Theory

This approach (Brunauer et al., 1938) employs the same basic assumption as the Langmuir theory, except that each layer of adsorbed molecules is assumed capable of supplying adsorption sites for the next layer. Since the derivation

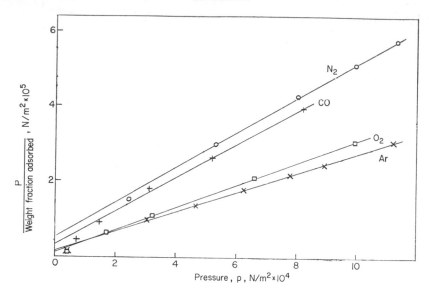

Fig. 7.4. Adsorption isotherms of various gases on charcoal at 90°K, plotted to determine the parameters of the Langmuir equation. From Brunauer and Emmett (1937).

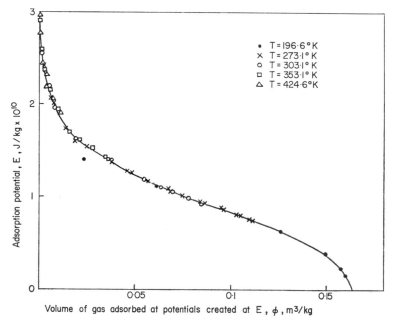

Fig. 7.5. Polanyi characteristic curve for the adsorption of carbon dioxide on charcoal From Berényi (1920).

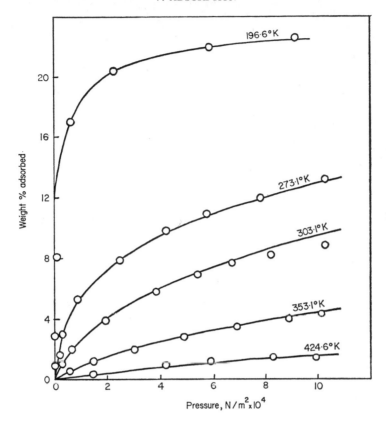

Fig. 7.6. Adsorption isotherms of carbon dioxide on charcoal, calculated from the characteristic curve of Fig. 5. From Berényi (1920).

of the isotherm is complex, only the end result is given here:

$$\theta = \frac{Cp}{(p_0 - p)[1 + (C-1)(p/p_0)]}$$

Where θ is still the fraction of single monolayer coverage, but is no longer restricted to values less than 1. p_0 is the saturation pressure of the adsorbate and C is equal to $\exp[(E_1 - E_2)/RT]$, where $(E_1 - E_2)$ is the difference between the heat of adsorption of the first layer and the heat of liquefaction. Young and Crowell (1962) present a detailed derivation of the BET isotherm, and review a substantial number of papers which rederive this equation, or modify it slightly.

The BET theory and its various extensions and modifications, has been most successful in representing a variety of adsorbate–adsorbent systems. It has also been adopted as the basis for reporting the surface areas of various adsorbents. (See Gregg and Sing, 1967.)

4. *Mixed Gas Adsorption Theories*

Other theories have been proposed, some of which account for the interaction of adsorbed molecules. This approach is particularly important in the representation of multi-component adsorption.

Myers (1968) reviews the literature on adsorption of mixtures with particular emphasis on recent developments. Prausnitz and co-workers have published papers on the thermodynamics of mixed gas adsorption (Myers and Prausnitz, 1965; Kidnay and Myers, 1966; Hoory and Prausnitz, 1967). Bussey (1966), uses statistical thermodynamics to extend the BET multimolecular gas adsorption theory to ideal mixtures. An equation of state which is used to predict mixed gas adsorption of hydrocarbons is presented by Payne *et al.* (1968), and mixed adsorption at cryogenic temperatures is discussed by Basmadjian and Cook (1964).

5. *Liquid Phase Adsorption Equilibrium*

Although widely used in industry, this phenomenon is not well understood. Part of this may be attributed to the complexity of liquid phase physics and part to the experimental difficulties of measuring the degree of adsorption of pure liquids.

Freundlich's empirical equation, $w = kc^n$, provides a good representation of most liquid systems, particularly for dilute solutions. As before, w is the weight fraction of solute adsorbed, while c is the solution concentration expressed as mass of solute per unit volume of solvent. k and n are experimentally determined constants. Mason (1957) and Karwat (1961) use this equation to correlate their data on the adsorption of contaminants from cryogenic liquids.

B. PROPERTIES OF ISOTHERMS

Some typical isotherms for propane on charcoal are shown in Fig. 7.7, from which it is evident that:

(i) Capacity increases with pressure.

(ii) Capacity decreases with temperature.

(iii) At low pressures, or high temperatures, an analogy with Henry's law is appropriate.

(iv) At high pressure and low temperatures, capacity seems to be a very modest function of pressure.

(v) All of the isotherms are convex with respect to the pressure axis.

The first two observations are true of all adsorption isotherms; and, while many isotherms are not in accord with the last three observations, they are valid for all the isotherms considered in this chapter.

It is apparent that capacity is a very strong function of temperature, and that therefore some effort should be made to ensure that adsorption is carried out at the lowest practical temperature. In the drying of compressed air, for example, it is always best to place the dryer down-stream of the compressor after-cooler and, if the air stream is to be further cooled, drying should be delayed until just above the ice point.

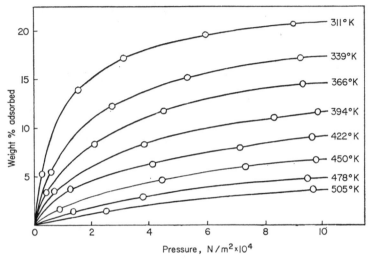

Fig. 7.7. Adsorption isotherms of propane on charcoal. From Ray and Box (1950).

Capacity will almost always be improved by operation at higher pressure, owing to the increased adsorbate partial pressure for a given concentration. It is seldom economical, however, to raise the operating pressure merely for this reason.

It is also apparent from the isotherm curves, that desorption can be accomplished either by increasing the temperature or by lowering the partial pressure. The choice between these alternatives is discussed in Section V.A.

IV. Adsorption Dynamics

An accurate mathematical description of the dynamic behaviour of a fixed-bed adsorber is extremely complex. There is no one general solution, but rather a number of solutions each based on a mathematical model embodying a set of simplifying assumptions. Accurate prediction or fixed bed adsorber performance requires the choice of the model most appropriate to the particular system being designed.

A. BASIC ASSUMPTIONS

Most models involve the following basic assumptions:

(i) A uniform adsorption equilibrium relationship exists throughout the adsorbent bed.

(ii) The initial concentration of the adsorbate is uniform throughout the adsorbent bed.

(iii) The adsorbent bed is initially at a uniform temperature, equal to that of the feed gas.

(iv) The feed gas has constant pressure, temperature, flow rate and composition.

(v) No radial gradients in temperature, flow rate or adsorbate composition exist in the bed.

(vi) The pressure gradient due to the flow through the adsorbent bed is negligible.

(vii) The change in flow rate due to the partial adsorption of the feed flow is negligible.

(viii) The effect of longitudinal despersion on the composition gradient is negligible.

In addition, further simplifications are commonly made concerning adsorption equilibria, diffusion of the adsorbate, and the effects of heat of adsorption. The complexities of the real system will be outlined before various approximate solutions are discussed.

B. ADSORPTION EQUILIBRIA

Most adsorption isotherms are determined for the two component, adsorbent-adsorbate system. But, in a fixed bed adsorber, at least one other component is always present. It is usually possible to use the concept of partial pressure to determine the effect of the presence of the extra component in the vapour phase, and various multi-component adsorption theories are available to describe the sorbed phase effects. Many mathematical models use the partial pressure model for the vapour phase and assume no adsorption of the other component.

A characteristic of most adsorbents which is ignored in some design procedures, is the so-called ageing effect. This phenomenon is most commonly due to the adsorption of trace quantities of components that are not adequately desorbed at normal reactivation temperatures. If provision is not made for an occasional extra hot reactivation, the build-up of such components can significantly reduce the capacity of the adsorbent. If the normal reactivation temperature is limited by the maximum allowable temperature of the adsorbent, this build-up cannot be removed. In this case the isotherm for the aged material should be used in the design of the bed.

C. DIFFUSION OF THE ADSORBATE

Diffusion may be significant at a number of different stages during the adsorption process. First, the adsorbate must diffuse from the bulk of the vapour to the surface of the pellet. This type of mass transfer resistance is common to a number of processes, such as catalysis and ion exchange, and has been correlated by a number of workers. See Nussey (1968) for a review. Diffusion into the adsorbent pellet occurs in the vapour within the pores, and in the sorbed phase on the walls of the pores. Nussey (1968) also gives a fine review of work on this topic. In some adsorbents, such as molecular sieves, diffusion in the sorbed phase also occurs within the crystals. Dworjanyn (1962) examined a number of ways to measure this type of diffusion.

If more than one component is adsorbed to a significant extent, each of these diffusion steps may be complicated by the counter-diffusion of the second component.

D. HEAT OF ADSORPTION

Adsorption, like condensation, occurs with the release of heat. The first result of this heat release is an increase in the temperature of the adsorbent pellet at the point where adsorption is taking place. If the adsorbent was initially at the same temperature of the feed gas, heat will now flow from the pellet to the process gas stream. This stream may then cool somewhat as it flows over the pellets down-stream of the adsorption point.

Some heat may flow out of the vessel through its walls. This effect is naturally more pronounced in small vessels which have greater surface area to adsorbent mass ratios. It is worth noting that because of this phenomenon, vessel diameter may have a significant effect on adsorber performance. It may therefore be impossible to predict the performance of a large vessel (which will behave almost adiabatically) by laboratory tests on a small, essentially isothermal adsorber.

The temperature rise resulting from heat of adsorption affects the performance of the adsorber principally because of the effect of temperature on equilibrium adsorption capacity, which can be estimated accordingly. The influence of temperature on the various diffusional processes is far less important. If the adsorbate is present in the feed only as a very small concentration, e.g. less than 0·1 %, the difference between adiabatic and isothermal operation will generally be negligible.

E. REVIEW OF MATHEMATICAL MODELS

The many different models have been extensively reviewed by Vermeulen (1958), Carter (1961, 1966) and Nussey (1968). Only some of the most immediately useful models are outlined here.

(a) *Rosen* (1952, 1954) assumed a linear isotherm, diffusional resistance from the bulk of the pellet, intra pellet diffusion, and isothermal operation. He developed an analytical solution, in the form of an integrated Hankel function in the complex plane, and presented numerical evaluations for a wide range of parameters. This model is capable of giving a detailed, accurate representation of fixed bed adsorber performance for systems in which the isotherm is truly linear. This is most liable to be the case for systems with a very dilute concentration for adsorbate, and relatively low feed pressure.

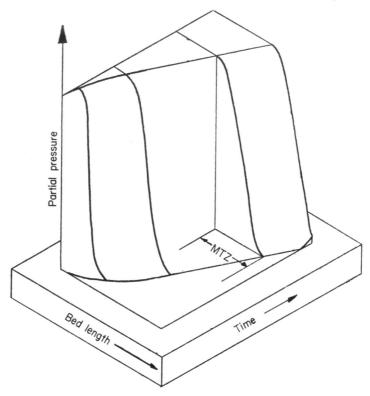

Fig. 7.8. Dynamic adsorption surface, illustrating a mass transfer zone of constant length (isothermal).

In addition to the isotherm, the two diffusional resistance parameters must be determined before Rosen's solutions can be used. The fluid film resistance parameter can be estimated from various correlations. The effective diffusivity of the pellets can be determined from small scale experiments. An alternative method, in which these two parameters are determined at the same time is described by Chandrasekharan (1962). This involves measuring break-through curves with an experimental adsorber, and comparing the shape of these curves with those calculated by Rosen. The values of the two diffusional resistance parameters are inferred to be those that result in the best fit of the data.

(b) *Michaels* (1952) was the first to suggest the use of the constant mass transfer zone (M.T.Z.) phenomenon, for the description of ion exchange systems. Treybal (1955) applied this concept to adsorber design. It has been observed (Bohart and Adams, 1920) that a constant pattern develops for systems with isotherms that are concave to the pressure axis. The more concave the isotherm, the sooner the constant pattern will be established.

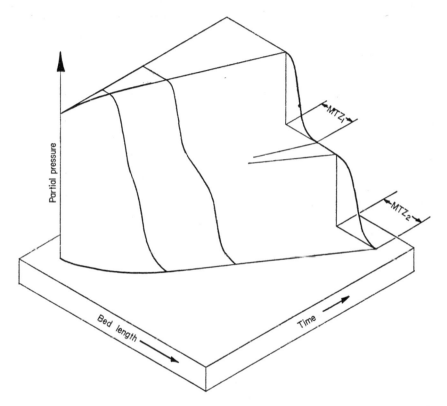

Fig. 7.9. Dynamic adsorption surface, for a non-isothermal process, illustrating the formation of two mass transfer zones moving at different rates.

This leads to a situation where a mass transfer zone of constant length passes along the adsorber at a constant speed. In front of the zone, the bed is free of adsorbate, behind it, the bed is saturated. This is indicated schematically in Fig. 7.8. In order to use this method, it is necessary to know the adsorbent capacity that is in equilibrium with the feed gas, and the length of the mass transfer zone. Both of these can be determined from small scale breakthrough curve experiments. It is, of course, necessary to be sure that the measured zone has in fact stabilized at a constant length.

(c) *Leavitt* (1962) extended MTZ concept to non-isothermal (adiabatic) systems. Under these conditions, two transfer zones may be formed. Figure 7.9 gives a schematic representation of this type of situation. Temperature and concentration profiles for a point in time are shown in Fig. 7.10. Carter (1966, 1968) and Amundson *et al.* (1965) have also presented analyses of the adiabatic problem.

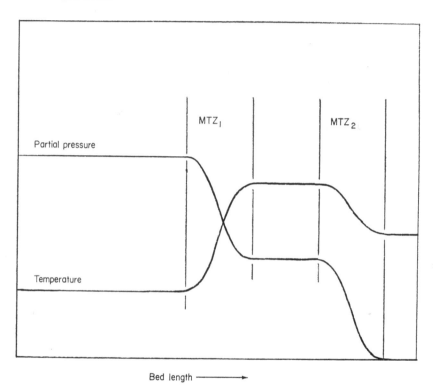

Fig. 7.10. Representative partial pressure and temperature profiles for a non-isothermal process with two mass transfer zones.

(d) *Cooney and Lightfoot* (1966) used the constant pattern approach to describe multicomponent adsorption of interfering solutes.

(e) *Other models.* It is relatively easy to write down the differential equations that govern the behaviour of even the most complex system. The real problem rises in the integration of these equations. Digital computers have made the numerical integration of differential equations more practical, and rather complex models of fixed bed adsorption have been treated in this way. Morton (1965), Wilson (1966) and Hall *et al.* (1966) are examples of this approach.

V. Reactivation Methods and Adsorption Cycles

Adsorption systems commonly include two adsorber vessels, of which one is on stream while the other is being reactivated. The time needed for reactivation frequently determines the choice of cycle time. Much of the cost of an adsorber system is associated in one way or another with reactivation. Yet reactivation is an aspect of adsorber design that has received relatively little attention in the literature. Mantell (1951) briefly discusses the reactivation of air and gas dryers, whilst sections of review papers have been given by Barry (1960) and Carter (1966).

A. CHOICE OF REACTIVATION METHOD

The choice between temperature or pressure cycling requires consideration of the following factors:

The minimum level of desorption is determined by the purity level demanded of the adsorber product. This requirement will, for instance, determine the partial pressure that a reactivating gas must have for any particular temperature. These considerations are illustrated in Fig. 7.11. Here it is assumed that adsorption takes place at T_1, with a partial pressure of P_1 in the feed gas. The weight percent adsorbed will be Q_1, so conditions in the saturated adsorbent may be represented by point A. If the allowable partial pressure of adsorbate in the effluent is P_2, then the adsorbent must be reactivated so that the quantity of adsorbate remaining after reactivation is less than Q_2. It is apparent that one way to meet this requirement would be to use a purge gas with a partial pressure less than P_2, and carry out the reactivation at the same temperature as for adsorption, T_1 (below point B). Another possibility would be to reactivate with feed gas with a partial pressure of P_1, and at a temperature exceeding T_6 (below point C). If, for example, T_5 is the maximum allowable temperature for reactivation, a purge gas with a partial pressure less than P_3 will have to be used.

Sometimes an essentially adsorbate-free gas can be used for reactivation purge gas. For example, pure nitrogen is often used to reactivate the silica-gel adsorbers which are used to remove hydrocarbons from liquid oxygen. More commonly, part of the feed stream is used, either before or after purification by adsorption.

If the reactivation gas is valuable, it may be desirable to condense out the desorbed component and recycle the purge gas, either to the process stream or to the reactivation system.

It usually takes longer to change the temperature of the adsorbent than it does to change the partial pressure of the adsorbate. For this reason, switching times of 15 min to 1 hr, can be used with pressure cycle reactivation, while periods of 8 or more hours are associated with temperature cycling. It will be shown later that short-period, pressure-swing, reactivation cycles are appropriate in applications where the adsorbate concentration in the feed is high (5 to 20%) and the feed pressure is moderate (1 to $4 \times 10^6 \text{N/m}^2$).

The precise value of the reactivation temperature is frequently determined either by the nature of the available heat source (for example, the pressure of the steam system), or by the maximum allowable temperature of the adsorbent.

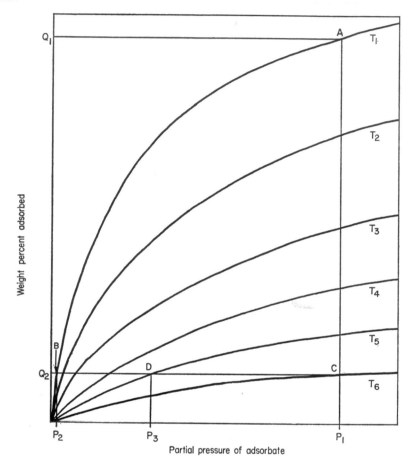

Fig. 7.11. Adsorption isotherms. Representation of various reactivation methods.

B. DESIGN OF REACTIVATION CYCLE

1. *Pressure Swing Reactivation*

(a) *Description of cycle.* Pressure-swing reactivation takes place in four steps:

(i) The vessel is taken off stream.

(ii) Depressurization of the vessel through the inlet end.

(iii) A small flow of purified purge gas is directed through the vessel from the outlet end. This purge is continued until the adsorbate that was deposited during the adsorption part of the cycle is removed in the effluent.

(iv) The vessel is repressurized.

(b) *Design of the reactivation process.* A simple model of the cycle may be based on the following assumptions:

 (i) The adsorber switch losses are negligible.

(ii) The quantity adsorbed at the bed inlet does not vary during the adsorption–desorption cycle.

(iii) The concentration of the adsorbate in the effluent is negligible.

The rate of adsorption will be $M_F X_F$ and the rate of desorption,

$$\frac{M_R X_E}{1 - X_E}$$

M_F and M_R are the molar flow rates of feed and reactivation purge gases respectively, and X_F and X_E are the mole fractions of adsorbate in the feed and purge effluent. Since these rates must be equal,

$$\frac{M_R}{M_F} = \frac{X_F(1 - X_E)}{X_E}$$

Assumption (ii) implies that the partial pressure of adsorbate at the bed inlet will be constant, i.e. $P_F \cdot X_F = P_R \cdot X_E$. By substitution,

$$\frac{M_R}{M_F} = \frac{P_R}{P_F} - X_F$$

This expression, which gives a first estimate of purge gas requirements, suggests that this type of reactivation will be most attractive where P_F and X_F are large. Sophisticated models have been developed which account for switch losses, but the benefits of more detailed calculation are limited by the difficulty of determining adsorbent requirements.

Adsorber design is made unusually complex by the significant heat effects that arise from the adsorption of large quantities of adsorbate. In addition, considerable temperature changes can occur as a result of isentropic expansion and compression during the reactivation step. In many applications, a further complications arises from the fact that more than one component must be removed from the feed gas. These difficulties usually mean that experimental methods must be used to determine the adsorbent requirements, which in turn affect the switch losses.

2. Temperature Swing Reactivation

(a) *Description of cycle.* This type of reactivation occurs in about six steps:

 (i) The vessel is taken off-stream.

(ii) Depressurization may or may not be carried out.

(iii) Heating of the bed is begun. Heating is usually accomplished by pre-heating the purge gas before it is passed through the bed. In order to keep the downstream end of the adsorber as free from adsorbate as possible, the purge gas is usually passed through the vessel in the reverse direction to the feed flow. The temperature of the adsorbent does not rise uniformly, instead a temperature wave front, analogous to the concentration wave front of the adsorption step, passes along the bed. As a result, a considerable period of time may elapse before the temperature of the purge gas effluent begins to rise. Because 100% reactivation is not required for the inlet portion of the bed, heating is frequently terminated before this temperature has stopped rising.

(iv) The purge gas heater is shut off and cool purge gas is allowed to flow through the vessel. This step is necessary to cool the adsorbent to near feed temperature so that its full capacity will be available when the reactivated adsorber is put on stream.

   It may be noted that the purge gas effluent temperature may continue to rise for a while after the reactivation heater is shut off. This temperature pulse is analogous to the concentration pulse obtained in chromatography.

   If a closed circuit reactivation system is used the purge gas will not be pure and it may be necessary to change the direction of the purge flow before cooling the adsorbent. This precaution will prevent contamination of the adsorbent at the bed outlet during cooling.

(v) Purging may be stopped when the effluent temperature begins to approach that of the feed gas. The nature of the process will determine the degree to which temperature equalization is important.

(vi) The vessel may now be prepared for use by repressurizing with feed gas.

(b) *Design of the reactivation process.*

(i) Total heat requirements. Heat must be supplied by the reactivation heater to fulfil the following requirements:

   Heating the adsorbent from adsorption to reactivation temperature.
   Supplying the integral heat of adsorption.
   Heating the adsorber and its insulation.
   Heating the associated piping, valves and their insulation.
   Heat loss through the insulation.
   Some heat also leaves the adsorber in the purge effluent.

(ii) Purge gas flow rate. The heat requirements will generally determine the total amount of purge gas needed and considerations of particle attrition will determine the flow rate at which this gas can be supplied. This rate, in turn, will determine how long the heating (and cooling) steps will take and will directly influence the time required for reactivation. It has been noted previously that reactivation time ultimately determines on-stream time requirements, and consequently, bed size. Since the heat requirements are obviously a function of bed size, it is apparent that good adsorber design will be an iterative procedure, in which the high capital cost of large adsorber vessels is balanced against the high energy cost associated with frequent reactivation. An indication of the nature of these considera-

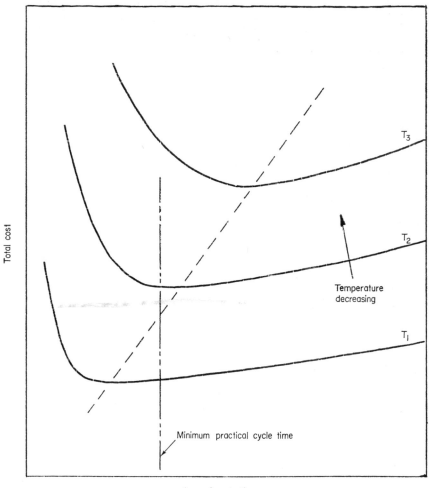

Fig. 7.12. Switching adsorber cycle time. The relationship between operating temperature and optimum cycle time.

tions for a cryogenic adsorber is given in Fig. 7.12. This figure illustrates that the optimum cycle time increases as the adsorption temperature decreases. This is due to the increasing cost of the refrigeration required during cool-down. A "minimum practical cycle time" is also indicated on the figure. This is usually either 4 or 8 hr, and is dictated by the complexity of the switching operation and the number of other activities that the plant operators have to carry out during an 8 hr shift. Fortunately, when the cycle time is set by this limitation, the penalty in total cost is usually quite small. This concept does not, of course, apply to automatically switched adsorbers. These can be designed for the optimum cycle time, and the actual operation can be readjusted to take full advantage of any overdesign.

VI. Adsorption System Design

A. PROCESS DESIGN

Previous sections have dealt in detail with the various aspects of adsorber design. The following table outlines a general procedure for the process design of an adsorber system. It is assumed that the feed conditions and flow rate are firmly established. Items (a) relate to the temperature reactivation. Items (b) relate to pressure reactivation.

Process Design Method for Switching Adsorbers
  (i) Select adsorbent material and pellet size.
 (ii) Calculate allowable velocity for upward flow of the feed stream, see Ledoux (1948), or use 70% of the velocity for which bed pressure drop equals bed weight.
(iii) Calculate cross-sectional area required.
 (iv) Calculate adsorbent requirements per unit time on the basis of equilibrium adsorption.
  (v) (a) Select a cycle time for temperature reactivation and calculate total adsorbent requirements based on 70% saturation.
      (b) Select a cycle time for pressure reactivation and calculate total adsorbent requirements based on 25% saturation.
 (vi) Calculate bed length. Check total bed pressure drop against process requirements. Bed lengths of less than 1·3 m, with L/D's less than 1·0 may suffer from flow maldistribution.
(vii) (a) Estimate vessel size and weight.
      (b) Estimate vessel and piping volumes.
(viii) (a) Select a suitable reactivation temperature.
       (b) Select a suitable reactivation pressure.
 (ix) (a) Estimate reactivation heating gas requirements. Assume valves and piping weigh $\frac{1}{2}$ of vessel weight. Assume that the insulation is heated to $\frac{1}{2}$ of the reactivation temperature. Remember that the heat of desorption must be supplied, as well as heat for adsorbent heating. To calculate the heat given up by the reactivation gas,

assume that its average exit temperature is $\overline{T} \equiv T$ initial $+$

$$\frac{T \text{ Reactivation} - T \text{ initial}}{4}$$

    (b) Estimate reactivation purge gas requirements.

(x) (a) Calculate the required reactivation gas velocity, assuming that about 40% of the reactivation time is available for the heating step.

    (b) Calculate reactivation purge gas velocity assuming 75% of reactivation time is available for purging.

(xi) Check that the bed pressure drop is not excessive with regard to adsorbent attrition or the pressure of the available reactivation gas.

(xii) It may be advantageous to alter the cross-sectional area, or to reverse the flow direction.

(xiii) It may be possible at this point to reject either temperature or pressure reactivation.

(xiv) (a) Recalculate from step v(a) using a more accurate method for estimating adsorbent requirements.

    (b) Conduct experiments to determine more accurately the requirements of adsorbent and purge gas.

In many situations, of course, it is not necessary to evaluate both methods of reactivation.

B. ADSORBER VESSEL DESIGN

The vessel and its associated piping should be designed so that there are no dead spaces which are inadequately purged during reactivation.

Liquid phase adsorbers must be designed with the liquid flow upward if there is any possibility of vapour entering the system during operation.

The inlet and outlet nozzles must be designed to ensure that the adsorbent is not damaged by excessive flow velocities. The nozzles should also be arranged so that maldistribution is minimized.

As with any vessel filled with pellets, some dusting will occur, especially early in the life of the bed. It is generally necessary to protect downstream equipment from this dust by the installation of filter elements. In some situations, it will be advantageous to incorporate the element in the vessel nozzle.

Provision must be made for filling and emptying the adsorbent pellets. The nozzles are also occasionally used for these operations.

Adsorbers that are to be reactivated by heating must be well insulated. In high pressure applications, where the vessel walls are liable to be heavy, the insulation is sometimes placed inside the adsorber. This reduces substantially the amount of heat that must be supplied during reactivation as the vessel walls do not have to be heated. Care must be taken, however, to ensure that the interior insulation does not provide a path for feed gas flow to by-pass the adsorbent.

VII. References

Amundson, N. R., Aris, R. and Swanson, R. (1965). *Proc. R. Soc.*, **A286**, 129.

Barry, H. M. (1960). *Chem. Eng.* **8**, 105.

Basmadjian, D. and Cook, W. H. (1964). Paper Q-1 of the Cryogenic Engineering Conference, Philadelphia.

Berényi, L. (1920). *Z. Phys. Chem.* **94**, 628.

Bohart and Adams (1920). *J. Am. Chem. Soc.* **42**, 523.

Brunauer, S. and Emmett, P. H. (1937). *J. Am. Chem. Soc.* **59**, 2682.

Brunauer, S., Emmett, P. H. and Teller, E. (1938). *J. Am. Chem. Soc.* **60**, 309. (For errata see Emmett and Dewitt (1941). *Industr. Eng. Chem.* **13**, 28.)

Brunauer, S. (1945). "The Physical Adsorption of Gases and Vapours". Clarendon Press, Oxford and Princeton University Press.

Bussey, B. W. (1966). *I. and E.C. Fund* **5**, 103–106.

Carter, J. W. (1961). *Brit. Chem. Eng.* **6**, 308.

Carter, J. W. (1966). *Trans. Inst. Chem. Eng.* **44**, T253–T259.

Carter, J. W. (1968). *Trans. Inst. Chem. Eng.* **46**, T213–T224.

Chandrasekharan, K. (1962). "A Study on the Performance of a Fixed-Bed Adsorber." Ph.D. Thesis, Imperial College, London.

Cooney, D. O. and Lightfoot, E. N. (1966). *I. and E.C. PDD* **5**, 25–32.

Dworjanyn, L. O. (1962). "Gaseous Diffusion Through Zeolites." Ph.D. Thesis, Imperial College, London.

Freundlich, H. (1926). "Colloid and Capillary Chemistry." Dutton, New York.

Gregg, S. J. and Sing, K. S. W. (1967). "Adsorption, Surface Area and Porosity." Academic Press, London and New York.

Hall, K. R., Eagleton, L. C., Acrivos, A. and Vermeulen, T. (1966). *Ind. and Eng. Chem. Fund* **5**, 212–223.

Hoory, S. E. and Prausuitz, J. M. (1967). *Chem. Eng. Sci.* **22**, 1025–1033.

Ishkin, I. P. and Burbo, P. F. (1940). *J. Appl. Chem.* **13**, 1022–1027.

Karwat, E. (1961). *Chem. Eng. Prog.* **57**, 41.

Kidnay, A. J. and Myers, A. L. (1966). *A.I.Ch.E. J.* **12**, No. 5, 981–986.

Langmuir, I. (1918). *J. Am. Chem. Soc.* **40**, 1361.

Leavitt, F. W. (1962). *Chem. Eng. Prog.* **58**, No. 8, 54–59.

Ledoux, E. (1948). *Chem. Eng.* **55**, 118.

Mantell, C. L. (1951). "Adsorption." McGraw–Hill, New York and London.

Mason, W. (1957). "Hydrocarbon Adsorption by Silica Gel in Liquid Oxygen and Liquid Nitrogen." Special Service Lab., Dow Chemical Co., Analytical Report.

Michaels, A. S. (1952). *Ind. Eng. Chem.* **44**, 1922.

Morton, E. L., Jr. (1965). "Analysis of Liquid–Phase Adsorption Fractionation in Fixed Beds." Ph.D. Thesis, Louisiana State University.

Myers, A. L. and Prausnitz, J. M. (1965). *A.I.Ch.E. J.* **11**, No. 1, 121–127.

Myers, A. L. (1968). *I. and E.C.* **60**, No. 5, 45–49.

Nussey, C. (1968). "A Study of the Transport Processes Relating to the Design of Fixed Bed Adsorbers." Ph.D. Thesis, University of Leeds.

Payne, H. K., Sturdevant, G. A. and Leyland, T. W. (1968). *I. and E.C. Fund* **7**, 363–374.

Polanyi, M. (1914). *Verh. Dtsch. Phys. Ges.* **16,** 1012.

Ray, G. C. and Box, Jr., E. O. (1950). *Ind. Eng. Chem.* **42,** 1315.

Rosen, J. B. (1952). *J. Chem. Phys.* **20,** 387–394.

Rosen, J. B. (1954). *Ind. Eng. Chem.* **46,** 1590–1594.

Treybal, R. E. (1955). "Mass-Transfer Operations." McGraw-Hill, New York.

Titoff, A. (1910). *Z. Phys. Chem.* **74,** 641–678.

Vermeulan, T. (1958). "Separation by Adsorption Methods", *Advances in Chemical Engineering* **2,** 147–208.

Wilson, K. B. (1966). "The Dynamic Behaviour of a Fixed Bed Adsorber." Ph.D. Thesis, Imperial College, London.

Young, D. M. and Crowell, A. D. (1962). "Physical Adsorption of Gases." Butterworths, London.

Chapter 8

Expanders and Pumps

J. E. BLACKFORD

Lucas Industrial Equipment, Liverpool, England

P. HALFORD

Petrocarbon Developments Ltd., Manchester ,England

D. H. TANTAM

BOC–Airco Ltd., London, England

I. Centrifugal Expanders

A. INTRODUCTION

The primary function of a cryogenic expansion turbine is the production of cold, that is to reduce the temperature of the gas passing through it. However the assessment of the performance of this type of machine should not be made on the basis of the actual temperature drop obtained compared with the isentropic drop obtained by expansion over the same ratio of pressures from the same inlet temperature, since even if the work output was zero there could still be a temperature drop due to throttle expansion. The efficiency is properly expressed as the ratio of the actual enthalpy drop to the isentropic enthalpy drop.

Although a brief description of various possible flow paths is given only the radial inward flow or centripetal turbine is considered in detail.

The flow through a turbo-machine is complex and can only be analyzed effectively by a three-dimensional treatment. Such a treatment is necessary for the design of efficient blading. In this section a one-dimensional flow assumption is made, this enables the basic principles of operation to be presented simply and can, when used in conjunction with experimentally determined loss coefficients, provide a sound basis for the estimation of the leading dimensions of the wheel and nozzle.

B. HISTORICAL

The first successful cryogenic expansion turbines, of which the author has knowledge, were produced in 1936 by Linde in Germany. They were radial inward flow machines having cantilever blades. Four groups of partial admission nozzles were arranged around the periphery of wheel. Efficiencies were expected to be about 65%, but the test value was only 53%.

At about this time Kapitza (1939), working in Moscow, suggested that a low temperature expansion turbine should resemble a water turbine rather than a steam turbine, due to the low kinematic viscosity of the working fluid. He described a machine, designed on the basis of this theory, having a wheel diameter of 8 cm running at 42,000 r.p.m., which achieved an efficiency of nearly 80%.

In 1942, the National Defence Research Committee of America sponsored work in this field and a turbine was designed by the Elliott Corporation and manufactured by the Sharples Corporation. The machine had a $6\frac{7}{8}$ in. diameter wheel running at 22,000 r.p.m., an efficiency of 82·5% was produced. Many of its design features have been incorporated in later machines.

During the 1950s rapid strides were made in the design of gas bearings, and these were applied to cryogenic turbines. Thus by 1958, the Lucas Company in England had developed a range of gas lubricated radial inward flow turbines for Petrocarbon Dev. Ltd., these are described in a paper by Beasley and Halford (1965).

Rotational speeds varied from 25,000 r.p.m. to 90,000 r.p.m. Gaseous nitrogen was employed as the bearing medium. A large number of these machines are now in service. A survey of this period has been given by Lady (1957).

A significant contribution has been made by Sixsmith (1959) and by Birmingham, Sixsmith and Wilson (1962) in the development of a high speed gas lubricated bearing. They produced a turbine having a wheel of only $\frac{5}{16}$ in. diameter rotating at 500,000 r.p.m. and yielding an efficiency of 50% when expanding Helium over a 20 : 1 ratio. The further development of this type of turbine has been undertaken by the British Oxygen Co., and a typical machine will be described later.

In 1960, Jekat, working for Worthington Corporation, produced an impulse turbine in which considerable attention was given to the mechanical design. Whilst the efficiency of this machine was not outstandingly high, the same constuctional principles were applied subsequently to a 50% reaction turbine having variable nozzles. This turbine produced an efficiency of 92%.

The La Fleur Corporation had, by 1964, produced a small gas lubricated turbine having conical bearings. This machine achieved 83% efficiency with a 3 in. diameter wheel running at 52,000 r.p.m.

Ruhemann (1965) reported on the operation of a helium expansion turbine (Lucas) which formed part of a prototype helium liquifier. The machine had gas lubricated bearings. The wheel diameter was $1\frac{5}{8}$ in. and at 70,000 r.p.m. produced an efficiency of 74%. The design was such that the central cartridge containing the rotor and bearings could be withdrawn without destroying the plant vacuum. Another interesting helium turbine was produced at that time by Linde in Germany, the bearing at the turbine end of the shaft was lubricated by helium gas, while at the other end conventional oil lubricated journal and thrust bearings were used. Both machines were operated with turbine inlet temperatures in the region of 20°K.

Recently, Linde have made significant advances in the field of high pressure turbines, with a machine capable of expanding air from 2,800 p.s.i.a. to 150 p.s.i.a. in a single stage. The wheel diameter is 3 in., the running speed 60,000 r.p.m., and an efficiency of 73% has been recorded. Although this is somewhat lower than that produced by an equivalent reciprocating expander, greater reliability and minimal maintenance requirements more than offset the difference.

C. REVIEW OF THEORY

1. *Description of Various Flow Paths*

The general flow path through a turbine is as follows: The gas enters the inlet (volute) casing or scroll where it is distributed to the nozzles. In the nozzles it is accelerated and directed at the correct angle into the turbine wheel where work is extracted from it. The gas usually passes from the wheel to a diffuser the purpose of which is to recover residual kinetic energy as useful pressure head., i.e. lowering the wheel exit pressure.

Flow paths may be divided into two main categories; the axial and the radial. Unfortunately, no generally accepted definitions exist. Two have been suggested:

(i) By the direction of the flow (in the meridional plane) of the gas passing through the rotor.

(ii) By the direction (again in the meridional plane) of the gas entering the rotor.

Typical wheel and nozzle arrangements are shown in Fig. 8.1 (a), (b) and (c).

Most turbines in cryogenic service are of the radial type, by definition (ii), and have wheels in which the blades are either entirely in the radial plane or are continued in the axial plane, and then curved to form the exducer section (Fig. 8.1(a)). The pure radial flow blades are often referred to as cantilever blades and are generally shrouded (Fig. 8.1(b)). The other types may be open or shrouded.

It may be seen that the flow in type (b) enters the blading radially and leaves it axially. By definition (i) this would be termed a mixed flow machine.

A further complication arises as some machines are constructed with their nozzles set on a conical surface so that the gas enters the rotor at an angle part way between the radial and the axial and are therefore mixed flow by either definition.

The cantilever type of turbine could be arranged so that the gas flows radially outward, the nozzles would then be located in the eye of the wheel. In practice this is never done because the centrifugal field would operate in the direction of flow, thus for quite small stage heads the nozzles would be required to run supersonically. The radial inward flow path on the other hand takes advantage of the pressure rise across the rotor in the reduction of nozzle expansion ratio and in many cases the nozzles can operate subsonically.

It may happen that the head to be utilized in a particular application exceeds that which can be handled efficiently in a single stage and it is necessary to compound stages. In such a case the axial path allows much easier flow from one stage to the next. The choice however may be made on considerations other than performance for the majority of extreme duties can be handled in a single stage radial machine of low specific speed provided a lower efficiency can be accepted. The poorer performance will be offset by lower initial cost.

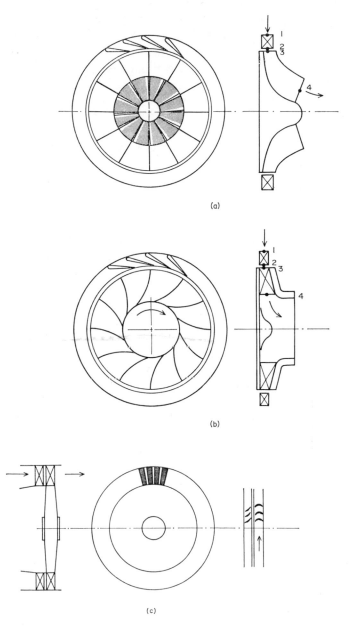

Fig. 8.1. Nozzle and wheel arrangements. Radial inward flow. (*a*) With exducer, (*b*) cantilever blades, (*c*) axial arrangement (single stage).

2. The Expansion Process on the T–S Diagram

The expansion process may be represented on a temperature–entropy diagram as shown in Fig. 8.2. The turbine duty is usually specified in terms of inlet pressure and temperature and outlet pressure, the outlet temperature being determined by the efficiency of the machine. The final choice of these parameters is made taking into account the outlet condition of the gas, usually chosen to be just above the vapour line.

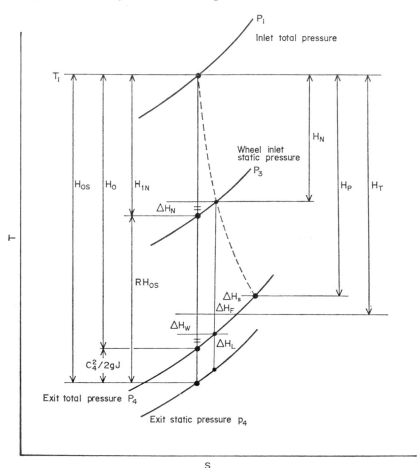

Fig. 8.2. The expansion on the temperature–entropy diagram.

In the absence of losses the expansion would take place at constant entropy and be represented by a vertical line on the diagram. The terminal point of this line usually lies in the vapour region. The actual expansion incurs various losses which have the effect of reducing the temperature drop (reheating effects) and is represented by the dotted line. The isobars shown on the diagram represent: total pressure at inlet, static pressure at wheel inlet, and total and static pressure at wheel outlet, the difference here being due to the equivalent of the velocity of the leaving stream.

At each of the points indicated on the diagram the enthalpy may be read off and it is in terms of enthalpy differences that the following three quantities are defined.

(a) *Loss*. The loss represented by the difference in enthalpy drop between an ideal expansion and an actual one is often referred to as a loss head although the units involved are those of energy. It is usual to relate this to the actual drop, the relationship being known as the loss coefficient ζ. This in turn is related to the velocity coefficient ϕ (ratio of actual to theoretical velocity) as follows:

For the nozzle

$$C_1 = \sqrt{(2gJH_{1N})} \quad \text{and} \quad C_2 = \sqrt{(2gJH_N)} \tag{8.1}$$

Now

$$\frac{C_2}{C_1} = \phi_n = \sqrt{\left(\frac{H_N}{H_{1N}}\right)} = \sqrt{\left(\frac{H_N}{H_N + \Delta H_N}\right)}$$

so

$$\phi_n^2 = \frac{1}{1 + \zeta_N} \quad \text{and} \quad \zeta_N = \frac{1}{\phi_N^2} - 1 \tag{8.2}$$

Where the suffix 1 denotes an ideal or theoretical condition and H is the (enthalpy) head.

In some publications losses are presented as pressure losses and their magnitude expressed in terms of pressure loss factors. Fig. 8.3 shows the relationship between the two methods.

Pressure loss factor is defined as:

$$= \frac{P_1 - P_2}{P_2 - p_2} \tag{8.3}$$

i.e. the pressure loss compared to the exit dynamic head.

(b) *Adiabatic efficiency*. This is defined as the ratio of actual enthalpy drop across the turbine to the drop obtained by an ideal expansion from the total inlet conditions to the outlet pressure, which may have a total or static value.

Since inlet velocities are low little error is introduced by taking the measured static value.

The *total-to-total efficiency* may be expressed as follows:

$$\eta_{TT} = \frac{H_0 - \Delta H_N - \Delta H_W - \Delta H_F - \Delta H_S}{H_0} = \frac{H_p}{H_0} \tag{8.4}$$

where H_0 is the total-to-total adiabatic drop and ΔH_N, ΔH_W, ΔH_F and ΔH_S

are the losses expressed as enthalpy heads associated with the nozzle, wheel, disc friction and seal leakage respectively. The numerator represents the head available as output power (H_p). The kinetic energy of the exit stream is regarded as recoverable in the diffuser and does not feature as a loss in the expression.

The *total-to-static efficiency* is obtained by comparing the power head with the total-to-static drop. Here the head associated with the leaving velocity C_4 is considered lost.

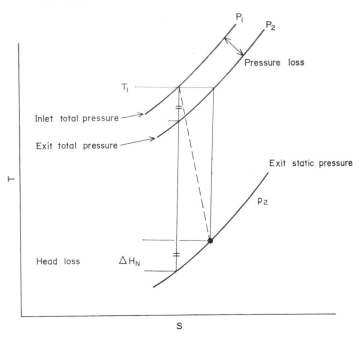

Fig. 8.3. Losses on the temperature–entropy diagram (nozzle).

The ratio $(C_4/C_0)^2$ (C_0 is the velocity equivalent of the adiabatic head.) frequently features in statements of performance and is known as the exhaust energy factor. It represents the proportion of total head leaving the machine as wasted energy. The total-to-total efficiency is related to the total-to-static efficiency as follows:

$$\eta_{TT} = \frac{H_p}{H_0} \quad \text{and} \quad \eta_{TS} = \frac{H_p}{H_0 + (C_4/C_0)^2 H_0}$$

so

$$\frac{\eta_{TT}}{\eta_{TS}} = 1 + \left(\frac{C_4}{C_0}\right)^2 \tag{8.5}$$

It is not possible in practice to recover all the leaving energy for apart from diffuser efficiency considerations the gas must possess a finite exit velocity.

The fitting of a diffuser often has a significant bearing on performance particularly on a high specific speed machine where exit velocities are large. Its effect is to reduce the exit pressure of the turbine wheel, which in turn increases the theoretical head across the turbine and enables a greater rate of cold production to be achieved. The small temperature rise in the diffuser does not off-set the increased cooling achieved by the wheel.

It is difficult to measure the exit temperature accurately due to variations across the duct, so frequently the drop is calculated from measured shaft power and a known throughput.

(c) *Reaction.* Turbines are usually designed so that only part of the pressure drop occurs in the stationary nozzles, whilst the remainder occurs in the rotor blades. A machine of this type is called a reaction turbine. The degree of reaction may be defined in a number of ways the most usual however is: The ratio of static head across the wheel to the total head across the machine.

The lower the degree of reaction the greater is the head expanded in the nozzles. If the degree of reaction is zero the machine is known as an impulse turbine.

Successful low reaction turbines have been produced, these have higher nozzle velocities for a given overall specification. The inlet velocity triangle gives rise to non-radial inlet blading and to reduced wheel speed. The penalty is greater loading of the rotor blades and increased turning within them both of which tend to reduce efficiency. It seems reasonable therefore only to employ such a design where the wheel speed would otherwise be above the safe limit.

If the wheel speed is reduced at a constant expansion ratio the centrifugal head is reduced, a reduction in nozzle exit pressure results, the head across the nozzle is increased as is the nozzle velocity and relative velocities in the wheel. This means that both reaction and throughput are dependent upon speed. The magnitude of the change in flow-rate depends on how near to the choked condition the nozzles are operating initially.

3. *Losses.*

Losses in turbomachinery can be classified as either hydraulic or miscellaneous. The hydraulic losses are: scroll and nozzle losses, wheel flow losses and leaving losses.

In cryogenic turbines, the scroll velocities are low and the associated losses can either be lumped with those of the nozzle or taken as equal to the inlet dynamic head. The losses in the nozzle arise from:

(i) Frictional effects.

(ii) Turning effects for the gas passing through the nozzle is required to undergo a rapid change in angular momentum. As a result a significant difference in velocity may exist on each side of the blades. The flow on the

16*

low pressure side will undergo a more rapid acceleration followed by a deceleration. This loss is reduced by increasing the number of nozzle blades.

As the expansion ratio of the nozzle is increased a point is reached where supersonic shock effects occur in the region beyond the throat and further losses are incurred. Flow deviation towards the radial takes place, however if this is taken into account simple nozzles can perform satisfactorily in the sonic region.

The most significant wheel losses occur from the formation of eddies produced by flow break-away in wheels so designed that low or zero velocities are present at the blade surfaces, particularly near the periphery of the wheel. The effect is outlined later. Such losses can be avoided by ensuring an adequate mean velocity and by causing the relative flow entering the wheel to do so at an incident angle to the blade, i.e. some slip is introduced (slip is defined as the ratio of the peripheral component of the absolute velocity entering the wheel to the tip speed). The amount required is calculated by a process similar to that giving the amount of slip occurring in radial compressors (Eck, 1953). If the incidence differs from this optimum value, incidence losses occur at the inlet to the wheel.

There are two other causes of loss within the wheel, one due to frictional effects as the gas flows down the wheel passages and the other due to a rapid change of direction from radial to axial.

The energy due to the velocity of the gas stream leaving the wheel is only partially recoverable and therefore a proportion at least must be considered as lost. The amount depends upon the efficiency of the diffuser.

The hydraulic or peripheral efficiency as it is sometimes called, is given by

$$\eta_h = \frac{H_{0_s} - \Delta H_N - \Delta H_W - \Delta H_L}{H_{0_s}} \tag{8.6}$$

where H_{0_s} = adiabatic head; ΔH_N = nozzle loss head; ΔH_W = wheel loss head; ΔH_L = leaving loss head.

The two major miscellaneous losses are due to disc friction and seal leakage. The former may be regarded as a direct reheating within the expansion process. The latter occurs across seals separating high and low pressure regions of the wheel and causes a quantity of gas, which has been merely throttled in its passage through the seal, to mix with the cooler exit stream.

A further loss known as ventilation occurs in machines operating with partial admission due to the flow pattern set up in the region of the closed-off nozzles.

The inflow of heat from the surroundings and the outflow of cold gas along shaft seals gives rise to losses which can be reduced to a negligible level by effective lagging, careful machinery design and the introduction of sealing gas at an appropriate pressure. Bearing power losses do not affect the thermodynamic efficiency and normally produce no appreciable reduction in output power.

4. *The energy equation of the wheel.*

The work which theoretically can be extracted from a turbine stage is given

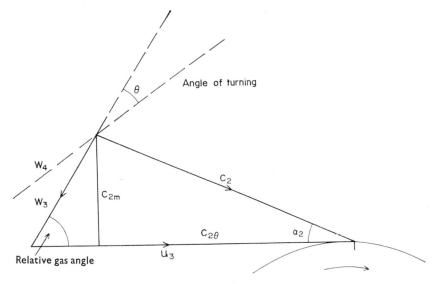

Fig. 8.4(*a*). Wheel inlet velocity triangle.

W = relative velocity
C = absolute velocity
C_θ = whirl component fluid velocity
u — peripheral wheel speed
Subscript 2—nozzle exit
Subscript 3—wheel inlet
Subscript 4—wheel exit.

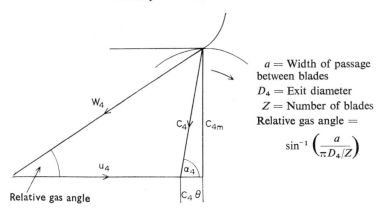

a = Width of passage between blades
D_4 = Exit diameter
Z = Number of blades
Relative gas angle =

$$\sin^{-1}\left(\frac{a}{\pi D_4/Z}\right)$$

(*b*) Wheel exit velocity triangle.

by the Euler equation which is derived by considering the change in moment of momentum of the gas in its passage through the wheel (see Fig. 8.4).

(It is necessary to take into account only the tangential component of the absolute velocity entering and leaving the wheel.)

If M is the weight of gas entering the wheel per sec the inlet moment of momentum is:

$$\frac{WC_2 \cos \alpha_2 r_3}{g} \tag{8.7}$$

similarly the exit moment of momentum in the general case is:

$$\frac{MC_4 \cos \alpha_4 r_4}{g} \tag{8.8}$$

This must be written minus or plus depending on whether the swirl at exit is with or against the direction of wheel rotation.

The power produced is

$$\frac{W\omega}{g}(C_2 \cos \alpha_2 r_3 \pm C_4 \cos \alpha_4 r_4)$$

Now $\omega r_3 = u_3 \omega r_4 = U_4$ so that the head which can be handled (in heat units).

$$H_T = \frac{C_2 \cos \alpha_2 U_3 \pm C_4 \cos \alpha_4 U_4}{gJ} \tag{8.9}$$

From equation 8.9 and applying the cosine law to the inlet and exit velocity triangles we get the basic Euler equation:

$$H_T = \frac{C_2^2 - C_4^2}{2gJ} + \frac{u_3^2 - u_4^2}{2gJ} + \frac{W_4^2 - W_3^2}{2gJ} \tag{8.10}$$

From Fig. 8.2 $RH_{0_s} = H_T - H_N + \Delta H_W + \Delta H_L$ from which with $H_N = C_2^2/2gJ$ and $\Delta H_L = C_4^2/2gJ$

$$RH_{0_s} + \left(\frac{W_3^2}{2gJ} - \frac{W_4^2}{2gJ}\right) - \Delta H_W = \frac{u_3^2 - u_4^2}{2gJ} \tag{8.11}$$

This means that the energy in the wheel after losses have been taken into account (relative energy entering the wheel minus the relative energy leaving the wheel plus the reaction head) expands against an additional loss-free head derived from rotational effects.

This is often claimed as one of the advantages of the radial inward flow path. The magnitude of this head, however, depends on the difference in peripheral speed $U_3 - U_4$ and is therefore more significant in low specific speed machines due to the greater radial extent of the blading.

The absence of swirl at exit is frequently taken as a design criterion, this condition means that C_4 has no tangential component so that equation 8.9 becomes:

$$H_T = \frac{C_{2\theta}U_3}{gJ} \tag{8.12}$$

where $C_{2\theta}$ is the peripheral component of nozzle velocity. From this it is clear that one of the major factors limiting the stage head is the maximum permissible tip speed, which is determined by wheel stress considerations.

5. *Velocity diagrams.*

Diagrams can be drawn representing the gas absolute and relative velocities and the blade peripheral or tangential velocity. Such diagrams are of assistance in the early stages of design in establishing blade angles and are useful in demonstrating the mechanism by which work is obtained from the expansion process. For these purposes it is assumed that the gas velocity is uniform across the flow path or can be represented by a mean value.

Figure 8.4 shows the general case inlet and outlet velocity triangles for a radial inward flow turbine. Where

        C represents the absolute velocity of the gas

        W represents the gas velocity relative to the wheel

        U represents the wheel or blade peripheral velocity.

The inlet triangle is located at a point on the outer diameter of the wheel, the outlet triangle at the inner diameter of the blades or in the circle which divides the exit annulus into two annuli of equal area for the case of a wheel with an exducer. The resultant of the nozzle velocity and the peripheral velocity, the relative velocity, dictates the blade angle at inlet although it is usual to design with some negative incidence (against rotation) as this reduces the blade loading in the region of the tip, that is the difference in velocities on either side of the blade.

It is necessary to make the distinction between blade and gas angles for apart from the incorporation of incidence at inlet both the nozzle and wheel exit gas angles are given more accurately by the sine formula than by the assumption that blade and gas angles are the same (see Fig. 8.4 (*b*)).

When simple nozzles are operated supersonically the exit flow will be deflected by a shock wave pattern and a decrease in gas angle measured from a radius will occur. When the inlet and exit diagrams are superimposed, the amount of turning in the wheel is clearly indicated. A small amount of turning, i.e. a small angular difference between the relative velocity vectors gives rise to efficient operation but requires a high degree of reaction which dictates a high blade tip speed.

Thus a gain in efficiency is made in a high specific speed machine which would be offset in a low specific speed machine by the larger disc frictional losses.

There are certain velocity relationships which affect the performance of an expansion turbine and while these are not directly related to velocity triangles they are conveniently considered under this heading.

Velocity ratio (V.R.). This is the ratio of wheel peripheral velocity U_3 to the theoretical velocity calculated from the adiabatic head H_0 (in energy units) V.R. $= U_3/C_0$ where $C_0 = \sqrt{(2gJH_0)}$ and is known as the spouting velocity. It is not an actual gas velocity.

The relationship is an important one since all types of radial inward flow turbines are found to produce maximum efficiencies over the range V.R. $=$ 0·65 to 0·70, see Fig. 8.5.

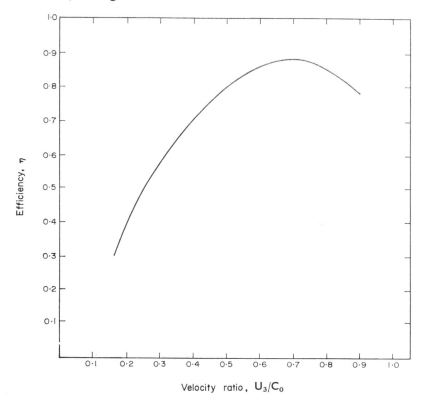

Fig. 8.5. Efficiency versus velocity ratio.

Two further useful expressions are the ratios of meridional inlet velocity and meridional exit velocity to wheel peripheral velocity, C_{3m}/U_3 and C_{4m}/U_3, respectively. The second expression is commonly given the symbol ϕ. The former governs the mean velocity at wheel inlet, its value being related to the number of blades. Too low a value will give rise to a reversal of flow on the pressure side of the blades. Too high a value of ϕ will produce an excessive amount of "waste" energy.

The relationship between these ratios, i.e. C_{4m}/C_{3m}, has a value generally between 1 and 1·25, that is the flow through the wheel is accelerated.

6. Machine Reynolds Number and Mach Number.

Mention must be made of these non-dimensional machine groups since they are frequently referred to in turbine literature. The groups differ from the more familiar local quantities which dictate flow regime, susceptibility to shock effects, choking, flow deflection and losses, in that their use is largely restricted to the prediction of probable machine performance.

The two groups are defined as follows:

$$\text{Machine Reynolds number} = \frac{C_0 D_3}{v}$$

Where C_0 is the spouting or theoretical velocity
$\quad D_3$ is the wheel diameter
$\quad v \quad$ is the kinematic viscosity of the fluid at the wheel inlet.

$$\text{Machine Mach number} = \frac{U_3}{C_s}$$

Where U is the blade tip velocity
$\quad C_s$ is the velocity of sound calculated at the wheel inlet condition (static temperature)

$$C_s = \sqrt{\left(\frac{2\gamma}{\gamma+1} gRT_1\right)}$$

Figure 8.6 shows the effect of Reynolds number and Mach number on efficiency.

The viscosity of gases at cryogenic temperatures is low so that most machines operate at high Reynolds numbers. Low Reynolds numbers occur when the wheel diameter is small or when the working fluid has a low density.

It should be appreciated that the machine Mach number has no significance as far as actual flow velocities are concerned. High Mach number machines may be designed in which flow velocities are entirely subsonic and vice versa.

7. The Coriolis Effect.

For the purpose of presenting a simplified theory conditions across the blade passage have been assumed uniform. This cannot be the case in practice for a driving torque is produced by the gas passing through the blading. This necessitates a pressure difference across the blades and consequently across the flow passages.

In order to appreciate the mechanism we have to consider the forces acting on a particle of gas passing through the wheel. Two forces are produced, firstly, the centrifugal force of magnitude $m\omega^2 r$ in a radial direction and secondly, the Coriolis force of magnitude $m2W\omega$ in an approximately tangential direction.

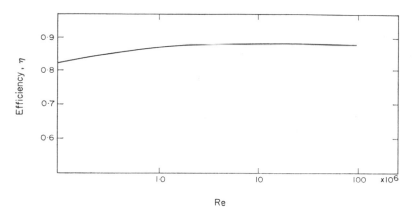

Fig. 8.6. (*a*) Efficiency versus Reynolds number.

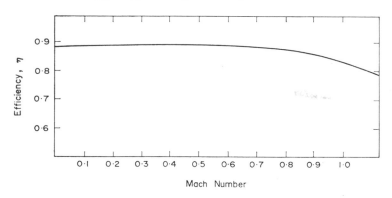

(*b*) Efficiency versus Mach number.

Where m is the mass of the particle

ω is the angular velocity of the wheel

r is the radial distance of the particle from the centre of the wheel

W is the velocity of the particle relative to the wheel.

These forces are balanced by pressure forces set up in the fluid. In the radial direction the pressure decreases from periphery to hub, while in the tangential direction the variation in pressure produces the driving torque, the value of which can be calculated by integrating an appropriate expression within the limits of radial extent of the blading. The "Coriolis head" is an advantage the radial turbine possesses over the axial turbine as this head can be considered to operate in a loss-free manner.

The assumption can be made that the gas entering the wheel has constant total pressure. The pressure variation across the flow passage then exists as the result of the difference of dynamic head, thus the velocity is lower on the pressure side of the passage and, if the average relative velocity through the wheel decreases below a certain level, flow reversal will occur on the trailing or pressure face of the blades. The wider the blade spacing the more likely is this phenomenon. If the boundary layer velocity decreases sufficiently breakaway will take place causing the formation of eddies in the main flow stream. The losses produced will be excessive and a low efficiency will result.

A high blade loading (i.e. a large amount of work per blade) gives rise to large velocity differences, and it is usual to choose sufficient blades to limit these differences and to ensure that the direction of flow is always positive. A very large number of blades is required in designs having low values of throughflow velocity particularly if the nozzle angle in Fig. 8.4(a) is small.

8. *The Concept of Specific Speed.*

It is possible to select from the operational and geometric parameters of turbo machines of a particular type a dimensionless group which will be found to have the same optimum value (i.e. the one corresponding to the conditions under which most efficient operation takes place) over a wide range of fluids, flow rates, speeds and heads. Specific speed is such a group but not the only one. It will be found also that the groups can be related by the use of simple geometric and gas relationships. In practice the groups are not non-dimensional since familiar rather than consistent units are used.

In the field of hydraulic turbines Power Specific Speed is used and defined as

$$N_s = \frac{NHP}{H^{5/4}} \tag{8.13}$$

Where

$$N_s = \text{specific speed}$$
$$HP = \text{horse power}$$
$$H = \text{head}$$
$$N = \text{r.p.m.}$$

It provides a means of classifying machinery and aids the selection of the type of turbine for given applications. For each of the three main types—Pelton Wheel, Francis Turbine and Kaplan Turbine—there is a different specific speed range for optimum performance.

In the field of cryogenic expansion turbines specific speed is defined as follows:

$$N_s = \frac{N(Q_4)^{\frac{1}{2}}}{H_0^{\frac{3}{4}}} \tag{8.14}$$

Where

N is the speed of the machine in r.p.m.

Q_4 the exit volume flow in ft³/sec

H_0 is the total-to-total adiabatic head in feet

Here the concept of specific speed presents no clear criterion for the selection of a particular type. It does, however, provide a means of assessing the probable performance of a particular design. Figure 8.7 indicates the trend and shows that, generally, high efficiencies are only obtained at specific speeds in the excess of 70.

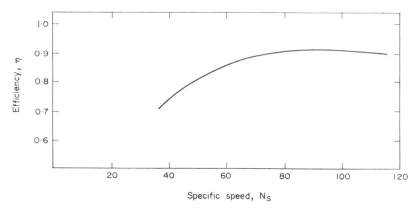

Fig. 8.7. Variation of efficiency with specific speed.

It can be seen from equation 8.14 that for a given specification (flow and head) a high specific speed dictates a high rotational speed. A limit is set in practice by turbine and compressor wheel stress considerations or by limits of bearing stability.

It is possible by means of the relationships

$$C_0 = \sqrt{(2gJH_0)} \qquad U_3 = \frac{\pi D_3 N}{60}$$

$$Q_4 = A_4 C_4 \quad \text{and} \quad \phi = \frac{C_4}{U_3}$$

to express the specific speed equation in the following alternative forms:

$$N_s = k \left(\frac{Q_4}{C_0}\right)^{\frac{1}{2}} \left(\frac{U_3}{C_0}\right) \left(\frac{1}{D_3}\right) \tag{8.15}$$

and

$$N_s = k(\phi)^{\frac{1}{2}} \left(\frac{U_3}{C_0}\right)^{\frac{3}{2}} \left(\frac{D_4}{D_3}\right) \tag{8.16}$$

where k is a constant; the hub is neglected.

Equation 8.15 indicates that for a given specification and for a selected velocity ratio the wheel diameter is inversely proportional to specific speed. In practice the optimum velocity ratio varies little over the useful range of specific speed so that a low specific speed design will have a large wheel, and thereby incur high disc frictional losses, while a high specific speed machine will have a relatively small wheel.

Equation 8.16 introduces a second ratio of velocities and, since the value of ϕ does not vary significantly over the range of specific speeds considered, indicates that the exit diameter of a high specific speed wheel will be large relative to its major diameter.

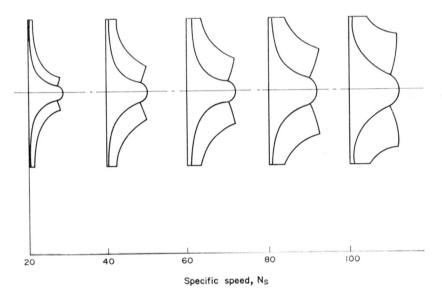

Specific speed, N_S

Fig. 8.8. Variation of wheel form with specific speed.

Figure 8.8 shows that the variation in wheel form with specific speed and for a constant wheel diameter. The form of a low specific speed wheel enables it to cope with high heads and low flow rates while at the other end of the range high flow rates can be accommodated.

If in addition to the above velocity ratios a hub/wheel diameter ratio is selected the leading dimensions of the wheel can be calculated directly from its specific speed. The point can be made that if it is desired to design a range of geometrically similar machines specific speed can provide the basis for a very effective "design by model" approach, provided they have identical velocity ratios.

9. *The Influence of Expansion Ratio and Fluid on Design.*

The two most significant variables of a design are the wheel diameter and the speed at which it should run. The effect of expansion ratio and working fluid on these values is considered on the following basis:

 (i) Geometric similarity of the wheels is maintained.

 (ii) The exit flow rate is constant.

(iii) The velocity relationships remain unaltered.

It can be seen from equation 8.16 that the specific speed is fixed by the above conditions (i) and (iii) so that, from equation 8.14, speed is inversely proportional to $H_0^{\frac{3}{4}}$ also from equation 8.15 re-arranged:

$$N_s = K\left(\frac{Q_4}{H_0^{\frac{1}{2}}}\right)^{\frac{1}{4}}\left(\frac{U_3}{C_0}\right)\left(\frac{1}{D_3}\right) \tag{8.17}$$

it follows that the wheel diameter is inversly proportional to the fourth root of the adiabatic head.

It is now necessary to see how the head varies with expansion ratio. For this purpose the efficiency is assumed $= 100\%$, i.e. the expansion is adiabatic and reversible. A particular exit temperature is chosen.

From the gas laws:

$$H_0 = C_p T_4\left(\left(\frac{P_1}{P_4}\right)^{(\lambda-1)/\lambda} - 1\right) \tag{8.18}$$

The value of H_0 for various expansion ratios using both air and hydrogen as gases of widely differing characteristics has been used with the relationships derived from the specific speed equations to give the plot illustrated in Fig. 8.9. The values being relative to those for air at a typical ratio of 4:1, see Table 8.I. It should be noted that since

$$C_p = \frac{\gamma}{\gamma-1}, \frac{R}{J} = \frac{\gamma}{\gamma-1} \frac{\mathscr{R}}{J \times \text{molecular weight}}$$

TABLE 8.I. A Comparison of Values at $r = 4$

Expansion ratio	Fluid	Wheel diameter	Speed	Gas or tip velocity
4 : 1	Air	1	1	1
4 : 1	Hydrogen	0·5	7·4	3·7

The head for a given ratio is inversely proportional to the molecular weight of the fluid. The influence of the adiabatic exponent is small.

In the case of the air turbine, large ratios make it necessary to run the wheel at reduced velocity ratios (U/C_0) due to wheel stress considerations and may also give rise to high local Mach numbers. Both effects tend to lower the efficiency of the machine. In practice high heads would be tackled at a reduced degree of reaction, i.e. nearer to the impulse type of turbine and the wheel geometry would be altered to give a backward curvature to the blades at inlet.

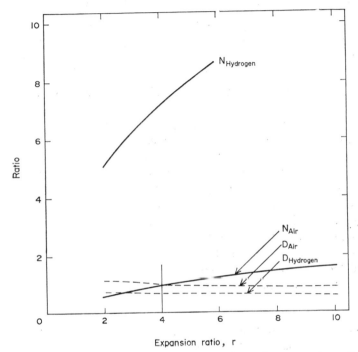

Fig. 8.9. A comparison of speeds and diameters for various expansion ratios.

The smaller wheel size in the case of hydrogen results from increased through-flow velocity $\propto H^{\frac{1}{2}}$. The tip speed will be proportionally high so that the maximum head which can be handled efficiently is dependent on the value of safe stress in the wheel.

The practical design of a hydrogen turbine with this limitation makes it unlikely that adverse Mach number effects will be encountered.

10. *The Presentation of Performance.*

It is important for the process designer to know the refrigeration capability of an expansion turbine for various operating conditions. Although it is the design point performance which is of prime importance, the off-design characteristics have to be acceptable. Nozzle grouping, or variable geometry nozzles may be incorporated to improve performance in these regions if flexibility of throughput is required.

Some of the relationships important in power turbine applications (e.g. torque–speed characteristics) have no significance in cryogenics where the design aim is to produce the maximum amount of cold for a specified throughput.

The most common presentations of performance therefore take the form of a plot of efficiency against various dimensional or non-dimensional parameters. These fulfil two main functions.

(i) They assist the designer in the choice of dimensional relationships for new designs.

(ii) They present relevant information about the capability of the turbine.

In the first category the most common plots are efficiency versus velocity ratio, specific speed, Mach and Reynolds number. Such curves are built up from test results taken from a variety of machines operating over widely differing conditions and are usefully compared with curves based on theory incorporating empirically determined loss factors. In the second category, the most important aspects are efficiency, flow and expansion ratio. The fixed nozzle machine operates at various inlet pressures and the most useful plot is that of efficiency and flow rate versus expansion ratio. When provision for throughput variation is incorporated the machine operates at a constant head and it is then necessary to show how efficiency changes with percentage flow. These curves relate to particular turbines the specifications of which are dictated by the process under consideration. Examples of curves of each type are shown in Figs. 8.5, 8.6, 8.7 and 8.20.

D. BEARINGS

Design charts for most types of bearing in common use are readily available. The purpose of this section is to indicate important design factors. The functions of the bearing system are:

(a) To support the rotor in the correct position relative to the static parts of the machine.

(b) To permit the rotor to run stably up to the design speed and beyond it by an acceptable overspeed margin.

(c) To withstand the axial and radial forces imposed on the rotor by the working fluid.

In order to perform these functions the bearing system must normally consist of two journal bearings, and two thrust bearings.

Bearings used in cryogenic turbines may be either oil or gas lubricated. The former type have been in use for many years and have proved extremely reliable. Some inherent disadvantages exist and these become more serious as the size of the turbine is reduced.

Gas-lubricated bearings are ideal for small turbines and their development in a form suitable for application to larger machinery is proceeding.

The following table outlines the advantages and disadvantages of each type:

Oil Lubricated Bearings

Advantages	*Disadvantages*
High axial loads can be carried.	Hazard of oil contamination. Need to maintain operating temperature within close limits. Resulting heat inflow in small turbines is unacceptable. Rotational speed is limited. Ancilliary equipment must be provided except where this already exists to lubricate the gearbox.

Gas Lubricated Bearings

Advantages	*Disadvantages*
Complete absence of oil contamination Can accept a wide range of operating temperature. Extremely high speeds can be achieved. Heat inflow can be minimized. Often no ancilliary equipment is required.	Relatively low load carrying capacity.

1. *Bearing Design Considerations*

The type of bearings will have been selected on the basis of such factors as size of machine, approximate size and operating speed of the rotor, the operating temperature level, and type of brake.

The general design procedure is as follows:

(*a*) Estimate from the specification the size and shape of the main rotor components and its rotational speed.

(*b*) Estimate the magnitudes of loads applied to the bearings paying particular attention to maximum values likely to be encountered in extreme operating conditions.

(*c*) Calculate the bearing dimensions to satisfy these loadings.

(*d*) Assess the mechanical layout of the proposed rotor and bearing system.

(*e*) Check the whirl frequencies.

The design process is essentially iterative, and while ideally the bearing system should permit complete freedom of thermodynamic design, and the machine layout should permit complete bearing design freedom, in practice it is sometimes necessary to alter the wheel design to avoid a speed limitation imposed by the bearings or to modify the bearing layout so that they may be more readily accommodated within the machine structure.

Two important factors must be borne in mind.

(i) The journal bearings are high speed lightly loaded components.

(ii) The thrust bearings are subject to considerable load variation since axial forces cannot be balanced completely at all operating conditions.

Thus for the journals the main consideration is the whirl onset speed while for thrust bearings the load capacity is the design criterion.

In general the greater load capacity of an oil-lubricated bearing, size for size, enables a larger factor of safety to be incorporated in the design, whilst the most vulnerable design in this respect is the self-acting gas-lubricated bearing. When it is used extreme care must be taken in the assessment to include all the load producing factors.

It is necessary to consider off-design operating conditions since in some cases a bearing is required to support a load before rotation commences, e.g. a turbine with a high exit pressure which must be brought into operation in parallel with an already operating machine. The direct forces which must be accommodated are generated by gas pressures acting on the wheels, by centrifugal imbalance effects and by gas inertia forces.

Besides the direct forces are others produced by misalignment. Thus the supporting structure can have an adverse effect on bearing performance. Where aerodynamic bearings are use it is frequently necessary to make these self-aligning.

The shaft of a turbo machine must be essentially rigid since deflection produces whirl effects similar to, but usually worse than, those of misalignment. This is particularly important when gas lubricated bearings are used, for the lack of damping makes it impossible to run through mechanical critical speeds. Where oil-lubricated bearings are used an additional requirement is to ensure an adequate flow of lubricant to establish satisfactory thermal conditions within the bearing.

Finally the bearings must be designed to give an acceptable operating eccentricity. This is important for seal leakage is a minimum where a high degree of running concentricity can be achieved.

2. Bearing Materials

The three primary requirements for bearing materials are:

1. That the components can be manufactured with the necessary degree of accuracy by the use of available machine tools or processes.
2. That mechanical tolerances will be maintained within acceptable limits over the operating life of the machine.
3. That the surfaces have an adequate degree of compatibility, that is they will not score or weld readily under a transient condition of excess loading which destroys the lubricant film.

It is equally important that the entire machine assembly should be geometrically stable, since cryogenic turbines have to operate over a range of temperature and any inbuilt stresses, or those resulting from differential contractions or large centrifugal effects, will be progressively relieved. The resulting movement may cause distortion of the bearings or of the general alignment of components within the machine.

There is little difference between the types of materials used in oil- and gas-lubricated bearings in that both normally employ a hard shaft in conjunction with a relatively soft bearing material such as a bronze.

The shaft surface may be nitrided or a hard material applied by a metalliza-tion process. The latter enables the shaft material to be chosen on the basis of its thermal conductivity or expansion. This is particularly important in gas-lubricated bearings where accurate clearances must be maintained throughout their wide working temperature range.

The hydrodynamic gas-lubricated bearing has a unique requirement in that it is necessary for the surfaces to operate under conditions of dry rubbing contact until sufficient speed difference exists between the surfaces to establish correct bearing action. A very satisfactory material for use under these conditions is a P.T.F.E. impregnated bronze, called D.U., manufactured by the Glacier Metal Co. (London, England), and now used extensively in tilting pad bearings. The necessary surface accuracy is normally obtained by lapping on a mandrel of the appropriate diameter although light machining is permissable.

3. *Thermal and Speed Effects—Oil-lubricated bearings*

The performance of an oil-lubricated bearing is, to a large extent, determined by the viscosity of the oil within it. In order to control its temperature within reasonable limits the high speed bearing must be fed under pressure with a sufficient quantity of oil to carry away the heat generated. For a given bearing the amount of heat produced is proportional to the speed of rotation of the shaft. The oil is fed into the bearing via a system of grooving designed to distribute the oil in a manner which will produce the most beneficial cooling effect. Too large a temperature rise in a bearing, besides reducing the viscosity, can produce serious thermal expansion problems as most bearing materials have high coefficients of thermal expansion. Clearances may be increased or distortion may occur due to differential movement.

Two serious speed effects may present themselves in poorly designed bearings, these are:

(a) *Whirl.* This is a form of instability generated by the fluid film and causes trouble when the whirl orbit speed of the shaft approaches half its rotational speed. This may be explained as follows:

Consider the centre of a rotating shaft fixed in an eccentric position relative to the bearing. The mean velocity of the fluid entering the converging wedge formed between the surfaces is equal to half the shaft peripheral speed. If the shaft centre is now caused to orbit in the direction of rotation by forces created within the bearing the position of the wedge will also move in this direction. When it does so at a speed near to the mean velocity of the fluid the bearing action will cease and the bearing will fail. Figure 8.10 illustrates this effect.

(b) *Reynolds number effect.* The concept of critical velocity of flow in a pipe is a familiar one. At a Reynolds number defined as

$$Re = \frac{Vd}{v} \simeq 2000$$

where V = The velocity of flow
d = Inside diameter of the pipe
v = Kinematic viscosity

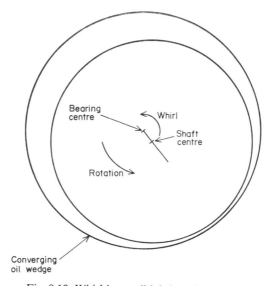

Fig. 8.10. Whirl in an oil-lubricated bearing.

turbulent flow commences with a sharp rise in resistance to shear and a consequent increase in losses. A similar relationship can be devised incorporating the bearing dimensions, or more particularly, the clearance, the shaft peripheral speed and the fluid viscosity.

Now because low viscosity lubricants are used in high speed bearings, which have relatively large clearances to permit adequate cooling flow, care has to be taken to ensure that the critical Reynolds number is not approached, otherwise the initiation of turbulence would give rise to greatly increased frictional losses and excessive heating of the bearing. The effect is much less likely to occur in the fully floating journal bearing (to be described) due to the smaller clearances and lower relative velocities in each film.

4. *Bearing Forms* (*Oil-Lubricated*)

The most common form of thrust bearing in general use is the fixed pad type. In this the annular stationary member is divided into six or more pads by radial grooves. The surface of each of the pads is formed so that the clearance between it and the rotating member converges in the direction of movement. The effect is achieved either with a step or a slope followed by a parallel portion as shown in Fig. 8.11.

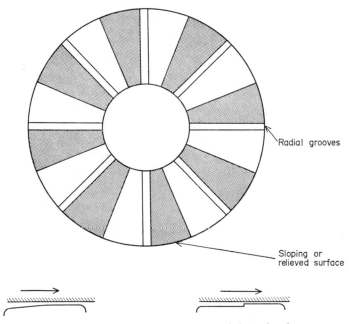

Fig. 8.11. The fixed pad oil-lubricated thrust bearing.

In some designs the entire static member is pivoted so that it may align more readily with the rotating part. In larger pad type bearings, the pads are individually pivoted so that they can assume an angular relationship giving the most satisfactory performance at each operational condition of load and speed.

Journal bearings may be of either fixed or pivoted pad type or cylindrical, having an oil admission hole and an associated system of grooving designed to control oil flow and to eliminate any tendency to whirl. By making the pads of the former type assymetrical the position of the shaft in its bearings may be adjusted to counteract the effect of rotor weight and achieve a near concentric attitude.

Another form of bearing used in high-speed machinery is the fully floating journal bearing in which the sleeve supporting the shaft is itself supported in a cylindrical bearing so that there is a hydrodynamic film between the shaft and the sleeve and another between the sleeve and the stationary bearing.

5. *Lubricating Oil Systems*

An adequate supply of pure lubricating oil at the correct temperature and pressure is essential for long turbine bearing life. The oil supply system generally consists of:

(i) An oil pump which can either be directly driven by the turbine shaft or separately by an electric motor. Directly driven pumps can act as power absorbers for small expansion turbines.

(ii) Oil filters. Sintered stainless steel filter elements are used to remove all particles larger than 5 microns.

(iii) Oil coolers. When the oil pump constitutes the load on the turbine then all the generated power must be dissipated as heat in the oil cooler.

(iv) Sump heaters. These are fitted to ensure that the oil viscosity is correct during start-up and during low-load operation.

(v) Vapour liquid separators. Process seal gas tends to become dissolved or entrained in the oil and must be separated.

(vi) Instrumentation. The oil temperature and pressure should be indicated at important points in the system. Alarms and automatic shut-down devices may also be incorporated to respond to shaft over-speed, low oil pressure or excessive bearing temperature.

When only electrically driven oil pumps are fitted, an emergency run-down system is fitted to protect the machine in the event of a pump failure. Such a system is illustrated in Fig. 8.12. It consists of an auxiliary oil receiver which is filled during start up by diverting a small flow of oil from the main oil circuit. When the vessel is full the oil feed to it is cut off by means of a float-operated valve. If a low oil pressure alarm condition should occur the power-operated valve in the feed line to the turbine closes, nitrogen is admitted to pressurize the auxiliary oil receiver and force the lubricating oil contained in it through the turbine bearings. The auxiliary vessel is sized to provide a supply of oil during the run-down period of the turbine. An auxiliary oil system is not generally fitted when the lubricating oil pump is driven by the turbine shaft.

E. GAS LUBRICATED BEARINGS

1. *Principles of Operation*

There are two main categories of gas lubricated bearing.

1. The self-acting or hydrodynamic, sometimes known as aerodynamic.

2. The externally pressurized or hydrostatic bearing, sometimes called aerostatic.

The first type develops "lift" as the result of viscous forces in the working fluid. The second requires a supply of pressurized gas from an external source.

2. *The Self-acting Bearing*

The self-acting bearing consists in its simplest form of a moving plate inclined relative to a fixed one. Pressure is built up by the viscous forces

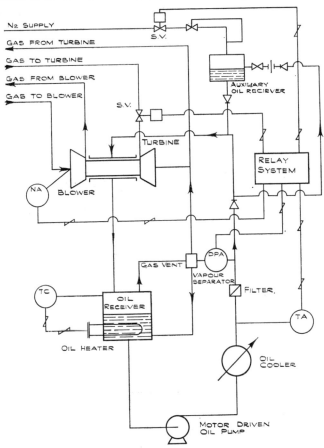

Fig. 8.12. Blower loaded expansion turbine with lube oil system. Alarms and safety trips are shown.

within the film wedge. The concept is familiar in oil lubrication, in gas lubrication the effect is greatly reduced due to the much lower viscosity, of the order of 1/1000 and also due to compressibility effects, that is as pressure rises through the bearing the density increases. The changing volume flow through the bearing modifies the pressure characteristic giving a lower load carrying capcity. If the bearing is operating at a high ambient pressure the compression ratio $P_{max}/P_{ambient}$ approaches unity, compressibility effects become insignificant and performance is improved.

Self-acting bearings may be subdivided into those of fixed geometry, the performance of which may be enhanced by a variety of surface configurations and those in which the non-rotating parts arc free to move relative to the rotating member and each other. The most usual form of this type is the tilting pad bearing which can be used for thrust or journal application but is more usually a journal bearing. The most popular self-acting thrust bearing is probably the spiral grooved configuration, see Fig. 8.13.

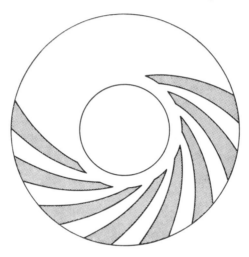

Groove configuration of fixed member

Direction of rotation of moving member

View on edge

Fig. 8.13. A spiral grooved thrust bearing (inward pumping).

A self-acting journal bearing possesses both converging and diverging wedges. The former acts as a compressor and the latter as a diffuser in which sub-ambient pressures occur. Gas flows into the bearing in this region and is expelled in the converging section. Because of the distribution of generated pressure shown in Fig. 8.14 the displacement of the shaft in the bearing is not

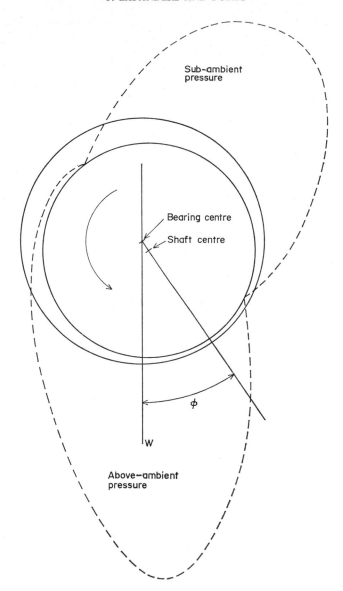

Fig. 8.14. Pressure distribution within a self-acting journal bearing.

in line with the direction of the applied load. The acute angle between the line of centres and the load line is known as the attitude angle, and this with the eccentricity (defined as the displacement of centres divided by the radial clearance) gives the shaft position within its bearing. A plot of attitude angle against eccentricity gives the attitude–eccentricity locus for the bearing and is important in whirl calculations where the radial spring stiffness must first be obtained by plotting radial force against radial film thickness and evaluating the slope of the curve at the chosen eccentricity ratio.

One of the most important parameters in hydrodynamic bearing design is the compressibility number

$$\Lambda = \frac{6\mu\omega}{P_a}\left(\frac{R}{C}\right)^2 \tag{8.19}$$

Where μ = gas viscosity
ω = angular frequency
P_a = ambient pressure
R = bearing radius
C = radial clearance

derived from basic lubrication theory. A large compressibility number indicates that pressure changes within the bearing are significant compared to the ambient pressure P_a. It may be shown that the load capacity is a function of the compressibility number and the eccentricity ratio.

Because the compressibility number is so useful in correlating the experimental results of various bearings it is frequently used as a base parameter in design charts from which load capacity, attitude angle and whirl threshold speed may be obtained for various eccentricity ratios.

3. *Externally Pressurized Bearings*

The simplest form of externally pressurized bearing consists of two concentric parallel plates, one of which has a central orifice through which pressurizing gas is fed to the clearance between them. The distance between the plates is variable with load but for a given supply pressure the upper plate would be supported a fixed distance above the lower one by a force equal to its weight derived by an integration of the pressure over its surface.

Increasing load will progressively reduce the clearance while the pressure within it will rise due to increasing restriction to maintaining equilibrium.

It is seen that the bearing has stiffness, the value of which may be obtained from a plot of load versus deflection. There is an optimum ratio of resistances to flow, i.e. an optimum feed orifice size for a given bearing geometry and clearance. If the feed orifice is too small the pressure drop across it will be excessive and little "lift" pressure will be available. If it is too large the pressure under the upper plate approaches supply pressure, the bearing becomes effectively a constant load device and possesses no stiffness. A hydrostatic journal bearing can be regarded as a number of bearings of the simple type described above, arranged around the shaft.

The practical form of thrust bearing is annular so that the rotor shaft may pass through it. An exception to this is one used on the small B.O.C. machine which consists of a circular plate, having a slight concavity, fed from a central orifice.

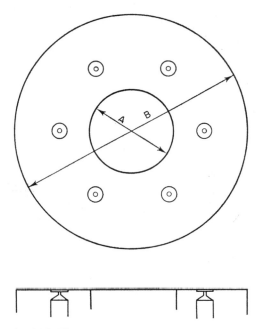

Fig. 8.15. The externally pressurized thrust bearing.

(a) *The hydrostatic thrust bearing.* This is fed by six or more orifices arranged on a pitch circle diameter given by $D = \sqrt{(AB)}$ where A and B are the inner and outer diameters of the bearing, see Fig. 8.15. The orifices may lead into shallow pockets or a system of grooving designed to improve the load carrying capacity by modification of the pressure distribution within the clearance.

The performance of a thrust, or journal bearing is expressed in terms of its load coefficient. Defined as the ratio of the actual load supported at a given clearance (or eccentricity in the case of a journal bearing) to the product of the plan area of the bearing and the available pressure drop.

A hydrostatic thrust bearing can produce values up to 0·4.

(b) *The hydrostatic journal bearing.* Gas is fed into the bearing via a number of orifices, the axial position of a ring of orifices being known as a plane of admission. Bearings may be single or double plane. If the latter, the planes are usually set at $\frac{1}{4}$ stations, i.e. $L/4$ from each end where L is the bearing length. Otherwise the admission plane is in the centre of the bearing.

17

The orifices may be plain drilled holes in the bearing wall in which case the restriction is the fringe area between the circumference at the end of the hole and the shaft surface, this is known as inherent compensation, or a pocket may be provided downstream of the orifice again in the bearing wall, this is known as orifice compensation.

Another form is the step bearing in which the gas is admitted between the two journal bearings and flows towards the outer ends where it encounters over the last 2–3 mm a region of reduced clearance in the form of a step—the type is reported to be extremely stable in operation (see Fig. 8.16).

4. *Instabilities.*

Three forms of instability can occur in gas-lubricated bearings:

  (i) Fractional speed whirl.
 (ii) Pneumatic hammer.
(iii) Lock up.

The first may be defined as the orbital motion of the shaft centre about the bearing centre. The term is generally applied to the motion of the shaft but bearings which are flexibly supported also can exhibit the phenomenon.

Pneumatic hammer is associated with both journal and thrust bearings of the hydrostatic type. It gives rise to rapid vibrations which may be audible.

Lock-up is the name given to a condition in which the hydrostatic pressure causes the shaft to be held firmly against the bearing wall.

The last two forms of instability can be avoided by careful design but fractional speed whirl will always be encountered if the rotational speed of the shaft is increased sufficiently. It is important that this type of whirl should not occur within the operating speed range of the rotor.

(a) *Whirl.* The plain cylindrical journal bearing is rarely used due to its low whirl threshold speed; some improvement may be made as in the oil lubricated case by admitting bearing fluid into the unloaded region of the clearance.

When fractional speed whirl takes place in a self-acting bearing the frequency at which failure occurs is close to half that of shaft rotation.

The behaviour of a shaft accelerated from rest in a hydrostatic bearing is of interest. The sequence of events is as described below.

Initially the whirl amplitude increases steadily due to a cyclic centrifugal force produced by residual inbalance. This is known as synchronous whirl. A speed is then reached which corresponds to the lowest natural frequency of the shaft bearing system, i.e. the frequency at which the shaft would oscillate if given a sharp radial disturbance from rest in its bearings. At this point the shaft motion changes from rotation about a geometric axis to rotation about a mass axis and is said to invert. As it does so a large resonant amplitude peak occurs, the magnitude depending primarily on the degree of balance of the rotating assembly.

Above inversion the rotor is no longer subject to centrifugal forces and the amplitude of whirl is considerably decreased.

As rotational speed is further increased, fractional whirl is encountered which, due to its rapidly increasing amplitude, if allowed to persist leads to complete failure of the bearings.

The frequency of this whirl is close to that of the synchronous whirl at inversion and its onset speed is some integral multiple of it, N. The whirl ratio is defined as $1/N$.

Now if the shaft is considered stiff in comparison with the bearing film the inversion frequency will be approximately

$$f \simeq \frac{1}{2\pi}\sqrt{\left(\frac{K}{M}\right)} \qquad \text{for cylindrical whirl}$$

and

$$f \simeq \frac{1}{2\pi}\sqrt{\left(\frac{KL^2}{4I_T}\right)} \qquad \text{for conical whirl}$$

where K = Bearing stiffness
M = mass of the rotor
L = distance between the bearing centres
I_T = transverse moment of inertia of the rotor.

The whirl onset speed will equal $(N/2\pi)\sqrt{(K/M)}$ or $(N/2\pi)\sqrt{(KL^2/4I_T)}$, depending on whichever mode of whirl has the lower frequency.

The whirl ratio can vary from $\frac{1}{2}$ to $\frac{1}{4}$ the radial load on the bearing being the primary determination factor.

The whirls described above are rigid body modes, their frequencies are calculated from the mass or inertia of a rigid shaft and the spring stiffness of the bearing film. Since this film possesses little damping it imposes no appreciable restraint on the shaft. For this reason the shaft must be considered a free-free beam and should be designed so that it does not whirl in this critical mode within the operating speed range of the rotor.

By far the most stable bearing form is the tilting pad type. This has three or four pads per journal. It suffers the disadvantage of complexity and the difficulty of setting the pivots to give the correct pad loading and a satisfactory alignment of the rotor within the machine, sometimes one or two of the pads are spring loaded. The journal equivalent of the spiral grooved bearing also produces a wide range of stable operation.

These forms can be used if the hydrostatic pressure available provides insufficient stiffness to achieve the minimum whirl onset speed, and boosting the supply is impracticable. Hydrostatic bearings with small clearances develop significant hydrodynamic effects which greatly enhance the stiffness of the bearing system at high speeds.

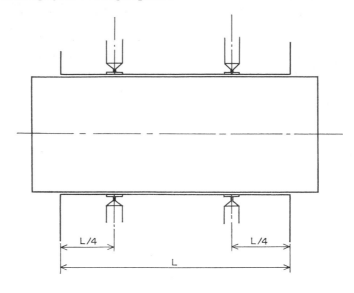

(a) Orifice compensation

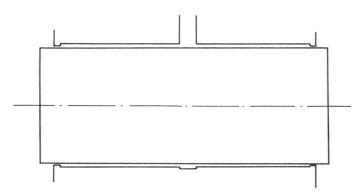

(b) The step bearing.

Fig. 8.16. Externally pressurized journal bearings.

It should be noted that while the whirling speed of a simple shaft-bearing system can be predicted with reasonable accuracy; that of the rotor cannot, for the passage of gas over the blading, and the swash action of the thrust bearings, may considerably lower the whirl onset speed of the journals.

(b) *Pneumatic hammer.* The factors governing air hammer and lock up may be said to be at opposite extremes of design possibility as for example, large clearances give rise to hammer while with small ones the shaft is liable to lock up.

The most significant consideration is feeder hole geometry or more specifically, the ratio of the total volume of feeder holes (or pockets downstream of the orifices) to the clearance volume. The inherently compensated bearing is for this reason more stable, but at the expense of stiffness. Bearings with high feed pressures are more prone to this instability.

(c) *Lock-up.* This is more likely to occur in bearings having a diametrical clearance less than 0·025 mm. and particularly if the bearing geometry is such that uneven flows are set up. This can result from poor manufacturing technique or differing feeder hole dimensions.

The inherently compensated bearing is more likely to exhibit lock-up than the orifice compensated type.

5. *Ultra High Speed Bearings.*

As the size of a machine is reduced, it is necessary to increase the rotational speed of the rotor. This is because the peripheral velocity is fixed by the stage head. Now it has been seen that as rotational speed is increased in a conventional bearing a point is ultimately reached where a catastrophic whirl condition is encountered. In other applications of gas lubricated bearings the mounting of the sleeve on rubber "o" rings enables the whirl period to be safely negotiated. This approach, however, is ruled out in cryogenic expanders on two accounts:

(i) The position of the rotor is no longer precisely determined.

(ii) They limit the range of temperature over which the bearings may be used.

Bearings may be stabilized by the application of a radial load generated for example by an assymetric pressure distribution created within it but this has the disadvantage of increasing the eccentricity and reducing the minimum operating clearance. Two types of bearing are suitable:

(i) The tilting pad type.

(ii) The tuned cavity type.

The major problem is that of producing a tilting pad bearing of small size. The solution used by Birmingham, Sixsmith and Wilson (1961, 1963) was to support the pads on beryllium copper strip hinges—so successful was this work that it allowed speeds of 12,000 r.p.s. using a $\frac{3}{16}$ in. diameter shaft.

With larger shafts at very high speeds the pads themselves become unstable and a method of controlling their motion has to be introduced.

The second method of stabilization relies on the use of several tuned cavities connecting with the bearing clearance; the frequency f at which the resonators operate is determined by their volume and the size of the restricting orifice used in accordance with the following relationship

$$f = \sqrt{\left(\frac{C_s^2 A}{LV}\right)}$$
(8.20)

Where C_s = sonic velocity
A = area of orifice
L = effective channel length
V = volume of the cavity.

Since the whirl action of the shaft is maintained by cyclic variations of pressure within the film the feeding in of variations of suitable phase relationship can cancel the whirl producing effect.

A particular cavity is capable of damping a range of frequency of oscillation. The damping range can be extended by incorporating chambers of several sizes within the same bearing or by varying the size of orifice used.

6. The Supply of Gas for Gas-lubricated Bearings.

The clearances employed in gas-lubricated bearings of the hydrostatic type range from 25 to 50 microns and this with the low environmental temperatures typical of the application make it necessary for the fluid used to conform to high standards of purity.

It is important that any impurity which will condense to form a liquid or freeze to form a solid is removed since the first cold surface to be encountered will be the bearing shell and it is likely that deposition will occur within the bearings.

It is normally recommended that the gas supplied should have a Dewpoint not greater than $-70°C$ for although bearings have operated satisfactorily at higher levels a factor of safety must be provided.

The presence of carbon dioxide is less serious for there is evidence to suggest that relatively large quantities of this contaminant can be tolerated where the bearing gas is supplied at ambient temperature. Acidic gases and hydrocarbons other than methane must be absent.

One method of providing a pure source of bearing gas is to use product nitrogen which is available on most air separation plants. Such nitrogen contains no water, carbon dioxide or other condensable impurity.

When helium is used as a bearing fluid for expansion turbines fitted in helium liquefiers or cryogenic refrigerators it must be processed to remove water, carbon dioxide, acidic gases, hydrocarbons (including compressor oil) and any air or nitrogen which may be present in the system. Such purification is carried out by adsorption using either activated carbon or molecular sieves at both ambient and liquid nitrogen temperatures.

Careful choice of bearing gas is essential where high purity gases are being expanded in the turbine. If, for example, Product Nitrogen must contain less than 10 p.p.m. of oxygen, and has to be expanded in a turbine to produce the necessary process refrigeration, air should not be used as a bearing fluid. Where no suitable stream is available extra seals must be provided to prevent the mixing of bearing gas discharge with the process stream.

F. MECHANICAL DESIGN CONSIDERATIONS

1. *Materials*

It is usual to manufacture the bearing housing and cold casings from austenitic stainless steel, since this has a low coefficient of thermal conductivity at cryogenic temperatures and has good ductility even at the extremely low temperatures encountered in helium applications. It can be cast readily into the necessary forms and has good stability. It also has a high coefficient of thermal expansion and is in this respect compatible with the material of the bearings which are often shrunk into their housings. The casing and bearing housing are best constructed of the same material as high thermally produced stresses, which would cause seal and possible bearing misalignment, are thereby avoided.

Stainless steel has the disadvantage of cost and weight, so the casings at the warm end of the machine can advantageously be of aluminium. In fact, medium sized turbines have been constructed using aluminium castings throughout at a small penalty on efficiency. 18/8 stainless steel is also frequently used as a shaft material since its low thermal conductivity is advantageous in limiting heat flow into the cold region of the machine. It is necessary to treat the surface to improve its bearing properties.

One of the disadvantages of the radial inward flow path is the tendency for foreign particles to accumulate in the space between the nozzles and the wheel, causing surface damage by erosion. In severe cases, the trailing edges of the nozzles have been completely worn away. The use of stainless steel nozzles reduces the rate of deterioration but the only satisfactory cure is the prevention of particle entry by filtration.

The turbine and compressor wheels can be produced by a variety of manufacturing techniques but aluminium alloy is used universally as the material, the major requirements for the duty being high strength and low weight. The necessary forms can be produced with sufficient accuracy by advanced casting methods. Where it is necessary to fix a shroud on to the wheel, the aluminium alloy H.E.9 can be used. The blades are machined on the wheel and the shroud joined on to them by a salt bath brazing process.

2. *Oil or Gas Lubrication*

The choice between oil- or gas-lubricated bearings is closely related to the method of power recovery, which is in turn largely a function of machine size. The dividing line occurs at approximately the power level of 100 kW.

Below this level it is usually not economical to recover the power. Above this level it is usual to connect the turbine to a generator or alternator through reduction gearing. Occasionally the loading device for a large turbine takes the form of a centrifugal compressor which is used to boost the pressure level in part of the plant circuit, but control becomes somewhat complex.

At low powers, significant advantages are to be gained by the use of gas-lubricated bearings, these advantages increase as the size of the machine decreases.

The application of gas-lubricated bearings to large machines is not at present generally accepted. There are two main reasons.

(i) Extremely satisfactory oil-lubricated machines have been produced.

(ii) The commercial risk of introducing a relatively untried form of turbo-machinery into a large plant is considerable.

The main advantage of gas-lubricated bearings is the elimination of oil contaminant, but wherever a gear-box is required complete elimination is impossible.

Additional advantages of gas lubrication of smaller cryogenic turbines are as follows:

(i) Considerable machine simplification is achieved particularly as regards sealing arrangements.

(ii) The installation is simplified for the machine can be bolted directly on to the cold box or preferably into a light box structure mounted upon it.

(iii) Lubricant temperature control is not required.

(iv) No speed limitation is imposed by the bearing system.

(v) Heat inflow is minimized in that the bearing at the cold end of the machine can operate at the natural temperature of its environment. For externally pressurized bearings treated instrument air can generally be used.

There are instances where oil lubrication is a logical choice even on small turbines. Where low molecular weight fluids are involved the power produced by the machine will be considerably higher than that of the same sized air turbine, and the loading requirement exceeds the capability of a single stage centrifugal compressor. If as a result an oil brake is used it is then sensible to use oil bearings. The brake may take the form of an oil pump which in turn supplies the bearings.

3. *Heat Inflow*

Heat in-leak may occur from any of these sources.

(i) The environment.

(ii) Heat generated within the machine.

(iii) Heat carried into the machine in the lubricant.

Heat entering the cold area does so mainly by conduction along the structural parts of the machine, but some may enter by radiation. The machine is usually supported by the part of its structure which is at the highest temperature, and the flow paths to the cold end are made as resistant to heat flow as possible by the use of minimum cross sectional areas consistent with strength. Materials having low thermal conductivity are used and heat insulating materials are incorporated wherever possible. The turbine is surrounded by insulation which may be either in the form of glass fibre or a low density powder contained by the box in which the machine is mounted.

The heat flow associated with the bearings becomes increasingly significant, relative to the refrigeration performance as the size of the machine is reduced.

4. *Sealing Arrangements*

(a) *Gas seals*. Gas seals are used to limit the leakage of process gas along the shaft or across a fully shrouded wheel. They also provide a means of adjusting the axial forces acting on the rotor in that they are used to define the areas over which high or low pressures act.

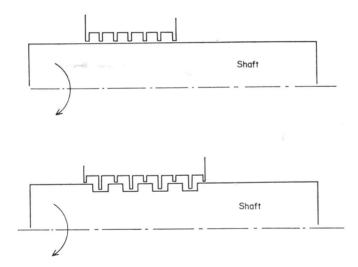

Fig. 8.17. Forms of labyrinth seal.

The most usual form of gas seal is the labyrinth (see Fig. 8.17) which introduces into the gas flow a number of restrictions and expansions, the effect of which is to dissipate the pressure head as kinetic energy rather than as net flow. For a given pressure differential the leakage flow will be lower than for a plain seal of equal diametrical clearance.

(b) *Oil seals*. Oil seals are required to seal the shaft at each end of the bearing compartment, but it is the seal at the turbine end of the arrangement which is the most important, for any leakage of oil into the process stream can have extremely serious consequences. This type of seal is necessarily more complex and consists usually of two labyrinths plus a means of preventing the direct impingement of oil onto the end of the seal. This is achieved by fitting one or more oil slinger rings onto the shaft. These throw the oil outwards onto the walls of the housing from which it runs into a drain.

The flow of oil-contaminated gas is prevented by supplying seal gas at a point halfway along the labyrinths, thus the flow of seal gas opposes any possible leakage. The seal gas pressure is set so that the pressure differential along the remainder of the labyrinths is small.

5. Types of Loading Device

If an alternator is used for loading the turbine it must be run at constant speed. A popular loading device at lower power levels is the centrifugal compressor. This has a power-speed law which is approximately

$$HP \propto N^3$$

thus when used on a machine of variable throughput the compressor discharge valve has to be adjusted if the maximum efficiency is to be obtained at all conditions and if overspeed is to be avoided with increasing flow. A reduction of load is effected by restricting the exit flow or by reducing the pressure and hence the density of the working fluid. It is possible to utilize the compressor to boost the pressure of the stream supplied to the turbine but this involves complex control sequences and is not commonly done.

If the turbine is of fixed nozzle geometry, and its throughput control is by inlet throttling, then quite efficient operation will be produced throughout the range at a fixed compressor setting. As the turbine head is reduced, the spouting velocity C_0 is reduced proportionally to $H_0^{\frac{1}{2}}$. Power is reduced due to the head reduction and the decreased throughput. The resulting lower speed gives a velocity ratio U_3/C_0 close to the optimum.

Because of its simplicity and ease of control the centrifugal compressor is ideally suited for the loading of small turbines. It has the additional advantage that it is capable of high speeds.

For small turbines whose work output exceeds that absorbable by a centrifugal gas compressor an electrical or oil brake may be used. The electrical device may be an eddy current brake or permanent magnet alternator, the latter having the advantage that heat is generated in an external load. Both can be used with gas lubricated bearings and have a form of rotor construction capable of withstanding large stresses.

6. Control of Throughput

Throughput control may be necessary for any of the following reasons: to give maximum refrigeration at startup, to provide the differing conditions dictated by gas and gas and liquid production, to adjust to a gradually deteriorating plant condition and finally to match a continuously varying product demand.

Four methods of control are available each showing particular advantages dependent on the above requirements.

(i) Interchangeable nozzle rings (and possibly wheels).
(ii) Inlet throttling.
(iii) Nozzle groups.
(iv) Variable geometry nozzles.

Interchangeable nozzles are used on small machines where the need for change occurs infrequently. If the percentage flow change is large, and efficiency is not to suffer, it is also necessary to change the turbine wheel.

The most usual form of control, for relatively small automatic plants, is the inlet throttle. This is certainly the simplest since only a diaphram actuated valve is used and can be readily integrated with the plant control system. The main disadvantage is the overall low efficiency of the method since the pressure drop across the valve generates no work and practically no temperature drop.

Nozzle grouping with multiple inlet valves has been found an extremely effective control method for large turbines. Turndown to 50% of the nominal throughput does not usually produce a fall of efficiency in excess of 5%. The method is more suitable for manual than automatic control. By carefully selecting the number of nozzles in each group it is possible to vary the throughput of the machine in small steps.

The loss in efficiency results from two main causes.

(a) *Ventilation losses.* Arise from the pattern of flow in the region of the closed nozzles.

(b) *Flow losses.* Occur in the wheel as the mean flow velocity is reduced due to regions of reverse flow and eddy losses.

By far the most effective form of throughput control is the variable geometry nozzle. This design also permits flows in excess of the nominal or design value. Mechanical complexity is inevitable and the components must be made with a high degree of accuracy. The nozzle efficiency is likely to be lower than that of a fixed nozzle due to leakage through blade clearances. It may be reduced by spring loading the side plates against the blades.

7. *Examples of Operational Cryogenic Turbines*

(a) *An oil-lubricated turbine.* A cross sectional view of an oil-lubricated turbine of advanced design is shown in Fig. 8.18. A significant feature of the design is the use of "cartridge" construction. The central portion of the turbine containing the rotor and bearings can be withdrawn without disturbing

the main structure or insulation of the machine. This greatly facilitates over-
haul and has an additional advantage in that forces generated by pipework
are not transmitted to the bearing assembly.

Fixed taper land bearings are used to carry the thrust load while the journal
bearings are of assymetrical three pad design. These allow the rotor to run
stably in a central position in the bearings, the sealing diameters then run
concentrically and seal leakage is a minimum.

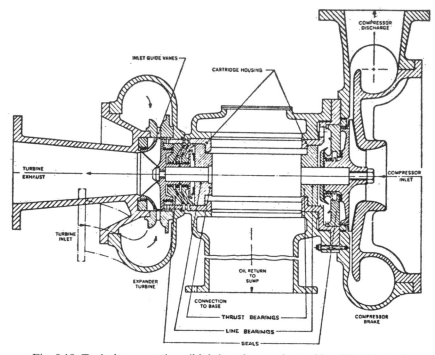

Fig. 8.18. Typical cross section oil-lubricated expansion turbine (Worthington).

It is extremely important that oil is confined to the bearing area as even
small amounts entering the process stream can constitute a hazard or make
necessary an extensive and costly clean out of the plant. External oil leaks in
the region of the plant are most undesirable and can be dangerous. The
problem is overcome in the design by the provision of labyrinth seals at each
end of the rotor plus an adequate drain system. Process gas leakage is prevented by feeding dry gas into the seals. The entire lubrication system is
hermetically sealed to eliminate any external leakage. Additional seals are
provided behind the turbine and compressor wheels to control the level of
axial force produced by gas pressures acting on these components. The
turbine wheel is fully shrouded thus leakage across the blade tops is prevented.
The machine is fitted with variable nozzles although these are not shown.

(b) *A gas-lubricated cryogenic turbine* (*variable nozzle geometry*). A cross sectional view of a gas-lubricated turbine is shown in Fig. 8.19. The centre section of the machine is formed from a single casting and is provided with strengthening webs (not shown) which prevent mis-alignment of the bearings under the loading of forces generated by the expansion bellows in the associated pipework. The larger end flange forms the back plate of the compressor

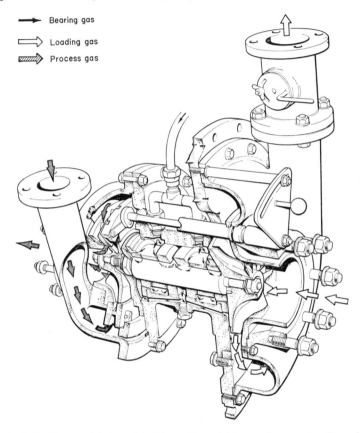

Fig. 8.19. The gas lubricated turbine with variable nozzle geometry (Lucas).

casing and is extended to provide a mounting flange which supports the machine in its insulated enclosure. The smaller end flange forms the back plate of the expansion casing and is provided with a spigot which locates the seal plate, the function of which is to separate the bearing area from the process stream.

Mounted on the seal plate is the nozzle assembly consisting of an upper and lower ring, the actuating ring and nineteen nozzle blades, spindles and actuating levers. The levers are clamped to the spindles on which the blades are fixed. The spindles are located in the upper ring and pass through the lower one.

An important feature of the design is the fixed outer portion of the blade which allows the support rings to be accurately positioned with respect to one another permitting the blade clearance to be reduced to a minimum. The

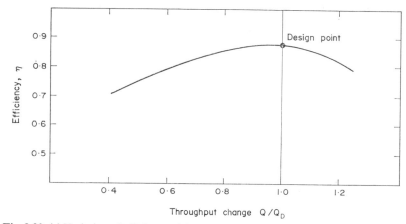

Fig. 8.20. (*a*) Variation of efficiency with change of throughput for a large variable nozzle turbine (at constant speed).

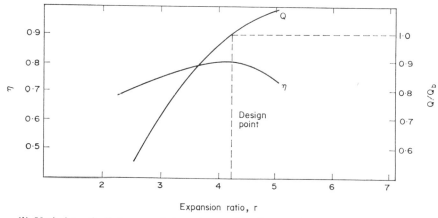

(*b*) Variation of efficiency and throughput with change of expansion ratio for a small fixed nozzle machine.

position of the blades is altered by movement of a quadrant at the compressor end of the machine which, via a crank and link arrangement, produces a small circumferential movement of the actuating ring and hence of the levers which are located in slots around it.

Efficiency, throughput, expansion ratio curves for such a machine, with and without variable inlet nozzles, is shown in Fig. 8.20.

The loading device takes the form of a centrifugal compressor, load being varied by means of a throttle valve attached to the casing exit.

The rotor support system consists of two journal and two thrust bearings all of the hydrostatic type. The direction of flow of the bearing gas is shown. The bearing exhaust is taken to the compressor where it may be used to purge the loading circuit. The rotor is of extremely simple construction having a

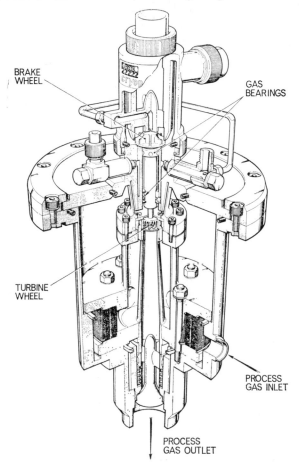

BRAKE
WHEEL

GAS
BEARINGS

TURBINE
WHEEL

PROCESS
GAS INLET

PROCESS
GAS OUTLET

Fig. 8.21. A B.O.C. Cryogenic expansion turbine.

single plain journal. The back face of the compressor wheel forms one thrust bearing member. At the other end a thrust plate is provided. The shaft is then reduced in diameter to pass through the seal plate.

Since with a variable nozzle machine it is important that the speed is reset after each nozzle adjustment a means of speed measurement has to be provided. A permanent magnet is set into the compressor wheel and produces a signal of sufficient strength to operate an electronic tachometer. The electro-magnetic transducer is mounted on the outside of the casing.

(c) *A miniature cryogenic expansion turbine.* Figure 8.21 shows the constructional features of a small cryogenic turbine manufactured by B.O.C. Ltd, England, and clearly indicates the elaborate precautions which have to be taken to minimize heat inflow by the extensive use of thin shells.

The outer casing is constructed so that the top cover allows the turbine body complete axial and radial freedom and thus the build-up of machining tolerances and the effects of differential contraction can be accommodated without straining the central structure.

The turbine body carries two hydrostatic journal bearings as in the larger machine described above but these have an unusual configuration which greatly increases the speed at which they will operate stably. Gas is fed to the bearings via the right-hand union assembly while the left-hand one conducts flow from the stabilizing devices.

The rotor is again of simple construction having at one end the loading compressor, the lower face of which forms one member of annular thrust bearing. Axial loads in the other direction are carried by a thrust bearing having a single central feed orifice. The lower end of the shaft is of reduced diameter and carries the turbine wheel, a radiation shield is fitted immediately above the wheel. The nozzles are formed in the cylindrical surrounding member. It should be noted that the nozzle flow is in the radial plane while the flow through the wheel is axial.

The process gas is led to the nozzle via the annular space between the diffuser tube and the inner casing. To ensure absolute cleanliness a process gas inlet filter is incorporated.

The spring loaded valve at exit facilitates the removal of the cold turbine by closing off the pipe as the inner assembly is withdrawn.

G. OPERATING PRECAUTIONS

In order to achieve long periods of trouble-free operation it is essential to ensure that the following precautions are taken.

(*a*) High standards of cleanliness must be observed when carrying out installation and maintenance work on expansion turbines. Any foreign matter such as dust, pipe scale and grease will block lubrication ports or score close clearance seals. Fitters should wear clean overalls and should carry out maintenance work in a "clean area". During transportation or storage expansion turbine components should be kept in clean, dry polythene bags containing dessicant packs to remove water vapour.

(*b*) Pipe work up-stream of expansion turbines must be treated to remove pipe scale, weld splatter and other foreign materials. Temporary line strainers should be fitted during initial plant start up. Inlet lines to expansion turbines must be blown out with clean nitrogen for long periods before the turbine is installed.

(c) Pipe work connected to expansion turbine cases should be designed to prevent the transmission of thermal stresses to the machine. It is generally necessary to fit expansion bellows to the inlet and exit gas connections at the turbine end. The bellows should be restrained to prevent the transmission of large to the turbine case.

(d) When atmospheric air is used as the working fluid in the loading compressor care should be taken to avoid the fouling of the blading by contaminants. The larger particles would pass through but oily substances and smoke particles will be deposited on the blading and case reducing efficiency. They can also dissolve in condensation on shut-down subsequently solidifying in seal rings and bearings. These contaminants can only be removed by extremely fine filtration capable of retaining 98% of particles larger than 1 micron. Such filters must be continuously monitored as an excessively high pressure drop will cause the machine to overspeed.

When a blower is loaded with air or waste gas containing water vapour problems can occur during an emergency shut down as the blower end will often drop in temperature due to thermal conduction to the cold end. This can result in ice formation in the thrust bearings and blower wheel, which will result in a seizure if an attempt is made to start under these conditions.

The most satisfactory fluid with which to load an expansion turbine compressor is a product gas from a low temperature plant. As an example oxygen enriched dry waste gas is used in Petrocarbon Developements Ltd. nitrogen plant, whilst product nitrogen is used in their oxygen–nitrogen plants. Many plants of this type are known to the authors which operate continuously for periods of 1–2 years without maintenance.

II. Reciprocating Expansion Engines

A. INTRODUCTION

Generally reciprocating expansion engines are employed when the inlet pressure and pressure ratio are high and when the volume is low. The inlet pressure to expansion engines used in air separation plant varies between 600 p.s.i.g. and 3,000 p.s.i.g., and the isentropic efficiencies achieved are from 82 to 87%.

A typical expansion engine is illustrated in Fig. 8.22. It consists of a cylinder of cast carbon or stainless steel inside of which runs an elongated piston fitted with piston and rider rings to prevent the leakage of gas into the crank case. The gas inlet and exit valves are controlled by push rods operated by arms fitted to the crank shaft. The piston drives the crank shaft via a cross head and a piston rod. A pulley is attached to the crank shaft to transfer the work via a belt drive to a suitable power absorbing device such as an electric generator.

18

The first expansion engine used in an air separation plant was built by G. Claude in 1902. It was a converted steam engine that used a leather cup as a piston seal. Unlubricated engines using leather cup packing have operated reliably with inlet pressures up to 600 p.s.i.g. and piston speeds of the order of 200 ft/min with efficiency of 60 to 70%.

Heylandt (1912) improved the expansion engine; he lengthened the cylinder and piston to enable the shaft seal to operate at ambient temperatures.

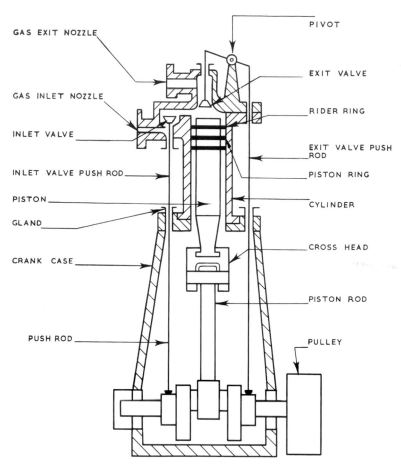

Fig. 8.22. Schematic diagram of a reciprocating expansion engine.

A cap made of a material having a low thermal conductivity was attached to the end of the piston in contact with the expanding gas, the clearance between the cap and the cylinder was relatively large, therefore the only friction was between the piston and the cylinder. Heat generated in the piston was transferred primarily to the atmosphere because of the low conductivity of the cap. Collins (1947) was responsible for the next major advance when he developed a highly efficient expansion engine for his helium liquefier. The expander was unique as the piston rods and push rods were flexible and kept in tension, which allowed thin stainless steel rods to be used and considerably reduced heat inleakage. To keep frictional forces to a minimum, the piston was free to align itself within the cylinder. A micarta sleeve was fitted round the piston to improve its wearing properties.

Land (1957) discussed the performance of expansion engines and quotes figures to illustrate the types and magnitude of the losses to be expected. W. A. Morain and J. S. Holmes of the Cooper Bessemer Corp. (1963) describe a mathematical model developed by them to study the performance of expansion engines, the implementation of this model on a digital computer and the verification of the results obtained on a prototype machine. Computer investigations were conducted to determine the effects of variations in clearance volume, compression ratio, valve loss and heat leakage.

Subsequently Morain (1967) describes the development of a large expansion engine with unlubricated cylinder design for hydrogen liquefaction service. Performance figures are given that illustrate the relationship between adiabatic efficiency, cut-off, developed power and mass flow rate.

B. ANALYSIS OF EXPANSION ENGINE PERFORMANCE

For minimum work the compression stage is isothermal. For an ideal reciprocating expansion engine the expansion stage is isentropic.

The ideal work for isentropic expansion of a perfect gas is:

$$W = H_1 - H_2 = C_p(T_1 - T_2) = T_2 C_p\left(\frac{T_1}{T_2} - 1\right) \qquad (8.21)$$

W = work done.

H = enthalpy

C_p = specific heat at constant pressure.

T = absolute temperature.

C_v = specific heat at constant volume.

$$\frac{T_1}{T_2} = \left(\frac{P_1}{P_2}\right)^{(\gamma-1)/\gamma} \qquad \gamma = \frac{C_p}{C_v}$$

Subscript 1 denotes inlet conditions and 2 denotes exit conditions.

$$W \text{ cycle} = \frac{\gamma}{\gamma-1} R T_1\left[\left(\frac{P_1}{P_2}\right)^{(\gamma-1)/\gamma} - 1\right] \qquad (8.22)$$

For a real gas the ideal work from isentropic expansion is given as follows:

$$W \text{ cycle} = \frac{n}{n-1}RT_1\left[\left(\frac{P_1}{P_2}\right)^{(n-1)/n} - 1\right] \tag{8.23}$$

where n is the constant in the equation $pv^n = $ constant.

The relationship between pressure and volume (the indicator diagram) for the idealized and actual compression cycles are shown in Fig. 8.23.

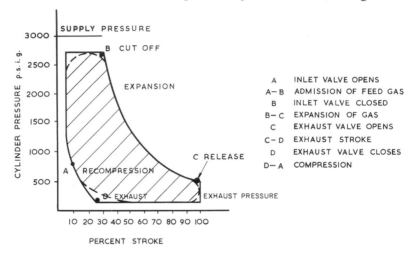

Fig. 8.23. Indicator (P.V.) diagram for ideal and actual expansion engine cycle.

1. Losses in Expansion Engines

There exist five principle sources of inefficiency:

(a) *Valve losses.* Because of the resistance to fluid flow of the cylinder inlet and exhaust valves, there will be a pressure drop across these valves, as the flow through the valves is isenthalpic this will reduce the ideal isentropic enthalpy drop.

Figure 8.24 illustrates these losses on a temperature-entropy diagram.

The actual isentropic drop is reduced from conditions 1 to 2, to conditions 1′ to 2′.

Equation 8.22 now takes the following form:

$$W \text{ cycle} = \frac{n}{n-1}RT_1\left[\left(\frac{P_1 - \Delta P_1}{P_2 + \Delta P_2}\right)^{(n-1)/n} - 1\right] \tag{8.23}$$

Where ΔP_1 and ΔP_2 are the losses through the inlet and exhaust valves, respectively.

(b) *Incomplete expansion.* Because the piston is only instantaneously at its extreme positions, the inlet and exhaust valves are opened and closed before and after the extreme positions of the piston.

Referring to Fig. 8.23 admission of gas to the cylinder takes place between points A and B (point B is known as the cut-off point). It is obvious that all of the gas is not introduced into the cylinder at the upper pressure P_1. This effect increases the losses due to isenthalpic expansion through the inlet valves. The expanded gas is exhausted from the cylinder between points C and D.

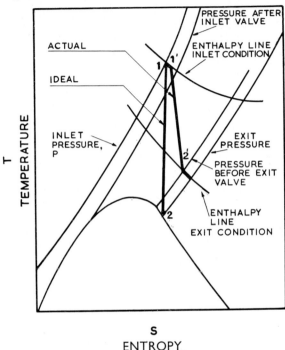

Fig. 8.24. Entropy temperature enthropy diagram for ideal and actual expansion strokes.

As C occurs before the extreme position of the piston, not all the available work in the gas is recovered. The exhaust valve is closed at point D. This means that a portion of the gas is adiabatically compressed from the exhaust to the inlet pressure. The energy required to compress this gas must be provided by the work done during the expansion stroke and this reduces the work available by an equivalent amount.

(c) *Heat leakage.* Heat transfer from the surroundings to the fluid during expansion reduces the enthalpy drop by the amount of heat transferred.

$$(H_1 - H_2) = W - Q \qquad (8.24)$$

where Q equals the amount of heat transferred.

(d) *Non-ideal expansion.* The friction that is generated between the piston rings and the cylinders generate heat which is transferred to the expanded fluid and so reduces the enthalpy drop.

(e) *Clearance losses.* To allow for unequal thermal expansion between the cylinder and the piston and connecting rod, it is not possible for the piston to travel the full length of the cylinder; a clearance must always be left between the outboard end of the cylinder and the piston when it is at its extreme position. The volume of gas thus trapped is approximately 5% of the swept volume of the cylinder. The gas in the clearance space is initially at the exhaust pressure and has to be compressed by the entering gas to the inlet pressure. This work of compression is not recovered as work done on the piston, and the resulting temperature rise reduces the overall refrigeration effect.

An expansion engine operating on an air separation plant would have the following losses, according to Land (1957):

	%
Valve losses	4
Incomplete expansion	2
Heat leakage	2
Non ideal expansion	6
Clearance	4
	──
Total losses	18
Efficiency	82

C. DESIGN FEATURES OF EXPANSION ENGINES FOR INDUSTRIAL APPLICATION

Expansion engines used in industrial liquefaction plant require a high degree of reliability as down time can be very expensive. All large industrial expansion engines are single acting to avoid complicated valve actuating mechanisms.

1. *Cylinder and Piston Design Features*

The piston seal used in the earliest expansion engines was a leather cup fitted to the top of the piston. It ran unlubricated at relatively low speeds in a cylinder fitted with a cast iron liner.

Expansion engines with oil lubricated cylinders are used for air separation service with exit temperatures down to −184°C. Such machines have force-feed lubrication of the bearings and cylinder which would be fitted with a high strength hardened replaceable cast iron liner with a honed microfinish. Piston and rider rings are normally of aluminium bronze. The mean piston speed would be up to 600 ft/min. They would be single acting with the gas inlet and exit points at the outboard end of the cylinder.

Expansion engines are designed to have a relatively high stroke to diameter ratio of approximately 2 : 1. They are fitted with long pistons made from a material having a low thermal conductivity such as austenitic stainless steel.

Because of these design features the outboard end of the cylinder operates at a temperature approaching the exhaust temperature whilst the inboard end tends to work at ambient temperature, being heated by circulating lubricating

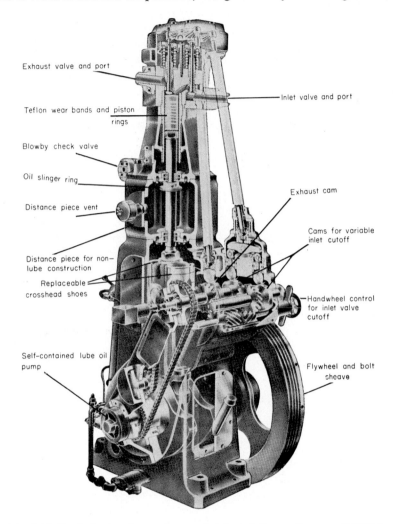

Exhaust valve and port

Teflon wear bands and piston rings

Blowby check valve

Oil slinger ring

Distance piece vent

Distance piece for non— lube construction

Replaceable crosshead shoes

Self-contained lube oil pump

Inlet valve and port

Exhaust cam

Cams for variable inlet cutoff

Handwheel control for inlet valve cutoff

Flywheel and bolt sheave

Fig. 8.25. Vertical expansion engine non-lubricated Cooper Bessemer E.X.N–1.

oil from the crank case. Some engines are provided with hot water jacketing around the inboard end of the cylinder and the valve operating gear to prevent the lubricating oil congealing. Some designs incorporate a shaft oil seal to prevent excess oil being carried into the cylinder from the crank case. Alternative machine designs do not require a gland between the crankcase and the cylinder but rely on the piston rings to provide the necessary sealing. Expansion engines fitted with oil-lubricated cylinders should be provided with oil filters or scrubbers in the gas discharge line to prevent oil droplets being carried forward into the plant where it would create blockages and would, in an air separation plant, present a serious hazard.

When gas exit temperatures below $-184°C$ are required or when the presence of oil in the process gas is not permissible, non-lubricated cylinders are fitted, as shown in Fig. 8.25. The piston rings and rider rings are then made from P.T.F.E. The cylinder is provided with a replaceable liner. All non-lubricated cylinders are fitted with a gland seal at the inboard end to prevent oil from the crank case entering the cylinder. Certain designs, such as the one illustrated, also incorporate a distance piece between the crank case and the cylinder, so dimensioned that the oil-wetted part of the shaft cannot enter the cylinder. The distance piece is provided with two sets of packings, one set of oil scraper rings at the crank-case end and a second set of packing rings fitted to the cylinder. It is not necessary to provide hot water jackets on non-lubricated cylinders.

Expansion engines designed for operation with exhaust temperature below $-212°C$ down to $-268°C$ have some or all of the following additional features:

  (i) Hollow pistons and piston rods made from K Monel to decrease heat transfer down the rod.

 (ii) Long distance pieces to reduce heat conduction.

(iii) Insulated spacer blocks at either end of the cylinder to reduce heat conduction.

(iv) Vacuum jacketed cylinder and exhaust line to minimize heat inleak.

Expansion engines operating at these low temperatures are generally provided with austenitic stainless steel cylinder heads, liners, distance pieces, piping and bolts.

2. *Valves*

Industrial expansion engines utilize spring-loaded poppet valves operated by push-rods. They are generally designed so that the inlet valve cut-off point can be varied from 0–50% of the stroke.

3. *Power Absorption*

The work is generally absorbed by one of the following methods:

(a) *Electric generator.* The expansion engine is coupled by means of belt drive to an electric generator. Some safety device must be fitted to automatically shut off the gas inlet valve in the event of a no-load condition. A no-load condition can be caused by belt breakage or slippage, or could be caused by a reduction in electrical load. The simplest safety device is a centrifugal governer as fitted to steam engines.

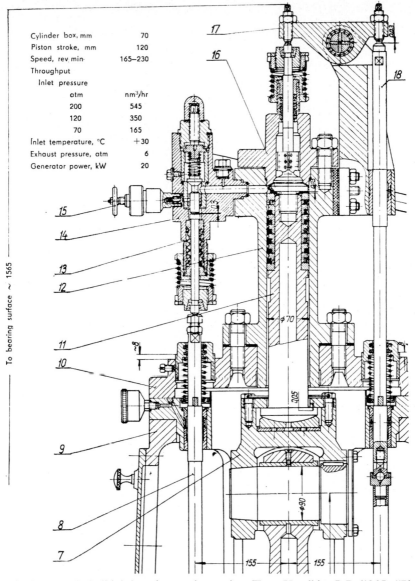

Cylinder box, mm	70	
Piston stroke, mm	120	
Speed, rev min	165–230	
Throughput		
Inlet pressure		
atm	nm³/hr	
200	545	
120	350	
70	165	
Inlet temperature, °C	+30	
Exhaust pressure, atm	6	
Generator power, kW	20	

To bearing surface ~ 1565

Fig. 8.26. Vertical oil-lubricated expansion engine. (From Usyulkim, I. P. (1966). "Plant and Machinery for the Separation of Air by Low Temperature Methods". Pergamon Press, London, with permission).

7. Crosshead
8. Exhaust valve push rod
9. Gear case
10. Cylinder base
11. Piston
12. Cylinder

13. Stuffing box ring
14. Inlet valve
15. Pressure lubricator
16. Exhaust valve
17. Valve rod rocker
18. Exhaust valve push rod

18*

(b) *Air compressor*. The work can be used to compress a portion of the feed air; the simplest method of achieving this is to fit the compressor piston on to the same rod, or alternatively to fit the expander piston to the same crank-shaft as the compressor.

(c) *Oil pump*. Small expansion engines are often coupled to lubricating oil pumps. The work is dissipated by pumping the oil to high pressure, throttling it through a valve, and cooling it in an oil cooler.

4. *Capacity Control of Expansion Engines*

(a) *Variable inlet pressure control*. The most common method of controlling the refrigeration output of an expansion engine is to control the pressure of the feed gas whilst keeping its temperature and the machines cut-off point constant. This changes the throughput of the machine as well as the expansion ratio. This technique is used in the Claude cycle for the production of gaseous oxygen. The plant is initially cooled down using an expansion engine inlet pressure of approximately 2,800 p.s.i.g. When the plant is on stream the compressor discharge, and hence the expansion inlet pressure, is reduced to about 600 p.s.i.g.

(b) *Variable cut-off control*. When it is necessary to maintain the inlet pressure, expansion ratio, inlet temperature and speed constant, and to operate at high efficiency under reduced load conditions, variable inlet cut-off is fitted to the machine. Variable inlet cut-off adjusts the time that the inlet valves are open during the expansion stroke and thus varies the throughput of the machine.

The advantage of variable cut-off is that it improves the efficiency under reduced load conditions whilst keeping the speed constant; it therefore finds application on large machines driving electrical generators.

(c) *Variable speed control*. If the inlet pressure, expansion ratio, inlet temperature and cut-off point are to be maintained constant, then to vary the throughput of the machine one must vary its running speed. To reduce the speed of a machine one must increase the braking effect of the power absorber. This is generally only practical when the power is being absorbed by an hydraulic pump.

(d) *Inlet throttle valve control*. This method is cheap to apply but is inefficient as the pressure drop across the inlet throttle valve cannot be recovered as useful work.

D. EXAMPLES OF COMMERCIAL EXPANSION ENGINES

1. *Expansion Engine for Air Separation Service*

I. P. Voyulkin of the U.S.S.R. (1965) describes a vertical expansion engine which is illustrated in Fig. 8.26. The machine has the following specification.

Maximum inlet pressure	2,900 p.s.i.g.
Exhaust pressure	73 p.s.i.g.
Flow	20,500 s.c.f.h.
Fluid	Air
Inlet temperature	30°C
Exit temperature	*ca.* −138°C
Generator power	20 kW
Cylinder bore	2·75 in. (70 mm)
Stroke	4·75 in. (120 mm)
Maximum speed	230 r.p.m.

The cylinder is force-feed oil lubricated.

2. *Expansion Engine for Hydrogen and Liquefaction Service*

W. A. Morain of Cooper Bessemer Corporation (1967) describes a large expansion engine designed for hydrogen liquefaction service. The machine had the following specifications when tested on hydrogen and helium (see Fig. 8.27).

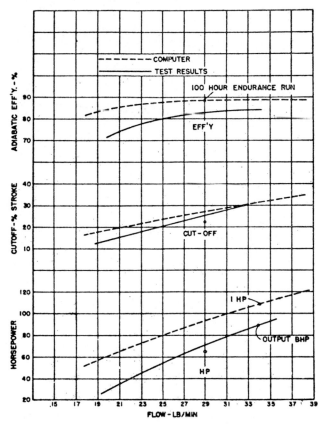

Fig. 8.27. Performance curves for Helium expansion engine pressure ratio 19·2 to 1 (Morain, 1967).

Fluid	Hydrogen	Helium
Maximum inlet pressure	1,935 p.s.i.g.	1,938 p.s.i.g.
Exhaust pressure	85 p.s.i.g.	88 p.s.i.g.
Flow	272,500 s.c.f.h.	198,000 s.c.f.h.
Inlet temperature	not given	not given
Exit temperature	−406·8°F (30°K)	−406·8°F (30°K)
Efficiency adiabatic	80%	80%
Cylinder bore	5 in.	
Cylinder stroke	10·5 in.	
Engine speed	300 r.p.m.	
Cylinder piston speed	525 ft/min.	
Variable cut-off % stroke	50	
Displacement at rated speed	35·8 c.f.m.	

The machine employs the standard crank-case, cross-head and valve gear of a well developed air expander, although hydraulic lifters are added to the valve operating gear. Conductive heat transfer is minimized by the use of long heat flow paths and material of low thermal conductivity. To reduce radiant heat transfer the cylinder and head are surrounded with a vacuum jacket containing superinsulation and all parts in the cold area are lined with radiation shielding material.

The cylinder, constructed of 304L stainless steel, is isolated from the cross head housing by means of 304L stainless steel distance piece and rigid high strength non-metallic spacers. A hollow Monel piston rod is employed to reduce heat loss by conduction. The piston is made from bronze and fitted with Teflon piston rings operating without lubrication in a cylinder liner which is dense chrome plated.

Stainless steel valves are provided with long stems, and two sets of packings, and a vented lantern ring, are fitted in the valve guides. Gas leaking through the lantern ring, and piston ring blow-by gas, amount to $\frac{1}{2}$% of the gas throughput of the machine. This gas is collected and returned to the low pressure section of the plant.

The outer surfaces of the cylinder head and distance pieces plus the inner surface of the diaphragm and cover plate are lined with a thin sandwich of several layers of aluminized Mylar.

The calculated heat inleakage from the surroundings is

Radiation	20 B.T.U./min
Gas conduction	20 B.T.U./min
Conduction along structural members	43 B.T.U./min
Total	83 B.T.U./min

Heat inleakage due to ring friction = 200 B.T.U./min

Performance curves for this machine are given in Fig. 8.27.

3. *Expansion Engines for a small Helium Liquefaction Service*

S. C. Collins (1957) describes a helium liquefier that uses three expansion engines operating with helium as a working fluid. One of the engines used is illustrated in Fig. 8.28.

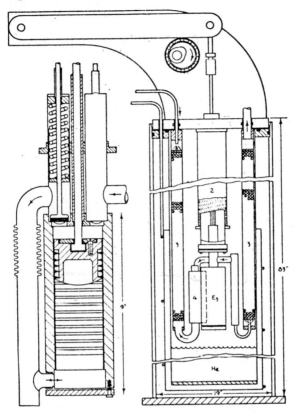

Fig. 8.28. Collins Helium Liquefier.

The running gear of the engine is mounted on top of the plate. It consists of a single shaft on which are keyed cams for controlling the motion of the piston and valves. Piston and valve pull rods are attached to the ends of walking beams which follow the cams. The engine cylinders are made from low carbon steel chromium plated internally. The piston is constructed from cast iron and fitted with leather piston rings.

4. *Valveless Piston Expansion Engines*

A small valveless expansion engine has been developed by Doll and Eder in 1962 and described by Becker, Doll and Eder (1968). The machine, which is used in helium liquefiers, avoids the use of valves and valve operating gear.

Two sizes of the machine, with working volumes of 20 and 50 cm³, respectively, have been described, one being illustrated in Fig. 8.29. Gas at 30 kg/cm² enters the cylinder via the annular inlet chamber in the cylinder and passes via the radial ports in the piston downwards through the central bore. The gas moves the piston upwards sealing off the inlet ports, expansion of the gas occurs moving the piston further until the exhaust ports are exposed. The gas is exhausted at lower pressure from the cylinder via the annular collecting ring. The action of the machine is best illustrated by reference to Fig. 8.30 taken from a paper by R. Kneuer and E. Turnwald of Linde A. G. (1968).

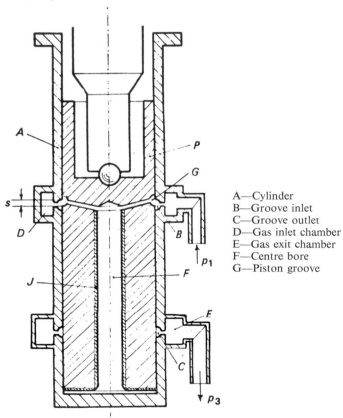

A—Cylinder
B—Groove inlet
C—Groove outlet
D—Gas inlet chamber
E—Gas exit chamber
F—Centre bore
G—Piston groove

Fig. 8.29. Valveless expansion engine.

Leakage of gas past the piston amounts to 5–10% of the engine throughout the leakage flow is designed to give stabilization of the piston within the cylinder and give complete gas lubrication by means of a number of grooves on the piston surface.

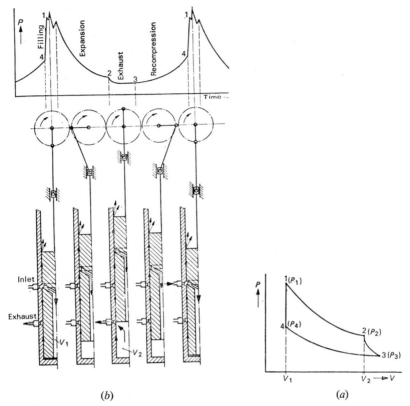

Fig. 8.30. Valveless expansion engine. (*a*) Plot of *P* versus *V*; (*b*) pressure in the working volume versus time.

The engine has been operated at speeds between 1,000 and 1,700 r.p.m. at efficiencies of approximately 70%.

5. *Bellows Expansion Engine*

An expansion engine using a metal bellows was described by Long and Simon in 1954. Development of this engine was restricted by fatigue failure of the bellows and difficulty in filling the clearance volume inside the convolutions of the bellows. J. L. Smith, Jr. (1969) describes a development of the metal bellows expansion engine which overcomes the problems of fatigue failure and clearance volume.

The advantages of the Bellows Expansion Engine are the complete elimination of gas leakage past the piston and the absence of lubricant in the cylinder. In the expansion engine described by Smith, see Fig. 8.31, a type 347 stainless steel bellows constructed by edge welding rings of sheet metal together was employed. This bellows has the advantage when compared with the formed bronze bellows previously used that it can be more tightly collapsed and thus has a much reduced clearance volume.

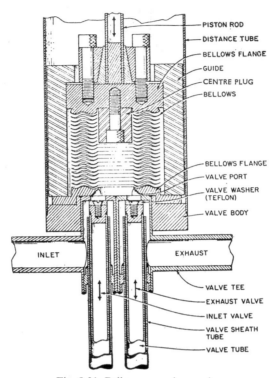

Fig. 8.31. Bellow expansion engine.

Both the piston rod and bellows are maintained in tension at all times. This is achieved by maintaining a pressure in the distance piece tube equal to the engine inlet pressure. The bellows has an outside diameter of 1·25 in. and inside diameter of 0·5 in., with a stroke 0·75 in. and a displacement of 0·44 in.³. The rotational speed is 500 r.p.m.

The valve tubes extend out of the bottom of the engine to a cam case whilst the stainless steel tubular piston rod 30 in long terminates in a sealed crank case.

Performance figures given in Smith's paper are as follows:

Inlet pressure	75 p.s.i.a.
Exhaust pressure	16·4 p.s.i.a.
Inlet temperature	28·5°K
Exit temperature	18·5°K
Adiabatic efficiency	77%
Flow	3·16 lb/hr helium
Refrigeration rate	20·8 W

The equipment was run without fatigue failure for 63·2 hr equivalent to $1·9 \times 10^6$ cycles of the bellows.

E. OPERATIONAL EXPERIENCE

Whilst expansion engines have proved reliable in service they suffer from the defects that are common to all reciprocating machinery, i.e. maintenance of valves, valve gear, piston rings and seals is necessary at regular intervals. Expansion engines that use oil as a cylinder lubricant must be fitted with down-stream filters to eliminate oil carryover. In air separation plants the oil hazard is particularly severe.

As the gas flow to and from an expansion engine is necessarily pulsating careful attention should be given to the fitting of pulsation dampers in the feed and exhaust lines in order to minimize the transmission of vibrations to the pipe work and adjacent items.

If start up problems are to be avoided it is necessary to ensure that a high standard of cleanliness is achieved during the installation period of the machine. Equipment and pipe work up-stream of the expansion engine must be treated to remove pipe scale, weld splatter and dirt. Before starting up the engine the inlet pipework should be blow out with compressed dry clean air or nitrogen. Suitable line filters should be fitted where appropriate in the gas feed line.

Care should be taken during plant operation to avoid any operational condition that would result in the precipitation of water, carbon dioxide or any other condensate within the expansion engine.

III. Liquid Pumps and Associated Systems

A. METHODS FOR TRANSFERRING CRYOGENIC LIQUID

When cryogenic liquids first came into wide use they were discharged from storage tanks by pressurization. The pressure was produced by taking some of the liquid from the bottom of the vessel, evaporating it through a coil using ambient heat, and applying the gas generated to the upper part of the vessel (see Fig. 8.32). The statification that occurs within the liquid adjacent to the surface enables this system to operate without adding appreciable heat to the body of the liquid. The pressure generated by evaporation is controlled by a valve either manually or automatically. In the course of time the pressure raising coil frosts up and the rate of gas generation diminishes.

In many modern distribution systems large flow rates must be sustained with high controllability and therefore, as in Fig. 8.33, mechanical pumps are used. An additional advantage is that higher pressures may be achieved,

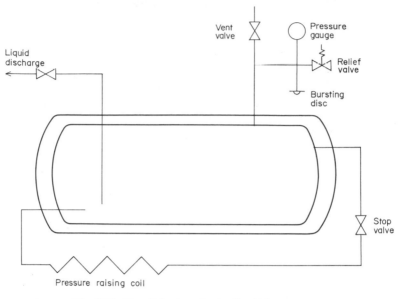

Fig. 8.32. Circuit for transferring liquid by pressure.

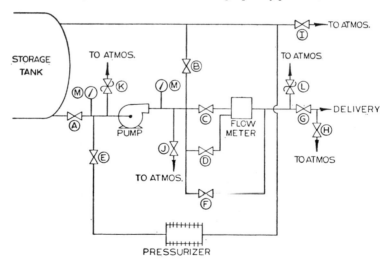

Fig. 8.33. Typical circuit for transferring liquid by pump.

A. Main suction valve
B. Priming valve
C. Meter inlet valve
D. Meter cool-down valve
E. Pressurizing valve
F. Meter bypass valve
G. Discharge valve

H. Bleed-off valve
I. Vent valve
J. Rapid priming valve
K. Relief valve (50 p.s.i.)
L. Relief valve (100 p.s.i. over rated system pressure)
M. Pressure gauge

as is required for charging high pressure receivers. Such a receiver, containing liquid under pressure, is a convenient way to store gas used for welding or cutting. By using a pump capable of operating against the receiver pressure the receiver may be recharged from a transporter or tanker without interrupting the supply of gas to the user.

It is important that the pump is reliable under the operating conditions of low temperature with a liquid at or near its boiling point where heat inleakage results in vapour formation and rapid volume expansion. Typical duties are listed in Table 8.II.

TABLE 8.II. Typical Duties for Industrial Pumping Systems

Application	Duty pattern	Head ft.	Flowrate Imp. gal/min
Bulk transfer, e.g. from storage to tanker	Intermittent over 3–4 hr twice a day.	1C0	100
Plant output to storage	Continuous	100	2C
Delivery to customer pressurized storage from tanker	Periods of $\frac{1}{2}$–1 hr several times on one journey	600	150
Bulk gas supply to grid, supplying street works or chemical plants	Continuous when operating otherwise maintained in condition of readiness	1,000–1,350	80
Gas cylinder filling	$\frac{1}{2}$ hr in each hr	6,000	3

The basic designs of pumps used for cryogenic liquid transfer fall into the normal categories of centrifugal, turbo and reciprocating. The important differences from normal liquid pumps concern the compatibility of the materials, bearings, seals and other components with the environment in which they operate. The environmental factors include the absence of conventional lubrication and the chemical nature of the liquid being pumped, e.g. liquid oxygen.

B. HYDRODYNAMIC PUMP PERFORMANCE

The simple theory for an outward radial flow hydrodynamic or centrifugal pump is summarized in Fig. 8.34. Liquid enters the eye of the impeller at A and leaves at the perimeter B.

The total head developed (Euler head) H_E is given by

$$H_E = \frac{V_{2w}u_2 - V_{1w}u_1}{g} \tag{8.25}$$

or

$$H_E = \frac{1}{2g}[(V_2^2 - V_1^2) + (u_2^2 - u_1^2) + (V_{r1}^3 - V_{r2}^2)] \qquad (8.26)$$

| Absolute kinetic energy | Centrifugal head | Head produced by change of relative velocity |

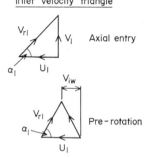

Fig. 8.34. Velocity triangles for a radial flow hydrodynamic pump.

The disadvantage of the hydrodynamic pump is the stringent relationship between head delivered and quantity throughput at the design speed. This characteristic is illustrated in Fig. 8.35 for pumps of the type use for tanker filling.

As has been shown in the case of cryogenic expansion turbines it is useful to define the Specific Speed (equation 8.14) in order to correlate optimal speeds and probable efficiencies of different pump geometries. Figure 8.36 presents a plot of efficiency against Specific Speed for a range of throughputs Q specified in gallons per minute. The units of H are feet, and those of N are r.p.m. The optimum efficiency is seen to occur for values of Specific Speed of about 3,000.

Head variation by alteration of speed is limited by the type of drive unit. In many cases a 3-phase induction motor will be used, and therefore the speed will be fixed. It is particularly important therefore that the customer specifies exactly the duty required of the pump.

For the most part the performance of centrifugal pumps on cryogenic liquids conforms to the standard design theory, but owing to the fact that the liquid is stored at or near its boiling point, cavitation must be avoided by ensuring the net positive suction head (NPSH) requirements are fully satisfied. Cavitation is the formation of vapour and the breakdown of ordered fluid flow in a rotodynamic pump, with a consequent reduction of efficiency.

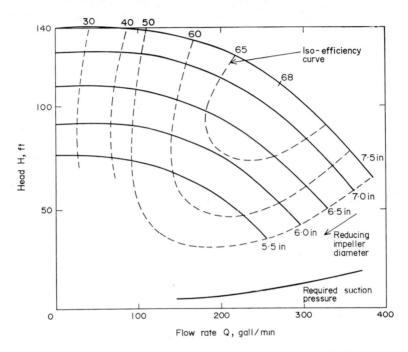

Fig. 8.35. Flow rate versus head curves for different sizes of cryogenic liquid pumps of the type used for liquid oxygen or nitrogen. The dotted curves are lines of constant efficiency. The rotational speed is 2,880 r.p.m.

1. Net Positive Suction Head

The Net Positive Suction Head (NPSH) is an extremely important factor in determining the performance of a cryogenic pump. A large proportion of poor running conditions, including lack of prime, emanate from the insufficiency of head at inlet to the pump. The minimum NPSH for a particular type of pump must be met.

A typical pump suction arrangement is shown in Fig. 8.37. The tank contains liquid at temperature T_{s_1}. The pressure in the gas phase is greater than the corresponding saturation pressure P_{s_1}. This excess pressure head is denoted by H_p. The tank is elevated to provide a static head of liquid h.

Thus the initial head available for moving liquid into the pump against losses, vapour pressure, etc., is

$$H_p + h$$

This is shown at point (1) both Figs. 8.37 and 8.38.

Figure 8.38 shows a typical vapour pressure diagram for a cryogenic liquid and the pressures displayed refer to those stated in diagrammatic form in Fig. 8.37.

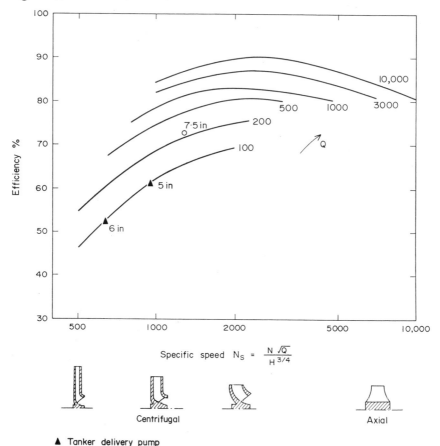

Fig. 8.36. Curves of efficiency against specific speed for a range of throughputs.

The following pressure losses occur:

(i) Pipe friction losses (hf), (1) to (2).

(ii) Loss of available head due to temperature rise, (2) to (3) resulting from friction losses in the system. Thus the available head is reduced by this amount. (Shown in Fig. 8.38 as h_{fv}.)

(iii) Heat inleak similarly causes a temperature and hence vapour pressure increase. (Shown in Fig. 8.38 as hq) (3) to (4).

(iv) The NPSH required by the pump manufacturer. This includes friction loss, dynamic head and acceleration pressure loss at vane inlet (H_{sv}). (4) to (5).

(v) An operational allowance (5) to (6).

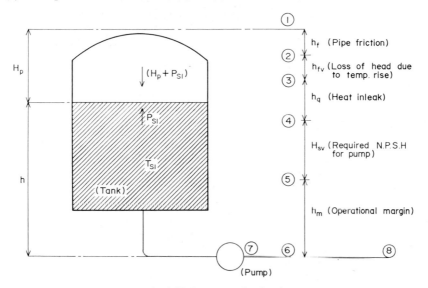

Fig. 8.37. Pump suction head.

It can be seen from Fig. 8.38 that the temperature increase in the liquid $(T_6 - T_1)$ has caused an increase in vapour pressure of the fluid from P_{s1} to P_{s6}. Thus if the system is allowed to accept any appreciable amount of heat either by poor lagging or continuous re-circulation of liquid then the available head will be markedly reduced.

Likewise, the tank pressurizing vapour will eventually reject so much heat to the liquid such that saturated conditions will prevail at some slightly higher temperature. The only head then remaining for flow will be the static liquid head h.

It is recommended that the operational margin h_m is sufficient to allow for this. On no account should point (4) lie in the vapour region of the curve.

Cryogenic pump suction pipes and discharge pipes are generally designed for relatively low velocities, particularly on pipes of 3 in. or less, to facilitate quick cool down. (2 to 8 ft/sec is common.) Friction losses are therefore usually a small factor except when little static head is available. The recommended pressure drop should not exceed 0·5 p.s.i. which is equivalent to approximately 0·5°C change in saturation temperature of liquid oxygen.

The heat inleak to pipelines varies with the type of insulation used but is not normally a major contributor to head loss.

The advantages of reducing the *required* NPSH for a pump are:

(1) It may be possible to use a cheaper and lighter vented tank.

(2) A reduction in structural costs will result from a ground level tank.

(3) No vaporizer unit is required for pressurizing the tank.

(4) Minimum start-up time for pumping.

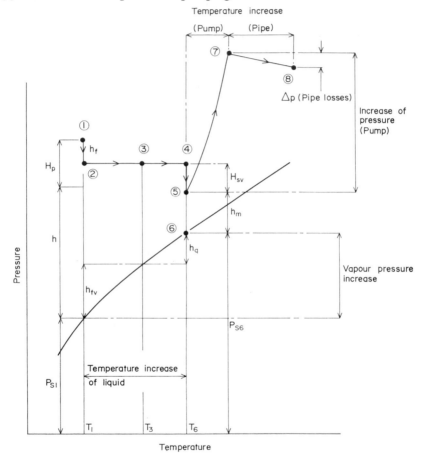

Fig. 8.38. Liquid vapour pressure.

The NPSH to a pump may be increased (depending on the requirement) by use of a small auxiliary boost pump, or by diverting some of the discharge into the pump suction line.

The increase in pressure of the liquid through the pump is illustrated by the line (5) to (7) in Fig. 8.38 Delivery system pressure loss and temperature rise of the fluid results in a terminal state (8).

2. *Cavitation*

The onset of cavitation for a range of pumps having the same rotor sizes but different speeds and profiles is demonstrated in Fig. 8.39. For a pump of low specific speed a reduction of the discharge head from A to B, with consequent flow increase, will cause cavitation at B, and for any flow and head combination to the right of the curve falling from B. For higher specific speed units, as the flow characteristics become more axial, the stable range of operation increases to points C, D, etc.

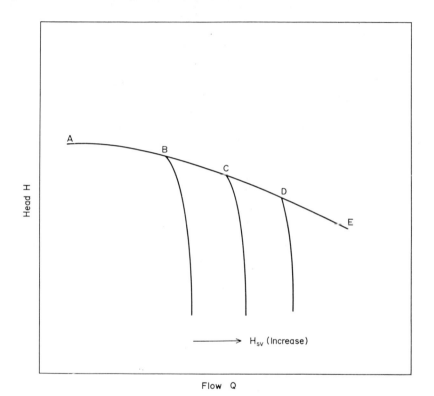

Fig. 8.39. Cavitation characteristics of low specific speed pumps.
AE—Normal characteristic curve.
B,C,D—Start of cavitation.

The controlling equations may be summarized as follows (referring to Figs. 8.34 and 8.37 for the various symbols):

$$H_{sv} = H_p + h - h_v = h_f + \frac{V_1^2}{2g} + \lambda \frac{V_{r1}^2}{2g} \qquad (8.27)$$

(absolute head at pump inlet) (friction loss plus dynamic head)

Therefore static head available

$$h_s = h - \left(h_f + \frac{V_1^2}{2g} + \lambda \frac{V_{r1}^2}{2g} \right) \qquad (8.28)$$

At cavitation

$$h_v = H_p + h_s$$

$$= H_p + \left[h - \left(h_f + \frac{V_1^2}{2g} + \lambda \frac{V_{r1}^2}{2g} \right) \right] \qquad (8.29)$$

Combining with equation 8.27

$$H_{sv} = \left(h_f + \frac{V_1^2}{2g} + \lambda \frac{V_{r1}^2}{2g} \right) \qquad (8.30)$$

Two useful variables for defining the cavitation regime are the Thomas Cavitation Parameter σ, given by

$$\sigma = \frac{H_{sv}}{H} \qquad (8.31)$$

and the Suction Specific Speed S, given by

$$S = \frac{NQ^{\frac{1}{2}}}{(H_{sv})^{\frac{3}{4}}} \qquad (8.34)$$

Although the normal cause of cavitation is the fact that the local liquid pressure falls below the saturation vapour pressure at the prevailing temperature, it can also be stimulated by a number of special effects. These include local intense acceleration effects, poor blade design and off-design running.

The vapour formed in cavitation is usually recompressed with consequent rapid recondensation in the pressure transforming region of the pump. It is this rapid compression, with the formation and destruction of the associated fluid wave, that determines the local erosion. (The alternating compression and suction at the material surface causes breakdown stresses.)

The formation of large quantities of vapour exerts a considerable effect on the developed head. The reason lies in the rather narrow passages between the blades—dictated by design requirements for the production of the given head. An appreciable amount of vapour in the impeller passages will quickly choke the flow. Machines of higher specific speeds, tending towards the axial type, have larger flow areas and are not so sensitive.

C. RECIPROCATING PUMP PERFORMANCE

Reciprocating pumps (Fig. 8.40) introduce pulsating flow which further aggravates the pressure drop on the suction side and in addition often necessitates the introduction of a pulsation damping vessel. This vessel adds to the complexity of the system and also increases the thermal mass to be cooled down and maintained cold. Reciprocating pumps generally have smaller flow rates than rotary types, and a compromise must be struck when determining the feed pipe diameter between the pressure drop and the heat leak into the system. Insulation is essential where the duty calls for pumping for short periods intermittently, e.g. when the pump forms part of a system for filling high-pressure gas cylinders. In these cases the suction piping and the pump cylinder are usually vacuum insulated or the pump is submerged in its own vacuum insulated liquid container. Reciprocating pumps are normally

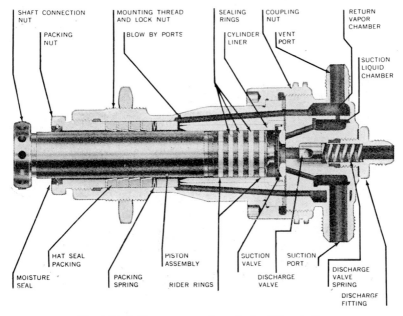

Fig. 8.40. A Plunger pump for cylinder filling duties.

restricted to static plant applications because of the size of the unbalanced loads which must be withstood. The layout of piping should be such that the gain in static head is greater than the loss of NPSH due to heat inleak and pressure drop. If this cannot be achieved due to local constraints well designed vapour separators must be installed near the pump section otherwise the loss of product and the wear on the pump will be prohibitive.

D. SYSTEM DESIGN AND OPERATION

Not until cool-down is complete should a pump be started. As premature starting causes mechanical wear and stress on pump components, the operator must be trained to recognize when satisfactory cool-down has been reached. The discharge side has little influence on the pump operation and the discharge valve can be opened into a warm system without affecting the pump other than imposing a back pressure. If a heavy pressure build-up is likely to occur, it is customary to operate in the following manner. Firstly, with all the valves systems closed, the pump is connected to the receiver and the receiver isolating valve opened. A non-return valve prevents the back flow of the liquid from the receiver to the storage vessel. The suction valve to the pump is then opened and the return valve to the storage container is opened slowly allowing the liquid to flow through the casing of the pump. When the pump has cooled to the appropriate temperature it is started and the discharge valve is opened slowly and the return valve to the tank is closed. If a flow-meter is required it is usual to instal this on the delivery side of the pump. It is vital that it be cooled down at the same time as the pump. Figure 8.33 shows the flow connections for this purpose.

1. *Insulation*

When a cryogenic liquid pump operates with a high load factor it should be insulated to minimize heat inleak. At start-up such a system will generate additional boil-off to cool the insulant. For intermittent pumping duties it is often preferable to work without insulation, and to allow the pump to frost over during the short period of use.

2. *Cooling Down Losses*

The trend in modern pump design is towards reduction in mass. In early installations centrifugal pumps were employed for low pressure discharge, turbo pumps for medium pressure discharge and reciprocators for very high pressures. The aim with turbo pumps was to obtain the required pressure with the minimum of stages thereby reducing the heat mass and weight of the pump. However, the turbo pump incorporates seals, particularly between suction and delivery, and the quality of these seals is vital for good operation. The potentially high wear rate arising from the fact that the cryogenic liquid is the only lubricant available poses a considerable challenge for the designer. He must provide for the maintenance of the correct clearances throughout the life of the pump and ensure correct alignment of the pump components when initially assembled, and after maintenance. This alignment must, of course, be maintained over the full temperature range of the pump.

The centrifugal pump is preferred because there are no internal low clearance seals. Recent development has been directed towards multi-stage centrifugals in place of the turbo units despite their more complicated assembly, higher cost and weight. The longer the shaft to accommodate the increased number of impellers the more bearings are required within the pump. Operationally the centrifugal pump has the added advantage over the turbo (or semi-positive type) in that the valve on the delivery side can be closed when running without building up severe hydraulic pressures. This avoids the need for relief devices which themselves require maintenance and are a likely cause of leakage. Furthermore if a centrifugal pump is called upon to operate under gas conditions at start-up, or during operation, severe gas pressures are not generated, although the bearings, especially if they are situated within the pump, may suffer from lack of lubrication and cooling.

E. BEARING DESIGN

Journal bearings have been the subject of considerable development (Hargreaves, 1960), and in early pump designs most bearings were of the plain or sleeve type. The success of these designs in service depended essentially on the selection of materials to avoid the effects of differential contraction on working clearances. As far as possible these materials should not only be compatible with the fluid being pumped but should withstand the effects of sustained relative motion. Owing to the marginal nature of the lubrication process, when using cryogenic liquids, and the practical likelihood of periods when the lubricant-coolant is not present, these component materials must be capable of tolerating considerable abuse for short periods. Usually the bearing shell is of bronze and the shaft of stainless steel, thus conforming to conventional plain bearing design but incorporating provision for the low temperature effects. The bearing shell and the housing assembly must accommodate the differential contraction without leading to dimensional difficulties or distortion, Otherwise seizure or mechanical failure could result. Modern bearings incorporate some form of non-metallic, low-friction, solid such as PTFE, thus enabling heavier bearing loads to be sustained while operating on low viscosity liquids such as liquid oxygen or liquid nitrogen. However, the non-metallic components have vastly different values of thermal-conductivity and coefficient of contraction. Bearing development has therefore concentrated on accommodating these non-metallic components in metallic housings in ways which minimize these incompatibilities. A general treatise on the subject is given by Hargreaves (1960).

1. Lubrication

A frequent cause of bearing failure is the blockage of the cryogenic liquid supply system with debris from the bearing or other extraneous matter. The use of filters is not necessarily a solution since these filters in turn become blocked and cause loss of liquid supply. As an alternative approach, ball or roller bearings have now been introduced. The cage designs and the materials employed allow such bearings to be used in cryogenic environments whilst retaining ball bearing steels to obtain the required surface hardness. The clearances must be designed to allow for the dimensional change during cooling down. The cage or separator materials are subject to sliding as opposed to rolling friction and so require particular attention for lubrication.

Here again P.T.F.E. has been used to advantage but the same problems of differential contraction must be allowed for.

One often neglected factor with cryogenic pumps is the damage caused by corrosion due to melting ice, or to internal condensation of moisture, at times when pumps are inoperative. Chemical attack in the region of bearings or seals will render them useless. It is therefore essential that dismantling and reassembly is carried out under controlled environmental conditions.

In a reciprocating pump there are areas of sliding contact, exposed to the cryogenic liquid, between the plunger and the bore. In early designs a solid plunger with a close fitting clearance was adopted because of the lack of knowledge of ring designs and materials. However since the satisfactory performance of the pump depends on the maintenance of these sealing surfaces, and therefore on the absence of wear, much attention has been given to aiding the lubrication by the introduction of non-metallic materials such as carbon or P.T.F.E. Correct alignment of the motion work is vital to the maintenance of these clearances. Recent designs incorporate a plunger with a hard surface finish, accurately ground to close tolerances, and cylinder liners fabricated from P.T.F.E. impregnated bronze, also machined *in situ* to obtain the close working clearance with the plunger. Since all the machining and fitting is carried out at ambient temperatures the "warm" dimensions and materials of the assembly must be such that the correct working clearances are obtained when the pump is cooled down. It is clear that the different materials of the plunger and cylinder results in varying clearances between them at all stages of cooling down and warming up. If a pump should be started before it is fully cooled down considerable wear will take place and greatly reduce the working life of the components.

Reciprocating pumps also include valves, and these suffer not only from the wear in the guides or other parts of the assembly subjected to sliding friction, but from the hammer of the valve on its seat. Since the choice of materials is limited by the cryogenic conditions, and in some cases by chemical compatability, it is difficult to produce a compromise which will withstand the impact loading without failure. Furthermore the valve port imposes a pressure drop in the suction line which not only affects the volumetric efficiency but can cause gas accumulation within the pump itself. One object of the early pump designs was to permit this gas to escape without impeding the free flow of supply liquid. It was, therefore, usual to find a header tank close to the suction side, communicating with a series of ports in the cylinder which were exposed by the piston. An improved solution employed a double piston system. The extra piston was used to pressurize the suction liquid, to condense the gas in the cylinder space, and to improve the volumetric efficiency. This, however, increased mechanical complexity and although successful designs are now in operation the trend has been towards dispensing with the suction ports in favour of conventional suction valves and discharge valves. Where suction valves are used it is important that the maximum area for flow is provided in order to reduce the resistance across the seat.

F. ATMOSPHERIC SEALS

In pumps by far the most difficult component to maintain is the seal between the cryogenic fluid and the atmosphere. This seal has two functions:

(1) To prevent loss of liquid and loss of pressure.
(2) To prevent the ingress of extraneous matter especially lubricating oil migrating along the shaft from the drive components.

In the case of centrifugal pumps the drive system is usually a direct coupled electric motor and the lubrication of the motor bearings and their satisfactory life is related to the efficient functioning of the cryogenic pump seal. If the seal should fail the leaking cryogenic liquid will cause a rapid reduction in

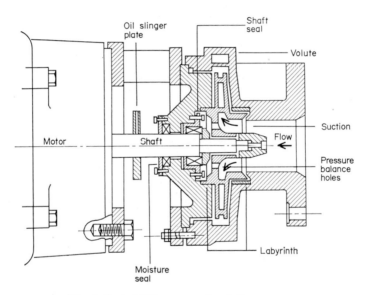

Fig. 8.41. A 7,200 r.p.m. tanker delivery pump.

temperature of the motor casing particularly in the region of the bearings thereby causing the hydrocarbon lubricant to solidify leading to rapid bearing failure. In the case of reciprocating pumps the drive is through a crank mechanism which usually runs in an oil bath, and in some cases, because the reciprocating pump works at low speed, a gear box similarly lubricated is also necessary. Particular attention must be paid therefore to maintaining the lubricant in the crank case and gear box in the right condition to ensure that the lubrication is effective. Not only will lubrication be impaired by the leakage of cryogenic liquid but it can also be affected by the conduction along the shaft of the "cold" from the low temperature components if the pump is maintained in a cooled-down condition while standing. When the pump is operating the heat generated in the motion work will establish the correct temperature gradient such that the frost line is about mid-way between the motion work and the pump. In all the designs the heat break between these two sub-assemblies is established by correct design incorporating low conductivity materials and transition pieces of the lowest cross section, and the longest length, with provision for the introduction of an ambient heat leak. A typical assembly is shown in Fig. 8.41.

Two design philosophies have been explored for seals on rotating shafts:

(1) the seal operates at cryogenic liquid temperatures and steps are taken to prevent atmospheric impurities condensing or solidifying on the seal;

(2) the seal is designed to operate at atmospheric or near atmospheric temperatures and a heat break is imposed between the seal and the cryogenic liquid.

In early designs the latter course was adopted and the rotary seal was accommodated at the outer end of an extension from the pump casing. Again the heat-break provided by the extension was not necessarily effective when the pump was stationary, and then further complication was introduced by the addition of an external source of heat, usually in the form of an electric heating tape wrapped around the seal housing. An electrical interlock is necessary to switch off the external heat when the pump is started otherwise the combined heat generated often causes overheating of the seal and its rapid failure. Furthermore this extension results in a longer overhang of the impeller from the motor bearing, thereby imposing even more severe requirements on the alignment of the assembly and often the need for further bearings.

In reciprocating pumps, most of which are single-acting, the plunger itself provides the seal between the cryogenic liquid and the atmosphere and accommodates the required temperature gradient. The seal must prevent the ingress of extraneous matter into the pump region. Care is therefore taken in the design (A typical assembly is shown Fig. 8.40) to ensure:

(i) that the plunger or piston rod seals are well aligned and tight;

(ii) that the seals on the oil lubricated drive are in good order. Since these are conventional, their design is based on well established principles;

(iii) an adequate distance between the pump and the drive, i.e. greater than the length of stroke to ensure that no part of the piston rod enters both the oil-lubricated region and the cryogenic pump;

(iv) a free venting area between the drive and the pump to avoid the mixing of vapours or liquids from both sources;

(v) an oil slinger or garter on the rod to prevent migration of oil along the rod. A major cause of seal failure is the damage sustained when ice formed on the surface of the rod enters the seal on start-up. This is most likely to occur when a pump is on intermittent operation or is kept cold for stand-by purposes.

Until the development of lip or mechanical seals for cryogenic operation packings such as asbestos rope impregnated with graphite were used. However this type of packing was insufficiently resilient to take up the differential contraction when the pump was repeatedly cooled down. The resulting leakage usually caused ice to form which in turn destroyed the seal. Adjusting the seal on each cool-down is operationally practical only when this occurs infrequently.

1. *Face Seals*

Face seals or mechanical rubbing seals do not operate satisfactorily when extraneous matter, particularly ice or water, is present. Success has been achieved by positioning the seal close to the cryogenic liquid with an additional restriction to prevent the back diffusion of atmospheric air and its impurities (as shown in Fig. 8.41). The design and installation of the mechanical seal must not only be capable of meeting normal operational loads but also of withstanding the build-up of ice between and around the mating faces. For instance when partial thawing occurs between one operation and the next ice forms in such a manner that it welds the rotating and stationary faces of the seal together. If the pump is then started it must be expected that mechanical failure of the bellows will occur, or sometimes both faces of the seal will be irreparably damaged.

Extensive development of face seals for cryogenic duties has focused attention on the following points:

(i) the materials in contact must be selected to give the best wear life;
(ii) the faces must be plain and of the correct surface finish;
(iii) the interface loading, from bellows and/or spring tension (adjusted on assembly) and differential pressure, must ensure the correct compromise between sealing and frictional forces at the working condition;
(iv) the alignment and concentricity must be correct;
(v) the faces must be parallel and normal to the axis of rotation;
(vi) the assembly must be free from constraints due to housing or differential contraction;
(vii) vibrations, self induced or externally imposed, must be damped to a safe level.

H. HIGH SPEED LOW MASS PUMPS

Faced with all these considerations the designer has tried to reduce the heat-mass of the pump while at the same time avoiding the need for bearings

19

within the pump. One way is to reduce the number of stages of the pump and make the distance between the pump and the motion work such that an overhung design will operate satisfactorily. This approach places a greater emphasis on the design of the motor bearing and the alignment of the pump and motor. To achieve the pressures required whilst reducing the number of stages, preferably to one, higher rotation speeds are required since the diameter of the impeller is governed by the flow rate. These higher speeds can be obtained by introducing gear-boxes, belt-drives or high frequency motors. Belt drives and Gear-boxes involve added cost, maintenance and space requirements, whilst belt-drives introduce side thrusts which must be accommodated by the drive system. It is therefore preferable to use a high frequency drive and a typical modern pump of this type is shown in Fig. 8.42.

Fig. 8.42. High speed pump exploded view.

A high speed system calls for careful bearing design, correct storage and highly skilled maintenance in a controlled environment. NPSH difficulties at the entry to the pump are also aggravated by the high speed. The risk of cavitation is normally reduced by the introduction of the inducer which comprises an archimedian spiral generating a small pressure rise upstream of the main impeller. Out of balance forces must be kept within the strength of the rotating components, and the high rotational speeds demand careful balancing after each assembly operation. In particular the adequacy of the attachment of the impeller to the shaft when the pump is at operating temperature is of paramount importance.

The usual combination of an aluminium impeller fitted to a stainless steel shaft provides a good compromise between the requirements of strength, low mass and low thermal conductivity at low temperatures. The stainless steel shaft by reason of its good tensile properties and low thermal conductivity provides the best heat break. The assembly is normally designed to provide a good margin below the whirling speed. Conversely if it is the intention to pass through the first critical speed the starting procedure must be carefully controlled.

1. Installation Principles

Working clearances and balance are affected by alignment, and it is therefore essential that no distortion occurs due to cool-down stresses. These stresses may arise within the assembly or be imposed from outside by piping or valves. High-speed pumps with light casings are particularly sensitive to these effects and it is generally necessary to make pipe connections to the pump through bellows. In general it is more convenient to use gear-boxes or belt drives for stationary pumps while for vehicle applications high frequency motors may be used, the special alternators being driven through gearing by the engine.

Such high speed units are particularly convenient for mobile duties where cooling down time and pay load considerations influence the overall economics of the operation.

H. PUMPING SYSTEMS SUBMERGED IN LIQUID

A number of significant advantages may be gained by submerging the pump in liquid so that the pump is always cold and ready for instant operation. Moreover liquid loss at cool-down is avoided. Since the pump remains cold stresses arising from differential contraction are minimized. The pump in the submerged condition has an open suction and the NPSH requirements are therefore reduced to the losses at entry to the pump itself. In practice the head required by the pump is a matter of inches water-gauge, and this combined with the height of the suction from the base of the tank represents a level of liquid in the tank that cannot be discharged. This is a disadvantage in mobile duties since it represents a loss of pay load.

For cryogenic liquids other than oxygen it is often possible to submerge the electric drive as well as the pump, as shown in Fig. 8.43. The following additional advantages arise:

(i) Absence of shaft seals between the cryogenic system and atmosphere.

(ii) Long drive shafts are avoided.

(iii) Mechanical and bearing design is less complicated.

Fig. 8.43. Fully submerged pump and drive.

(iv) Resistance losses in electric motor are reduced by the effect of low temperature on the material.

(v) The assembly can be withdrawn easily through the top of the storage vessel. This is particularly advantageous with underground storage or ship tanks.

(vi) Low weight and increased pay load or chassis space.

This system, being fully sealed, is particularly well suited to the pumping of combustible or toxic substances, e.g. liquid hydrogen, LNG or liquid methane. The small ohmic heat input from the electric motor is usually more than offset by the savings in heat inleak which would otherwise occur along the pump shaft and housing.

Although submerged pumps with remote electric drive are in service for liquid oxygen they lose much of their appeal owing to complexity, weight, and maintenance difficulties associated with the long drive shaft and rotating seal. Shaft flexibility inevitably introduces problems due to whirling which require the incorporation of intermediate bearings. These bearings in turn require a continuous supply of the cryogenic liquid for cooling and lubrication. It is very difficult to ensure that under no circumstances will the pump be started or run without this supply being effective. Hence failures readily occur.

The submerged pump installed in its own small tank, permanently coupled to the main liquid storage, is attractive for stand-by duties and gives near-instantaneous operation for varying periods at any time.

I. SAFETY

For hazard-free service, close attention must be given to safety from the initial design, through all stages of manufacture and installation, to all operating procedures. This subject is dealt with in detail in Chapter 11, and a useful guide to good practice has been issued by the British Cryogenics Council (1970). The following points are worthy of special study in relation to cryogenic pumping systems and associated equipment.

A standard of cleanliness must be laid down and enforced to ensure that no extraneous matter above the specified quantity, size and composition can be assembled into the pump, or enter it, before it is fully sealed in its installed position. To achieve this aim consistently it is essential that the correct facilities and tools are provided, and used, and that all skilled and unskilled labour who handle the pump or its components, are adequately trained for the part they play. Disregard for the rules by any one person may result in a hazardous situation. It is therefore essential that careful inspection is carried out at vital stages in the sequence, from design to operation.

It is advisable to arrange for packaging in sealed containers, preferably transparent, so that the storemen have no need to expose the parts to atmospheric contamination in order to identify them, and so that clean degreased items or whole assemblies will not deteriorate before use.

Clearances within the pump must be in accordance with design when warm as there is no other satisfactory way of ensuring that the running clearances when cold are correct. Inspection must check that alignment, etc., is correct and the installation is to specification and does not include strains through bad fitting.

To ensure cleanliness it is customary to adopt the procedure that, when a pump or seal failure occurs, the complete assembly is replaced by a fully tested, sealed unit, and that maintenance and repair is carried out off-site under controlled conditions by the approved staff.

The controls on the drive system must enable remedial action to be taken manually or automatically in the event of difficulties arising. Where pressure build-up from boiling cryogenic liquid can occur fully automatic safeguards must be installed and periodically inspected and checked.

In addition to all these precautions the pumping systems must fail safe in the event of the unexpected happening. Broadly, the following protective measures must be incorporated in the system and its drive:

(i) Overload protection on the drive system, which automatically shuts down the system.
(ii) "Loss of prime" fusing, usually by the load dropping below a predetermined value, prevents the continued operation of the pump under conditions which can lead to mechanical damage. An over-ride with an automatic time limit is necessary for start-up.
(iii) Emergency stop buttons placed at strategic points.
(iv) An isolating valve, automatic or manual, for quick shut-off in the event of a sudden loss of liquid pressure downstream of the tank outlet, which indicates a failure of the pump or joints.

J. MATERIALS OF CONSTRUCTION

When liquid oxygen is to be pumped the materials of construction must be chosen not only for their low temperature properties but also to minimize the combustion hazard. This subject is dealt with at length in the British Cryogenic Council Safety Manual (1970). Broadly speaking the properties which matter are the heat of oxidation and the thermal conductivity. Copper alloys are good, ferrous alloys bad and aluminium alloys are tolerable under certain circumstances.

IV. References

Beasley, S. A. and Halford, P. (1965). International Advances in Cryogenic Engineering, Vol. 10, pp. 22–39. Plenum Press, New York.

Becker, H., Doll, R. and Eden, F. X. (1968). Proc. Second Int. Engineering Conference (ICEC2) pp. 9–12. Brighton.

British Cryogenics Council (1970). "Cryogenics Safety Manual—A Guide to Good Practice." *Inst. Chem. Engrs.* B.C.C., London.

Claude, G. (1902). *Proc. Acad. Sci. Paris* **134**, 1568.

Collins, S. C. (1947). Proceedings of the 1956 Cryogenic Engineering Conference, Vol. 2. Paper A2, pp. 8–11.

Eck, B. (1953). "Ventilatoren". Springer–Verlag, Berlin.

Kapitza, P. (1939). *J. Phys. U.S.S.R.* **1**, 7–28.

Kneuer, R. and Turnwald, E. (1968). Proc. 2nd. Int. Cryogenic Eng. Conference (ICEC2), pp. 12–16. Brighton.

Lady, E. R. (1957). *British Chem. Eng.* **2**, 128–130.

Land, M. L. (1957). Proceedings of the 1956 Cryogenic Engineering Conference, Vol. 2, Paper G2, pp. 250–260. N.B.S., Boulder.

Long, H. M. and Simon, F. E. (1954). *Applied Sci. Res.* **4**, 237.

Mann, A., Sixsmith, H., Wilson, W. A. and Birmingham, B. W. (1963). Proceedings of the 1962 Cryogenic Engineering Conference, Vol. 8. Paper D6, pp. 228–235. N.B.S., Boulder.

Morain, W. A. (1967). Proceedings of the 1956 Cryogenic Engineering Conference, Vol. G2, **12**, 585–594. N.B.S., Boulder.

Morain, W. A. and Holmes, J. S. (1963). Proceedings of the 1962 Cryogenic Engineering Conference, Vol. 8. Paper D6, pp. 228–235. N.B.S., Boulder.

Ruhemann, M. (1965). *In* "International Gas Turbine Engineering Handbook" (La Fleur, ed.).

Smith, Jr., J. L. (1969). "Advances in Cryogenic Engineering", Vol. 12, pp. 595–601.

Sixsmith, H. (1959). "The Theory and Design of Gas-lubricated bearings of high stability", Proc. 1st. Int. Symposium on Gas-lubricated bearings, ACR-49. Office of Naval Research, Washington.

Strass, W. (1959). *Kaltetechnik*. **11**, 5.

Voynitkin, I. P. (1965). *In* "Plant and Machinery for the Separation of Air by Low Temperature Methods" (English Edition) (Ruhemann, ed.). Pergamon Press, London.

Chapter 9

Superconductivity

R. HANCOX

Culham Laboratory, Culham, Abingdon, England

J. A. CATTERALL

National Physical Laboratory, Teddington, England

I. Introduction

A. OCCURENCE OF SUPERCONDUCTIVITY

Soon after Kamerlingh Onnes succeeded in liquefying helium in 1908, he began investigating the electrical conductivity of metals at very low temperatures, and the effect of impurities on their resistivity. His experiments on mercury produced a completely unforeseen result, since the resistance did not decrease smoothly as expected but dropped suddenly to zero at a temperature of 4·1°K, in the manner shown in Fig. 9.1. The condition of the mercury below 4·1°K was described, for obvious reasons, as "superconducting" and 4·1°K was called the "critical temperature" of mercury (T_c). Further examination showed that superconductivity could be destroyed by the application of a sufficiently strong magnetic field, when the superconductor returned to its normal resistive state. The minimum field required to accomplish this was the "critical field" H_c.

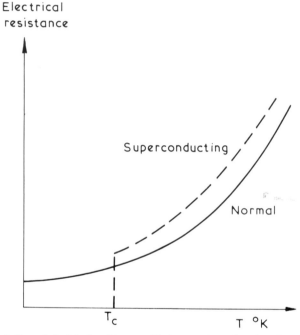

Fig. 9.1. Variation of electrical resistance with temperature, showing the critical temperature T_c for a superconductor.

Since the time of Kamerlingh Onnes many metals, alloys and compounds have been found to behave like mercury, and about a thousand materials are now known with critical temperatures ranging from 0·01°K to 20·9°K (Roberts, 1964). In the past few years there has been a growing interest among scientists and engineers in the phenomenon of superconductivity, owing to the possibility of constructing various devices using superconductors in which there is no consumption of electrical power because there is no resistance. Some of these potential or actual uses are described later in this chapter.

1. *Elements*

Thirty-five elements have been found to be superconductors, and Table 9.I lists the values of some critical temperatures and fields. This table also

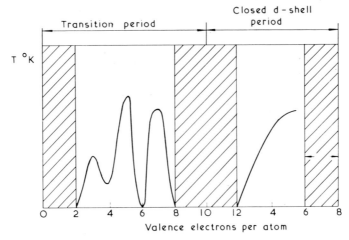

Fig. 9.2. Variation of the critical temperature with valence electrons per atom (*e/a*) in the Periodic Table (Matthias, 1957).

TABLE 9.I. Properties of Superconducting Elements

Periodic table group number	Element	$T_c °K$	Type	$H_c(0)$ (gauss) $0°K$	H_{c1} (gauss) $4·2°K$	H_{c2} (gauss) $4·2°K$
3A	La(α)	4·90	1			
	La(β)	6·10		1,600		
4A	Ti	0·39		100		
	Zr	0·55		47		
	Th	1·37		150		
5A	V	5·41	2		430	820
	Nb	9·2	2		1,390	2,680
	Ta	4·48	1	805		
6A	Mo	0·92		98		
	W	0·01				
	U	0·68				
7A	Tc	7·75	2		1,180	1,450
	Re	1·70		193		
8A	Ru	0·49	1	66		
	Os	0·66	1	65		
	Ir	0·14				
2B	Zn	0·85	1	53		
	Cd	0·55	1	29		
	Hg(α)	4·15	1	415		
	Hg(β)	3·95		339		
3B	Al	1·19	1	102		
	Ga	1·09	1	55		
	In	3·4	1	285		
	Tl	2·39	1	170		
4B	Sn	3·72	1	305		
	Pb	7·19	1	803		

distinguishes between type I and type II superconductors (defined more fully in section I.D). Figure 9.2 shows that there is a variation of the critical temperature in a regular manner in the Periodic Table in terms of valence electrons per atom (e/a) of the element. Three peaks occur at 3, 5 and 7, and the existence of such a relationship, which was first pointed out by Matthias (1957), is qualitatively explained by the microscopic theory of superconductivity.

2. *Solid-Solution Alloys*

When a superconducting element is converted to an alloy by the addition of a solute, the values of the critical temperature and field are changed. The critical temperature is related to the number of valence electrons per atom

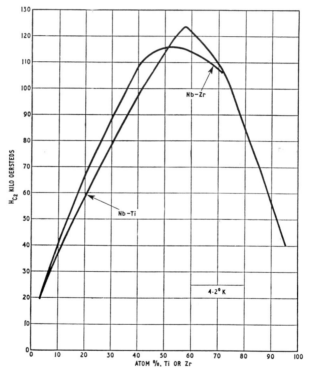

Fig. 9.3. Variation of the upper critical field H_{c2} with composition in niobium titanium and niobium zirconium alloys at $4.2°$K (Coffey *et al.*, 1965).

in a manner similar to that above for elements, but the peaks in the critical temperature are established more accurately. Hulm and Blaugher (1961) examined solid solution alloys of neighbouring body-centred-cubic elements in the same row of the Periodic Table, and found peaks at values of 4.7 and 6.4 valence electrons per atom. Other factors can affect the critical temperature in solid solutions, however. Ferromagnetic solutes depress the critical temperature extremely rapidly, ordering can change the critical temperature and so also can precipitation. Precipitation of a second phase causes complex effects on the critical temperature, particularly if the second phase is also superconducting. The measured critical temperature will depend on the size and distribution of the second phase, and the quantity of solute remaining in the matrix.

The upper critical field H_{c2} of type II superconductors can be increased by alloying, and type I superconductors can be converted to type II. The variation of H_{c2} in the niobium–titanium and niobium–zirconium alloy systems is shown for example in Fig. 9.3.

3. Metallic Compounds

The various compounds which are known to be superconducting include a large number of different crystal structures. However one particular crystal structure, the A15 or "β–W" structure shown in Fig. 9.4, is of particular

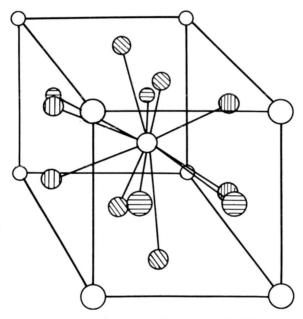

Fig. 9.4. The A15 or β–W crystal structure, for compounds of the composition A_3B. B atoms unshaded, A atoms shaded.

interest since the highest known superconducting transition temperatures of 18 to 20°K are found among compounds such as niobium tin (Nb_3Sn) with this type of lattice. The interstitial compounds of the sodium chloride type, such as niobium nitride and niobium carbide also have high transition temperatures, and other notable structures are the "sigma-phase" and the "alpha manganese" lattices. Table 9.II lists values of the critical temperature for various compounds, and the effect of the number of valence electrons per atom is again evident as illustrated in Fig. 9.5.

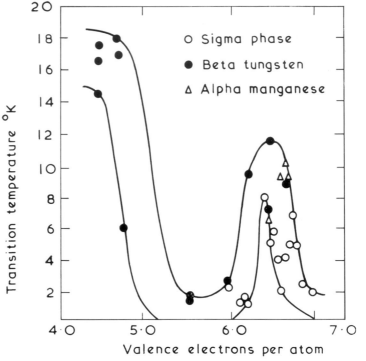

Fig. 9.5. The variation of the critical temperature with valence electrons per atom for three different structures (Blaugher and Hulm, 1961).

4. *Semi-conductors and Other Compounds*

A number of semi-conductors have recently been found to be superconducting although with fairly low critical temperatures. For example, germanium–tellurium has a critical temperature of about $0.25°K$. The theory of such behaviour has been discussed by Cohen (1964). The sodium tungsten bronzes (Na_xWO_3) have a critical temperature of $0.5°K$ (Raub *et al*., 1964), and several intercalation compounds of graphite with alkali metals (e.g. C_8K) are also superconducting (Hannay *et al*., 1965).

B. PHYSICAL PROPERTIES OF SUPERCONDUCTORS

1. *The Critical Temperature*

Below the critical temperature, recent measurements have placed the resistance of a superconductor as less than 10^{-21} ohm cm compared with a value

TABLE 9.II. The Critical Temperatures Shown by Various Crystal Structures

Structure type	Compound	$T_c°K$
Beta–tungsten	Nb_3Sn	18·45
	V_3Ga	16·8
	Nb_3Ga	14·5
	Nb_3Al	17·1
	V_3Si	17·0
Sigma	Nb_5Pt_3	4·2
	Nb_3Ir_2	9·8
	Cr_2Ru	2·0
Sodium chloride	NbN	15·5
	MoC	11·3
	NbC	10·5
	ZrN	10·0
Alpha–manganese	Re_3Mo	9·8
	Re_3W	9·0
	Re_3Ta	6·8
B.C. Tetragonal	Mo_3P	5·3

of 10^{-9} ohm cm for high purity copper at $4·2°K$. In other words, the resistance is zero as far as can be ascertained by experiment. The value of the critical temperature in a superconducting material is extremely dependent on the chemical purity of the specimen. Traces of impurity, particularly ferro-magnetic materials, can reduce the critical temperature or even suppress superconductivity completely. Molybdenum, for example, was found not to be a superconductor until all traces of iron had been removed.

One method of measuring the critical temperature is to pass a very small current through a superconducting wire, allow the temperature to rise and record the temperature at which a detectable resistance first appears. The value can only be defined accurately if the transition to the normal state occurs over a narrow temperature interval as in Fig. 9.6. In practice this interval may be less than $0·01°K$, but poor specimens containing metallurgical defects such as those produced by cold-working, give much broader transitions.

2. The Critical Field

Apart from the absence of electrical resistance, there is a second and equally fundamental property of a superconductor. If a specimen is placed in a magnetic field at room temperature, flux will penetrate the material and the internal and external flux densities will be identical. If the same specimen is now cooled in the presence of the magnetic field the flux is expelled from the interior of the specimen as the temperature passes through the critical temperature. This exclusion of flux is very nearly complete, but there is a thin

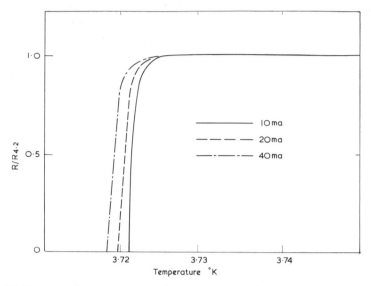

Fig. 9.6. Fraction of resistance restored with temperature in pure tin, as a function of the measuring current (De Haas and Voogd, 1931).

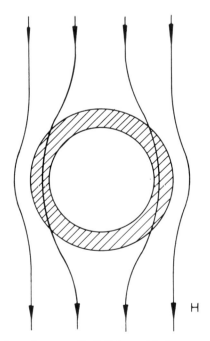

Fig. 9.7. The penetration of a magnetic field in an ideal superconductor (not to scale).

penetration layer of the order of 10^{-5} to 10^{-6} cm at the surface, as shown on an exaggerated scale in Fig. 9.7. This behaviour of superconductors towards magnetic fields is known after its discoverer as the "Meissner effect" and it represents almost perfect diamagnetism, since in the well-known relation

$$B = H + 4\pi M \qquad (9.1)$$

between the flux density B inside a material, the external field H, and the

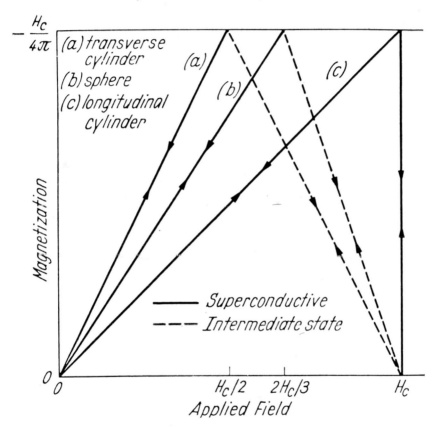

Fig. 9.8. Variation of the magnetization with magnetic field for various superconducting ellipsoids showing the "intermediate state" field region.

magnetization M, the magnetic susceptibility $k\,(= M/H)$ is equal to $-\frac{1}{4}\pi$ when the internal flux density is zero. (C.G.S. Electro-magnetic units are used for the numbered equations.) The exclusion of the flux is accomplished by circulating superconducting currents which are induced in the surface layer of the specimen.

The magnetic behaviour of superconductors divides materials into two classes or types, known as type I and type II. The difference between these types is discussed later but, briefly, type I is characterized by a single critical field H_c whilst type II is characterized by a lower (H_{c1}) and an upper (H_{c2}) critical field.

If the external magnetic field is too large, the screening currents will be unable to prevent flux penetration, and the material will return to the normal or non-superconducting condition. The minimum field required for penetration to occur is the critical field which is found to depend upon the temperature through the approximate relation

$$H_c \simeq H_0 \left\{ 1 - \left(\frac{T}{T_c} \right)^2 \right\} \tag{9.2}$$

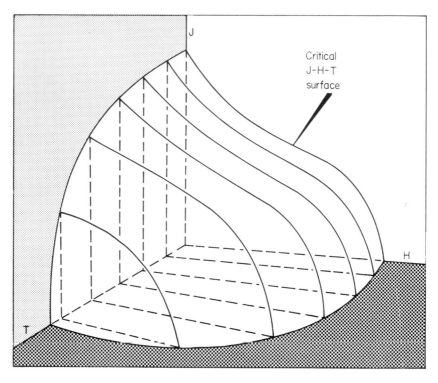

Fig. 9.9. Superconducting region bounded by a critical J–H–T surface. The superconducting region is below the surface, the resistive region above (Benz, 1965).

In a type I material the magnetization at the critical field falls from its large value $-H_c/4\pi$ to the very small value of the normal metal, and gives characteristic triangular magnetization curves, as shown in Fig. 9.8. The drop in magnetization at the critical field is abrupt only in favourable circumstances such as in a long thin cylinder parallel to the field. For other shapes or configurations the "demagnetization factor" causes a more gradual flux penetration. The region shown by the dotted lines in Fig. 9.8 is known as the "intermediate state", since the material here contains both superconducting and normal phases existing together.

3. The Critical Current

It is found that there is an upper limit to the current which can be passed through a type I or type II superconductor without driving the material back into its normal condition. There is a close relation between this maximum or "critical current" (I_c) and the magnetic properties. A current I in a round wire, for example, generates a magnetic field at the surface equal to $2I/10R$ gauss, and in a type I material when this field equals the critical field, superconductivity is destroyed. This behaviour is known as the "Silsbee effect" and the current required to accomplish it is the critical current. The more complex situation in a type II material will be considered in Section III. Since the critical field decreases with increase in temperature, so the critical current also decreases with increase in temperature. In fact, the critical current, the critical field and the critical temperature are all related in the manner shown in Fig. 9.9. Between the origin and the surface of the figure the material is fully superconducting, but outside the surface resistance appears.

4. The Penetration Depth

The penetration of a magnetic field into a superconductor has already been illustrated in Fig. 9.7. The strength of the field decreases inwards from the surface according to a relation

$$H = H_s \exp(-x/\lambda) \qquad (9.3)$$

where H_s is the surface field, and λ is a characteristic length known as the "penetration depth". The value of this penetration depth has been measured in thin films, wires and colloids. In tin, for example, it is $5 \cdot 0 \times 10^{-6}$ cm.

5. The Specific Heat

In the absence of a magnetic field the superconducting transition involves no latent heat change but it does produce a discontinuity in the specific heat as shown in Fig. 9.10. A change of this type is called a phase change of the second kind. In the presence of a magnetic field a latent heat of transition appears and the change then becomes one of the first kind.

6. Mechanical Effects

The critical temperature of superconductors can be changed slightly by tension, and in tin, indium and mercury it is increased (Sizoo and Onnes, 1925; Kan et al., 1948). Pressure has the opposite effect, and very high pressures (20,000 atm) can cause bismuth, which is not normally a superconductor, to become superconducting with a critical temperature of 7°K

(Chester and Jones, 1953). Plastic deformation of type II superconductors can radically alter their electrical and magnetic properties. The critical field of tin and tantalum is increased by tension. Below the critical temperature there is a change in the elastic constants of tin (Landauer, 1954) and of lead, niobium and tantalum (Alers and Waldorf, 1962). Some early X-ray work on lead revealed no apparent change in the lattice parameter, although thermodynamic theory predicts discontinuities in the thermal expansion.

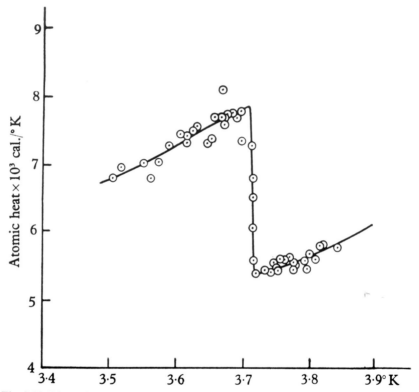

Fig. 9.10. Discontinuity in the atomic heat of tin at the critical temperature (Keesom and Kok, 1932).

7. Thermodynamic Relationships

At the critical field of a type I superconductor the free energies of the normal and superconducting phases are equal, i.e.

$$G_s(H_c) = G_n(H_c) \tag{9.4}$$

The free energy of a superconductor in a magnetic field which is less than the critical field is greater than the free energy in zero field by the amount of energy required to keep out the magnetic flux from the interior. This energy of flux exclusion is equal to the area under the magnetization curve. Consequently

$$G_s(H) = G_s(0) - \int_0^H M \, . \, dH \tag{9.5}$$

It has already been shown that $M = -H/4\pi$, so that

$$G_s(H) = G_s(0) + \frac{H^2}{8\pi} \tag{9.6}$$

The normal metal is only very weakly magnetic and $G_n(H_c) \simeq G_n$ and is independent of field. Writing G_s for $G_s(0)$ and using equations 9.5 and 9.6 we find the free-energy difference between the normal and superconducting phases per unit volume to be

$$G_n - G_s = \frac{H_c^2}{8\pi} \tag{9.7}$$

Differentiating equation (9.7) with respect to temperature gives the entropy difference

$$S_n - S_s = \frac{-H_c}{4\pi} \cdot \frac{\partial H_c}{\partial T} \tag{9.8}$$

Since $\partial H_c/\partial T$ is always negative, S_s is always less than S_n. The superconducting phase is consequently a state of higher order.

Differentiating (9.8) with respect to temperature gives the difference in specific heat per unit volume

$$C_s - C_n = \frac{H_c T}{4\pi} \cdot \frac{\partial^2 H_c}{\partial T^2} + \frac{T}{4\pi} \cdot \left\{ \frac{\partial H_c}{\partial T} \right\}^2 \tag{9.9}$$

C. THEORY OF SUPERCONDUCTIVITY

The present "microscopic" theory of superconductivity was evolved in the 1950s and culminated in 1957 with the "BCS" theory by Bardeen, Cooper and Schrieffer (1957). Before this time, the most successful approaches had been through phenomenological theories, and the best-known of these are the two-fluid model of Gorter and Casimir (1934), to account for the thermal properties, and the London theory (1950), to account for the electromagnetic properties.

1. *The Two-Fluid Model*

Essentially the two-fluid model proposes that below the transition temperature conduction electrons are divided between two groups of energy levels separated by an energy gap. A fraction $1-x$ of the electrons occupies the lower group of levels and is described as "condensed" or superconducting, and the residue x remains "uncondensed" or normal. The quantity $1x-$

varies from unity at zero temperature (when all electrons are superconducting) to zero at the transition temperature. Zero electrical resistance would, of course, be observed at all temperatures below the transition temperature since the superconducting electrons would "short out" the normal ones.

The condensation energy of the superconducting phase relative to the normal metal is $-\beta(1-x)$ where $\beta = V_m H_0^2/8\pi$ and V_m is the specific volume.

If f_n is the free energy of the normal phase and f_c that of the condensed phase, the free energy f per unit volume is

$$f = x \cdot f_n + (1-x) \cdot f_c \tag{9.10}$$

Gorter and Casimir assumed that the free energy of the electrons obeyed a relation

$$f = x^\eta \cdot f_n + (1-x) \cdot f_c \tag{9.11}$$

where $f_n = -\gamma T^2/2$ and γT is the electronic specific heat.

The value $\eta = \frac{1}{2}$ fits the experimental observations best, and the form of f_n and f_c is chosen to agree with the temperature dependence of the entropy. The system is in equilibrium at a given temperature when $(\partial f/\partial x)_T = 0$, and under these conditions

$$\beta = \gamma T_c^2/4 \tag{9.12}$$

and

$$x = (\gamma T^2/4\beta)^2 = (T/T_c)^4 \tag{9.13}$$

Expressions for the entropy S and specific heat C per unit volume may now be obtained by substituting in the free-energy equations and differentiating with respect to the temperature

$$S = \gamma T^3/T_c^2 \tag{9.14}$$

$$C = 3\gamma T^3/T_c^2 \tag{9.15}$$

One of the successes of the theory was the prediction of the way in which the penetration depth γ varies with temperature. In the two-fluid model the number of superconducting electrons n_s is proportional to $1-x$, and for a value $\eta = \frac{1}{2}$ we have $1-x = 1-(T/T_c)^4$. The London equations show that $\lambda^2 \propto 1/n_s$ and consequently the temperature-dependence of the penetration depth is

$$\lambda = \lambda_0/\{1-(T/T_c)^4\}^{\frac{1}{2}} \tag{9.16}$$

This is in agreement with observation.

2. *The London Theory*

The equations arising out of the theory proposed by F. and H. London relate the current density in a superconductor to the electric and magnetic fields. In the formulation of the theory a distinction is made between the current due to superconducting electrons j_s and the normal current j_n, where $j = j_s + j_n$.

It is assumed that the normal current j_n obeys Ohm's Law $j_n = \sigma E$ where σ is the electrical conductivity.

The equations peculiar to the theory are those which relate j_s with the electric and magnetic fields. There are two of these equations

$$\Omega \, \mathrm{curl} \, j_s = -B/c \qquad (9.17)$$

$$\Omega(\partial j_s/\partial t) = E \qquad (9.18)$$

where Ω is a constant whose value varies with the superconductor. The two equations describe the Meissner effect and the zero resistance respectively. They are combined with Maxwell's equations and lead to differential equations for the magnetic field and current density. From these it is also possible to derive the penetration depth

$$\lambda^2 = \Omega c^2/4\pi \qquad (9.19)$$

The London equations determine the fields not only in the interior of a macroscopic specimen but also in the region of depth λ close to the surface. For large specimens the characteristic feature of these equations is that they decay exponentially into the interior of the specimen, so that field and current density are negligibly small except within a distance λ from the surface. The equations have been solved for some simple special cases such as plates, cylinders and spheres in uniform magnetic fields and for cylinders carrying a current. They have also been generalized by von Laue and Ginzburg for anisotropic bodies. The behaviour of superconductors in a high-frequency field may be obtained by replacing static fields with time-dependent fields in the equations and by making assumptions about the behaviour of the normal conductivity σ.

Some limitations of the London theory came to light when measurements of the penetration depth in alloys of indium in tin, by Pippard and others, showed that the penetration depth increased when the electron mean free path l in the alloy decreased. Such a relationship lies outside the London model, and the concept of a "range of coherence" ξ was introduced by Pippard. This concept permits a spatial variation of the superconducting order parameter ψ (where $\psi^2 = n_s$, the number or density of superconducting electrons) over distances of the order of 10^{-4} cm, and modifies the London equations since the penetration depth now becomes

$$\lambda = \lambda_L(\xi_0/\xi l)^{\frac{1}{2}} \qquad (9.20)$$

where ξ_0 is the range of coherence in the pure material and ξ_l that in an alloy having a mean free path l.

The two quantities are related by

$$1/\xi_l = 1/\xi_0 + 1/\alpha l \qquad (9.21)$$

where α is a constant.

3. *The GLAG Theory*

In the phenomenological theory of Ginzburg and Landau (1950) an order
parameter ψ was introduced. Equations were then derived relating this order
parameter to the superconducting free energy and the applied magnetic
field. There are three parameters in the theory which need to be determined
experimentally, λ_0, H_c and κ, where λ_0 is the penetration depth, H_c the bulk
critical field and κ is defined by the relation

$$\kappa^2 = [2(e^2/h^2 c^2)H_c^2 \, . \, \lambda_0^4 \tag{9.22}$$

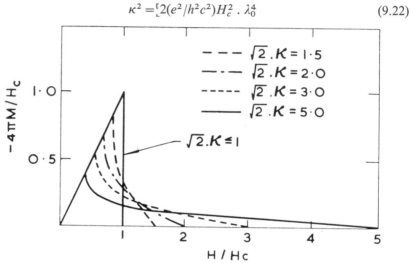

Fig. 9.11. The effect of variation in the Ginzburg–Landau constant κ on the magnetization curve.

The parameter κ is related to the surface energy (described later) since if
κ is less than $1/\sqrt{2}$ the surface energy is positive (type I superconductor) and
if κ is greater than $1/\sqrt{2}$ the surface energy is negative (type II superconductor). Abrikosov (1957) studied theoretically the behaviour of type II superconductors in a magnetic field and distinguished two critical fields H_{c1} and
H_{c2}. The effect of changes in κ on the magnetization curves are shown in
Fig. 9.11 and are discussed further in terms of flux penetration later in this
chapter. This theory of superconductivity is now known as the GLAG theory
after the four Russian workers, Ginzburg, Landau, Abrikosov and Gor'kov.
The theory has also established relationships between the critical fields and
the parameter κ as follows:

$$H_{c2} = \sqrt{2} \, . \, \kappa \, . \, H_c \tag{9.23}$$

and also

$$H_{c1} = H_c \ln \kappa + 0.08)/\sqrt{2}\kappa \tag{9.24}$$

From the practical point of view, equation (9.23) shows that to obtain a
high value of the upper critical field, a high value of κ is necessary.

4. *The BCS Theory*

It was not until 1957 that a formal theory relating the microscopic or electronic properties of a superconductor to its superconducting behaviour was first put forward. The difficulty of formulating a theory has been due to the small energy change involved in the transition itself, about 10^{-8} eV/atom, and in the mathematical methods required to handle the various interactions between the electrons. The first step forward towards the presently accepted theory was taken by Fröhlich in 1950 when he proposed an interaction between electrons much smaller than the normal Coulomb repulsive interaction. Individual electrons were postulated as causing polarization of the ionic lattice, and this polarization reacted back on the other electrons leading to a small attractive interaction between them. It was shown that if a certain inequality was satisfied the interaction led to a removal of some of the electrons from the Fermi surface to a spherical shell of slightly greater momentum. This "energy gap" could be identified with that occurring in the two fluid model. The theory predicted that the critical temperature would be related to the ionic mass M and this was subsequently discovered experimentally as the "isotope effect" where

$$T_c \propto 1/M^{\frac{1}{2}} \tag{9.25}$$

The idea of an attractive "electron-phonon" interaction has since been developed by Bardeen, Cooper and Schrieffer.

The basic idea of the BCS theory is that the electron–lattice interaction produces stable pairs of electrons. The electrons forming such pairs are derived from those at or near the Fermi surface. Physically, this process may be visualized as one in which the first electron moving through the lattice is surrounded by a cloud of phonons which, provided certain conditions regarding the relative magnitudes of electron and phonon energy are satisfied, gives a positive screening charge slightly greater than the charge on the electron. The second electron is consequently attracted by a net positive charge and this attraction must overcome normal Coulomb repulsion for pair formation to be stable and superconductivity to occur.

The theory shows maximum stability of the pairs if the two electrons have opposite spin and momentum. Correlation between all the pairs exists, and it takes a finite amount of energy to break a pair, corresponding to the energy gap. At temperatures above absolute zero some pairs dissociate into normal electrons, and one can identify two fluids, the normal electrons which can be scattered, and the correlated pair electrons which cannot be scattered.

The BCS theory gives the critical temperature T_c as

$$kT_c = 1\cdot14\,\hbar \,.\, w_{ph} \,.\, \exp\left(-1/N(E)_F \,.\, V\right) \tag{9.26}$$

where $\hbar w_{ph}$ is the phonon energy, and accounts for the isotope effect,

$\quad N(E)_F$ is the density of states at the Fermi surface,

$\quad V \quad$ is an electron phonon interaction constant.

The quantity $N(E)_F$ in this equation explains in a qualitative way the empirical observations of Matthias, since $N(E)_F$ is related to the electron/atom ratio. Good accounts of the BCS theory are given by Lynton (1962) and Cooper (1960), and more detailed and more recent accounts in the book edited by Parks (1969).

D. TYPE I AND TYPE II SUPERCONDUCTIVITY

The properties of a superconducting material cause it to fall into one of two classes, known as type I or type II. From the experimental point of view, these two classes are most readily distinguished through their magnetization behaviour. Typical magnetization curves are shown in Fig. 9.12. It is readily apparent from these curves that, in a type I superconductor, magnetic flux is completely excluded up to the critical field H_c, beyond which complete penetration takes place. In a type II superconductor however, flux begins to penetrate at a lower critical field H_{c1}, and penetration is only completed at an upper critical field H_{c2}.

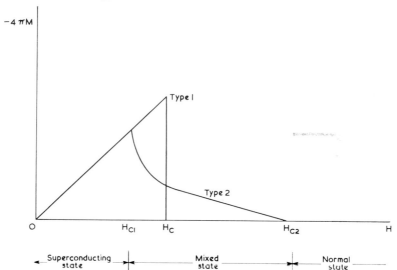

Fig. 9.12. Magnetization curves of type I and type II superconductors.

This different behaviour may be understood in a qualitative way through the "surface energy" concept. When a magnetic field is excluded from the interior of a superconductor the rise in magnetic energy is $H^2/8\pi$ per unit volume. The field is, nevertheless capable of penetrating the surface to a depth equal to the penetration depth. Consequently if a superconductor were to break up into superconducting regions of dimensions less than the penetration depth, separated by thin normal regions, the magnetic field would be able to spread into the interior of the material, since it would be able to penetrate both the normal and the superconducting phases. The energy of field exclusion would be recovered. Such a situation would always arise but for the presence of a surface-energy term. Since there is an increase in the surface area between the two phases if this mixing occurs, it is only stable if the increase in surface energy does not exceed the reduction in magnetic energy.

1. Flux Penetration in Type I Superconductors

Although in ideal circumstances flux penetrates abruptly at the critical field H_c, it has already been shown in Fig. 9.8 that, for certain shapes of specimen, flux penetrates over a range of fields to give an intermediate state structure. The intermediate state can also be produced by a current flowing in a wire if the current is large enough to exceed the value given by Silsbee's Rule. In this intermediate state the material consists of both normal and superconducting regions in a fairly characteristic array, and the formation and behaviour of the intermediate state under various conditions of magnetic field and current have been studied by a number of workers.

In observing the nature of the intermediate state experimentally, use is made of the fact that the normal regions have magnetic flux in them whilst

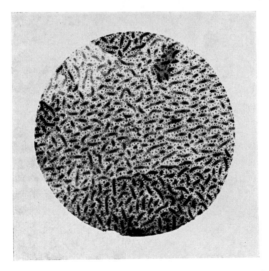

Fig. 9.13. Intermediate state pattern on a pure lead disc, axial magnetic field cycled from 0 to 650 Oe and 400 Oe. Magnification $\times 20$ (Sarma and Wilcockson, 1969).

the superconducting regions do not. Thus, a magnetic contour map of the surface of the material may be drawn by using a magneto-resistance probe as a detector. Alternatively, the material may be dusted with ferro-magnetic powder to produce a form of "Bitter pattern" in which the powder decorates the magnetic regions. Again, the effect of the magnetic field regions on the rotation of the plane of polarized light in cerium phosphate glass may be used to demonstrate visually both static and dynamic effects in the intermediate state. A very comprehensive range of photographs of intermediate state structures is shown in the article by DeSorba and Healy(1964); and an example by Sarma and Wilcockson (1969) is shown in Fig. 9.13.

Observations show that the laminar model of alternating superconducting and normal regions suggested by Landau is in fact modified near the surface into a corrugated structure, and that the structures actually observed also depend upon the field and temperature history of the specimen, the dynamics of the flux motion, the specimen shape and the degree of superheating or supercooling. The intermediate state structure of type I superconductors is affected by the presence of a transport current and, when this current is perpendicular to the applied field, the superconducting-normal boundaries tend to align in a direction perpendicular to the current flow. The extent of alignment depends on the magnetic field, and on the strength of the current, which must be sufficient to cause the flux to move. This effect is demonstrated in Fig. 9.14.

Although the intermediate state in type I superconductors has some relationship with the mixed state in type II superconductors, the principle difference is that the intermediate state is a macroscopic structure with a laminar separation of the order of $0 \cdot 01$ cm. This separation is much coarser than the mixed state structure, which in high fields can have spacings of only 10^{-5} cm. A more detailed discussion of the intermediate state in general has been given in the article by Livingston (Parks, 1969).

2. Flux Penetration in Type II Superconductors

As shown in Fig. 9.12 the magnetization curves of type II superconductors show two critical fields, H_{c1} and H_{c2}. When the applied magnetic field lies between these two critical fields, magnetic flux penetrates the specimen and it can be shown both theoretically and experimentally that this flux is present as an array of flux tubes parallel to the field, each flux tube containing one quantum of flux ($\phi_0 = hc/2e = 2 \times 10^{-7}$ gauss cm^2). Each flux tube consists of a normal core surrounded by a screening current vortex and under ideal conditions, the array of flux tubes is triangular as shown in Fig. 9.15. Very approximately, the normal core has a radius comparable to the coherence length ξ, whilst the screening currents occupy a distance comparable with the penetration depth λ. As the field is increased, the number of flux tubes increases until at the upper critical field the normal regions overlap, and superconductivity is quenched. The regular array of tubes or vortices has been demonstrated experimentally by Träuble and Essmann (1968) and Sarma and Wilcockson (1968), Fig. 9.16.

The magnetization of a type II superconductor often shows marked hysteresis effects which causes some modification to the ideal curve and an extreme example of such behaviour appears in Fig. 9.17. The existence of

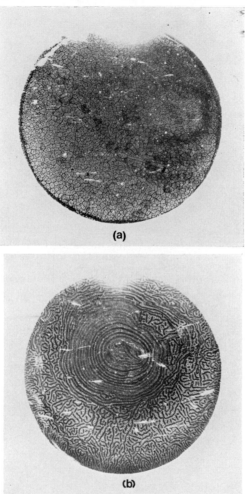

(a)

(b)

Fig. 9.14. Intermediate state pattern on a pure lead disc carrying a radial current,
(*a*) with current of 12·5 A and axial magnetic field cycled from 0 to 650 Oe to 360 Oe. Magnification ×20.
(*b*) with current of 20 A, and axial magnetic field cycled from 0 to 650 Oe to 400 Oe. Magnification ×20. (Sarma and Wilcockson, 1969.)

such hysteresis means that the motion of flux is hindered within the material as the external field is changed, resulting in gradients of flux density and the presence of screening currents (Fig. 9.18). Many experiments have confirmed

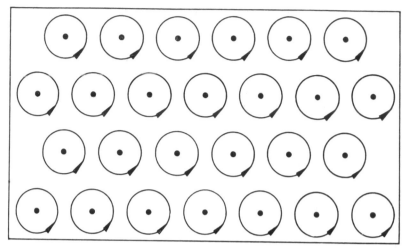

Fig. 9.15. Flux lattice in a type II superconductor for $H_{c1} < H < H_{c2}$, H passing into the paper. Dots are normal cores, circles are screening currents.

that it is the existence of certain structural defects in the material which restrict the easy passage of flux and "pin" the flux tubes in position. Dislocation arrays, grain boundaries, precipitates of second phases and voids are all effective in this respect. The more effective the defect, and the larger their

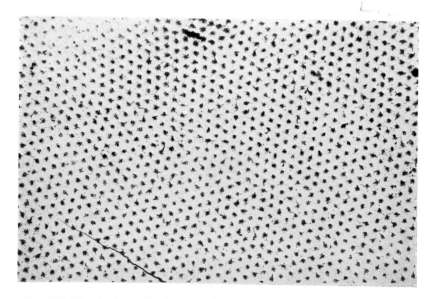

Fig. 9.16. Flux lattice in Pb–6 a/o In alloy. Lattice spacing 9,000 Å (Sarma and Wilcockson, 1968).

concentration, then the larger is the area of the hysteresis loop. The flux tubes can be unpinned from their pinning sites by either a rise in the temperature, the presence of a sufficiently large flux gradient or when a sufficiently large transport current is passed through the specimen. The presence of defects in the material which are able to pin flux is of great practical significance, since they enable large currents to pass through the superconductor without power loss. This subject is discussed in detail in Section III.

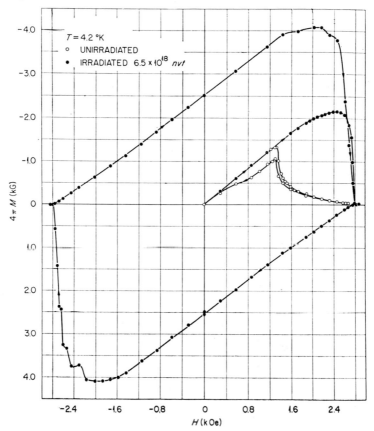

Fig. 9.17. Initial magnetization cycle and subsequent hysteresis loop for a neutron-irradiated sample of niobium (Sekula and Kernohan, 1968).

If the applied magnetic field is less than the lower critical field of a type II superconductor all flux is excluded, and its behaviour resembles a type I material. If the field is greater than the upper critical field all superconductivity in the bulk of the material is quenched but Saint-James and de Gennes (1963) predicted that, if the magnetic field was parallel to the surface, then a surface superconducting layer or sheath would be stable up to a field H_{c3}

equal to $1.695 H_{c2}$, and this has been confirmed experimentally. The total current carried by the surface sheath is small however, and appears to have little practical significance.

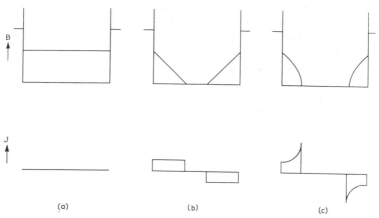

Fig. 9.18. Distribution of field B and current density J through the cross section of type II superconductor (Livingston, 1964).

(*a*) Defect free material (note the surface field step).
(*b*) Material containing defects, with constant critical current.
(*c*) Material containing defects, with field-dependent critical current.

II. Practical Materials

The most useful superconductors for technological purposes have proved to be of the type II variety. Materials in this class have the highest values of critical temperature, field and current, and these three parameters are found to be affected either by alloying or metallurgical treatment. The following sections describe in more detail the properties of type II or "high-field" superconductors.

A. EFFECT OF ALLOYING ON CRITICAL FIELD AND TEMPERATURE

The magnitude of the upper critical field is important in practical applications since it determines the maximum operating field of any device. It has been shown that the upper critical field is related to the Ginzburg–Landau parameter κ through equation 9.23, so that an increase in κ causes an increase in the upper critical field. Goodman (1962) derived the following relation between κ and the electronic structure of an alloy:

$$\kappa = \kappa_0 + (7.5 \times 10^{-6} \gamma_{\frac{1}{2}} \rho) \tag{9.28}$$

where κ_0 is the value in the pure metal, κ that in the alloy, γ is the coefficient of the electronic specific heat and ρ the normal state resistivity. Consequently increasing the resistivity by alloying will increase κ and the upper critical field. The effect has been demonstrated experimentally in lead-base alloys, for example Druyvesteyn et al. (1964).

The values of the upper critical field for various type II superconductors are shown in Table I and the effect of alloying in the niobium–titanium and niobium–zirconium systems is shown in Fig. 9.3. Alloying offers the possibility of increasing the critical field still further in certain compounds and, although most ternary alloys of the A15 variety have shown a deterioration in superconducting properties, the compound of niobium, aluminium and germanium (Matthias et al., 1967) has a critical field of 410 kgauss at 4·2°K. Hake (1967) has suggested that it may also be possible to measure κ_0 in these compounds and, under these circumstances, a maximum value of the upper critical field at 0°K could be predicted. The predicted values are shown in Table 9.III.

TABLE 9.III. Comparison Between the Measured and Predicted Maximum Values of Upper Critical Field

Compound	$H_{c2}(0)$ κgauss measured	$H_{c2}(max)$ κgauss predicted
Nb_3Sn	245	880
V_3Si	235	850
V_3Ga	210	810
NbN	153	250

The effect of alloying on the critical temperature may be deduced from equation 9.26. Although the quantities in the right-hand side of this equation cannot be predicted for any special case, measurements of low-temperature specific heat which gives $N(E)_F$ can demonstrate how they vary in particular systems. The product $N(E)_F \cdot V$ must be large for a high T_c to be achieved but, in practice, a given alloying element usually alters $N(E)_F$, w_{ph}, and V simultaneously.

The A15 materials have high values of $N(E)_F$ and low values of V (Geballe et al., 1966). The value of T_c is also highly dependent upon the stoichiometry and degree of order, the achievement of the exact stoichiometric ratio and maximum degree of order favouring the highest critical temperatures.

B. EFFECT OF DEFECTS ON THE CRITICAL CURRENT

The presence of metallurgical defects is necessary in type II superconductors for these materials to be able to carry any useful current. The experiments of Heaton and Rose–Innes (1963) on niobium–tantalum alloy wire demonstrate clearly that fully annealed material is very poor in this respect.

The critical current density which can be supported is directly dependent on the density and strength of pinning sites, and the variation of the critical

20

current density with field depends on the variation of the strength of these pinning sites with field. The pinning force depends on the nature of the defect and its size with respect to the flux tubes. The flux tube has a core whose radius is comparable to the coherence length and maximum pinning occurs when the defect is of a similar dimension. In addition, the pinning force on a flux tube depends upon the number of other filaments in the neighbourhood. It is clear therefore that the variation of the pinning force and the critical current density with field may be complicated and, in some cases, the net pinning force can increase with field (Livingstone, 1966), over certain ranges of field. The more usual effect of increasing field is, nevertheless, to reduce the critical current.

The defects in the material which are responsible for pinning flux can be divided into four categories, as suggested by Campbell *et al.* (1968), Livingston (1968).

1. *Point Defects*

Point defects are represented by voids, small particles of second phase or small clusters of different kinds of defect. The introduction of point defects into a material is accomplished by standard metallurgical techniques. For example voids are produced by sintering, precipitates of second phases by suitable cycles of heat treatment, and clusters by radiation damage. The effectiveness of point defects in pinning flux has been realized since the classical work of Livingston, who produced precipitation in various lead-base alloys and related the microstructure to the amount of magnetic hysteresis which was present. It has also been shown, mainly by Livingston and his co-workers, that the amount of pinning depends on the nature of the precipitate itself, as well as on its size and spacing. The greater the difference between the electronic structure of the matrix and the precipitate, the greater is the pinning force. Consequently the pinning force increases with the presence of either superconducting, normal or non-conducting precipitated phases respectively. Ferromagnetic particles can also provide pinning, the strength of the pinning depending upon whether the magnetic axis of the precipitate is parallel or anti-parallel to the flux contained in the flux thread.

Voids are a special case of non-conducting defects, and increasing the number of voids initially increases the flux pinning as shown, for example, by sintered niobium (Catterall and Williams, 1967), although a maximum in the critical current eventually occurs since the amount of superconducting material available decreases at the same time as the number of voids increases.

It has been mentioned earlier that the optimum size of a pinning site is the coherence length. However, as far as the optimum spacing between sites is concerned this is a more difficult quantity to specify, since the ability of a flux tube to be pinned depends upon the forces exerted on it by neighbouring flux tubes, in addition to those exerted by the pinning site itself. Since forces between flux lines depend on the number and spacing of the lines, the optimum separation of sites for pinning is a field dependent parameter.

2. *Line Defects*

The principle line defect occurring in materials is the dislocation, and it has been known for some time that cold-working a type II superconductor can increase its critical current density. A low-temperature heat-treatment of cold-worked material, which re-arranges the dislocations into a cellular

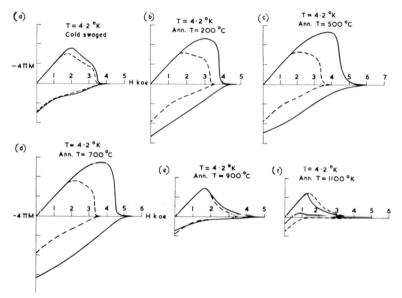

Fig. 9.19. Effect of annealing temperature on the magnetization curve of cold-swaged niobium at 4·2°K (Williams and Catterall, 1966).

pattern is even more effective, and this has been demonstrated by Narlikar and Dew-Hughes (1966) and Williams and Catterall (1966). In this work the dislocation structures as revealed by electron microscopy were related to the low-temperature properties of niobium (Fig. 9.19). The structure of heat-treated commercial niobium–titanium alloys has been examined by Neal (1969), who finds that, for a given field, the pinning force varies inversely with the cell size of the microstructure. The interpretation proposed is that pinning is due to the cell walls, and that as the cell size decreases the number of pinning sites increases, causing an increase in the total pinning strength.

It has been suggested by various workers that the low temperature heat-treatment of cold-worked material not only re-arranges the dislocation networks but also causes the cell walls to be decorated with precipitates of either interstitial compounds (NbC, TiC) or decomposition phases. Such precipitation could enhance the flux pinning due to the cell walls, through a local reduction in the electronic mean-free path and an enhancement of the Ginzburg–Landau parameter κ.

3. *Surface Defects*

Two-dimensional defects common in crystalline materials are grain boundaries, phase boundaries and boundaries between transformed and untransformed regions such as in martensitic, twin or allotropic transformations. For example, a proportionality between the critical current density and the surface area between phases has been demonstrated in eutectic alloys by Levy *et al.* (1966) and Campbell *et al.* (1968), and a dependence on grain size has been shown by Hanak and Enstrom (1966). It is also possible that the high critical currents obtained in niobium tin may be in part due to a partial martensitic transformation which is known to occur at low temperatures.

4. *Volume Defects*

Large voids or precipitates would be expected to be relatively ineffective as pinning sites since their dimensions will be much larger than either the coherence length or the flux line separation distance. Pinning will be mainly due to the surface between the defect and the matrix.

C. METALLURGICAL PRODUCTION OF PRACTICAL MATERIALS

From the practical point of view the materials which have proved most useful in constructing devices such as magnets have been either the body-centred-cubic alloys of niobium–48% titanium, or the cubic A15 compounds, especially the compound niobium tin. There are other compounds with high critical temperatures, in particular the interstitial sodium chloride type compounds shown in Table 9.II, but no methods have been developed either for assessing their properties in a reliable fashion or of producing them by a process suitable for commercial production.

The body-centred-cubic alloys have proved extremely useful for the construction of magnets with field strengths up to 100 kgauss. The niobium zirconium alloys remain superconducting up to 80 kgauss whilst the niobium titanium alloys reach 120 kgauss. The crystal structure accounts for their ductile behaviour and, although niobium zirconium alloys were the first to be used for winding magnets, the superior ductility of the niobium titanium alloys combined with their higher critical field has rendered the former alloys obsolete. The metallurgical processing of niobium titanium into wire is accomplished by the standard processes of extrusion in sealed cans followed by rod and wire drawing. The requirements of thermal and magnetic stability (discussed later in this chapter) result in the niobium titanium wires being marketed for most purposes clad with high-conductivity copper. The copper is incorporated early in the process, usually before the extrusion stage, and the niobium titanium is arranged within the copper in the desired fashion at this time. Subsequent extrusion and drawing leaves the original arrangement relatively undisturbed. Depending upon the requirement of the user, many configurations of wires or filaments in copper rod, strip, tube, cable or braid can be produced. Figure 9.20 shows some examples. In order to achieve a high current density in the niobium titanium alloy, it is customary to give the cold-drawn wire a short anneal at 400°C.

Fig. 9.20. Composite conductors consisting of niobium titanium alloy wires in copper (courtesy Imperial Metal Industries Ltd.).

The compound niobium tin has superior superconducting properties to niobium titanium in terms of critical current, field and temperature. The compound is, however, extremely brittle and its use required the development of special production techniques. In essence, the methods which have been established produce a layer on a substrate which is in strip form. If the layer is thin ($\simeq 10~\mu$m) it has sufficient flexibility for the strip as a whole to be wound into a magnet without cracking. There are two alternative processes by which such strip is produced. The first was developed by Hanak et al. (1964) and involved reducing a mixture of gaseous niobium chloride and tin chloride with hydrogen. If the proportions of the gases are correct, the compound Nb_3Sn is formed directly, and is deposited on to a heated stainless steel ribbon. Control of the rate at which the ribbon passes through the reaction chamber gives the required thickness of the deposit. Copper for stabilization can be incorporated by soldering strips on each side to form a sandwich. The stainless steel substrate gives the tape useful mechanical strength.

The second process involves coating a niobium ribbon with a layer of tin, and heat-treating the ribbon to form niobium tin by diffusion of tin into the niobium. Again, the thickness of the layer can be controlled through the rate of reaction. Copper can also be soldered on for stability and stainless steel for strength.

Typical curves of the current density of niobium titanium, niobium zirconium and niobium tin are shown in Fig. 9.21.

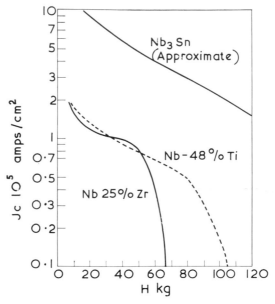

Fig. 9.21. Critical currents of niobium–25% zirconium, niobium–48% titanium and niobium–tin.

D. FUTURE DEVELOPMENT OF MATERIALS

There are three major problems in the development of superconducting materials, namely, raising critical temperatures, reducing the cost and minimizing the alternating current loss.

1. *The Critical Temperature*

It appears that most cases of superconductivity in metals arise from the phonon mechanism, as described in the BCS theory. The highest known temperatures occur in the A15 compounds and, after considerable experimentation, the alloy $Nb_3(Al_{0.8}Ge_{0.2})$ has been found with a critical temperature of $20.9°K$. In most cases, however, adding a third component to Al5 structures depresses their critical temperature and, although some further small increase above $20.9°K$ might be expected, it seems unlikely that this figure will be exceeded by any great margin. For the phonon mechanism it has been estimated that 25 to $30°K$ is the maximum that can be expected. For any progress to be made beyond this point some other mechanism of superconductivity must be found, and two or three possibilities have been discussed (Ginzburg, 1968).

One suggestion is that interaction between electrons can occur through "excitons" instead of phonons. An exciton is an excitation of the electronic structure, and spin waves and plasma waves are examples of such excitations. It is possible that one way of inducing an interaction of this kind and of enhancing the critical temperature is to sandwich a thin ($=20$ Å) layer of superconductor between two layers of a suitable dielectric. Excitons propagating in the dielectric induce electron–electron interaction in the superconductor. This possibility has not yet been checked experimentally, and it is likely to prove difficult to establish. A further difficulty in using such an arrangement in any practical device is that in order to carry a reasonable total current a large number of sandwiches in parallel would be necessary.

A second suggestion by Little (1964) involving one-dimensional chains of organic molecules now seems theoretically unsound. In this model the interaction considered is between pairs of electrons in a conjugated chain of the form shown in Fig. 9.22, where R represents some easily polarizable side

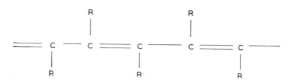

Fig. 9.22. Proposed superconducting organic molecule.

chain. Little's view was that the existence of excited electronic states in these side chains can cause an attractive interaction between pairs of free electrons moving along the conjugated chain, the magnitude of the interaction being of the order of 1 or 2 eV. Consequently electron pairs will be formed and, according to the degree of bond localization in the chain, transition temperatures of several hundred degrees Kelvin could be possible. However, the difficulty is that in a one-dimensional superconductor of this kind, thermodynamic fluctuations compete with the ordering process and prevent superconductivity from occurring. Whether such difficulties apply to a three dimensional array of chains in close proximity is still not clear.

A third method of enhancing the critical temperature arises within the framework of the BCS phonon mechanism, and gives rise in turn to three possibilities which are concerned with small crystallites or thin films. Firstly, Parmenter (1967) has suggested that, in very small superconducting particles, the BCS equations must be modified owing to the discrete nature of the electron states in k (or momentum) space. Such a modification gives a value of critical temperature in which

$$\ln (T_c/T_{cb}) \propto 1/a^3 \qquad (9.29)$$

where T_{cb} is the bulk transition temperature and a the particle size. A second possibility lies in producing sandwiches of metal and dielectric or making dirty films in which the crystallites are separated by oxide layers. The Coulomb repulsion between pairs of interacting electrons which are separated by the dielectric or oxide layer is affected more than the phonon interaction, resulting in an enhancement of the critical temperature. A third mechanism is in the reduction of the mean phonon frequency near the surface of small particles, giving an increased value of the BCS coupling constant V and leading to

$$\ln (T_c/T_{cb}) \propto 1/a \qquad (9.30)$$

This mechanism, however, produces other effects such as a reduction in $N(E)_F$ which tend to offset the increase in V. Its value is therefore uncertain.

To summarize, the greatest hope of raising the critical temperature lies in achieving small increments in the critical temperature of the A15 type compounds. The thin film sandwiches may give some success, but utilizing them in practical devices may prove difficult.

The objective of raising critical temperatures of superconductors is to raise the operating temperature of any device which employs them. A higher operating temperature reduces the refrigerator power, and the device is consequently more economic to use. In general, superconductors are not able to carry very large currents at temperatures above about $\frac{1}{2}T_c$, so that to operate in liquid hydrogen (20°K) a superconductor with a critical temperature of 30 to 40°K would be necessary.

2. The Cost of Material

For very large machines such as motors and generators, the superconducting material can form about one-third of the total cost. Superconducting cables are unlikely to be economically attractive for the transmission of electrical power unless they have a lower capital cost for the same power load than conventional cables. Consequently, attempts to reduce the cost of superconducting material are of great importance. To some extent, this reduction can be achieved by developments in the metallurgical processing of existing materials, and by raising critical current densities since less material is then required for a given total current. The cheapest superconductors are lead based alloys, but these alloys do not compare favourably in terms of superconducting properties with niobium and its alloys. Hindley and Watson (1969) however, have shown that reducing the physical dimensions of superconductors sufficiently enhances the properties of lead alloys so that this approach may provide a route towards lowering material costs.

3. The Alternating Current Loss

In order to minimize losses in superconductors which occur with alternating currents and fields it is necessary to operate below H_{c1} in type II, or below H_c in type I superconductors. In this "Meissner state" region losses are dependent on the degree of roughness of the surface of the material, smooth surfaces showing the lowest losses. For present design of transmission lines therefore, it will be necessary to produce very smooth surfaces on niobium by a continuous process. Later sections of this chapter discuss twisted filamentary conductors for use in pulsed magnets, but it appears necessary to transpose the individual filaments for operation at frequencies higher than a few cycles per second. This requirement will pose considerable metallurgical problems.

III. Dynamic Behaviour of Superconductors

We have seen in the previous sections that practically useful currents can be carried by superconductors in magnetic fields up to at least 200 kgauss, and that there is a range of materials suitable for engineering applications. The behaviour of these materials in practice, however, is very complex and requires an understanding of their dynamic behaviour as well as their static characteristics discussed so far. This is particularly important for type II materials which can exist not only in the fully superconducting or normal states but may exhibit flux motion in the mixed state, retaining their superconducting properties yet having some resistance. This behaviour leads to the hysteresis observed in magnetization curves, is the cause of power losses in alternating current devices, and leads to instabilities and degradation which can result in the failure of superconducting magnets.

A. FLUX MOTION AND THE CRITICAL STATE MODEL

In a completely homogeneous type II superconductor in applied fields above the lower critical field H_{c1} fluxoids can move freely within the material, being subject only to their own mutual interaction. The result is a regular array of fluxoids without any field gradients or bulk currents. If the material is not homogeneous, however, but includes strains, dislocations, grain boundaries, inclusions or variations in composition then these structural defects act as pinning sites which limit the motion of fluxoids, and their behaviour is no longer reversible and field gradients and bulk transport currents can exist.

20*

Such barriers to flux motion exert a pinning force which may be overcome by a combination of the long range Coulomb interactions of the fluxoids and thermal activity within the material. The resultant motion is thermally activated flux creep (Anderson, 1962) which is very slow and can only be observed by precision measurements such as those of Kim, Hempstead and Strnad (1962) who reported a decay of the magnetic field trapped in a niobium zirconium tube of about 14 gauss/decade.

The driving force for this flux motion is the Lorentz force $F_l = J \times B$ where J is the average macroscopic current density and B the local magnetic induction, and the pinning force can be related to the free energy of the superconductor $H_c^2/8\pi$. Assuming that the effectiveness of a pinning site of dimension d is only a fraction p of the available energy difference between the superconducting and normal states, the total energy of the barrier to flux motion in the presence of a transport current is

$$E_b = p(H_c^2/8\pi)d^3 - J \times Bd^4 \qquad (9.31)$$

and due to thermal activation the rate of motion of flux through such a barrier will be

$$R = R_0 \exp(-E_b/kT) \qquad (9.32)$$

where R is a characteristic vibration frequency which is of the order 10^{-6} to 10^{-10}/sec.

Several consequences follow from equations 9.31 and 9.32 of which the most important is that, unlike a type I superconductor in which perfectly persistent currents flow, the current density in a type II superconductor decays with time. The rate of change is of the form

$$J \times B \propto \text{constant} - (kT/d^4) \log t \qquad (9.33)$$

which is logarithmic with time and therefore soon becomes imperceptably slow. It follows that, after any change of the external magnetic field, any transport current will rapidly redistribute itself so that the current density falls to its original level. This level is the commonly observed critical current density J_c and is a property of the structure and composition of the super-conductor.

From equations 9.31 and 9.32 it is seen that this critical current density is given by

$$J_c \times B = pH_c^2/8\pi d - qkT/d^4 \qquad (9.34)$$

where p and q are constants determined, respectively, by the effectiveness of the pinning sites and the exact definition of the quasi-static critical current. Neglecting any field dependence of p it follows that the critical current density is expected to vary inversely as the local magnetic field, and this relationship has been found to be true for many materials. Furthermore, the critical current density decreases with increasing temperature, and it will be seen in later sections that this is the basic cause of the instabilities observed in bulk superconductors and superconducting magnets. Both these dependencies are illustrated in Fig. 9.23 which shows measurements of the critical current of a niobium–titanium wire.

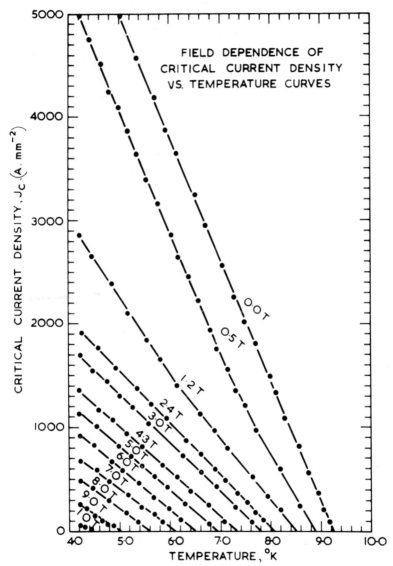

Fig. 9.23. Variation of the critical current density of a niobium–titanium wire with temperature and field (Hampshire *et al.*, 1969).

If the bulk transport current in an irreversible type II superconductor is increased above the level at which thermally activated flux creep is observed, such that the Lorentz force becomes comparable to the pinning force, more rapid motion of the flux lattice occurs known as flux flow. In this state the superconductor retains most of its superconducting properties but exhibits an electric field and hence a resistance. Above the critical current the voltage increases linearly with current and with a slope which is independent of the critical current. Measurements at various applied magnetic fields show that the slope is proportional to the field and becomes equal to the normal state resistivity at the upper critical field H_{c2}. Thus the differential flux flow resistivity is a property of the material and is given by

$$\rho_f = \rho_n(H/H_{c2}) \qquad (9.35)$$

The mechanisms which produce the viscous damping of this flux flow are complicated. Kim, Hampstead and Strnad (1965) have discussed several possible alternatives but the details of their precise nature are not fully understood. It is certain, however, that flux motion in a type II superconductor in the mixed state is always associated with the generation of an electric field. This can be illustrated by two novel experiments. Firstly, the d.c. transformer of Giaever (1965) in which flux motion is induced in a strip of superconductor by means of a closely coupled primary strip carrying a current greater than its critical current. An electric field is observed in the secondary strip equal to that caused by the flux flow in the primary strip and, if several secondary strips are connected in series, a multiplication of the voltage is obtained. The second experiment is that of Lowell, Munoz and Sousa (1968) who produced flux motion in the absence of a transport current by the application of a thermal gradient and again an electric field is observed, perpendicular to the direction of flux motion. It appears therefore that, whatever the precise mechanisms, flux motion is always accompanied by an electric field and is a dissipative process.

It has been shown that in an irreversible type II material we may distinguish two types of flux motion. Firstly, when the driving force is small compared with the pinning force thermally activated flux creep occurs, which is very slow. Secondly, when the driving forces are much larger flux flow occurs which is a much more rapid redistribution of flux. In materials with a very high density of pinning sites, and which are therefore of greatest practical importance, the range of current over which these effects occur is quite small, and one may define more clearly a critical current density above which flux motion is rapid and easily observable and below which it is negligible. This is called the critical state and is the equilibrium situation where the rate of change becomes so slow as to be unimportant within the time scale of most experiments.

The implication of the critical state model is that wherever current has been duced to flow within the bulk of an irreversible type II superconductor, the local current density will always be equal to the critical current density. This model was first proposed by Bean (1962) and has been used successfully to describe both hysteresis and alternating current losses in type II super-conductors. The principle is illustrated in Fig. 9.24 which shows the variation of field and current profiles within a slab of superconductor with an external magnetic field applied parallel to its surface as the magnetic field is cycled from zero, through a maximum, and back to zero. When the magnetic field is first applied flux begins to enter through the surface of the slab. Near the

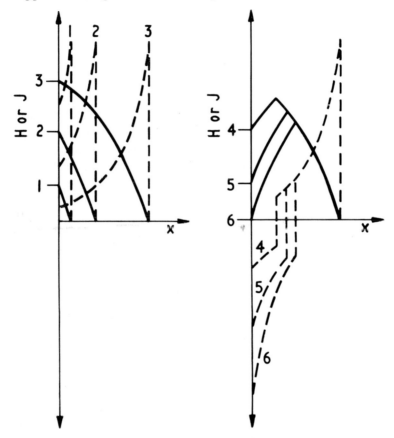

Fig. 9.24. Stages in the penetration of magnetic field and current into a type II super-conductor, illustrating the critical state model. (a) External field rising, (b) falling. Full curves represent the field profiles and broken curves the current density profiles.

surface the critical state is established and the current density is equal to the critical current density, but only partial penetration occurs and no current flows in the central regions. Since the magnetic field is lower inside the material than at its surface, the critical current density is highest at the inner boundary. As the external field is raised flux flows through the existing critical state region and the depth of penetration increases until current has been induced throughout the slab.

Continuing with the same example, as the applied field is reduced flux again moves from the outer regions first, generating new critical states with current flowing in the reverse direction, so that current is now flowing in both directions simultaneously. Further reduction of the field causes the boundary separating these regions of opposite current directions to move inwards, until finally, with no external magnetic field, the total currents flowing in either direction are equal. The slab is therefore left with a remnant trapped flux and circulating currents which will continue to flow indefinitely.

The existence of such circulating currents is easily demonstrated in most superconducting magnets. If, after operation, the current is reduced to zero and the supply disconnected a residual magnetic field is still observed in the bore of the magnet. Although the field integral $\int H . dx$ along the axis is zero, as required by Maxwell's equations, the magnetic field at the centre of the coil may still be several hundred gauss. Similar effects have been observed on a much smaller scale with a short length of niobium zirconium wire by Iwasa and Williams (1966).

Because of its wide application in calculations involving dissipative processes in type II superconductors it is important to appreciate the limitation of the critical state model and the type of problem in which it may be safely used to describe the complex events which occur in superconductors when flux motion is involved. The principle assumption is that the behaviour of a superconductor can be characterized by a critical current density. As has already been seen, this is not strictly true since the current density is always decaying slowly due to flux creep and therefore the measured critical current density in any experiment depends on the time scale of the experiment or the rate of change of the fields involved. This is paricularly important in alternating current loss calculations since in this case the superconductor is in a continual state of flux flow. Nevertheless, when the density of pinning sites is high, the transition from flux creep to flux flow is sufficiently rapid that, in practically useful materials with high critical currents, the critical current density varies by less than 0·1 % for the frequency range 0·1 c/sec to 1 kc/sec. This may not be true however for low current density materials and then the critical state model becomes a poor approximation.

The critical state model also fails at low applied magnetic fields. At fields below the lower critical field H_{c1} flux does not enter a type II superconductor and its behaviour is reversible. Even above H_{c1} when the material is in the mixed state surface currents will persist and surface barriers may influence the motion of flux. The model is therefore only applicable to a material with a high Ginzburg–Landau constant K and a high bulk critical current density compared with its surface current density. Again, this is true for most materials of interest in engineering applications, provided the applied magnetic fields exceed about 5 kgauss.

Finally, the critical state model gives a good description of the dynamic behaviour of an irreversible type II superconductor provided the real variation of critical current density with magnetic field and temperature is used. In practice this complicates the mathematical application of the model to the extent that analytical solutions are impossible, and therefore simple approximations have to be used. The simplest such approximation, proposed by London, is that the current density should be considered independent of the local magnetic field. This approximation is used in most alternating current loss calculations, and will be used in the following sections in describing the growth of instabilities in flux motion. An alternative approximation proposed by Kim, which is a more general form of the relationship shown in equation 9.34 is that the critical current density is proportional to $1/(B+B_0)$ where B is the local magnetic induction and B_0 is a constant. This fits more closely the variation with field observed in practice.

B. INSTABILITIES IN FLUX MOTION

Magnetic flux may move through an irreversible type II superconductor and, in doing so, generates an electric field and dissipates energy. Since the rate at which heat can diffuse through the material is orders of magnitude slower than the velocity of flux motion, the temperature will increase and the current density which can be supported by the superconductor will fall. This situation is unstable and catastrophic motion of flux can occur which, if unchecked, can drive the superconductor normal. Such instabilities, known as flux jumps, are common events and have been a major problem in the development of superconducting magnets and other devices.

The behaviour of flux jumps is illustrated in Fig. 9.25 which shows measurements of the field inside a hollow niobium–tin cylinder as the external field is increased and then decreased. If the superconductor behaved in a stable manner the internal field would vary smoothly once the critical state had been established across the full wall thickness of the cylinder, but in practice after partial penetration of flux an instability occurs and the diamagnetism of the superconductor disappears. If the cylinder is allowed to cool and then the field increased again the same cycle is repeated several times, smooth flux flow being followed by a flux jump at fairly regular intervals.

In the following sections this behaviour will be considered in more detail, and different approaches to the calculation of the stability criteria will be reviewed. The simple case of a one-dimensional instability in semi-infinite geometry will be discussed in detail since it shows most clearly the main parameters which control the growth of flux jumps.

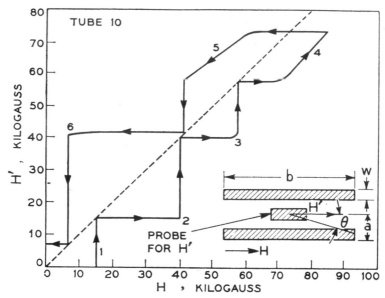

Fig. 9.25. Magnetization of a hollow cylinder of type II superconductor. H is the externally applied magnetic field and H^1 the field measured inside the cylinder (Kim *et al.*, 1963).

1. *The Stability Criterion in a One Dimensional System*

The simplest approach to the calculation of the limit of stable flux flow is the isothermal approximation used by Lange (1966), in which the system is tested for stability by applying a small increase of temperature and seeing whether the energy dissipated by the resultant flux flow is greater or less than the thermal energy required to produce the change. The two energies are equal at the limit of stability, and beyond this limit a thermal runaway may occur. This approach neglects any temperature gradients inside the superconductor and introduces instead a large thermal barrier at the surface. For simplicity the variation of critical current with magnetic field is neglected.

The stability criterion calculated in this way gives a critical field

$$H_s = \sqrt{(12\pi C T_0)} \tag{9.36}$$

where C is the product of specific heat and density of the superconductor and T_0 is a characteristic temperature defined by $(\delta J_c/\delta T)_H = -J_c/T_0$.

This critical field H_s is the largest change in external magnetic field for which flux motion inside the superconductor will remain completely stable. Using typical values for niobium tin at 4·2°K this limit of stability is found to be as low as 5 kgauss, which explains why flux jumping is so prevalent.

The Lange criterion is only approximate because it neglects the poor thermal conductivity of the superconductor and assumes isothermal conditions. A better representation is to assume a negligible thermal conductivity and use an adiabatic calculation. Using the same semi-infinite plane geometry and critical current density independent of magnetic field, the effect of a small change of external field is calculated. Several approximate solutions of this problem have been published, but Swartz and Bean (1968) have obtained an exact solution using a second order differential equation to describe the flux motion and solving for the limit in which complete penetration occurs. The result is a limiting field change, as before, below which flux jumps do not occur.

$$H_s = \sqrt{(\pi^3 C T_0)} \qquad (9.37)$$

This expression only differs from the previous result by a factor $\sqrt{(\pi^2/12)}$, showing that within the limitations of the model used both approaches to the calculation give a similar result.

Equations 9.36 and 9.37 give the field change at which flux motion begins to become unstable, but no indication of the extent to which the instability will grow, or whether the superconductor will be driven normal. To do this another limiting field change H_f can be calculated which corresponds to a flux jump of sufficient magnitude to heat the superconductor to its transition temperature which is given by

$$\frac{H_f^2}{8\pi} = \int_T^{T_c} C \, dT \qquad (9.38)$$

These two limiting fields can be evaluated and are shown in Fig. 9.26 for the specific case of niobium–tin, assuming that the specific heat increases with the cube of the temperature. At low temperatures flux motion is obviously very unstable but there is insufficient energy in the flux jumps to quench the superconductor, whereas at temperatures close to the transition temperature the superconductor will be driven normal as soon as an instability occurs.

The situation during the growth of a flux jump has been studied in greater detail by Wipf (1967) who showed that during the course of a flux jump an additional energy term must be added to the equation to represent the viscous damping of the motion. For field changes slightly greater than H_s there is limited stability with incipient flux jumps which are quenched before they grow to an observable size, and a new limit H_j must be defined at which complete collapse occurs. One effect of this damping is that the limit H_j includes a term involving the rate of change of the applied magnetic field and proportional to $\log(dH/dt)$, which is in good agreement with experimental observations.

2. General Behaviour of Flux Instabilities

Whilst the full criterion for flux jumping H_j cannot be evaluated explicitly, it is closely related to the simple limit H_s which can therefore be used to describe the main features of flux instabilities and highlight the important parameters which determine the onset of flux jumping.

The first of these is the slope $\delta J_c / \delta T$ of the relationship between the critical current density and temperature represented by T_0. The very existence of instabilities is due to the fact that for all single phase superconductors this slope is negative, and it follows that if the critical current density could be made to increase with increasing temperature all flux motion would be stable. This principle has been tested by Hart and Livingston (1968) using a lead–indium–tin alloy suitably heat treated so that, over a limited range of magnetic fields and temperature, the critical current density did increase

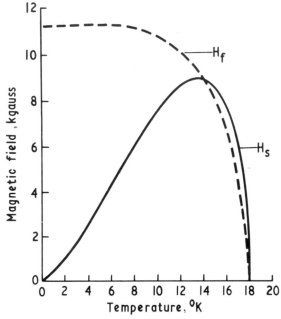

Fig. 9.26. Temperature dependence of the limit of stability H_s and the field required for the flux jump to drive a superconductor normal H_f (Swartz and Bean, 1968).

with temperature and stable behaviour was observed. Unfortunately such materials are rare and none has yet been found with a sufficiently high critical field or critical current density to be of practical importance, and therefore other methods of stabilization must be used.

The second important parameter which determines the stability limit is the specific heat of the superconductor. This has been verified by tests with a thermally isolated niobium tin cylinder at various temperatures, in which not only were flux jumps observed but also the thermal capacity and the critical current density were measured (Hancox, 1965b). Agreement with theory was quite good; an increase of specific heat by a factor 10 was accompanied by an increase in the stable screening field by a factor 3·2. It has also been demonstrated that allowing liquid helium into porous niobium tin increases the stability limit by an order of magnitude, due to the very large specific heat of the liquid helium.

Finally it is seen that the critical current density does not influence the limit of stability. This is only true in the simple semi-infinite case considered here, but again it has been verified experimentally by showing that with a wax impregnated niobium tin cylinder quenched in a range of background fields a variation by a factor of six in the critical current density only caused a slight change in the field change which produced flux jumps (Hancox, 1965a).

3. *Calculation of the Stability Criteria for a Round Wire*

So far the simplest geometry has been used for the calculation of the limit of stability for flux motion. In practice, however, the more interesting case is when the superconductor is in the form of wire or strip wound into a coil, and stability is required when both magnetic fields and transport currents are involved. The first approach to this problem is to approximate a single layer of round wire in a long solenoid to a sheet of superconductor whose thickness is equal to the wire diameter. The sheet carries a current and also experiences an externally applied magnetic field large enough for complete flux penetration. Since the thickness of the sheet D is now quite small the isothermal approximation may be used, and following the same argument as before it can be shown that the system is stable up to the full critical current of the sheet if

$$4\pi J^2 D^2 \leqslant 3CT_0 \qquad (9.39)$$

As before, stability depends on the characteristic temperature T_0 and the specific heat of the superconductor, but now the critical current density and the wire diameter are also involved. If stability is required at high current densities the diameter of the superconductor must be kept small, and it will be seen later that this is the basis of intrinsic stabilization in high current density magnets.

Equation 9.39 can be extended by calculating the stability limit for any current density, J averaged over the whole cross-section, less than the critical current density J_c in which case the limit becomes

$$4\pi(J_c^2 + 3J^2)D^2 = 12CT_0 \qquad (9.40)$$

This region of stable behaviour lies in the high field section of a superconducting solenoid, which explains the fact that particularly unstable materials can often be used effectively in a high field magnet when they are quite unsuitable for use in lower fields. This is illustrated in Fig. 9.27 by the behaviour of a coil of niobium tin ribbon when energized inside a larger magnet. With low background fields severe degradation was observed but at higher fields the operating current increased until it approached the critical current of the material. This high field stabilization may be used to optimize the design of a magnet using two or three different materials graded to match their critical currents and stability characteristics to the local fields which they experience in the magnet.

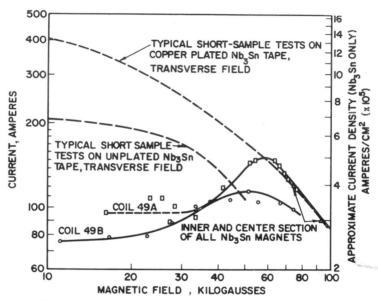

Fig. 9.27. Critical currents and current densities for (a) unplated and copper plated niobium tin tapes and (b) two coils built from similar tapes (Schrader, Freedman and Fakan, 1964).

Since the low field regions of a magnet are the least stable it is necessary to calculate whether any flux jumps which occur will be large enough to drive the superconductor normal and at what current level the superconductor will quench. The limiting case is when the highly diamagnetic critical state collapses to the point at which the transport current flows uniformly through the whole cross-section of the conductor (Hancox, 1966). Using the isothermal approximation as before and neglecting any variation of the critical current density with magnetic field, this gives

$$\pi\left\{J_c^2 - J_d^2 + 6J_d^2 \ln\left(\frac{J_c}{J_d}\right)\right\}D^2 = 6\int_{T_1}^{T_2} C\, dT \tag{9.41}$$

where J_c and J_d are the critical current densities at the temperatures T_1 and T_2, before and after the transition.

The limits defined by equations 9.40 and 9.41 are shown in Fig. 9.28 for a typical niobium titanium wire, and form a stability diagram which indicates regions of stability, partial flux jumping and quenching for a wire when used in a magnet. The current at which the magnet quenches is the lowest unstable

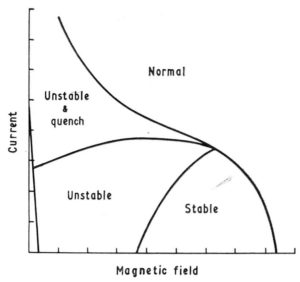

Magnetic field

Fig. 9.28. Stability diagram for a round superconducting wire when used in a solenoid. The boundaries below the critical current curve correspond to equations (9.40) and (9.41).

current in the relevant field range generated by the winding and, in some cases, this may be well below the critical current of the wire.

As with the stability limits calculated in previous sections the simplifications required to obtain a tractable mathematical model together with a lack of adequate low temperature data means that the stability diagram is not suitable for precise quantitative predictions of the behaviour of specific magnets. It does, however, act as a very useful basis for further discussion of the causes of degradation and methods of stabilization in wire wound magnets.

4. The Stability Criteria in a Magnetically Damped Conductor

Damping of the flux motion has been neglected so far since the electrical diffusivity of a superconductor is always much greater than its thermal diffusivity. There are some situations, however, in which the damping of flux motion will be significant such as multi-filament conductors incorporating many fine strands of superconductor in a copper matrix or strip wound coils in which superconducting ribbon is clad with high conductivity copper. It is therefore necessary to discuss the opposite extreme in which it is assumed that flux motion is heavily damped and the temperature is controlled entirely by thermal conduction. The stability criteria in this situation were originally derived by Hart (1968).

As before a slab of superconductor is considered of thickness D, with an externally applied magnetic field parallel to the surface which is sufficient to cause complete flux penetration. Stability is studied by considering the result of a small temperature disturbance δT, and the system is considered dynamically stable if the temperature decays back to its original value. Since the flux motion is heavily damped the original magnetic field distribution is maintained during this temperature excursion and so, although the critical current density is reduced, the actual current density cannot change appreciably, resulting in a rate of heat generation $\dot{q} = J_c^2 \rho \delta T / T_0$ where ρ is the average electrical resistivity. If the thermal impedence at the interface between the superconductor and its surroundings is small the temperature is determined by the thermal conductivity K together with the boundary condition that the temperature at the surface of the superconductor is constant. The solution of the diffusion equation is a time dependent Fourier series, and the requirement for stability is that none of the terms grow with time, which leads to the criterion

$$J_c^2 D^2 \leqslant \pi^2 K T_0 / \rho \qquad (9.42)$$

In the simpler case of a semi-infinite block of superconductor without transport current this simplifies to

$$H^2 \leqslant 4\pi^4 K T_0 / \rho \qquad (9.43)$$

which is similar to the result obtained by Chester (1967) for the stability of a laminated structure in a perpendicular magnetic field.

If the surface thermal barrier is more important than the bulk conductivity of the superconductor, the corresponding stability criteria are obtained in terms of the surface conductivity h

$$J^2 D \leqslant 2h T_0 / \rho \qquad (9.44)$$

$$H \leqslant 4\pi h T_0 / J\rho \qquad (9.45)$$

As in previous criteria it is seen that stability is influenced by the temperature dependence of the critical current density, and that reduction of the dimension of the conductor perpendicular to the magnetic field will allow stability to be obtained at higher current densities. Improved stability can also be obtained by increasing both the thermal and the electrical conductivities of the system by the addition of a normal material such as copper.

C. DEGRADATION AND STABILIZATION OF SUPERCONDUCTING MAGNETS

The instabilities in flux motion described in the last section have serious consequences in the design and operation of superconducting coils. When wires of niobium–zirconium and niobium–tin capable of carrying high currents in useful magnetic fields were first fabricated in 1961 it appeared that superconducting magnets generating 50 to 100 kgauss without any power dissipation were at last a practical reality, but early experience proved disappointing and magnet development was painfully slow.

Two major problems were encountered in early superconducting magnets. Firstly, it was generally found that as the current was increased the coil suddenly became normal at a current well below that expected from critical current measurements on a short sample of wire. This effect, termed degradation, was neither predictable nor reproducible but was a local effect occurring randomly in various parts of the coil. Secondly, such normal regions propagated through the coil and unless the current was rapidly reduced excessive heating occurred resulting in permanent damage to the coil. This latter problem was overcome by plating the wire with copper, thus reducing the normal state resistance by between one and two orders of magnitude, which was sufficient to prevent damage in a small coil. This question of protection will be dealt with in greater detail in section D.

The problem of degradation was only slightly reduced by copper plating of the wire and quenching currents were still typically 30% to 60% of the critical current of the superconductor. Furthermore, degradation appeared to depend on coil geometry, became more serious in larger coils and was worst in coils wound from materials with a high critical current. Whilst all these effects are now only of historic interest, they delayed the development of superconducting magnets for several years and, at times, doubts were raised as to whether the problems were so fundamental as to be insoluble.

During this early development of superconducting magnets, several clues were discovered to the complex process involved in the quenching of a coil which lead in due course to a greatly increased understanding of their behaviour. For example, it was often reported that coils were "noisy" and that the terminal voltage during current changes was not steady but showed a a continuous series of transients corresponding to discontinuous changes of inductance, which we can now associate with unstable flux penetration into the wire. It was also observed that degradation could be induced in some short samples of wire by field changes rather than by current changes, and that this was most pronounced in low magnetic fields and in materials with a high critical current density. Combining these observations, Iwasa and Montgomery (1965) were able to induce degraded performance in short lengths of wire subjected to a swept magnetic field by the application of a small superimposed field pulse. Their results, shown in Fig. 9.29, indicated that the pulse only served to initiate unstable flux motion, and that there was a minimum transport current above which instabilities, once established, caused sufficient heating to quench the wire. This threshold current was related to the quenching current of a degraded coil, but since neither the thermal environment nor the magnetic history were identical to conditions in a coil the two currents did not correspond exactly.

It is now generally agreed that the principle cause of degradation in superconducting coils is unstable flux jumping. The situation may be confused,

however, by the effect of wire movement. Within a coil the wire is subject to
large magnetic forces and slight mechanical movement can occur relative to
the local magnetic field which is equivalent to rapid flux motion through the
conductor and can cause quenching. This may cause "training" in degraded
coils, so that the quenching current increases with successive quenches until
the final operating current is well above the initial quenching current. In such
a situation it is uncertain whether after training no further movement takes

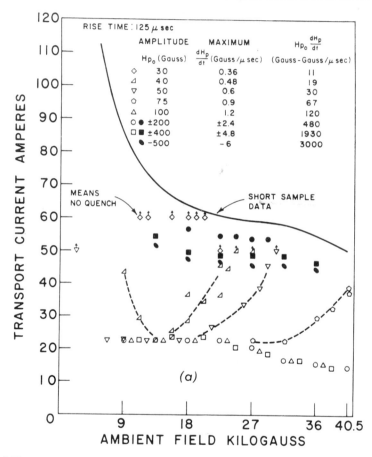

Fig. 9.29. Transport current in a niobium–zirconium wire just sufficient to cause quenching
on the application of a pulse of specified rise time, polarity and amplitude. The sample
was 15 m of non-inductively wound unplated wire (Iwasa and Montgomery, 1965).

place or whether the winding simply reaches some equilibrium in which
small movements still trigger flux jumps. Comparison of the performance
of coils which have been tested both before and after impregnation with epoxy
resin suggest that in unimpregnated coils wire movement is still important.

A further factor which complicates the situation in coils is the presence of reversed or rotating fields. In most short sample tests uni-directional fields are used, but it has been shown that degradation can be induced in a wire by exposing it to a small reverse field before applying the field in the normal direction (Taquet, 1965) and small coils tested in an external magnetic field show severe degradation when they experience reversed fields. Since in a large superconducting coil the changing diamagnetism of the superconductor can cause the kernal, or zero field point, to move as the current is increased a reverse field of several hundred gauss can exist, although it is not certain that this effect is large enough to be significant.

Finally there is still the possibility that, due to the close packing of the wires in a coil, a flux jump at any point will trigger further flux jumps in adjacent regions and that by some co-operative process instabilities which are individually too small to quench the coil may grow to a sufficient size to cause degradation. The coupling between flux jumps will be both magnetic and thermal and it has been suggested that this interaction is the cause of a size effect which results in greater degradation in large coils. No satisfactory analysis of this complex situation is available since, with increased understanding of the cause and the growth of single flux jumps, most thought has been given to the question of stabilization and the elimination of degradation rather than to the understanding of degraded coils.

1. Methods of Stabilization

A wide variety of methods of stabilizing superconducting coils have now been developed. Several have evolved empirically and been proved by experience, others are the direct result of the application of stabilization theory. Most approaches are applicable to particular types or sizes of coil, and the choice of method depends on many factors, including the degree of reliability which is required.

In order to categorize the various stabilization techniques they are generally divided into two groups depending on whether or not normal regions are allowed to exist in the winding. In the first group the philosophy is to allow the formation of normal regions provided that they do not grow to a size which interferes with the performance of the coil or causes it to quench. The behaviour of such limited normal regions can be analyzed on the basis of steady state theory, and the resulting criterion is often referred to as cryostatic stabilization. Alternatively, the existence of these normal regions can be regarded as a transient phenomenon and this leads to the concept of a minimum propagating current below which they will not spread but rather will contract and disappear. This latter approach is usually experimental rather than analytical, but the principles are similar in both cases.

The second group of stabilization techniques requires a fuller understanding of the mechanism of flux jumping and the way in which it can lead to the formation of localized normal regions. The principle is that instabilities should either be completely eliminated or at least so limited that they are prevented from driving the superconductor normal. There are several approaches to this type of stabilization including enthalpy stabilization, intrinsic stabilization and dynamic stabilization. These techniques are based on the analysis of unstable flux motion considered in Section B and their validity has been demonstrated in a variety of magnets. All of these methods of stabilization will be considered in detail in the following sections. In general those in the first group, particularly cryostatic stabilization, are most applicable in large magnets operating at modest current densities, i.e. 2,000 to 10,000 A/cm², where a high degree of reliability is required. Those in the second group allow magnets to be operated at much higher current densities, i.e. 10,000 to 50,000 A/cm² and this makes protection more difficult and limits their use to smaller magnets.

2. Cryostatic Stabilization

In magnets in which the appearance of normal regions is allowed a fully stable winding may be defined as one in which the normal regions disappear again for all currents up to the designed maximum operating current. This design current is usually, but not necessarily, the critical current of the superconductor in the region of highest magnetic field. By definition a fully stabilized magnet does not exhibit degradation. For a preliminary consideration of this type of stabilization the nature of the disturbance causing the appearance of the normal region is unimportant, although it will be seen later that a simple understanding of the mechanism of degradation is necessary if catastrophic quenches are to be avoided in large magnets.

The principles of full stabilization were first outlined by Kantrowitz and Stekly (1965). The superconductor is backed by a normal material of high electrical conductivity, such as copper or aluminium, which is exposed to liquid helium to obtain good cooling of the composite conductor. To achieve stabilization it is necessary that the cross-section of normal material and the proportion of its surface exposed to the helium are sufficient that with all the current flowing in the normal material the temperature of the superconductor is still below its critical temperature. In this situation the superconductor cannot be permanently driven normal and after any disturbance the current will transfer back into the superconductor.

Kantrowitz and Stekly demonstrated their principle by building a 12·5 cm bore 40 kgauss coil using a conductor with nine niobium zirconium wires embedded in a copper strip. The ratio of the cross-sectional areas of copper and superconductor was 27 : 1, which was sufficient to stabilize the coil at its critical current of 710 A. Not only was the coil free from degradation and training but it could be operated at currents up to 820 A in the current sharing mode, i.e. with critical current flowing in the superconductor and the excess current flowing resistively in the copper. The appearance of resistance in the coil in this situation occurred gradually and there was no sudden transition to normality. A further example of a cryogenically stabilized winding is the bending magnet shown in Fig. 9.30, built at the Rutherford High Energy Laboratory.

Fig. 9.30. Cryostatically stable bending magnet built at the Rutherford High Energy Laboratory.

A simple analysis of the behaviour of fully stabilized coils was given by Stekly and Zar (1965), assuming a constant heat transfer coefficient for the interface between the normal material and the liquid helium. If a fraction f of the total current I is flowing in the normal material the voltage per unit length of conductor is $V = \rho I f / A$ where ρ and A are the resistivity and cross-sectional area, respectively, of the normal material. The temperature difference between the conductor and the helium is then $T - T_h = VI/hP = \rho I^2 f / hPA$ where h is the heat transfer coefficient and P the perimeter of the conductor exposed to the helium. The critical current of the superconductor is assumed to fall linearly with increasing temperature from I_c at liquid helium temperature T_h to zero at the critical temperature T_c, so that the maximum current that can flow in the superconductor is

$$I_s = I_c\{1 - (T - T_h)/(T_c - T_h)\}.$$

Introducing a stability parameter α such that $\alpha = \rho I_c^2 / hPA(T_c - T_h)$ yields the following relationships

$$f = \frac{(I/I_c) - 1}{(I/I_c)(1 - \alpha(I/I_c))} \qquad (9.46)$$

$$V = \frac{\rho I_c\{(I/I_c) - 1\}}{A 1 - \alpha(I/I_c)} \qquad (9.47)$$

The variation of the voltage V is shown in Fig. 9.31 for a range of values of the stability parameter α. Two types of coil behaviour can be distinguished, depending on whether α is less than or greater than unity. If α is less than

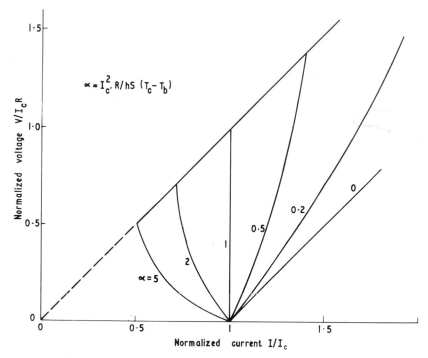

Fig. 9.31. Normalized voltage–current characteristic of a cryostatically stabilized conductor, assuming a constant surface heat transfer coefficient (Stekly and Zar, 1956).

unity, corresponding to a large cross-section of normal material and good heat transfer to the liquid helium, the function f is always single valued, changing smoothly from zero for all currents below the critical current to unity at very high currents. This corresponds to full stabilization for all currents up to the critical current and, if the conductor is driven normal, it will recover and the current will transfer back to the superconductor. For currents above the critical current stable current sharing is obtained with a smooth transition from superconducting to resistive behaviour.

If the stability parameter α is greater than unity the function f may be double valued, being either zero or unity. Any intermediate value is unstable in a constant current system. This situation corresponds to under stabilization and the result of a temporary disturbance will be that all the current is forced into the normal material causing such an increase in temperature that the superconductor cannot recover. Stability is still possible, however, at currents below $I_c/\sqrt{\alpha}$ where the parameter f is single valued.

The description of stability given above is very much simplified, because the coefficient of heat transfer to the liquid helium is not constant but a function of the heat flux, with a discontinuity as the mechanism of heat flow changes from nucleate boiling to film boiling. With increasing heat fluxes nucleate boiling with its relatively good heat transfer and small temperature differences occurs up to a maximum flux Q_n, at which point there is a transition to film boiling with much larger temperature differences. With decreasing heat fluxes the process is not reversible but film boiling continues to a lower flux Q_f before a transition back to nucleate boiling is possible. The values of Q_n and Q_f lie in the range 0·1 to 1·0 W/cm², and are dependent on the surface condition and orientation of the interface between the conductor and the liquid helium.

A simple analytical solution of the stability equations is now no longer possible, but the heat transfer data may be used to compute a revised plot of the voltage shown in Fig. 9.32. Gauster and Hendricks (1968) have compared, measured and computed voltage-current characteristics for a series of cabled niobium titanium composite conductors with copper to superconductor ratios varying from 27 : 1 to 3 : 1, and have shown that they are in good agreement. Furthermore the measurements from fourteen samples were all consistent with a single heat transfer characteristic. It may be concluded, therefore, that this type of calculation is accurate enough for use in the design of magnets, and that the heat transfer characteristic of a conductor measured in an environment identical to that to be used in a magnet is an adequate basis for design.

The behaviour of a conductor can now be characterized by three currents— the critical current of the superconductor, the take-off current at which the steady state current sharing fails to be reversible, and the recovery current at which the voltage returns to the reversible characteristic. Because the surface temperatures which exist during film boiling are comparable to the transition temperatures of the commonly used superconducting alloys niobium zirconium or niobium titanium, for these materials the take-off

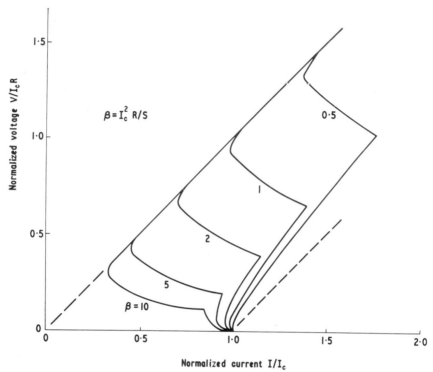

Fig. 9.32. Normalized voltage–current characteristic of a cryostatically stabilized conductor, taking into account the non-linear surface heat transfer coefficient.

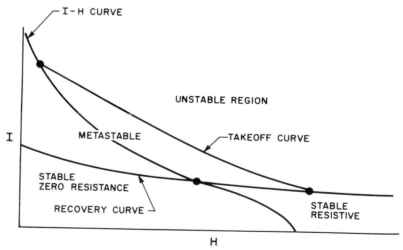

Fig. 9.33. Diagram indicating the modes of behaviour of a cryostatically stabilized conductor (Stekly *et al.*, 1968).

current corresponds to the transition from nucleate to film boiling and the recovery current is closely related to the minimum heat flux required to maintain film boiling. For superconductors with much higher transition temperatures such as niobium tin, however, these relationships are not so simple. In either case because of non-linearities in the heat transfer characteristics it is sometimes possible for the take-off and recovery currents to be double valued and, in these cases, the lowest value is of greatest practical importance.

The relationship between the critical, take-off and recovery currents of any particular composite conductor will change with the magnetic field, and Fig. 9.33 shows the various modes of operation which are possible. For cryogenic stability the conductor in a coil must always be in the stable-zero resistance region, with the operating current below both the critical current and the recovery current. In order to optimize the design of magnets the recovery current should be as high as possible and several detailed studies have been made of its dependence on the shape and size of the cooling channels and of the effect of the surface finish of the conductor. With a clean copper surface the minimum film boiling flux may be as low as $0\cdot15$ W/cm² but, with suitable surface treatment, it may be increased to $0\cdot6$ W/cm² (Butler $et\ al.$, 1969). With this limitation, and the requirement to provide adequate cooling channels for liquid helium the overall current density in a cryostatic stabilized magnet is usually in the range 2,000 to 5,000 A/cm².

So far in the analysis of cryostatic stability the size of the composite conductor or of the superconducting filaments within it have not been considered. In a more detailed analysis, however, both must be taken into account. Firstly, the temperature has been assumed to be uniform across the conductor but this may not be true and the superconductor, which has a poor thermal conductivity, may experience a thermal gradient in the current sharing mode. There is a characteristic radius of the superconducting filament above which this effect is important (Stekly $et\ al.$, 1968), which is

$$r_c = \sqrt{[K(T_c - T_h)/\rho J_c J_s]} \qquad (9.48)$$

where K is the thermal conductivity of the superconductor and J_c and J_s are respectively the critical current density of the superconductor and the current density in the normal conductor when it carries the critical current. The effect of the size of the superconducting filament on the voltage–current characteristic in the limit of an extremely well cooled conductor ($\alpha = o$) is shown in Fig. 9.34.

For a round filament of radius greater than $\sqrt{8}\ r_c$ the function becomes double valued at the critical current, implying that such a conductor will be unstable no matter how well it is cooled. A $0\cdot025$ cm diameter filament of niobium titanium operating in a field of 50 kgauss would show such an instability when used in a composite conductor with a copper to superconductor ratio of 2 : 1, and even in a highly stabilized conductor with a copper to superconductor ratio of 50 : 1 the filament size should not exceed 0·125 cm diameter.

Another assumption implied in the stability criterion was that the calculation was purely a steady state balance between the power dissipated in the composite conductor and the cooling of the liquid helium. This is only true if the energy dissipated in the initial disturbance forming the normal region is small, and this may set a maximum overall size to the composite conductor.

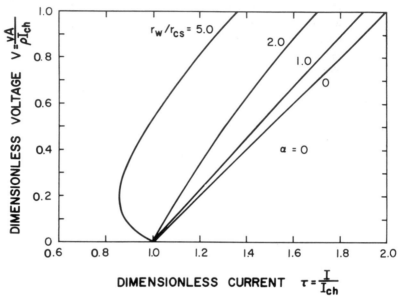

Fig. 9.34. Normalized terminal characteristics for a well cooled conductor ($\alpha \rightarrow 0$) as a function of the size of the superconductor (Stekly *et al.*, 1968).

No precise criterion exists yet for this limitation but it may be of importance in a very large magnet using wide strip, such as is used in bubble chambers. In such a conductor the current distribution will be very non-uniform due to the magnetization currents along the edges of the strip, and if the superconductor is uniformly distributed across a strip of width D the maximum energy which can be released by the collapse of these currents in a flux jump is $Q \approx \pi D^2 J^2 / 4$ where J is the critical current density averaged over the whole cross section of the conductor. If this energy is sufficient to vaporize all the liquid in a cooling channel adjacent to the conductor the disturbance might be serious enough to prevent subsequent recovery of the normal region. Assuming the cooling channel occupied 20% of the winding volume the condition for this to happen would be $JD \approx 2 \cdot 6 \times 10^4$, so that for an average current density of 4,000 A/cm² in the conductor its width should not exceed 6 cm. Whilst this type of failure has not yet been observed, it must obviously be taken seriously in the design of large magnets. Flux jumps of this type also represent large mechanical disturbances and, for this reason also, the width of the conductor should be limited.

3. *Transient Cryogenic Stabilization*

The analysis of cryostatic stabilization assumed steady state conditions with heat flow only across the boundary between the conductor and liquid helium, and recovery of a normal region occurring simultaneously at all points. In practice however there is also heat flow along the conductor at the ends of the normal region and the boundary between the normal and superconducting regions may move along the conductor. The velocity at which this boundary propagates will vary with the current, and may be positive (corresponding to growth of the normal region) or negative (corresponding to recovery) and at some particular current will be zero. This current at which the normal region is static is known as the minimum propagating current and can be used as a design criteria in superconducting coils.

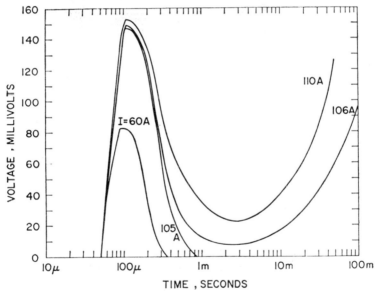

Fig. 9.35. Voltage traces for a conductor which is temporarily driven normal with a heat pulse (Iwasa *et al.*, 1969).

If a reasonable length of normal region is assumed so that each end may be treated separately, the minimum propagating current can be defined in terms of the Stekly stability parameter, and is

$$I_m = I_c\{(1+8\alpha)-1\}/2\alpha \qquad (9.49)$$

This current is greater than the cryostatic recovery current $I_c/\sqrt{\alpha}$ for all values of α greater than 1·0, although in practice the increase is only 10 or

21

20%. The main advantage of this approach is that, since the normal regions are only of limited length and disappear rapidly, only small cooling channels are required for the liquid helium, and higher overall current densities can be achieved.

In addition to the effect of longitudinal thermal conduction, recovery of a short length of conductor may also be improved due to the transient nature of the processes involved. A considerable advance in the understanding of these processes is due to the observations of Iwasa *et al.* (1969) on the variation of the resistance of a normal region. Measurements show, Fig. 9.35, that at currents only slightly above the minimum propagating current the normal region almost completely recovers after it has formed before becoming resistive again. This is due to the temporal variation of the nucleate heat transfer coefficient which may be as high as $2 \cdot 5$ W/cm^2 for periods up to one milli-second before a transition to film boiling occurs (Jackson, 1969). Thus the complete recovery of a normal region depends on whether the surface dissipation falls faster than the time dependent heat flux required to cause a transition from nucleate to film boiling and, if this is not the case, the normal region will begin to grow again as film boiling is established.

Although few of the factors which affect transient recovery are sufficiently well known to permit the accurate design of a conductor, a great advantage of the principle of transient stability is that the minimum propagating current can easily be measured on a short length of conductor (Cornish, 1966). In this way it is possible to develop empirically a composite conductor which, in a given magnetic field and thermal environment, will have any required minimum propagating current. The measurement is made by generating a small normal region by the application of a thermal or magnetic pulse, and observing propagation along the conductor by means of a series of potential tappings.

4. *Enthalpy Stabilization*

In the two methods of stabilization so far considered the formation of normal regions in the winding has been allowed and a parallel low resistance path provided for the current during the recovery to the superconducting state. The alternative approach to stabilization is to prevent the appearance of normal regions, with the advantage that the large cross-section of normal conductor is not required and higher overall current densities can be achieved. Several methods of doing this have been used, although none of them have been developed to the point where exact criteria can be quoted which have been proved by widespread use.

Historically the first method was enthalpy stabilization (Hancox, 1968) and, although it was quickly superseded by better alternatives, it is included here because it was the first indication that the improved understanding of unstable flux jumping could be applied to the stabilization of practical magnets. It follows directly from equation 9.41 which gives the energy dissipated in a flux jump in a round wire and the enthalpy required to prevent it from driving the coil normal. Since the specific heat of both the superconductor and copper cladding are small at $4.2°K$ (about 1.0 mJ/cm^3°K) it was suggested that the addition of some other material of higher specific heat such as indium (7.3 mJ/cm^3°K) or lead (9.1 mJ/cm^3°K) would give stability. It was also necessary to subdivide the superconductor to improve heat transfer, and to encapsulate the coil to prevent wire movement. A small 25 kgauss coil built in this way with lead plated niobium titanium wire operated reliably at its critical current of 162 A, corresponding to a current density of 12,500 A/cm^2, at $4.2°K$ in either liquid or gaseous helium. A similar coil using the same conductor without lead plating showed severely degraded performance.

The significance of this approach was not only that an improved overall current density was obtained, but that the coil was no longer dependent on good thermal contact between the conductor and liquid helium so that the cooling channels in the winding were no longer necessary. Furthermore, it had often been observed that the performance of degraded coils improved as the temperature was raised, and this is now seen to be due to the rapid increase of specific heat with temperature.

The principles of enthalpy stabilization have also been applied by Hale and Williams (1968) to coils wound with 0.23 cm wide niobium–tin ribbon, and it was found that the use of interleaving between layers of a high specific heat material such as cadmium was more effective than the more commonly used copper. In the case of strip wound coils, however, the energy involved in flux jumping is so much larger than in wire wound coils that it is unlikely that the coil could be fully stabilized in this way.

5. *Intrinsic Stabilization*

Equation 9.41 shows that a superconducting wire can be stabilized not only by increasing its enthalpy but also be reducing its diameter so that the energy involved in the collapse of the magnetization currents becomes very small. This implies superconducting filaments with diameters of 50 μm or less compared with the commonly used sizes of 250 μm or more. Because of the difficulty of using such fine wires, this approach did not at first seem very attractive but the development of multi filament co-processed niobium titanium composite conductors has now made it feasible.

The behaviour of a multi filament conductor is more complex than that of a single core wire because it is time dependent. As before, a two-dimensional model will be considered in which a layer of conductors is replaced by a uniform sheet conductor, and to simplify the calculation further it will be assumed to consist of only two superconducting sheets of thickness d separated by a normal conductor. In such a composite conductor magnetization currents in the superconductor which must also flow through normal material at the ends of the conductor, will decay with time. If the length of the conductor is l the decay time constant will be $\tau \approx 4\pi l^2/\rho$ and therefore these currents will be significant only if the rate of change of external field $\dot{H}$ is sufficient to induce the currents faster than they decay, i.e. if $\dot{H} < 4\pi J_0 d/\tau$. Thus there is a characteristic length

$$l_c = \sqrt{2J_c d\rho/\dot{H}} \qquad (9.50)$$

such that a conductor of length much greater than l_c will support magnetization currents and will behave as if it were a single large superconducting wire, whereas a conductor of length much less than l_c will only support circulating currents within the individual filaments. In the latter case the behaviour of the conductor is determined by the behaviour of the independent filaments and, if these are small enough, the conductor will be stable.

For a typical niobium titanium multi filament conductor with filaments of 50 μm diameter and current density $3 \cdot 10^5$ A/cm^2 embedded in copper of resistivity $2 \cdot 10^{-8}$ Ωcm and the external field varying at a rate of 30 gauss/sec the characteristic length is 10 cm. This is so short that it would be expected that the behaviour of the conductor in a coil would always be determined by the overall conductor size and could not be improved by the use of small filaments. This difficulty can be overcome, however, by twisting the conductor (Smith *et al.*, 1968) with a pitch less than the characteristic length so that the length of any magnetization current loop is limited and the currents decay sufficiently rapidly to allow intrinsic stabilization with small filaments.

The effectiveness of small filaments and a twisted conductor have been demonstrated by Iwasa (1969). Figure 9.36 shows magnetization measurements made on three non-inductively mounted conductors each 110 cm long and 750 μm diameter. Sample A was a single core wire and showed a rate independent magnetization and unstable flux jumps when a small field pulse was applied. Sample B contained 131 superconducting filaments uniformly distributed over the cross-section of the conductor and showed a rate dependent magnetization which decayed with a time constant of 27 sec when the field was held constant. Finally sample C was an identical 131 filament conductor which had been twisted with a pitch of approximately 1 cm and showed only a small magnetization which was rate independent and was stable against flux jumping. The principle of intrinsic stabilization has also been demonstrated in several small coils in which copper to superconductor ratios as low as $1 \cdot 1 : 1$ have been used and overall current densities over 50,000 A/cm^2 have been successfully obtained.

The simple solution of twisting a multi filament conductor to break up the diamagnetic current loops is only strictly correct if all the filaments are equally spaced from the longitudinal axis of the superconductor. Where the filaments are uniformly distributed over the cross-section of the conductor they should really be transposed so that all current paths are identical. In a co-processed conductor this is impossible and twisting must be accepted as an adequate approximation unless the overall diameter of the conductor is too large or the

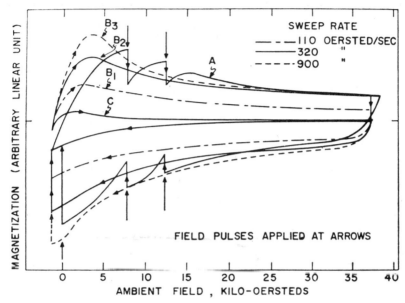

Fig. 9.36. Magnetization of three types of composite conductor (Iwasa, 1969).

(a) single superconducting core.
(b) multifilament conductor, untwisted.
(c) multifilament conductor, twisted.

field gradient generated by the magnet is too great so that conditions change appreciably across the diameter of the conductor or within its twist pitch. Smith *et al.* have given the following approximate criteria for the conditions under which twisting will be acceptable,

$$\frac{\Delta H \text{ across composite}}{\text{Mean } H \text{ across composite}} < \frac{d}{D} \qquad (9.51)$$

$$\frac{\Delta H \text{ along one twist}}{\text{Mean } H \text{ along one twist}} < \frac{d}{D} \qquad (9.52)$$

where d is the diameter of an individual filament and D the diameter of the complete composite. Up to 360 filaments can be used without greatly increasing the cost of the conductor so that d/D is limited to about 0·03. If larger diameter conductors are required several smaller multi filament conductors can be formed into a larger transposed conductor.

A further limitation to the use of simple twisting to give intrinsic stabilization is the possible instability of flux which must move across some filaments because not all current paths are identical. This instability is similar to the one-dimensional case considered in section B with damping of the flux motion due to a high conductivity material surrounding the superconducting filaments. For a typical niobium titanium and copper composite the maximum overall diameter of a simple twisted conductor should be 0·1 to 0·2 cm but there are no experimental measurements to confirm this result or indicate the accuracy of the calculation.

An application in which intrinsic stabilization is of particular interest is in pulsed or low frequency alternating current magnets where filamentary conductors must be used to reduce the power dissipation. In a pulsed synchrotron magnet, for example, the rate of change of magnetic field $\dot{H}$ may be as high as 60 kgauss/sec but, with the filaments in a copper matrix, the characteristic length would be only 0·2 cm and it would be impossible to achieve sufficient twisting to satisfy the stability criterion. This difficulty can be overcome by the use of a high resistivity matrix such as cupro-nickel, with a resistivity of 20×10^{-6} Ωcm, which increases the characteristic length to several centimetres. Such conductors have no low impedance path in parallel with the superconductor and introduce serious problems of coil protection, but it may be possible to obtain a compromise solution by the use of a tripartite composite with a high resistivity layer around each filament and a copper matrix.

At higher frequencies intrinsic stabilization offers the only hope of stabilizing alternating current conductors of type II superconductors carrying current in their bulk, but for these applications it will be necessary to develop filamentary systems with an insulating matrix or with individual filaments insulated from a conducting matrix.

6. *Dynamic Stabilization*

The last method of stabilization to be considered in detail is that resulting from magnetic, or flux, damping. The importance of damping the unstable motion of magnetic flux was realized at a very early stage in magnet development, and one advantage of the use of copper plating on niobium–zirconium wire was that it not only provided an alternative current path and increased the minimum propagating current, but that it also limited sudden changes of field in the wire and reduced the severity of flux jumping. This effect was increased by the use of copper foil between layers of a solenoid but, in neither case could the flux damping be fully effective because of the poor electromagnetic coupling between the copper and the superconductor. With the introduction of multifilament conductors the coupling was improved and in the last section it was seen that the criterion for the maximum diameter of such a conductor was determined by flux damping. Its greatest importance, however, is in strip wound niobium tin magnets where the superconductor is in the form of a thin layer and in intimate contact with a backing strip and flux damping becomes more effective than any other method of stabilization.

The behaviour of strip wound coils is complicated by the fact that the criteria for stability varies through the volume of the winding. When the direction of the magnetic field is parallel to the surface of the strip, flux motion is limited to the thickness of the superconducting layer and is generally intrinsically stable or enthalpy stabilized. When the direction of the magnetic field is perpendicular to the surface of the strip, however, the flux moves across the strip and stabilization must rely on the damping of flux motion by the copper. This is in general the least stable region of a magnet and the criteria derived in Section B show that good thermal conduction across the strip and into the liquid helium are both essential for stability. The criteria can be written in terms of a maximum allowed field component perpendicular to the surface of the tape, and experience with strip wound solenoids shows that this is correct. For example, with 1·25 cm wide niobium-tin tape capable of carrying 300 A at 100 kgauss and clad with 25 μm of copper on either side this limiting field component is about 30 kgauss. This allows 100 kgauss solenoids to be built relatively easily with overall current densities of 20,000 to 25,000 A/cm^2, but solenoids generating fields much higher than this are more difficult and require additional stabilization in the end discs to deal with the increased radial field component.

The exact thickness of copper which is sufficient to stabilize any particular winding must be established empirically since the stability criteria are not accurate enough to yield precise results, and because in practice there is also a rate dependent effect so that more copper is required if the field is to be increased rapidly. However, if additional copper is needed it may either be an integral part of the composite strip or else wound into the coil in parallel with the superconducting strip. No electrical or thermal connection is necessary since the essential function of the normal material is to damp the flux motion, although for protection purposes it is usually preferable to have the extra copper in direct contact with the superconducting strip.

An important limitation in strip wound coils is the thickness of the superconducting layer, because of the poor thermal conductivity of the superconductor. This limit is given by Hart (1968) as

$$J^2 d_s^3 < \frac{\pi^2 K_s T_0 d_n}{\rho} \tag{9.53}$$

where d_s and d_n are the thicknesses of the superconducting layer and the normal material, K_s is the thermal conductivity of the superconductor, and ρ the resistivity of the normal material. If the critical current density in the super-conductor is typically 10^6 A/cm^2 and the other parameters for a niobium-tin tape are as in the previous example this corresponds to a superconductor thickness of 6 μm or a current of 360 A in a 1·25 cm wide tape. Since this thickness only varies as the cube root of d_n/ρ this limitation cannot be over-come by the use of more stabilizing material without serious effect on the overall current density of the winding. This means that high current tapes, which are much cheaper to produce than an equivalent length of low current tape, will always be subject to instabilities when the direction of the magnetic field is perpendicular to the surface of the tape and cannot be used in some regions of a winding.

D. PROTECTION OF SUPERCONDUCTING MAGNETS

Whilst there is now a wide range of techniques for stabilizing magnets, the possibility of a sudden transition from the superconducting to the normal state in part of the winding and the consequent release of a considerable proportion of the stored magnetic energy as heat into the magnet and the surrounding liquid helium cannot be ignored. Such a failure may arise be-cause of a mechanical fault within the magnet itself, such as movement of a conductor or a breakage due to excessive mechanical forces, or may be the result of an external disturbance. Whatever the cause, precautions must be taken to allow for the discharge of the energy in the system as rapidly as possible in a controlled and safe manner. The time available for this discharge will vary with the size and construction of the magnet, but may be as short as 50 msec in a small solenoid or several minutes in a large winding such as the magnet for the Argonne National Laboratory bubble chamber which is designed with a time constant of 400 sec. In any case the energy must be removed fast enough to prevent the temperature of any normal region reach-ing a level which will damage the superconductor, and in practice this is taken to be about 0 to 100°C.

The simplest method of discharging the energy in a coil is to disconnect the power supply when a fault is indicated and allow the current to decay through a protective resistor which is permanently connected across the winding. When the superconductor goes normal the current transfers to the normal substrate and the temperature rises as the power dissipation heats the con-ductor. Since the helium will be boiled off rapidly the temperature rise will be governed by the enthalpy of the composite conductor and is determined by the energy balance $\int_0^\infty j^2 \mathrm{d}t = \int_{\theta_0}^{\theta} m(C/\rho)\mathrm{d}\theta = f(\theta_m)$ where θ_0 and θ_m are the initial and final temperatures, and C and ρ are averaged over the cross section of the conductor. If the normal region is only a small part of the whole winding the current will decay exponentially with a time constant $\tau = L/R$, and the maximum allowable current density in the substrate can be evaluated in terms of this time constant and the function $f(\theta)$.

$$J < \{2f(\theta)/\tau\}^{\frac{1}{2}} \qquad (9.54)$$

If, further, the normal substrate to superconductor ratio is not too small the contribution of the superconductor to the total enthalpy of the conductor can be neglected and the temperature rise in the winding can be related directly to the coil current, the cross-sectional area of the normal substrate and the decay time constant.

$$\theta_m = f^*(I^2\tau/A^2) \qquad (9.55)$$

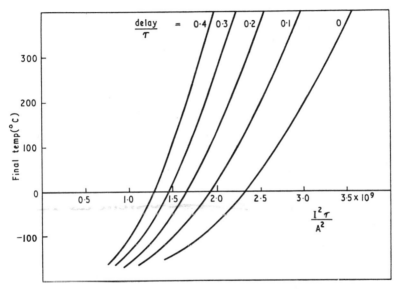

Fig. 9.37. Temperature rise due to resistive heating of a conductor during current decay.

This function is plotted in Fig. 9.37 for copper, and from this generalized curve the rate at which the energy must be extracted from a coil to limit the temperature rise to any predetermined level can be calculated. When this time constant is very small the delay in detecting failure and disconnecting the power supply may be significant and therefore curves are also given which allow various delays to be taken into account. Similar sets of curves have been calculated for other substrate materials (Maddock and James, 1968), but it is found that, from the point of view of protection, copper with its good conductivity and high enthalpy is clearly the best choice.

In practice the decay of current in a magnet can be faster than expected from the time constant calculated from the coil inductance and the resistance of the protective shunt due to propagation of the normal region within the coil. This propagation will not only be along the wire as was assumed in the discussion of the minimum propagating current but can be in two or three dimensions depending on the construction of the magnet. If these propagating velocities are known, the time dependent resistance of the coil and the consequent decay of current can be computed. Figure 9.38 shows a comparison

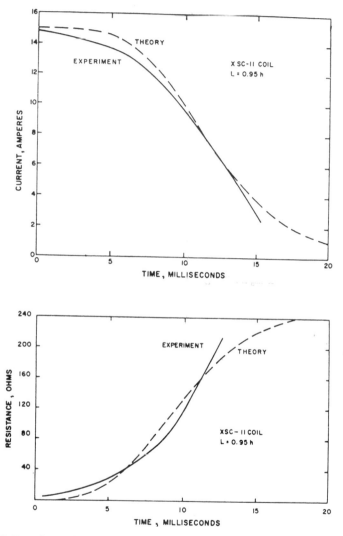

Fig. 9.38. Experimental and theoretical variations of current and resistance during decay due to the appearance of a normal region in a coil (Stekly, 1963).

between the observed decay of current and build up of resistance in an unprotected niobium–zirconium magnet and the computed current and resistance assuming three dimensional propagation (Stekly, 1963).

In very large magnets the stored magnetic energy is so great that safety becomes an important factor in magnet design. Such magnets will generally be cryostatically stable because this gives protection against the widest variety of possible failures, and the restriction which it imposes on the overall current density is not very important. Maddock and James (1968) have considered the optimization of current and current density in such coils and the relationship between the stabilization and protection requirements. The criterion for full stabilization may be written in the form

$$J < sI^{-\frac{1}{3}} \text{ where } s = (Q_cK/\rho)^{\frac{2}{3}} \tag{9.56}$$

J is the current density in the substrate when the superconductor is in the normal state. Similarly the criterion for protection derived above may be written in terms of the magnetic energy and the maximum terminal voltage

$$J < pI^{\frac{1}{2}} \text{ where } p = \{f(\theta_m)V/E\}^{\frac{1}{2}} \tag{9.57}$$

These limits are shown graphically in Fig. 9.39 for several values of s and p and it can be seen that there is an optimum operating current which gives a maximum current density, corresponding to the minimum necessary quantity of copper substrate. This optimum current is given by

$$I_{opt} = \frac{(Q_cK/\rho)^{4/5}}{\{f(\theta_m)V/E\}^{3/5}} \tag{9.58}$$

and increases with increasing stored energy in the magnet. Thus for a 100 Mjoule magnet with a maximum terminal voltage of 20 kV the best operating current would be around 400 A, but for a 10,000 MJ magnet it would be about 6,000 A. The optimum overall current density can also be calculated but, since it depends on the size of cooling channels and other design parameters, it is not so easily expressed analytically.

In practice the optimum current calculated by this method is not necessarily the best solution, and other factors must be taken into account in arriving at a practical design. The forces in the coil system must be considered and, if these are large, it may be necessary to use hardened copper or structural materials such as stainless steel in the conductor so changing the effective electrical and thermal properties. The cost of fabricating and winding the coils may also effect the final choice.

E. ALTERNATING CURRENT LOSSES

Although there are several applications for superconductors involving steady currents, particularly the generation of large volumes of magnetic field, the development and use of superconductors would be much further advanced if they could be used for alternating current applications. Such applications are limited however by the energy dissipation which occurs with changing fields, and even though the losses are orders of magnitude smaller than would be obtained with the corresponding cross-sectional area of copper at room temperature, they are still sufficient to hinder the economic application of superconductors to the generation and distribution of electricity.

The problem of these alternating current losses can be dealt with in two categories. Firstly, losses in a type II material in a magnetic field above H_{c1} in which flux penetration into the bulk of the superconductor occurs. The

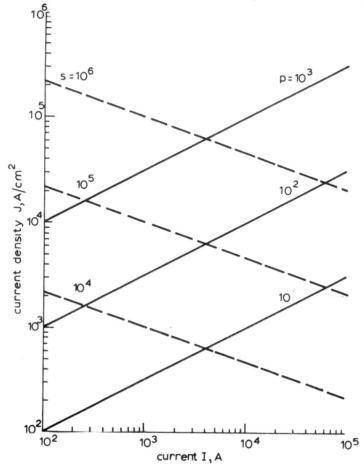

Fig. 9.39. Optimization of current density in a cryostatically stabilized magnet as determined by the requirements of stabilization and protection (Maddock and James, 1968).

theory in this case is based on the critical state model, and the losses can be calculated to within a factor two or three. Secondly, type II materials in fields below H_{c1} or type I materials where only surface currents are involved. In principle these materials should shown no losses but, in practice, due to surface roughness and slight field penetration small, but significant, dissipation is observed. This surface loss cannot be calculated and experimental measurement is the only means of determining the suitability of such materials for a particular application.

The correspondence between alternating current losses in a type II super-conductor measured at 60 and 600 c/sec by a helium boil-off technique and the hysteresis observed in quasi steady state magnetization curves has been convincingly demonstrated by Hart and Swartz (1963). Both magnetization and loss measurements were obtained with V_3Ga powders of various sizes and the agreement between the results is close enough to show that the basic mechanism involved in alternating current losses is the flux pinning which produces hysteresis in the material. It follows that at fields well above the lower critical field H_{c1} it is possible to calculate the losses in a type II super-conductor from the critical state model. If the applied field H_m is less than the field H_p required to fully penetrate the superconductor the loss per unit surface area is

$$Q = H_m^3/24\pi^2 J_c \qquad (9.59)$$

and if the applied field is greater than the field required for penetration the loss is

$$Q = 2D^2 H_m J_c(1 - 2H_p/3H_m) \qquad (9.60)$$

The transition between these two conditions is seen in the results of Chant, Halse and Lorch (1970) shown in Fig. 9.40, which shows the measured and calculated losses for a niobium zirconium strip.

Use of the critical state model breaks down when the applied magnetic field is close to the lower critical field H_{c1} so that surface currents become significant in comparison to the bulk currents in the superconductor. An extension of the critical state model has been proposed by Dunn and Hlawiczka (1968) which takes these surface currents into account, but because of the difficulty in practice of separating the contributions of bulk and surface currents at fields close to H_{c1}, a simpler empirical equation is more often used of the form

$$Q = \frac{(H - \Delta H)^3}{24\pi^2 J_c} \qquad (9.61)$$

where ΔH is a magnetic field comparable, but not equal, to H_{c1}. This equation is only approximate because the surface currents are field dependent, but the two parameters ΔH and J_c can be adjusted to give reasonable agreement with experimental results over a limited range. This is illustrated in Fig. 9.41 which shows experimental measurements of the loss in two samples of niobium with different surface finishes and the calculated loss using equation 9.61 with $J_c = 1 \times 10^5$ A/cm² and $\Delta H = 1,300$ gauss.

At fields below the lower critical field in a type II superconductor, or below the critical field in a type I material, some losses are still observed even though

only surface currents are induced. Figure 9.42 summarizes many of the
measurements that have been made of the surface loss of niobium, and the
variations between different results indicates the dependence of the loss on
surface treatment of the specimens. These losses can be attributed to field

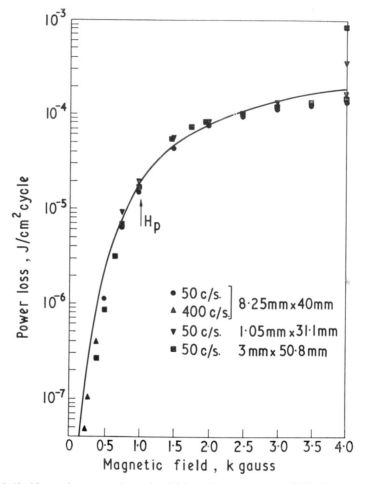

Fig. 9.40. Alternating current losses in niobium zirconium strip at fields above and below
the penetration field, H_p. The points represent experimental results from three different
samples and the curve is calculated from equations (9.59) and (9.60), assuming J is $3 \cdot 3 \times 10^5$
A/cm² (Chant et al., 1970).

enhancement at surface irregularities and to trapped magnetic flux in surface
layers. Measurements on lead cylinders and radio frequency cavities have
shown considerable reductions in losses when the specimens were cooled
under field free conditions in which the earth's magnetic field was cancelled
by compensating coils.

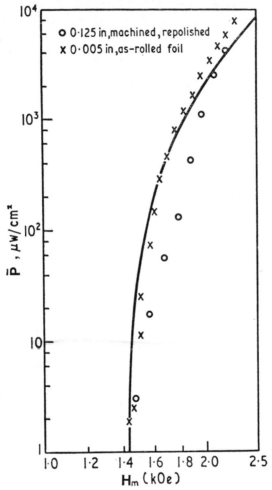

Fig. 9.41. Alternating current losses in niobium at applied magnetic fields above H_{c2}. The points are experimental measurements on two different samples and the curve is calculated from equation (9.61) (Easson and Hlawiczka, 1968).

IV. Applications of Superconductors

Whilst the understanding of the physical behaviour of superconductors has improved rapidly, the application of superconductivity in electrical engineering has been slow. This is because, despite the obvious advantage of zero resistance, operation at very low temperatures involves high additional costs. Improvements in the efficiency and reliability, and reductions in the capital cost, of refrigerators is slowly tipping the balance in favour of superconducting devices, but the expense and inconvenience of low temperature operation will always limit the use of superconductors in engineering applications. Another major disadvantage of superconductors is the power loss associated with alternating or changing currents and fields. This limits the alternating current applications to type I materials operating at relatively low magnetic fields, or

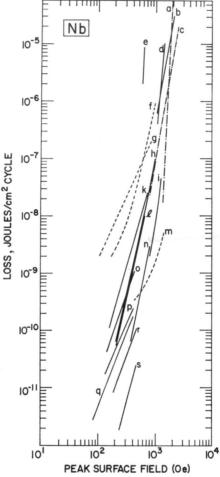

Fig. 9.42. Summary of reported alternating current losses in pure niobium at fields below H_{c1} (Wipf, 1968).

type II materials at very low frequencies. Finally superconducting materials such as niobium titanium and niobium tin will always be more expensive than their conventional counterparts copper or aluminium.

These advantages and disadvantages are illustrated in some of the more promising fields of application in the following section. The examples chosen represent large engineering applications since, in these cases, comparison with existing conventional machinery is possible. Small scale and electronic devices are not considered since most of these are specialized and use the unique properties of superconductors, such as flux quantization, which have no conventional counterpart. Excellent reviews of such devices have been given by Newhouse (1964), Mercereau (1968) and Kamper (1969).

A. APPLICATIONS IN HIGH ENERGY PHYSICS

The increased size of modern accelerators requires a very high power consumption if conventional magnets are used, and there is considerable economic incentive to introduce superconducting magnets. The first application of superconductivity was to the magnets of liquid hydrogen bubble chambers where many of the cryogenic problems had already been solved. This is being followed by the introduction of superconducting beam handling magnets. Superconducting accelerators are already being considered.

Fig. 9.43. Half of the 4·8 m diameter magnet for the Argonne National Laboratory bubble chamber during construction (Purcell, 1968).

1. Magnets for Bubble Chambers

Magnets already in operation have designed fields of 18 kgauss for the Argonne National Laboratory 3·7 m diameter chamber (Purcell, 1968) and some 30 kgauss for the Brookhaven National Laboratory 2·1 m chamber

22

(Brown *et al.*, 1968), whilst a 70 kgauss magnet has been designed for use with the proposed Rutherford Laboratory 1·5 m bubble chamber (Clee *et al.*, 1968).

In all these designs cryostatically stable conductors have been used because reliability and safety are of prime importance and the current density need not be particularly high. The Argonne magnet shown in Fig. 9.43, for example, which used 45,000 kg of composite conductor and has a stored energy of 80 MJ, has an overall current density in the winding of only 775 A/cm². The conductor contains six strands of niobium titanium superconductor which carry a total of 2,200 A in a copper strip 5·0 cm × 0·25 cm with a copper to superconductor ratio of 24 : 1, and for full stability a surface heat transfer coefficient as low as 0·13 W/cm² with only 5% of the surface exposed is adequate. Whilst the capital cost of this magnet, which was $2·4 m, was comparable to the cost of a conventional copper magnet, its running costs are estimated to be $400,000 per annum less due to savings in the power required to energize the winding.

Undoubtedly the successful operation of such large magnets, which have been very conservatively designed, has demonstrated the feasibility and advantage of superconducting magnets where large volumes of high magnetic field are required.

2. *Beam Transport Magnets*

The use of superconducting windings to replace conventional copper coils in beam transport magnets is clearly attractive, and several prototype magnets have been constructed. Bending magnets with twice the field and quadrupole focusing magnets with four times the field gradient of their conventional counterparts are possible, and developments in intrinsic stabilization might allow further improvements. Cost estimates show that a bending magnet will have a comparable cost to a conventional copper magnet, that a quadrupole will be cheaper, and that running costs will be reduced in both cases.

An example is the quadrupole magnet (Cornish, 1968) built for use at CERN. The rectangular conductor consists of 16 strands of niobium titanium in copper and operates at 820 A in liquid helium. Transient cryogenic stabilization is used, allowing an overall current density of 11,000 A/cm² with a maximum field of 49 kgauss in the winding. The quadrupole gives a field gradient of 5·7 kgauss/cm with a useful aperture of 5·0 cm and a length of 70 cm.

3. *Superconducting Synchrotrons*

The maina dvantage of a superconducting synchrotron is the reduced overall size of the machine due to the higher magnetic field at which it could be operated (Smith and Lewin, 1967). For a superconducting machine to be possible, however, there must be a considerable reduction of the alternating current losses associated with type II materials. Present-day accelerators require a pulsed magnetic field rising to its peak in about 1 sec, but the hysteresis loss in most conductors in this situation would be sufficient to quench the winding. The most promising solution is the use of intrinsically stabilized conductors in which the fine superconducting filaments reduce the loss and a high resistivity matrix allows rapid flux penetration into the bulk of the conductor. There are also serious engineering design problems to be overcome such as the problems of accuracy and homogeneity of magnetic fields, restraint of large mechanical forces, and the effects of radiation on both the superconducting material and the cryogenic losses. Should the optimism concerning the development of pulsed magnets not be justified an alternative, but less attractive, approach to the building of superconducting accelerators would be the construction of a fixed field alternating gradient machine.

4. *Superconducting Lineac*

A possible alternating current application of superconductors is the resonant cavities of a linear electron accelerator. The 20 GeV accelerator at Stanford University (Fairbank and Schwettman, 1967) operates at a peak power of 2,000 mW and since such high powers cannot be provided continuously the machine is operated with a 2 μsec pulse and duty cycle of 1 in 1,000, but this introduces several problems such as a magnified effective background in counting experiments and a poor homogeneity in the beam energy due to temperature changes. The use of superconducting cavities cooled by super-fluid helium would overcome these difficulties. To allow continuous operation and also overcome inefficiency of the refrigerators requires an improvement in the Q of the cavities by a factor of 10^6, but this is theoretically possible and a Q of 10^{10} should be obtainable in a lead cavity operating at temperatures below $1\cdot85°$K. Early experiments showed, however, that the Q of an unloaded cavity was only about half the theoretical expectation and did not increase further as the temperature was lowered. For example an unloaded lead cavity at $1\cdot85°$K, resonating at 2,856 Mc/sec, had a Q of 5×10^9.

A major limitation on the Q of such cavities was found to be the effect of trapped magnetic flux, due either to background fields or thermo-electric currents during cool down. Thus the Q of a lead cavity carefully cooled in a field below $0\cdot03$ gauss was 5×10^9 compared with 1×10^9 for the same cavity cooled in a field of $1\cdot5$ gauss. Another major problem in operating high Q cavities was loading and breakdown due to electric fields, which was particularly severe with soft materials such as lead which can be damaged easily.

Following successful work on a range of small cavities the Stanford team built a $1\cdot5$ m long section of an accelerator with 19 lead-coated cavities operating at 952 Mc/sec, which produced a 7 μA beam of 6 MeV electrons. Prospects for further development appear good, particularly with the development of niobium coated cavities with a Q comparable to lead cavities but which are less susceptible to breakdown. Niobium with a resistance ratio of 20,000 is obtainable, and an unloaded cavity has operated in the X band with an unloaded Q of 5×10^{10}, and a loaded cavity has operated with a Q of 2×10^9 and an electric field of 210 kV/cm.

B. APPLICATIONS IN THE GENERATION AND DISTRIBUTION OF POWER

The second area in which the application of superconductivity offers an exciting and profitable challenge is in the generation and distribution of electrical power, since quite modest improvements in efficiency or reductions in capital costs would yield very large savings. On the other hand, since any new components must comply with specifications based on the use of existing conventional equipment, the opportunity for exploiting any novel features of superconducting systems is limited.

1. *Power Transmission*

Several studies of transformers (Lorch, 1969; Harrowell, 1970) have concluded that in order to limit the magnetizing current an iron core is essential and that this determines the overall size, shape and weight of the system. Whilst the use of superconductors in a transformer might overcome the losses due to dissipation in the windings, the thermal losses through the terminals represent an inescapable refrigeration problem and the cost of the refrigerator could be comparable to the capitalized copper losses. Thus from capital cost considerations it appears unlikely that superconducting transformers will become economically attractive until there is a drastic reduction in refrigeration costs or new superconductors become available with higher operating temperatures.

Demonstrable savings in capital costs or, more probably, running costs are expected in superconducting cables compared with their normal equivalent. This will be particularly noticeable in high current or high power transmission systems. Garwin and Matisoo (1967) considered a 100 GW link operating at 200 kV d.c. and transmitting power for 1,000 km and concluded that it would be cheaper to instal than a conventional system and would have negligible running costs. A cross-section of their proposed cable is shown in Fig. 9.44 and consists of two 4 cm conductors of niobium tin in a common liquid helium container. Refrigerators to cool the line would be stationed every 20 km, and helium circulation pumps and vacuum pumps every 0·5 km. Other studies of direct current links have shown that superconducting systems become competitive for ratings above about 5,000 MW.

Secondary distribution systems in urban areas, where underground cables are essential, might offer an alternative application for superconducting cables. In this case the system would have to carry alternating current since the cost of d.c./a.c. convertors at the terminals would be prohibitive. The electrical design parameters and characteristics of such cables have been discussed by Cairns *et al.* (1969) and specific designs of 750 MVA 50 c/sec, 3 phase cables have been published by Wilkinson (1968) and, in more detail, by Rogers and Edwards (1967). The latter authors proposed the use of three concentric conductors, each comprising a thin layer of niobium to carry the

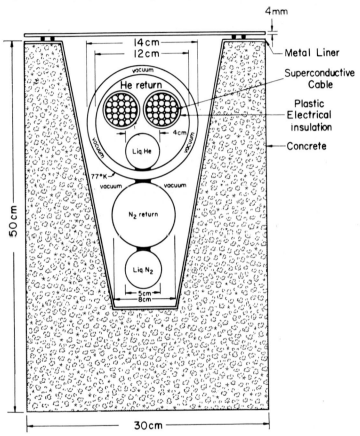

Fig. 9.44. Cross-section of a possible arrangement for a 100 GW superconducting transmission line (Garwin and Matisoo, 1967).

current under the normal load conditions and a backing of high purity aluminium to carry higher currents under fault conditions. The surface current density in the niobium was limited to 375 A(rms)/cm by the losses expected in the niobium (10 μw/cm^2) and this in turn fixed the size of the inner conductor which was 10 cm diameter. In order to reduce dielectric losses to an acceptable level (tan $\delta < 10^{-5}$) the conductors were vacuum insulated, and vacuum was also used for thermal insulation to a liquid nitrogen cooled heat shield. The estimated heat loads at 4°K with this construction were 40 W/km due to alternating current losses in the superconductor and 60 W/km through the thermal insulation, giving a total of 100 W/km and hence a running cost roughly half that of a conventional cable. The capital cost, however, was higher than for a normal cable.

2. *Motors and Generators*

The advantages of superconductors in a rotating machine are that an increase in either the current density in the winding or of the flux density can reduce the size and weight of the machine. Because of the alternating current loss in a type II material it is only possible to envisage the use of the superconductor in windings carrying steady currents.

The machine in which these conditions are most easily satisfied is the homopolar motor. Additional advantages of this type of machine are that there is no torque reaction on the field system and no armature reaction. The problems of supporting the magnet with low heat losses and stabilizing it against rapid field fluctuations are thus simplified. Homopolar machines are essentially low voltage, high current devices and have particular advantages where high torques at relatively slow speeds are required. Machines with ratings above 1 MW are expected to be economically competitive in applications such as ship propulsion units, steel rolling mill drives and winding machines. Similarly homopolar generators will find many uses in electrochemical processes.

A 3,250 h.p. prototype homopolar motor has been described by Appleton (1969) which runs at 200 r.p.m. with a terminal voltage of 450 V. The magnet for this machine generated a maximum flux density of 35 kgauss in a coil of 240 cm internal diameter and used 5,000 kg of niobium titanium–copper composite conductor. The winding was cryogenically stable and operated at 4·4°K in liquid helium at an average current density of 2,550 A/cm^2 with edge cooling of the conductor only. In order to strengthen the copper it was partially work hardened by the introduction of 3 % cold work, giving a yield stress of 12 kg/cm^2 with only a small change in electrical resistance.

An alternative use of superconductors in rotating machines would be in the field winding of an alternator. Here also the superconductor would be generating a quasi-steady magnetic field, although some alternating components might be induced. Preliminary studies suggest that the overall size of an alternator could be reduced by up to 50 %, with some saving in capital costs without too much change in the electrical characteristics of the machine.

V. References

Abrikosov, A. A. (1957). *Soviet Physics–JETP* **5**, 1174.
Alers, G. A. and Waldorf, D. L. (1962). *IBM. J. Res. Devel.* **6**, 89.

Anderson, P. W. (1962). *Phys. Rev. Lett.* **9**, 309–311.

Appleton, A. D. (1969). *Cryogenics* **9**, 147–157.

Bardeen, J., Cooper, L. N. and Schrieffer, J. R. (1957). *Phys. Rev.* **108**, 1175.

Bean, C. P. (1962). *Phys. Rev. Lett.* **8**, 250–253.

Benz, M. G. (1965). General Electric Research and Development Report 1965, No. 65-RL-4041 M.

Blaugher, R. D. and Hulm, J. K. (1961). *Phys. Chem. Solids* **19**, 134.

Brown, D. P., Burgess, R. W. and Mulholland, G. T. (1968). Proceedings of the Summer Study on Superconducting Devices, Brookhaven, pp. 794–814.

Butler, A. P., James, G. B., Maddock, B. J. and Norris, W. T. (1969). C.E.R.L. Report RD/L/R 1566.

Cairns, D. N. M., Minors, R. H., Norris, W. T. and Swift, D. A. (1969). Proc. Conf. on Low Temperatures and Electric Power, London.

Campbell, A. M., Evetts, J. E. and Dew–Hughes, D. (1968). *Phil. Mag.* **18**, 313.

Catterall, J. A. and Williams, I. (1967). *J. Less Common Metals* **12**, 258.

Chant, M. J., Halse, M. R. and Lorch, H. O. (1970). *Proc. Inst. Elec. Eng.* **117**, 1441–1447.

Chester, P. F. (1967). *Rep. Progr. Phys.* **30**, 561–614.

Chester, P. F. and Jones, G. O. (1953). *Phil. Mag.* **44**, 1281.

Clee, P. T. M., Thomas, D. B. and Trowbridge, C. W. (1968). Proceedings of the Summer Study on Superconducting Devices, Brookhaven, pp. 815–827.

Coffey, H. T., Hulm, J. K., Reynolds, W. T., Fox, D. K. and Span, R. E. (1965). *J. Appl. Phys.* **36**, 128.

Cohen, M. L. (1964). *Rev. Mod. Phys.* **36**, 240.

Cooper, L. N. (1960). *Am. J. Phys.* **28**, 91.

Cornish, D. N. (1966). *J. Sci. Instrum.* **43**, 16–20.

Cornish, D. N. (1968). Proceedings of the Summer Study on Superconducting Devices, Brookhaven, pp. 880–885.

De Haas, W. J. and Voogd, J. (1931). *Leiden Commun.* **214C**.

DeSorbo, W. and Healy, W. A. (1964). *Cryogenics* **4**, 257.

Druyvesteyn, W. F., van Ooijen, D. J. and Berben, T. J. (1964). *Rev. Mod. Phys.* **36**, 58.

Dunn, W. I. and Hlawiczka, P. (1968). *J. Phys. D.* **1**, 1469–1476.

Easson, R. M. and Hlawiczka, P. (1968). *J. Phys. D.* **1**, 1477–1485.

Fairbank, W. M. and Schwettman, H. A. (1967). *Cryogenic Eng. News* **2**, 46–49.

Gauster, F. G. and Hendricks, J. B. (1968). *I.E.E.E. Trans.* **MAG-4**, 489–492.

Garwin, R. L. and Matisoo, J. (1967). *Proc. I.E.E.E.* **55**, 538–548.

Geballe, T. H., Matthias, B. T., Remeika, J. P., Clogston, A. M., Compton, V. B., Maita, J. P. and Williams, H. J. (1966). *Physics* **2**, 293.

Giaever, I. (1965). *Phys. Rev. Lett.* **15**, 825–827.

Ginzburg, V. L. and Landau, L. D. (1950). *Zhur. Eksp. i Teor. Fiz.* **20**, 1064.

Ginzburg, V. L. (1968). *Contemp. Phys.* **9**, 355.

Goodman, B. B. (1962). *IBM J. Res. Dev.* **6**, 63.

Gorter, C. J. and Casimir, H. B. G. (1934). *Phys. Z.* **35**, 963.

Hake, R. R. (1967). *Appl. Phys. Lett.* **10**, 189.

Hale, J. R. and Williams, J. E. C. (1968). *J. Appl. Phys.* **39**, 2634–2638.

Hampshire, R., Sutton, J. and Taylor, M. T. (1969). Proc. Conference on Low Temperatures and Electric Power, London, pp. 69–76.

Hanak, J. J., Strater, K. and Cullen, G. W. (1964). *RCA Review* **25**, 342.

Hanak, J. J. and Enstrom, R. E. (1966). Proc. 10th Conf. Low. Temp. Phys., Moscow, 2B, 10.

Hancox, R. (1965a). *Phys. Lett.* **16**, 208–209.
Hancox, R. (1965b). *Appl. Phys. Lett.* **7**, 138–139.
Hancox, R. (1966). Proc. 10th Int. Conf. Low Temp. Phys., Moscow, Vol. 2, pp. 43–46.
Hancox, R. (1968). *I.E.E.E. Trans.* **MAG-4**, 486–488.
Hannay, N. B., Geballe, T. H., Matthias, B. T., Andres, K., Schmidt, P. and Macnair, D. (1965). *Phys. Rev. Lett.* **14**, 225.
Harrowell, R. V. (1970). *Proc. I.E.E.* **117**, 131–140.
Hart, H. R. (1968). Proceedings of the Summer Study on Superconducting Devices, Brookhaven, pp. 571–600.
Hart, H. R. and Livingston, J. D. (1968). Proc. 11th Int. Conf. on Low Temp. Phys., St. Andrews, pp. 869–872.
Hart, H. R. and Swartz, P. S. (1963). Progress Report No. 1, A.F.33(657)-11722, pp. 19–26.
Heaton, J. W. and Rose-Innes, A. C. (1963). *Appl. Phys. Lett.* **2**, 196.
Hindley, N. K. and Watson, J. H. P. (1969). *Phys. Rev.* **183**, 525.
Hulm, J. K. and Blaugher, R. D. (1961). *Phys. Rev.* **123**, 1569.
Iwasa, Y. (1969). *Appl. Phys. Lett.* **14**, 200–201.
Iwasa, Y. and Montgomery, D. B. (1965). *Appl. Phys. Lett.* **7**, 231–233.
Iwasa, Y. and Williams, J. E. C. (1966). *Appl. Phys. Lett.* **9**, 391–392.
Iwasa, Y., Weggel, C., Montgomery, D. B., Weggel, R. and Hale, J. R. (1969). *J. Appl. Phys.* **40**, 2006–2009.
Jackson, J. (1969). *Cryogenics* **9**, 103–105.
Kamper, R. A. (1969) *Cryogenics* **9**, 20–25.
Kan, L. S., Sudovstov, A. I. and Lazarev, B. G. (1948). *Zhur. Eksp. i Teor. Fiz.* **18**, 825.
Kantrowitz, A. R. and Stekly, Z. J. J. (1965). *Appl. Phys. Lett.* **6**, 56–57.
Keesom, W. H. and Kok, J. A. (1932). *Leiden Commun.* **221e**.
Kim, Y. B., Hempstead, C. F. and Strnad, A. R. (1962). *Phys. Rev. Lett.* **9**, 306–311.
Kim, Y. B., Hempstead, C. F. and Strnad, A. R. (1963). *Phys. Rev.* **129**, 528–535.
Kim, Y. B., Hempstead, C. F. and Strnad, A. R. (1965). *Phys. Rev.* **139**, A1163–72.
Landauer, J. K. (1954). *Phys. Rev.* **96**, 296.
Lange, F. (1966). *Cryogenics* **6**, 176–177.
Levy, S. A., Kim, Y. B. and Kraft, R. W. (1966). *J. Appl. Phys.* **37**, 3659.
Little, W. A. (1964). *Phys. Rev.* **134A**, 1416.
Livingston, J. D. (1964). General Electric Research and Development Report 1964, No. 64 RL 3810.
Livingston, J. D. (1966). *Appl. Phys. Lett.* **8**, 319.
Livingston, J. D. (1968). Proceedings of the Summer Study on Superconducting Devices, Brookhaven.
London, F. (1950). "Superfluids." Vol. 1. John Wiley, London.
London, H. (1963). *Phys. Lett.* **6**, 162–165.
Lorch, H. O. (1969). *Cryogenics* **9**, 354–361.
Lowell, J. Munoz, J. S. and Sousa, J. B. (1968). Proc. 11th Int. Conf. on Low Temp. Phys., St. Andrews, pp. 858–861.
Lynton, E. A. (1962). "Superconductivity". Methuen, London.
Maddock, B. J. and James, G. B. (1968). *Proc. I.E.E.E.* **155**, 543–547.
Maddock, B. J., Carter, C. N. and Barratt, P. B. (1967). Proc. 2nd Int. Conf. on Magnet Technology, Oxford, pp. 533–542.

Matthias, B. T. (1957). "Progress in Low Temperature Physics", Vol. 2, p. 138. Interscience, London.

Matthias, B. T., Geballe, T. H., Longinotti, L. D., Corenzwit, E., Hull, G. W., Willens, R. H. and Maita, J. P. (1967). *Science* **156**, 645.

Mercereau, J. E. (1968). *In* "Superconductivity in Science and Technology" (M. H. Cohen, ed.), pp. 63–76. University of Chicago Press.

Narlikar, A. V. and Dew-Hughes, D. (1966). *J. Mat. Sci.* **1**, 317.

Neal, D. F. (1969). Imperial Metal Industries Ltd. Fourth Report. Ministry of Technology Contract KJ/5H/31/CB7/8B2.

Newhouse, V. L. (1964). "Applied Superconductivity." J. Wiley and Sons, New York.

Parks, R. D. (1969). "Superconductivity." Dekker, New York.

Parmenter, R. H. (1967). *Phys. Rev.* **154**, 353.

Purcell, J. R. (1968). Proceedings of the Summer Study on Superconducting Devices, Brookhaven, pp. 765–785.

Raub, C. J., Sweedler, A. R., Jensen, M. A., Broadston, S. and Matthias, B. T. (1964). *Phys. Rev. Lett.* **13**, 746.

Roberts, B. W. (1964). *In* "Progress in Cryogenics", Vol. 4 (Mendelssohn, K., ed.). Academic Press, London and New York.

Rogers, E. C. and Edwards, D. R. (1967). *Electl. Rev. Lond.* **181**, 348–351.

Saint-James, D. and de Gennes, P. G. (1963). *Phys. Lett.* **7**, 306.

Sarma, N. V. and Wilcockson, A. (1968). International Research and Development Co. Ltd., Ministry of Technology Contract KJ/5H/26/CB78B. Report No. IRD. 68-5, 1968.

Sarma, N. V. and Wilcockson, A. (1969). *Phil. Mag.* **20**, 539.

Schrader, E. R., Freedman, N. S. and Fakan, J. C. (1964). *Appl. Phys. Lett.* **4**, 105–106.

Sekula, S. T. and Kernohan, R. H. (1968). *J. Appl. Phys.* **39**, 2516.

Sizoo, G. T. and Onnes, H. K. (1925). *Leiden Commun.* **1806.**

Smith, P. F. and Lewin, J. D. (1967). *Nucl. Instrum. Meth.* **52**, 298–308.

Smith, P. F., Wilson, M. N., Walters, C. P. and Lewin, J. D. (1968). Proceedings of the Summer Study on Superconducting Devices, Brookhaven, pp. 913–919.

Stekly, Z. J. J. (1963). *Adv. Cryogenic Eng.* **8**, 585–600.

Stekly, Z. J. J. and Zar, J. L. (1965). *I.E.E.E. Trans.* **NS-12**, 367–372.

Stekly, Z. J. J., Thome, R. and Strauss, B. (1968). Proceedings of the Summer Study on Superconducting Devices, Brookhaven, pp. 748–764.

Strnad, A. R., Hempstead, C. F. and Kim, Y. B. (1964). *Phys. Rev. Lett.* **13**, 794–797.

Swartz, P. S. and Bean, C. P. (1968). *J. Appl. Phys.* **39**, 4991–4998.

Taquet, B. (1965). *J. Appl. Phys.* **36**, 3250–3255.

Träuble, H. and Essmann, U. (1968). *Phys. Stat. Solidi.* **25**, 373.

Wilkinson, K. J. R. (1968). *Nature (Lond.)* **219**, 1317–1319.

Williams, I. and Catterall, J. A. (1966). *British J. Appl. Phys.* **17**, 505.

Wipf, S. L. (1967). *Phys. Rev.* **161**, 404–415.

Wipf, S. L. (1968). Proceedings of the Summer Study on Superconducting Devices, Brookhaven, pp. 511–543.

Chapter 10

Instrumentation

H. TOWNSLEY

BOC–Airco Cryogenic Plant Ltd., London, England

A. H. COCKETT

The British Oxygen Co. Ltd., Edmonton, London, England

I. Introduction

Although instrumentation on cryogenic plants is basically similar to that of chemical and petrochemical processes, there are special factors. In flow measurement, for example, liquids are normally at or near their boiling point. Other problems are encountered when measuring level in totally enclosed thermally insulated vessels. In cryogenic research measurements of temperature can be made with a precision unattainable at higher temperatures, but the lower the temperature the fewer the properties that are available for the measurement. Recourse must be had to methods based on fundamental thermodynamics. Automation of research apparatus can give rich rewards in productivity.

This chapter reviews current methods of measuring and recording temperature, flow, liquid level, pressure and chemical analysis. Calibration, accuracy and reliability are discussed. Both laboratory and plant instruments are considered.

II. Temperature Measurement

A. TEMPERATURE SCALES

A temperature scale may be based on any property of a body which changes with temperature: the linear dimensions of a solid, the volume of a liquid or gas, the resistance of a metal or semiconductor, the refractive index and so on. Practical scales for many purposes use the apparent expansion of liquids (mercury, alcohol) in glass. For each such scale reference values of the property at two fixed temperatures such as the freezing and boiling point of water define a fundamental interval which can be subdivided into an arbitrary number of smaller units, conveniently a hundred, to fix the size of a degree. However, agreement of different thermometers at the fixed points would not necessarily imply agreement at intermediate temperatures since most properties do not change linearly with temperature. Careful selection of property is therefore required to give a scientifically acceptable scale, and ultimately the scientifically preferred scale is one that is independent of the properties of any particular substance.

B. ABSOLUTE SCALE

Experiment shows that close agreement between different thermometers is obtained when the property employed is the pressure of one of the so-called "permanent gases" in a container of constant volume, or the volume measured at constant pressure. The kinetic theory of gases gives a sound theoretical backing to these observations since it predicts that for a perfect gas the product of pressure and volume for a fixed mass of gas is proportional to the temperature, and under the conditions used in thermometry the permanent gases are only slightly imperfect. Nevertheless for precision thermometry the imperfections are significant and whereas the perfect gas obeys the relation $PV = RT$, the properties of real permanent gases like air, nitrogen, hydrogen and helium must be expressed by more complicated relations such as the virial equation

$$PV = RT(1 + B/V + C/V^2 + \ldots) \tag{10.1}$$

Here B and C are the 2nd and 3rd virial coefficients and are functions of temperature but independent of pressure. This equation can be used to extrapolate the product PV to a limiting value at zero pressure and for real gases this limiting value, $\lim (PV)$, obeys the equation for a perfect gas $\lim (PV) = RT$ with R a universal constant.

The absolute or "perfect" gas scale of temperature is therefore defined by measurements of the low pressure limit of the product PV at the melting point of ice θ_i and boiling of water θ_s and making the difference between them $100°$. Then from the equation of state for a perfect gas $PV = R\theta$ we have

$$\lim (PV)_i = R\theta_i$$
$$\lim (PV)_s = R\theta_s = R(\theta_i + 100)$$

whence the ice point temperature is

$$\theta_i = 100 \frac{\lim (PV)_i}{\lim (PV)_s - \lim (PV)_i} \tag{10.2}$$

TABLE 10.I. Agreement Between Gas Thermometer Scales at 50°C

Gas	Difference from constant volume helium thermometer, °C	
	Constant volume thermometer	Constant pressure thermometer
Helium	0	−0·001
Hydrogen	+0·002	+0·003
Nitrogen	+0·009	+0·031

When observations made using hydrogen, nitrogen and helium are corrected in this way almost exact agreement is found, as shown in Table 10.I (Roberts and Miller, 1951; Holborn and Otto, 1926). The agreement between the different perfect gas thermometers ensures that θ_i is the same whatever gas is used; its numerical value is internationally agreed as $273 \cdot 15°K$ in accordance with the best available experimental data (C. R. des Séances, 1967).

However this scale is of more fundamental significance than would be apparent from the above argument, as may be seen on consideration of the second law of thermodynamics. This asserts that "no heat engine operating in a closed cycle can transfer heat from a reservoir at a lower temperature to one at a higher temperature" and leads to proof that no engine can have a higher efficiency than a Carnot engine operating between the same heat reservoirs. The Carnot efficiency is dependent only on the temperature T_1 and T_2 of these reservoirs and can be shown to be equal to $(T_1 - T_2)/T_1$ where $T_1 > T_2$. It can further be shown that if a perfect gas is used as the working substance in such an engine $T_1/T_2 = \theta_1/\theta_2$ and if $T_s - T_i$ is made $100°$ the two scales are identical. The scale so defined is the absolute thermodynamic or Kelvin scale of temperature which is independent of the properties of any material substance.

On this scale there is a temperature at which the volume of a perfect gas or the heat Q_1 taken in is zero and this is the absolute zero of temperature. The existence of such a temperature follows from the second law of thermo-dynamics, but the third law shows that it is unattainable in a finite number of steps. On the Celsius scale it is $273 \cdot 15°C$ below the melting point of ice and definition of this point as $273 \cdot 15°K$ fixes a scale in which $\theta_s - \theta_i = 100°$.

C. INTERNATIONAL PRACTICAL TEMPERATURE SCALE

While the absolute thermodynamic scale is defined in terms of the properties of a perfect gas or an ideal Carnot engine it is necessary for practical purposes to use more conventional thermometers yet to ensure that the temperatures deduced from them are as close as possible or necessary within the context of the measurement to the absolute scale. To this end the International Bureau of Weights and Measures has agreed the IPTS which from $13 \cdot 8°K$ to $900°K$ $(630°C)$ is based on measurement of the resistance of a standard platinum thermometer calibrated according to a specified procedure at $0°C$, $100°C$ and two other fixed points. For use above $0°C$ only one of these other points, the melting point of zinc, $419°C$, is required; for low temperature measurements a measurement at the boiling point of oxygen is also required. Again for practical reasons the observation at $0°C$ is carried out at the triple point of water, which is $0 \cdot 007°C$.

In the 1927 and 1948 International Temperature Scales (Burgess, 1928) the relation between resistance and temperature was given by the Callender equation 10.6 for temperatures above $0°C$:

$$t = 100 \frac{R_t - R_0}{R_{100} - R_0} + \delta\left(\frac{t}{100} - 1\right)\frac{t}{100} \tag{10.3}$$

where δ is a constant and a first approximation is given by $t_p = 100(R_t - R_0)/(R_{100} - R_0)$ which can be used for t in the second term. For temperatures below $0°C$ van Dusen's modification was used

$$t = 100 \frac{R_t - R_0}{R_{100} - R_0} + \delta\left(\frac{t}{100} - 1\right)\frac{t}{100} + \beta\left(\frac{t}{100} - 1\right)\left(\frac{t}{100}\right)^3 \tag{10.4}$$

For temperatures below $90°K$ no simple relation between temperature and resistance had been found, nor was a single relation applicable to all platinum thermometers. National standards laboratories established their own standards and working thermometers were calibrated against them.

These formulae were found to give temperatures differing by a few hundredths of a degree from the Kelvin scale and the IPTS68 scale introduced a new approach as well as extending the range down to $13 \cdot 8°$K. This scale defines temperatures from $13 \cdot 81°$K to $273 \cdot 15°$K by the relation

$$W(T_{68}) = W_{CCT-68}(T_{68}) + \Delta W(T_{68}) \qquad (10.5)$$

TABLE 10.II. Values of the Function $W_{CCT68}(T_{68})$ at Selected Temperatures

Temperature T_{68} °K	$W_{CCT68}(T_{68})$	Temperature T_{68} °K	$W_{CCT68}(T_{68})$
14	0·00145973	150	0·49861135
20	0·00427780	160	0·54026792
30	0·01716768	170	0·58171423
40	0·04171968	180	0·62296972
50	0·07537756	190	0·66404996
60	0·11455312	200	0·70496694
70	0·15649541	210	0·74573026
80	0·19958212	220	0·78634756
90	0·24298315	230	0·82682531
100	0·28630201	240	0·86716894
110	0·32936765	250	0·90738309
120	0·37212331	260	0·94747152
130	0·41456709	270	0·98743642
140	0·45672033	273	0·99940199

where $W(T_{68})$ is the ratio $R(T)/R(273 \cdot 15)$ for the thermometer at temperature T_{68} and $W_{CCT68}(T_{68})$ is a standard resistance ratio given at selected temperatures in Table 10.II. A fuller table is given in Comptes Rendus des Séances (1948) and a tabulation sufficiently detailed to allow interpolation to an accuracy of 10^{-4}°K is available from the Bureau International des Poids et Mesures. The deviations $W(T_{68})$ are determined from calibration at specified fixed points and the use of interpolation formulae. The range between $13 \cdot 81°$K and $273 \cdot 15°$K is divided into four parts for which the limits, calibration points and interpolation formulae are given in Table 10.III. In each range the constants in the interpolation formula are deduced from the observed resistance ratios and the derivative at the highest calibration temperature in the range of the deviation function for the next higher range.

Agreement between the international and absolute scales within the limits quoted presupposes that the purity of the platinum is very high, as defined by restricting the value of R_{100}/R_0 to $1 \cdot 3925$.

Calibration at the points given in Table 10.III and interpolation by the IPTS formulae allows measurement of temperatures exactly (by definition) on the IPTS and as closely as is at present possible on the absolute or Kelvin

TABLE 10.III. Low Temperature Calibration Points

Range °K	Calibration points	Tempera-ture °K	Interpolation formulae
13·81–20·28	Triple point of hydrogen Temperature 17·042°K, vpH$_2$ = 25 cm Hg Boiling point e-hydrogen	13·81 17·042 20·28	$\Delta W(T_{68}) = A_1 + B_1(T_{68})$ $\quad + C_1(T_{68})^2 + D_1(T_{68})^3$
30·28–54·361	Boiling point of e-hydrogen Boiling point of neon Triple point of oxygen	20·28 27·102 54·361	$\Delta W(T_{68}) = A_2 + B_2(T_{68})$ $\quad + C_2(T_{68})^2 + D_2(T_{68})^3$
54·361–90·188	Triple point of oxygen Boiling point of oxygen	54·361 90·188	$\Delta W(T_{68}) = A_3 + B_3(T_{68})$ $\quad + C_3(T_{68})^2$
90·188–273·15	Boiling point of oxygen Boiling point of water	90·188 273·15	$\Delta W(T_{68}) = A_4(t_{68}) + C_4 t_{68}$ $\quad (t_{68} - 100°C)$ where $t_{68} = T_{68} - 273·15K$

TABLE 10.IV. Accuracy and Reproducibility of the Primary and Secondary Fixed Points

Fixed point	Temperature °K	Estimated uncertainty °K	Reproducibility °K
Triple point of equilibrium hydrogen	13·81	0·01	±0·003
17·042°K point	17·042	0·01	±0·003
Boiling point of equilibrium hydrogen	20·28	0·01	±0·003
Boiling point of neon	27·102	0·01	±0·003
Triple point of oxygen	54·361	0·01	±0·003
Boiling point of oxygen	90·188	0·01	±0·003
Triple point of water	273·16	Exact by definition	±0·0002
Boiling point of water	100 [a]	0·005	±0·001
Freezing point of tin	231·9681 [a]	0·015	±0·003
Freezing point of zinc	419·58 [a]	0·03	±0·003

[a] Temperature °C

scale. The accuracy with which this is reproduced is of course not better than the accuracy with which the fixed points are known on it, which is given in Table 10.IV.

Between 5°K and 13·81°K there is so far no internationally agreed scale, but below 5°K the scale is based on the vapour pressure of helium. The relation between pressure and temperature has been derived from all published vapour pressure measurements and related thermodynamic data independently by van Dijk and Durieux (1955) and Clement and Logan (1955) in Washington. The former workers used the latent and specific heats of helium and the theoretical equation

$$\ln P = i_0 - \frac{L_0}{RT} + \tfrac{5}{2} \ln T - \frac{1}{RT} \int_0^T S_l \, dT + \frac{1}{RT} \int_0^P V_l \, dP + \varepsilon \qquad (10.6)$$

in which

$$i_0 = \ln (2\pi m)^{\frac{3}{2}} k^{\frac{5}{2}} / h^3$$

and

$$\varepsilon = \ln \left(\frac{PV}{NRT} \right) - 2B \left(\frac{N}{V} \right) - \tfrac{3}{2} C \left(\frac{N}{V} \right)^2$$

Lo = the latent heat of vaporization at $0°K$

S_l and V_1 = the molar entropy and volume, respectively, of the liquid

m = the mass of a He^4 atom

B and C = second and third virial coefficients in the equation 10.1 used to extrapolate the isotherms to zero density to determine the temperature from gas thermometer measurements. Other symbols have their usual meaning.

Clement and Logan used all the best available measurements of vapour pressures using temperatures obtained from secondary thermometers calibrated against gas thermometers. The resulting values were in agreement within $0·004°$ and in 1958 the scales were combined to give the 1958 He temperature scale (Brickwedde et al., 1960), which has a maximum uncertainty of $±2$ millidegree between $1·0$ and $4·5°K$; above this the uncertainty is a little greater.

D. INDUSTRIAL SCALES

Much industrial instrumentation uses the Celsius or Kelvin scale of temperature but in U.S.A. the Fahrenheit and Rankine scales are also used.

The Fahrenheit scale has 180 divisions between the reference points of melting ice and boiling water, which are at 32 and $212°F$ respectively. Some British measurements are still based on the Fahrenheit scale, e.g. the British Thermal Unit, but with increasing standardization and the use of SI units the Celsius and Kelvin scales will predominate.

The Rankine scale is the absolute Fahrenheit scale with the same size units as degrees Fahrenheit but with zero degrees corresponding to absolute zero temperature as in the case of the Kelvin scale.

Zero degrees Rankine is equal to minus $459·69$ degrees Fahrenheit.

The Rankine scale is little used in industry and reference is made to it for interest and comparison only.

E. CONVERSIONS

Conversions from any one to another of the scales referred to can be performed by simple addition and/or multiplication. For conversion from Kelvin to Celsius subtract 273·15 and from Rankine to Fahrenheit subtract 459·69.

Other conversions are:

(a) Celsius to Fahrenheit
$$°F = (°C \times \tfrac{9}{5}) + 32$$

(b) Fahrenheit to Celsius
$$°C = (°F - 32) \times \tfrac{5}{9}$$

(c) Kelvin to Rankine
$$°R = K \times \tfrac{9}{5}$$

(d) Rankine to Kelvin
$$K = °R \times \tfrac{5}{9}$$

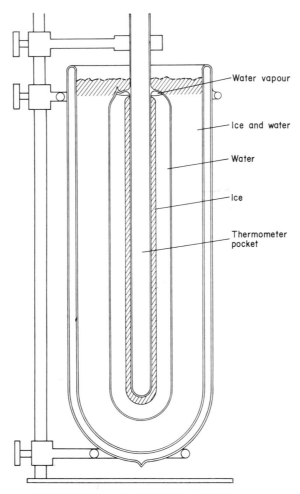

Fig. 10.1. Apparatus for triple point of water.

1. The Triple Point of Water

A constant temperature equal to that of the triple point of water can be realized with high precision in a sealed glass cell of the type shown in Fig. 10.1, containing only water of isotopic composition similar to that of ocean water and its vapour. The cell has an axial well for the thermometer. The triple point is attained whenever part of the water is frozen and the ice is in equilibrium with liquid and vapour. At a depth h below the liquid–vapour surface the pressure is greater than the triple point pressure by an amount equal to the hydrostatic head and the temperature is given by $t_{68} = 0.01°C - 7 \times 10^{-6} h°C$ where h is in centimetres.

Triple point cells can be purchased complete from laboratory furnishers and instrument makers or can easily be constructed by a competent glass blower, but then they have to be filled. To fill, the water is distilled at low pressure into an evacuated cell after previous distillation from potassium permanganate solution to remove dissolved solids and other contamination. The filling tube is sealed off under vacuum.

To prepare a cell for calibration of a thermometer a thick layer of ice is formed around the thermometer well by cooling from within using either solid CO_2 or liquid nitrogen as refrigerant. Insertion of the warm thermometer produces a thin layer of water adjacent the well giving a new liquid–solid interface there. During the first few hours after this preparation the temperature measured in the well rises rapidly by a few millidegrees probably due to changes in the crystal structure. Thereafter if the cell is kept in an ice bath the temperature can be maintained constant to $\pm 0.0001°K$ for several months. Differences between cells from different sources prepared and used in this way should not exceed $0.002°K$. Since the liquid and solid are relatively transparent it is necessary to shield the cell from radiation to prevent errors in the temperature of the thermometer.

Variation in the isotopic content from that of oceanic water, which contains the greatest concentration of heavy isotopes in natural waters to that of continental surface water decreases the temperature by up to $0.00025°K$ (Comptes Rendus des Séances, 1948).

2. The Boiling Point of Water

Calibration at the boiling point of water is usually carried out by the dynamic method using a hypsometer with the thermometer immersed directly in the saturated vapour. The temperature may be deduced from the barometric pressure with corrections if necessary for any excess pressure within the hypsometer. For highest precision it is preferable to use a closed system in which the boiler and manometer are connected to a manostat filled with air or helium.

The thermometer must be shielded from radiation from surfaces at temperatures different from that of the vapour in which it is immersed and the usual design of hypsometer achieves this by surrounding it with shields also immersed in the vapour. When equilibrium is attained the temperature of the thermometer is independent of the depth of immersion, of the time elapsed and of the rate of heat supply. If electrical heating is used a variable transformer is useful to adjust the heat input.

The IPTS68 uses the relation

$$t_{68} \,^{\circ}C = 100 + 28 \cdot 0216 \left(\frac{P}{P_0} - 1 \right) - 11 \cdot 642 \left(\frac{P}{P_0} - 1 \right)^2 + 7 \cdot 1 \left(\frac{P}{P_0} - 1 \right)^3 \quad (10.7)$$

to calculate the temperature from the vapour pressure P.

3. The Freezing Point of Zinc

A reproducible temperature at the freezing point of zinc can be obtained by slow cooling of the molten metal contained in a graphite crucible with a graphite lid within a silica tube in a metal block furnace. The crucible must be made of very pure (99·999%) artificial graphite and is conveniently about 5 cm diameter with an axial thermometer well deep enough to allow sufficient immersion of the thermometer to eliminate errors due to heat conduction down the leads. The silica tube should contain an inert atmosphere and the surface of the metal should be covered with graphite powder.

TABLE 10.V. The Effect of Submergence on the Boiling Point

Liquid	Temperature rise due to 10 cm head of liquid at normal boiling point, °K
Oxygen	0·105
Nitrogen	0·058
Neon	0·039
Hydrogen	0·0023
Helium	0·0013

In calibration the metal is melted and allowed to cool to the liquid temperature then a thin layer of solid is produced around the well by removing and cooling the thermometer or inserting a silica rod for about 30 sec before replacing the thermometer. The purity of the sample is adequate if the melting range does not exceed 0·001°K.

4. Low Temperature Boiling Points

The temperature of a boiling liquid may be determined from its vapour pressure and the simplest calibration arrangement is to immerse the thermometer in a bath of the liquid and measure the vapour pressure, which may be simply the barometric pressure. If the gas is to be recovered, as would be helium or neon, or is inflammable and must be conveyed out of the building (as hydrogen) there will be a pressure drop between the bath and the atmosphere. In such cases a simple manometer filled with oil of low vapour pressure can usually be used to give a sensitive measure of the difference by which the barometric reading must be corrected.

This vapour pressure gives the temperature at the surface; at the level of the thermometer the temperature may be higher due to the hydrostatic head of liquid corresponding to the depth of immersion. Whether this increase is fully attained depends on the vigour with which the liquid is stirred, so there is some uncertainty here in the actual temperature. The full temperature rise due to a 10 cm immersion for commonly used liquids at their normal boiling points is shown in table 10.V; at lower temperatures where the rate of change of temperature with pressure is greater, the temperature rise will be greater; e.g. for liquid helium at 2°K it is 15 mK.

Precise calibration at these low temperature boiling points is therefore best carried out by the static method in which the test thermometer is immersed in a bath of liquid and the temperature is deduced from the vapour pressure over the same liquid in a vapour pressure thermometer. The temperature of the bath may be kept constant by a manostat. The two thermometer bulbs must be at the same level in the liquid and close together to ensure equality of temperature, and the vapour pressure thermometer must not be in thermal contact with the liquid surface. Another arrangement is to insert the test thermometers in a hole drilled in a block of pure copper; a similar hole with sealed connection to a manometer can be used for the bulb of the vapour pressure thermometer. To ensure complete elimination of errors due to uncertainty in the depth of immersion a good arrangement is to allow the liquid to wet only the top of the block by surrounding the sides and base with a vacuum enclosure.

5. *Liquid Oxygen Point*

Calibration by the static method is particularly important in the case of oxygen which superheats by several tenths of a degree if left to boil freely in an open dewar. The effect may be reduced by stirring with a stream of pure oxygen gas if it is necessary to determine the temperature from the barometric pressure. Commercially available liquid oxygen is quite satisfactory for the liquid bath but very pure oxygen is required for the vapour pressure thermometer—as will be described.

For temperatures near the normal boiling point, i.e. vapour pressure near one atmosphere, the temperature may be calculated from the pressure by the relation:

$$T_{68}K = 90 \cdot 188 + 9 \cdot 5648\left(\frac{P}{P_0} - 1\right) - 3 \cdot 69\left(\frac{P}{P_0} - 1\right)^2 + 2 \cdot 22\left(\frac{P}{P_0} - 1\right)^3 \quad (10.8)$$

TABLE 10.VI. Oxygen Vapour Pressure–Temperature Table. (From the data of J. Hoge)

P_{mm}	Temperature °K										
	0	10	20	30	40	50	60	70	80	90	100
100	74·471	75·077	75·640	76·167	76·661	77·125	77·565	77·984	78·384	78·767	79·134
	0·606	0·563	0·527	0·494	0·464	0·440	0·419	0·400	0·383	0·367	0·353
200	79·134	79·487	79·826	80·152	80·468	80·774	81·071	81·359	81·639	81·911	82·175
	0·353	0·339	0·326	0·316	0·306	0·297	0·288	0·280	0·272	0·264	0·257
300	82·175	82·432	82·683	82·927	83·165	83·397	83·624	83·847	84·065	84·279	84·488
	0·257	0·251	0·244	0·238	0·232	0·227	0·223	0·218	0·214	0·209	0·205
400	84·488	84·693	84·894	85·091	85·284	85·474	85·661	85·845	86·026	86·204	86·379
	0·205	0·201	0·197	0·193	0·190	0·187	0·184	0·181	0·178	0·175	0·172
500	86·379	86·551	86·721	86·888	87·052	87·214	87·374	87·532	87·688	87·842	87·994
	0·172	0·170	0·167	0·164	0·162	0·160	0·158	0·156	0·154	0·152	0·150
600	87·994	88·144	88·292	88·438	88·582	88·724	88·865	89·004	89·141	89·277	89·412
	0·150	0·148	0·146	0·144	0·142	0·141	0·139	0·137	0·136	0·135	0·133
700	89·412	89·545	89·677	89·808	89·938	90·064	90·190	90·315	90·439	90·562	90·683
	0·133	0·132	0·131	0·130	0·128	0·126	0·125	0·124	0·123	0·121	0·120
800	90·683	90·803	90·922	91·040	91·157	91·273	91·388	91·502	91·615	91·727	91·838
	0·120	0·119	0·118	0·117	0·116	0·115	0·114	0·113	0·112	0·111	0·110
900	91·838	91·948	92·057	92·165	92·272	92·378	92·483	92·587	92·691	92·794	92·896
	0·110	0·109	0·108	0·107	0·106	0·105	0·104	0·104	0·103	0·102	0·102
1000	92·896	92·998	93·099	93·198	93·297	93·396	93·494	93·591	93·687	93·782	93·877
	0·102	0·101	0·099	0·099	0·099	0·098	0·097	0·096	0·095	0·095	0·095
1100	93·877	93·972	94·066	94·159	94·252	94·344	94·435	94·526	94·616	94·705	94·794
	0·095	0·094	0·093	0·093	0·092	0·091	0·091	0·090	0·089	0·089	0·089
1200	94·794	94·883	94·971	95·059	95·146	95·232	95·317	95·402	95·486	95·570	95·654
	0·089	0·088	0·088	0·087	0·086	0·085	0·085	0·084	0·084	0·084	0·083

or Table 10.VI based on the data of Hoge may be used with linear interpolation to give temperature on the 1948 IPTS.

6. Liquid Neon Point

Calibration at the boiling point of liquid neon can follow the same procedure but the neon is usually contained in a dewar surrounded by a heat shield at about 77°K, and the dewar is connected to a manostat to maintain constant pressure, therefore temperature, and to a gas recovery system.

Liquid neon can be obtained from commercial suppliers. The price is relatively high because of the high price of the gas and the high density (i.e. large volume of gas per unit volume) of the liquid. However a generous allowance is usually made if the gas is recovered and returned in a pure state and this reduces the net cost of a calibration. Liquid neon can also be produced from gaseous neon by heat exchange with liquid hydrogen if this is available, or less efficiently by heat exchange with liquid helium. Heat exchangers for these processes must be designed with particular care to prevent excessive undercooling of the liquid which would result in blockages since the triple point is only 3°K below the normal boiling point. Neon for the vapour pressure bulb can be obtained by evaporating some from the liquid bath, or obtained separately as pure gas from a commercial supplier. The temperature is deduced from the vapour pressure by the relation (Holborn and Otto, 1926):

$$T_{68}K = 27 \cdot 102 + 3 \cdot 3144 \left(\frac{P}{P_0} - 1 \right) - 1 \cdot 24 \left(\frac{P}{P_0} - 1 \right)^2 + 0 \cdot 74 \left(\frac{P}{P_0} - 1 \right)^3 \quad (10.9)$$

7. Liquid Hydrogen Point

Calibration at the liquid hydrogen point can also follow this procedure except that the gas is not recovered but vented to external atmosphere to prevent any possible accumulation of a flammable mixture within the building. (Any mixture with air in the composition range from 5 to 80% hydrogen is flammable.) Precautions must also be taken to prevent condensation of solid air in the liquid; a closed cryostat with vent pipe actively venting gas ensures that this is achieved, but an inactive vent due to very efficient insulation and low boil off may not.

To obtain the defining temperature of 17·042°K the temperature of the bath is reduced by pumping until the vapour pressure falls to 25 Torr.

Liquid hydrogen can be produced *in situ* by means of a miniature or other liquefier or may be purchased from commercial suppliers.

For temperatures from 13·8°K to the normal boiling point the temperature on IPTS68 is calculated from the relation:

$$\log \frac{P}{P_0} = 1 \cdot 711466 - \frac{44 \cdot 01046}{T_{68}} + 0 \cdot 0235909 T_{68} - 0 \cdot 000048017 T_{68}^2 \quad (10.10)$$

8. *Liquid Helium Point*

Calibration at the liquid helium point may be carried out *in situ* in a helium cryostat. Since there are no interfering contaminants (except He^3 which is not likely to be present in sufficient concentration to affect the results) the temperature may be obtained from the vapour pressure over the bath using the 1958 tables. Correction may be made if necessary for the hydrostatic head due to immersion in the liquid. However the static method is preferable and in more general use. The procedure follows that already outlined for other liquids of low temperature boiling points.

Liquid helium may be obtained from a laboratory liquefier or purchased from commercial suppliers.

9. *Carbon Dioxide Point*

A secondary calibration point that is sometimes convenient to use, although not one of the defining points of the IPTS, is the one atmosphere sublimation point of carbon dioxide, the temperature being about 195°K. Since carbon dioxide is solid special care is required to ensure that the true equilibrium temperature is obtained. Solid carbon dioxide broken into small pieces and used to fill a dewar quickly attains a temperature considerably below 195°K because the partial pressure of the gas is well below 1 atm, the interspaces containing much air. As the solid vaporizes and displaces the air the temperature rises and may ultimately reach the equilibrium figure. To expedite the process and ensure sufficient continuous vaporization to prevent back diffusion of air a small heating coil should be immersed at the bottom of the dewar.

Solid carbon dioxide can be produced in small quantities as snow by the Joule Thomson effect on allowing gas to escape from a cylinder, but it is usually much more convenient to obtain it in blocks from commercial suppliers.

Temperatures in the range 194 to 195°K may be deduced from the pressure in a vapour pressure thermometer using the relation (Meyer and van Dusen 1933; Giauque and Eagan, 1937)

$$T_{68} = 194 \cdot 674 + 12 \cdot 264\left(\frac{P}{P_0} - 1\right) - 9 \cdot 15\left(\frac{P}{P_0} - 1\right)^2 \qquad (10.11)$$

G. THERMOMETERS

1. *Gas Thermometers*

(a) *Constant volume precision gas thermometer.* The constant volume gas thermometer is the standard by which the thermodynamic temperature scale is reproduced and some recent work is outlined by Preston–Thomas and Kirby (1968), Preston–Thomas and Bedford (1968) and Barber and Horsford (1965). It is a device for measuring the pressure P of a gas in a bulb of volume

V at the temperature concerned. The measuring equipment is normally at room temperature so the bulb must be connected to the pressure gauge by a capillary tube which is at some intermediate and perhaps variable temperature. Its volume must therefore be small, as should also the gauge, although both these requirements conflict with the need to make precise pressure measurements. Precision measurements of pressures up to a few atmospheres are usually made with a mercury manometer which should be of fairly wide bore (say 2 cm) to avoid capillary depression errors but on the other hand it should not be too wide otherwise changes in the mercury meniscus shape can introduce significant volume errors in the gas space above it. Some difficulties here may be avoided by inserting a diaphragm transducer used as a null detector although this gives a limit to sensitivity due to the finite resolution of the transducing system. It also introduces some uncertainty in the volume of the system. At low temperatures where the pressure also could be low the mean free path of the gas molecules is comparable with the dimenions of the thermometer. (At ambient temperature and a pressure of 10^{-2} mm Hg the mean free path is 1 cm. It is inversely proportional to pressure and varies approximately as $T^{1.2}$.) Then a small bore capillary introduces errors due to the thermomolecular pressure effect. If T_1 and T_2 are temperatures of the bulb and manometer respectively and P_1 and P_2 the corresponding pressures, the relation $P_1/P_2 = \sqrt{(T_1/T_2)}$ applies for short connecting tubes. For more complicated situations the theory has been discussed by Knudsen (1927) and by Keesom (1942). For precise work it may also be necessary to consider adsorption on the walls of the bulb although the effect is small.

Gas thermometers are usually filled with helium to about 1 atm or perhaps 1,000 Torr at 0°C but for low temperature work this filling pressure may be used at some lower temperature, say 4°K, in order to give higher pressures at the low temperature of use. The upper limit to the filling pressure is set by the requirement that below the critical temperature the total pressure must be less than the vapour pressure at the temperature concerned. Although normally helium is the gas used, in the past many measurements have been made with nitrogen and hydrogen in the temperature ranges for which they are suitable.

Since even helium is imperfect, especially at low temperatures, it is necessary whatever gas is used to correct for imperfections. At the low pressures used in thermometry this is done with sufficient accuracy by the virial equation 10.1 already discussed, limited to the second virial coefficient B.

Numerical values for this coefficient can be obtained from Brewer (1967) and Mann (1962). A detailed procedure for making the corrections is given by Roberts and Miller (1951).

(b) *Laboratory gas thermometer.* A less precise form of gas thermometer suitable for many experiments on semi-conductors, such as the Hall effect, etc. for which an accuracy of about 3 % in the temperature is adequate, may be made by coupling the measuring bulb to a standard Bourdon gauge with a range from 0 to about 2 atm, Fig. 10.2, as suggested by Simon in Ruheman (1937). At the lowest temperatures this acts as a vapour pressure thermometer changing

Fig. 10.2. Simon gas thermometer.

to a gas thermometer at the temperature at which all the liquid has evaporated. The scale may be made to cover a range extending up to about 100°K without overloading the pressure gauge when the bulb is at ambient temperature. The sensitivity for any selected temperature range may be adjusted by adjusting the relative volumes of the bulb and the gauge. If we regard the gas as perfect and fill to a pressure p_0 at temperature T_0, the pressure p when the bulb of volume v is at temperature T is given by

$$\frac{pv}{T} + \frac{pV}{T_0} = \frac{p_0 v}{T_0} + \frac{p_0 V}{T_0} \tag{10.12}$$

where V is the gauge volume. Writing this as

$$\frac{1}{T}(vT_0) + V = \frac{p_0}{p}(v + V)$$

shows that the volumes can be found by calibration at two known temperatures, e.g. by immersing the bulb in liquid helium and liquid oxygen. The low temperature sensitivity is increased by making the gauge volume large in comparison with the bulb volume so that only at low temperatures is the majority of the gas in the bulb. Then at high temperatures it is insensitive to changes in bulb temperature although sensitive to changes in gauge temperature. Such a thermometer may be accurate to about $\frac{1}{20}°$K at temperatures below 30°K. On the other hand if the bulb volume is made large equation 10.12 indicates an almost linear relation between p and T and the thermometer sensitivity is more or less independent of the temperature being measured.

G. K. White (1961) has described a modification of this version in which the bulb is connected to a mercury manometer arranged to have the high pressure meniscus brought to a given fiducial mark for every reading, thus making the instrument a constant volume thermometer. Provided that the ambient temperature is maintained constant this arrangement allows readings of temperature correct to 0·05°K between 55 and 90°K and to 0·01°K from 1·8 to 4·2°K. At other temperatures it is probably correct to within the limits of the reading accuracy.

Serious errors can be introduced if the bulb is at some temperature near, say, 40°K and the capillary traverses the liquid helium bath at 4·2°K. In this case a variable but relatively large proportion of the gas is in the capillary and either a correction should be made for this or some arrangement made to raise the temperature of the capillary above that of the bulb and, more important, that of the bath.

(c) *Industrial gas thermometers.* In industrial instruments the measuring system consists of a hollow metal bulb, a connecting capillary tube and a pressure sensitive device which can be a Bourdon tube, helix, capsule, etc. The complete system is filled with gas, sealed and the pressure sensitive element connected, via suitable linkages, to an indicating pointer or recording pen. The temperature range covered depends on the filling gas but can be wide compared to some other types of instrument. With nitrogen filling the overall range of measurement is about 73 to 800°K. If helium is used the minimum temperature can be 10°K or less. Compared to other types in general use the response of the gas thermometer is slow but, with a suitably sized bulb and measuring system, accuracy to within 1% of scale range is achievable.

TABLE 10.VII. Liquids for Vapour Pressure Thermometers

Liquid	Approximate temperature range, °K	Data source reference
Helium 3		
Helium 4	2·3–4·5	Brickwedde *et al.* (1960)
Hydrogen	13·8–21·2	Clement and Quinnel (1954)
Neon	21·0–28·0	Grilly (1951)
Nitrogen	63·0–80·0	Armstrong (1954)
Oxygen	74·0–93·0	Hoge (1950)
Methane	92·0–116·0	Mathews and Hurd (1946)
Nitric Oxide	107·0–123·0	Stull (1947)
Carbon Tetrafluoride	121·0–145·0	Stull (1947)
Ethylene	141·0–184·0	Michels and Wassenaar (1950)
Carbon Dioxide (Solid)	173·0–198·0	Brickwedde *et al.* (1960)
Hydrogen Sulphide	184·0–213·0	Clark *et al.* (1951)
Ammonia	206·0–246·0	Stull (1947)
Sulphur Dioxide	226·0–270·0	Rynning and Hurd (1945)

Bulbs are usually fairly large especially for narrow temperature ranges and the ratio of bulb volume to capillary volume may be up to 100 to 1. The difference in height between the bulb and the measuring element has no measurable effect on the indication of temperature.

2. *Vapour Pressure Thermometers*

(a) *Vapour pressure thermometry.* The vapour pressure over a liquid is a sensitive function of temperature over a range from below its boiling point up to its critical point. Provided the relation between pressure and temperature is known with sufficient accuracy a vapour pressure thermometer can therefore be used to give very precise temperature measurements. High precision in the pressure measurement can be obtained by using a mercury manometer and an accurate cathetometer. A wide bore manometer may be used since a large gas volume is of no consequence provided there is always some liquid in the bulb. Suitable cathetometers normally have a maximum range of 100 cm giving a maximum pressure measurement of 1,000 Torr. This defines the upper temperature limit of such a thermometer. If the precision of the pressure measurement is 0·02 Torr the precision of the temperature measurement is much better than the accuracy with which the boiling point is known on the absolute thermodynamic scale. For example at the normal boiling point of oxygen the pressure changes about 8 cm/°K so temperature changes

TABLE 10.VIII. Vapour Pressure of Liquid Helium in microns (10^{-3} Torr)

T,°K	0·0	0·1	0·2	0·3	0·4	0·5	0·6	0·7	0·8	0·9
0	—	—	—	—	—	0·016	0·281	2·288	11·445	41·581
1	120·0	292·2	625·0	1208·5	2155·4	3599·0	5689·9	8590·2	12466·1	17478·2
2	23767·0	31428·0	40465·0	51012·0	63304·0	77493·0	93733·0	112175·0	132952·0	156204·0
3	182073·0	210711·0	242266·0	276880·0	314697·0	355844·0	400471·0	448702·0	500688·0	556574·0
4	616537·0	680740·0	749328·0	822411·0	900258·0	983066·0	1071029·0	1164339·0	1263212·0	1367870·0
5	1478535·0	1595437·0	1718817.0	—	—	—	—	—	—	—

of 0·00025 °K can be detected; compare this with the figures given in Table 10.IV.

Vapour pressure thermometers can be used to cover a wide range of temperatures from about 60°K to ambient by choosing a suitable range of liquids. Assuming the use of a cathetometer and mercury manometer with a pressure range 50 to 1,000 Torr. Table 10.VII shows that in principle it is possible to cover the range from 63 to 270°K completely with the substances listed although some are hardly likely to be used. Only limited parts of the

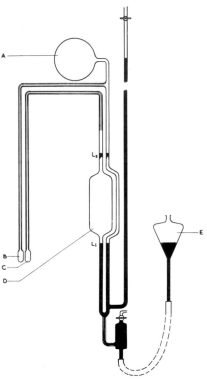

Fig. 10.3. Apparatus for testing purity of liquids.

range from 2 to 63°K can be covered; suitable liquids do not exist. The substances most commonly used in vapour pressure thermometers are helium, neon, hydrogen and oxygen, which are easily obtained in a pure state and for which the relation between pressure and temperature is well known. This information is not available for some of the hydrocarbons listed because of the difficulty of preparing them in a sufficiently pure state.

The criterion of purity for thermometry is that there be no effect due to impurity on the vapour pressure and this can be checked by use of the double bulb thermometer shown in Fig. 10.3. In one limb there is a large reservoir D, which can be filled with mercury or with the gas whereas the other contains a reservoir A always filled with the gas.

After filling with the selected gas and bringing the bulbs B and C to a temperature at which they contain some liquid the pressures in the two limbs are equal at L1 as indicated by the differential mercury manometer. On raising the mercury level to L2 by raising the reservoir E most of the gas in the first limb is condensed in the bulb and any impurity will give a change in pressure. In the other limb hardly any extra condensation occurs so there is no change in pressure and the differential mercury manometer gives a direct indication of the presence of impurity. Apart from the actual difference in pressures produced the performance of the instrument through the condensation process gives a good guide to the purity. If the gas is pure it condenses rapidly and the two limbs come to equilibrium within a few minutes. If the gas is impure condensation and equilibration progress over a long period and the differential pressure is extremely slow to stabilise.

A vapour pressure thermometer measures the temperature of its coldest point and therefore it is necessary to ensure that no part of the thermometer is at a lower temperature than the bulb. In particular it is important to insulate, for example by a vacuum or gas-filled jacket, any part of the stem traversing the refrigerant bath, so allowing heating by conduction from the warm end.

To ensure that the gas is air free the manometer is best filled by distilling the mercury into it at low pressure, after thorough evacuation of the thermometer and filling system. After further evacuation the bulb can be filled from the gas cylinder or other source of supply.

(b) *Liquids for vapour pressure thermometry.* All the substances listed in Table 10.VII and particularly the permanent gases can be obtained in a high state of purity from commercial suppliers. This is usually by far the most convenient method of obtaining them but laboratory preparation is also possible. Figgins (1965) has described a method of filling from a bath of liquid. Some comments on the individual gases follow.

Oxygen. A good laboratory method of preparation is by electrolysis of an air-free electrolyte, since the gas then contains scarcely any impurity except hydrogen which is easily removed by fractionation at 90°K. Another much used procedure is by heating potassium permanganate, but oxygen so produced always contains a lower boiling impurity which is probably nitrogen or argon or both desorbed from the permanganate at reaction temperature even after prolonged evacuation. Commercial cylinder oxygen may have similar impurities so both require purification by a much more careful fractionation at 90°K than the product of electrolysis.

At pressures near 1 atm the temperature on the 1968 scale is given by equation 10.8. At other pressures from 10 to 100 cm of mercury (100–1,000

Torr) the temperature on the 1948 scale dan be obtained from Table 10.VI, based on the data of Hoge. Corrections to IPTS68 are given in Comptes Rendus des Séances (1948).

Hydrogen. Hydrogen exists in two forms, ortho-hydrogen and para-hydrogen stable at room temperature and low temperature respectively, and at intermediate temperatures coexisting in proportions varying with and determined by the temperature. The two have slightly different boiling points so it is important to know and prefereable to select the concentration of each in the thermometer. The most convenient practice is to expedite conversion to the para form in the thermometer by inserting a small amount of a suitable catalyst, e.g. ferrous hydroxide, in the bulb. Since the boiling point of hydrogen is so far removed from that of helium and the atmospheric gases the liquid is readily obtained pure and may be evaporated to fill the thermometer if gas of high purity is not available.

The temperature on the 1968 scale can be deduced from equation 10.10.

Helium. Helium is readily obtained in a high state of purity in cylinders, and except for vaporization of liquid, no other source exists. The temperature on the 1968 scale can be obtained from the tables of Clement, van Durieux *et al.*, summarized in Table 10.VIII and given in detail by Brickwedde *et al.* (1960).

Carbon Dioxide. Commercial supplies of carbon dioxide whether gas or solid are usually of lower purity than required for vapour pressure thermometry, some nitrogen or air being present. Purification can be effected by condensing to solid in a suitable vessel and removing the contaminant by evacuation.

The temperature can be deduced from the pressure using the data of Meyers and van Dusen (1933) or equation 10.11 can be used.

Organic Liquids, Freons. Supplies of these materials are commercially available and in many cases the manufacturers can give vapour pressure temperature relations. Sources of information are also given in Table 10.VII but it must be remembered that the figures are for the particular liquid that the experimenter used and are given in terms of what are now obsolete temperature scales.

(d) *Industrial vapour pressure thermometers*. For industrial low temperature applications the bulb used is comparatively small, sufficient only to contain the small amount of liquid used at the lowest temperature in the range.

The pressure acts on a sensing element of the same type as in the gas thermometer and operates a pointer or pen. Because of the logarithmic relationship between pressure and temperature the scale is cramped at the lower end and open at the upper end.

Vapour pressure thermometers have a faster response than gas-filled instruments and, with suitably sized bulbs, can be used for narrow span measurement. The indicated temperature is not unduly affected by changes of temperature along the capillary but, in systems where liquid enters the capillary, it is necessary to allow for the liquid head when the bulb and measuring element are not on the same level. For above ambient or "cross ambient" temperature ranges the amount of liquid used is greater than for low temperature measurement although the type of system is the same.

A.
    BULB AT HIGHER TEMPERATURE THAN MEASURING SYSTEM WITH LIQUID FILLED CAPILLARY.
DIFFERENCE IN ELEVATION BETWEEN BULB AND MEASURING SYSTEM MUST BE TAKEN INTO ACCOUNT WHEN CALIBRATING

B.
    BULB AT LOWER TEMPERATURE THAN MEASURING SYSTEM WITH VAPOUR FILLED CAPILLARY

Fig. 10.4. Vapour pressure systems for above and below Ambient conditions.

Typical vapour pressure systems are shown in Fig. 10.4; (A) shows a system in which the bulb is at a higher temperature than the rest of the system with a liquid filled capillary. (B) shows a system used for low temperature measurement with a small amount of liquid in the bulb and the capillary filled with vapour.

H. RESISTANCE THERMOMETERS

1. Standard Platinum Resistance Thermometer

The platinum resistance thermometer is the official standard for reproducing the IPTS (1968) scale which it does exactly provided it is constructed to meet the specified requirements of purity and freedom from strain, and calibrated by the prescribed methods described above. Impurities affect the slope dR/dT of the resistance-temperature curve which is therefore a measure of their concentration, so the 1968 scale specifies an upper limit of $1\cdot3925$ for

23

R_{100}/R_0, the ratio of the resistances at 100°C and 0°C, respectively. For high reproducibility the thermometer must be constructed from annealed wire, suspended in a strain-free manner in a sealed protecting sheath and re-annealed after construction.

The wire diameter is usually from 0·005 to 0·010 in. and of such length that the resistance is 25 ohms at 0°C, giving a fundamental interval $R_{100}-R_0$ of 10 ohms and a slope of approximately 0·1 ohms/°C. Two leads of thicker platinum are attached to each end to make a four-terminal resistor and the thermometer sealed off in a sheath of glass, silica or platinum under a small pressure of dry air or oxygen. For low temperature work some helium may be added but oxygen is essential to preserve the condition of the surface and with it the reproducibility of the resistance.

Callendar's early design of a loose winding on a mica mandrel was for long the standard and most reliable form. An early successful departure was the freely suspended helical coil of Barber (1950) but coiled coils have since been developed (Stimson, 1955). Straight platinum wire bifilar wound on a mica cross and other bifilar windings have also been used, but the monofilar coil designs are of low enough self inductance to make the complication of bifilar winding generally unnecessary. It is, of course, a convenient way of bringing both ends of the wire to the same end of the thermometer.

Thermometers to several of these designs are available as commercial products, from a number of suppliers, e.g. Fig. 10.5. They can be obtained in either metal or glass sheaths, about 1 cm o.d. and 45 cm long, the thermometer occupying about 5 cm at the end. The sealed thermometer can also be obtained without protective sheath for insertion directly into experimental sheaths or apparatus.

For use below 90°K the thermometers are usually smaller than those for higher temperatures, a typical design being in a platinum sheath about 3 mm diameter and 20 mm long with a glass pinch seal.

Since the standard thermometer is constructed as a four-terminal resistor its resistance can be measured by any of the conventional methods. There is no need to discuss these in detail here, it suffices to mention that suitable instruments include the vernier and Diesselhorst otentiometer,p the Smith's difference bridge and other precision bridges (Smith, 1912; Hall, 1929). With all of these instruments the resistances of the leads are eliminated so that changes in them do not affect the precision of the measurement. Stray thermoelectric e.m.f.s are eliminated by reversing the measuring current and joining copper leads to the platinum in an isothermal volume, preferably that where the temperature is being measured or in a room temperature region shielded from draughts. For a temperature measurement to 0·01°K the resistance must be measured to 0·001 ohm, the actual resistance being in the range 5–25 ohms for temperatures between 70°K and 273°K.

Self balancing a.c. bridges are available with digital read-out and facilities for direct connection to computer inputs. One found very satisfactory for thermometry is based on an a.c. inductance bridge (Wolfendale, 1969) in which the ratio arms are precision inductances, which can be made to high accuracy by control of the geometry. A precise and stable standard resistance is required and the effect of frequency on the apparent resistance must be taken into account especially if the thermometer and standard resistance are calibrated by d.c. methods. For the thermometer the effect is usually negligible a change of less than 1 part in 2×10^6 at 375 Hz has been measured. For wire wound standard resistances, however, it may be significant, a not unreasonable figure being 10 parts in 10^6 at the same frequency, in agreement with estimates based on the geometry of the winding. For solid state resistors it is due to the large thermoelectric power of the material which introduces an appreciable Peltier e.m.f. during d.c. calibration but none at moderate a.c. working frequencies. Consequently it changes most rapidly at low frequencies and measurements at moderate frequencies, say from 40 Hz to 400 Hz cannot be extrapolated to zero frequency. For a small solid state resistor the author found that there was a difference of 50 parts in 10^6 between the d.c. value and that at 40 Hz, and Hill (1968) at the NPL measured a further change of 11 parts in 10^6 at 480 Hz, the measuring current being 5 mA.

Errors in temperature measurement due to the self heating effect of the measuring current are minimized by using low currents; from 1 to 3 mA is a satisfactory value above 70°K. The effects can be estimated by making measurements with two different currents, i.e. two different rates of power dissipation and, if necessary, extrapolating to zero power.

2. Industrial Resistance Thermometers

The resistance thermometer is one of the most common industrial instruments for measuring temperature in low temperature systems. For this purpose the Callendar–van Dusen equation (Burgess, 1928) which is equivalent to

$$R_t = R_0(1 + \alpha t + \beta t^2 + \gamma t^3) \tag{10.12}$$

can be approximated by the linear relation

$$R_t = R_0(1 + \alpha t) \tag{10.13}$$

Practical use is made of this resistance–temperature relationship in resistance thermometry. Nickel may also be used as the resistance element but platinum is in more general use.

A change of resistance will occur at constant temperature if the wire is subjected to strain (the principle of strain gauges). Therefore, it is essential to ensure that, when manufactured, the resistance winding is in a strain-free condition as in the standard platinum resistance thermometer already discussed. It is normally annealed to minimize initial strains.

The self-heating effect, which is proportional to the square of the measuring current times the resistance, must be kept low. For mechanical protection the

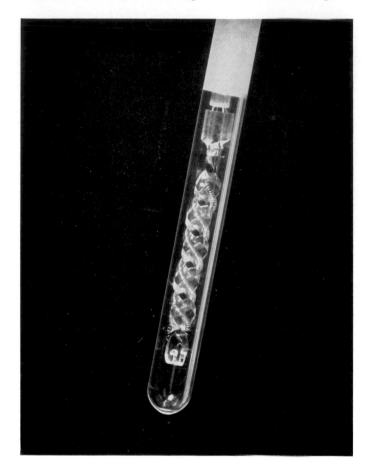

Fig. 10.5. Precision platinum resistance thermometer.

element must be covered by a sheath which, for low temperature measurement, is made of metal. For fast response this sheath should be of light construction consistent with the requirement for mechanical protection.

The various methods of constructing industrial resistance elements include:

(a) Wound on flat mica, insulated on the outside with mica and inserted in a high thermal conductivity metal sheath which is afterwards rolled flat. This type gives rapid response to temperature changes.

(b) As (a) but with a steel sheath. This gives better mechanical protection but has a slower response.

(c) Platinum wire wound on a steatite former, glazed and baked and encapsulated in a metal protective sheath.

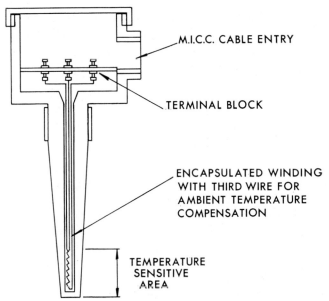

Fig. 10.6. Typical resistance thermometer as used in cryogenic applications.

Type (c) is a mechanically robust assembly and is suitable for use in cold boxes where accessibility is limited and the unit must have a high degree of electrical and mechanical reliability even at the expense of speed of response. For high velocity process streams some additional mechanical protection is sometimes necessary.

A typical resistance thermometer assembly is shown in Fig. 10.6. This comprises resistance element, protective metal sheath and terminal head with tapped hole to receive a gland for mineral insulated metal sheathed cable. Mineral insulated metal sheathed cable (M.I.C.C.) offers good mechanical properties at low temperature where thermoplastic insulants such as P.V.C. would become brittle and crumble. The cables from the individual thermometers are normally brought through the cold box casing via suitable glands at a convenient common point. A gland plate is usually provided for this purpose and the cable ends are then taken to a multiway terminal box fixed to the outside of the cold box. From the terminal box the cable runs to the measuring instruments can be made in conventional multicore cable.

When manufactured for industrial low temperature applications resistance thermometers are checked at one or more of the three fundamental fixed points, i.e. the temperatures of liquid oxygen, melting ice and boiling water, respectively. Usually only the two lower temperatures are used and accuracies within 0·1°C should be achievable. In practice, because of conduction and radiation effects between the measuring element and the pipe or vessel wall the accuracy is reduced and the error can be up to 1°C. The average industrial potentiometric type indicator has errors of less than 1 %of scale so for a measuring range of 100°C (or 100°K) the maximum overall error would be within 2°C. This compares very favourably with other industrial temperature measuring instruments but because of the need for mechanical robustness the speed of response may be several minutes in extreme cases. Where mechanical strength is not an essential requirement the element can be designed to have a time constant of only a few seconds.

Compared with the types of thermometer previously described the main advantages of the resistance thermometer are:

(a) Small size measuring element.
(b) No capillary tube which could be damaged (if a connecting cable is broken a repair is easily effected).
(c) Damaged elements can be changed in the field without having to change the complete thermal system.
(d) Where several measuring points are involved it is simpler and cheaper to run one multicore electrical cable than several metal capillaries each with bulb on one end and instrument on the other.
(e) By using multiway switches, or several individual switches one instrument can be used for multiple resistance thermometer installations.

For general industrial applications, then, the resistance thermometer is a more suitable temperature measuring instrument than either the gas or vapour pressure thermometers.

3. Other Resistance Thermometers

The 0·25 W Allen Bradley carbon resistor has proved very popular as a low temperature thermometer since it is cheap, easily available and robust. A

TABLE 10.IX. Typical Resistance Variation of a Carbon Resistor

Temperature °K	Resistance ohms
4	1,231
20	159
78	85
100	79
290	68

wide range of nominal resistance values is available giving scope for selection of the best value for the temperature range of interest. The variation of resistance with temperature for a typical specimen is indicated in Table 10.IX. They are insensitive to magnetic fields and not affected by the measuring current except for a small self-heating effect.

Thermal mass can be reduced and heat transfer increased by grinding off the outer protective ceramic coating. They are very useful in calorimetry below 20°C, where they are very sensitive. Selected pairs can be used differentially to give directly the temperature difference between two bodies.

They can be used as four terminal resistors in any of the conventional bridge or potentiometric circuits, the latter generally being preferred. It is usually reported that the thermometers are somewhat unstable, changing calibration each time they are warmed to ambient temperature, although they can easily be recalibrated to $\pm 0.5\%$ against the vapour pressure of the cryostat liquid. However Weinstock (1968) and Hudson (1968) report them stable to at least 0.1% over a period of two years and 20 cycles between room temperature and liquid helium temperature. The relation between resistance and temperature from about 1 to 20°K usually used is that of Clement and Quinnel (1952):

$$\log R + K/\log R = A + B/T$$

where R is the resistance at temperature T, and A, B and K are constants. However, Roach (1968) found better accuracy resulted from adding a term in $(\log R)^2$ to the left hand side.

For this temperature range the standard thermometer is the arsenic doped germanium crystal which is commercially available as a four terminal resistor (Gabelle *et al.*, 1958). The germanium is in the form of a thin rigid sheet enclosed in a sealed capsule containing helium gas, the diameter of the capsule being about 2 mm and its length 20 mm. Thermometers can be supplied with N.B.S. calibration or uncalibrated. They are reproducible to about $0.0001°K$ at 4°K and are usable from 1 to 100°K. Measuring circuits have been described by Wolfendale (1969).

I. THERMOCOUPLES

Thermocouples are very convenient thermometers where only a moderate absolute accuracy is acceptable or where low thermal capacity or a precisely located measuring element is needed. They are also used to give a direct and often sensitive measurement of a temperature difference. The electrical energy dissipated in them is negligible so they contribute little to the heat load and experience no temperature rise due to self-heating. The thermoelectric power, dE/dT, diminishes progressively with temperature for those normally used at higher temperatures, but for temperatures above 90°K they are quite satisfactory. The copper–constantan combination is often used, as is also the chromel–alumel one. White (1933) described potentiometric techniques for accurate measurement of the small e.m.f.s produced but self-balancing instruments make the measurements much easier to carry out.

Scott (1941) has examined a large number of copper constantan thermocouples made from American materials and show how calibration at three temperatures can give good accuracy. An accuracy of about 1 % in t can be obtained from standard tables and 0·1°K by using a deviation curve or computing a table from the relation

$$E = at + bt^2 + ct^3 \qquad (10.14)$$

for temperatures between 70–273°K. The constants a, b and c are deduced from measurements at 90 and 195°K against the vapour pressures of O_2 and CO_2, respectively, and at about 230°K against a standard platinum resistance thermometer. Scott's table was based on the mean of his many observations and is calculated from equation 10.14 with

$$a = -38\cdot476 \quad b = -0\cdot046080 \quad c = +3\cdot93 \times 10^{-5}$$

where $E\ \mu V$ is taken as positive for current flowing from constantan to copper at the reference junction although the temperature of the measuring junction $t°C$ is negative.

An abstract is given in Table 10.X with, in addition, some figures for the thermoelectric power dE/dt.

Carefully installed thermocouples constructed from selected lengths of homogeneous wire can make quite precise and permanent secondary thermometers as Giauque et al. (1927, 1937) and Aston (1941) have shown. The wires should be mounted strain-free and selected for homogeneity by testing for thermal e.m.f.s. A simple test is to connect the two ends of a wire to a potentiometer and draw the wire through a tube immersed in liquid nitrogen. A satisfactory specimen will show an e.m.f. not exceeding 1–2 μV throughout its length. Inhomogeneities may be eliminated by connecting several wires in parallel or by connecting several thermocouples in series in which case the additional advantage of higher sensitivity is obtained, at the expense of possible increase in insulation problems. This arrangement is particularly useful in measuring small temperature differences as in calorimetry.

For lower temperatures, especially in the range 4–20°K, other combinations have been developed based on the effect of trace impurities on the electrical properties of metals. A particularly useful one consists of a silver or chromel wire and an alloy of gold + 0·03 atomic per cent of iron (Berman and Huntley, 1963), which has a thermoelectric power of about 15 $\mu V/°K$ down to 20°K falling to 10 $\mu V/°K$ at about 4°K. Some figures are given in Table 10.XI. Particular care must be used in constructing and installing this type of couple because of the low concentration of iron, which can easily be lost by oxidation if the junction is overheated during the soldering operation. Equally important, especially at low temperatures, is good thermal contact between the junction and the experimental apparatus. If electrical contact between the two is permissible indium solder has been found very satisfactory.

TABLE 10.X. Thermal e.m.f.s and Thermo-electric Powers of Common Thermo-couple Combinations. Temperature of Reference Junction 273·15°K (0°C).

Temperature of measuring junction °C	Chromel–Alumel		Copper–Constantan		Gold+0·03 atom % Iron–Chromel P	
	e.m.f. e mV	de/dt $\mu V/°C$	e.m.f. e mV	de/dt $\mu V/°C$	e.m.f. e mV	de/dt $\mu V/°C$
0	0·0	39	0·0	38	0·0	20·0
−10	0·39	38	0·380	37	0·200	19·6
−20	0·77	37	0·751	37	0·396	19·5
−30	1·14	36	1·112	36	0·591	19·5
−40	1·50	36	1·463	35	0·786	19·4
−50	1·86	35	1·804	34	0·980	19·2
−60	2·21	34	2·134	33	1·172	19·0
−70	2·55	32	2·454	32	1·362	18·7
−80	2·87	32	2·763	31	1·549	18·5
−90	3·19	30	3·061	30	1·734	18·1
−100	3·49	29	3·348	29	1·915	17·8
−110	3·78	27	3·622	27	2·093	17·5
−120	4·05	27	3·886	26	2·268	17·3
−130	4·32	25	4·137	25	2·441	17·1
−140	4·57	24	4·376	24	2·612	16·8
−150	4·81	22	4·602	23	2·780	16·5
−160	5·03	21	4·815	21	2·945	16·3
−170	5·24	19	5·016	20	3·108	16·0
−180	5·43	17	5·203	19	3·268	15·8
−190	5·60	15	5·377	17	3·420	15·6
−200	5·75				3·500	

A combination of pure silver and pure aluminium gives a constant though small thermoelectric power of about 2·5 $\mu V/°K$ from 80 to 273°K and can be constructed from wire or strips evaporated on to an insulating substrate (Bailey *et al.*, 1969). Joints can be soldered using 90% indium–10% silver and 50% indium–50% tin for the silver and aluminium wires, respectively. The thermocouple has been found reproducible to at least 0·1°K at 77°K on repeated cycling to room temperature.

Other thermocouple combinations suitable for low temperature measurements have been studied by Powell and Sparks (1968), who have published tables and graphs of the thermoelectric functions.

A reference junction at some known temperature is required for measurement of absolute temperature. Melting ice or a triple point of water cell gives a reference at 0°C but then a large e.m.f. has to be measured to a microvolt or

TABLE 10.XI. Thermal e.m.f.s and Thermoelectric Powers of Au+0·03 Fe–Chromel
P Thermocouple at Temperatures below 100°K

(a) Temperature of reference junction 77·3°K

Temperature of measuring junction °K	Thermal e.m.f. e μV	Thermoelectric power de/dT $\mu V/°K$
10	940	15·4
20	786	14·2
30	644	13·5
40	509	13·0
50	379	13·3
60	246	13·8
70	108	14·5
80	− 37	

(b). Temperature of reference junction 4·2°K

Temperature of measuring junction °K	Thermal e.m.f. e μV	Thermoelectric power de/dT $\mu V/°K$
2·0	− 28	12
3·0	− 16	13
4·0	− 3	14
4·2	0	14
5·0	+ 11	14
6·0	25	14
7·0	39	15
8·0	54	15
9·0	69	16
10·0	85	16
11·0	101	16
12·0	117	16
13·0	133	16
14·0	149	15
15·0	164	15
16·0	179	15
17·0	194	15
18·0	209	15
19·0	224	15
20·0	239	15
20·4	245	

better. The total e.m.f. can be reduced by using a bath of boiling oxygen, nitrogen, hydrogen or helium as refrigerant according to the range of temperature in which the measurements are to be made. The temperature of the bath must be measured separately—either from its vapour pressure or by means of an immersed vapour pressure or standard platinum thermometer.

Thermoelectric powers for a number of couples useful at low temperatures are given in Table 10.X.

1. *Industrial Thermocouples*

In earlier types of thermocouple measuring equipment the reference temperature was maintained at 0°C by the use of an ice bath. This, in its simplest form, is a Dewar flask filled with melting ice and with the cold junction immersed. Obviously this method of maintaining a temperature of 0°C is subject to errors from many causes, amonst them being the purity of the water used, excessive water formed in the bottom of the flask as the ice melts, and the immersion length of the cold junction. It is also necessary to drain off water and replenish with ice from time to time. A unit which automatically maintains the reference temperature at ice point with a claimed accuracy of ±0·01°C is now available. Although, for most practical applications, such accuracy is unnecessary, with the increasing trend towards computers for on-line process control even marginal improvements in accuracy are desirable and may be economically necessary.

The most common rare metal thermocouple is platinum–platinum rhodium and base metal combinations are iron–constantan, copper–constantan and alloys of chromium nickel–aluminium nickel (chromel–alumel).

For most accurate results the leads connecting the thermocouple to the measuring instrument should be as short as possible and of the same materials as the couple. In the case of rare metal couples this is expensive and to a less extent this may also apply with base metal couples. Compensating leads having similar thermoelectric characteristics to the couple materials are used to reduce the overall installation cost.

Copper–constantan thermocouples are probably the most widely used for cryogenic applications and are suitable for temperatures down to about 73°K. For temperatures approaching absolute zero couples of copper–gold alloy are used. These are brittle and the alloy is too unstable to be soldered or brazed. Consequently great care is necessary when making and installing them.

A convenient method of making thermocouples for cryogenic applications inside cold boxes is by using mineral insulated thermocouple wire. Various methods of forming the couples are shown in Fig. 10.7. The leads can be brought out through the cold box casing in an unbroken length and thence to a locally mounted instrument or, where the distance is within practical limits, to a control room. Where the distance is too great for an unbroken run to be practical the leads can be taken to a terminal box located on or near the cold box casing.

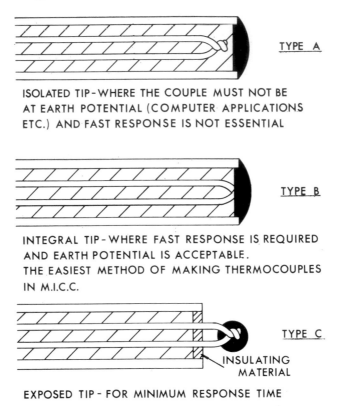

TYPE A

ISOLATED TIP-WHERE THE COUPLE MUST NOT BE
AT EARTH POTENTIAL (COMPUTER APPLICATIONS
ETC.) AND FAST RESPONSE IS NOT ESSENTIAL

TYPE B

INTEGRAL TIP-WHERE FAST RESPONSE IS REQUIRED
AND EARTH POTENTIAL IS ACCEPTABLE.
THE EASIEST METHOD OF MAKING THERMOCOUPLES
IN M.I.C.C.

TYPE C

INSULATING
MATERIAL

EXPOSED TIP - FOR MINIMUM RESPONSE TIME

Fig. 10.7. Methods of making thermocouples using mineral insulated copper covered thermocouple cable.

With base metal thermocouples used in the range 73 to $400°K$ accuracy of measurement is within $\pm 3\%$. With rare metals the accuracy is within $\pm 2\%$. The errors in the measuring instrument must also be allowed for. Although basically less accurate than resistance thermometers or filled systems thermocouples have the advantages of robust construction and fast response. If a M.I.C.C. thermocouple is formed with an exposed tip response times of less than one second can be achieved.

J. OTHER THERMOMETERS

For temperatures below about $1°K$ which can be obtained by adiabatic demagnetization of a dilute magnetic salt the susceptibility of the salt is a good index of temperature, since over a useful range Curie's law holds and $\chi = c/T$ to a close approximation. However at the lower limit of the range significant departures occur and a magnetic temperature T^x is first deduced from the relation $\chi = c/T^x$. The relation between T and T^x can be obtained calorimetrically as shown by Simon in Ambler and Hudson (1955). The thermometers are made from compressed powder or granules and Kurti and Simon (1938) and others have shown how to correct for the self demagnetization field and the filling factor. The difference between T and T^x is of the order of a few millidegrees at 0·4°K and a few hundredths at 0·1°K, but near 0·03°K T^x may be more than twice T. A very useful salt, which at temperatures below 0·1°K acts as both refrigerant and thermometer, is cerium–magnesium nitrate, and it obeys Curie's law down to 0·006°K (Mess et al., 1968; Zimmermann and Abeshouse, 1968; Niesen and Huiskomp, 1968; Hudson, 1968; Kurti, 1968).

The susceptibility may be measured ballistically or by an induction bridge, and the induction coils are usually immersed in the low temperature region.

Another phenomenon being investigated for low temperature thermometry is the thermal noise in a resistor for which the mean square voltage is proportional to the absolute temperature (Wagner et al., 1968).

III. Pressure Measurement

Pressures in low temperature plant can normally be measured by standard instruments at room temperature by connecting one to the other by a gas-filled capillary. Bourdon gauges cover a wide range of accuracies from the roughest to those requiring 0·1% accuracy for pressure ranges from 1 atm to several hundreds. For precise absolute measurements a pressure balance may be used and of course mercury manometers are frequently employed. More sensitive manometers using low density, low vapour pressure liquids like vacuum oil or dibutyl phthalate may also be used. Measurements on oxygen must be made with oil-free gauges, and where a range of gases is being used it is wise to use oil-free gauges throughout. Special difficulties arise with pressure balances. The use of a mercury transmitter cannot necessarily be relied upon to retain the oil in the balance. A diaphragm separator can be used.

For measurements on cryogenic liquids the pressure is transmitted partly by liquid and partly by gas with a meniscus located at some intermediate point. If the pressure is above the critical there will be no meniscus. For precise measurements corrections must be made for the hydrostatic head of liquid and perhaps even for that of the gas since at low temperature and high pressure gas densities are relatively high.

As mentioned in discussing gas and vapour thermometry, if the pressure to be measured is low there may be significant errors due to the thermo-molecular pressure effect for which corrections are required.

At very low pressures some workers have used a Pirani gauge installed in the low temperature bath, calibrated *in situ*. This increases the heat dissipation and therefore the liquid evaporation rate, but the extent of this effect is usually insignificant.

For some experiments unbonded strain gauges have been found satisfactory though subject to zero shift. They in fact measure the changing diameter of the container.

A. INDUSTRIAL PRESSURE MEASUREMENT

1. *Problems*

The main problems arising are:

(1) Formation of liquid in vertical sections of impulse tubing causing incorrect readings. This situation can be prevented by careful installation of impulse piping to ensure that where vertical legs are unavoidable they are in a "warm" part of the installation. Where space is limited inside the cold box the necessary length of horizontal tube can be achieved by forming a flat horizontal loop.

(2) Formation of solids, e.g. solid carbon dioxide, frozen water droplets, etc. causing blockages in impulse tubing.

In a well designed and correctly operated plant the build-up of solids between maintenance shut downs is insignificant. The effect of any abnormal build-up can be minimized by using tubing of sufficiently large bore and avoiding sharp corners or restrictions.

Where there is a likelihood of CO_2 build-up larger bore pipes up to 12 mm bore should be used with blanked off tees on the outside of the cold box for "rodding out".

(3) The effect of low temperature on the measuring element or manometer fluid. If a pressure measuring element is subjected to very low temperature the elasticity may be affected resulting in incorrect measurement. This can be overcome by locating the measuring unit in a warm part of the plant and by employing a sufficient length of connecting tubing to keep the instrument temperature within the manufacturer's recommended limit.

2. *Types of Measuring Device*

The type of measuring device used depends on the magnitude and range of pressure to be measured, ranging from slack diaphragms for very low pressures to bourdon tubes and pistons for high pressures.

(a) *U-tube manometers.* The simplest form of pressure measuring device is the U-tube manometer. Its accuracy is normally limited only by the accuracy with which the difference of height can be measured.

A variation of the U-tube manometer is the single limb manometer, shown in Fig. 10.8. This is, in fact, a double limb manometer with one limb much shorter and fatter than the other. When calibrating the indicating scale allowance must be made for the small fall in level of the wide limb. As an

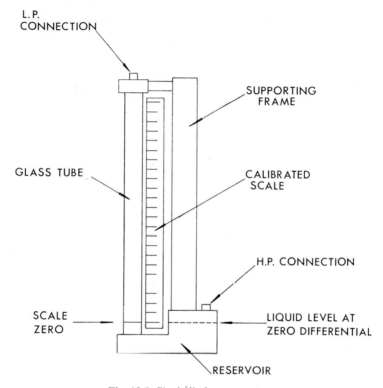

Fig. 10.8. Single limb manometer.

example assume that the area of the wide limb is nineteen times the area of the narrow limb. A fall of 1 mm in the level of the former will correspond to a rise of 19 mm in the level of the latter, the pressure difference being 20 mm. Thus, if the zero mark is made opposite the common level with no differential applied, then for the scale to be calibrated in cm water, each cm division will, in actual fact, be only 0·95 cm. For corresponding areas a and A, respectively, each division on the scale will be the required value multiplied by $A/A+a$. The manometer liquid may be chosen according to the pressure differential to be measured.

(b) *Diaphragms.* For very low pressures which are measured in centimetres of water the diaphragm type of element is used. Where the maximum pressure is of the order of fifteen centimetres water gauge or less the diaphragm is generally of synthetic rubber or plastics material. Metal diaphragms are normally used for pressures in excess of fifteen centimetres water gauge, the most common materials being phosphor bronze and stainless steel. With certain instrument designs these limits do not apply.

Diaphragms are also used for differential pressure measurement, the deflection of the diaphragm being transmitted through linkages, to the indicating pointer or transmitting device.

The diaphragm type measuring unit can be designed to withstand high overload pressures. An example of this is the Schafer pressure gauge the principle of which is illustrated in Fig. 10.9.

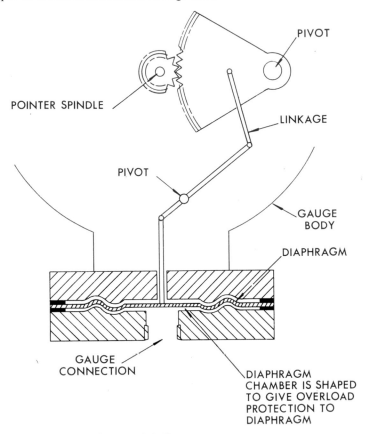

Fig. 10.9. Schaffer type pressure gauge.

(c) *Bourdon tubes.* Most pressure gauges for pressures from three hundred millibars upwards use the Bourdon tube as the measuring element. Materials used for the Bourdon tube may be phosphor bronze, K Monel or stainless steel. Accuracy of Bourdon tube pressure gauges for industrial use is $\pm 1\%$

although, where a higher degree of accuracy is required, this can be within $\pm\frac{1}{4}\%$. Gauges should be recalibrated at intervals of 6 months, especially if they are used at near full scale deflection.

(d) *Bellows.* For measurement of gauge pressure one end of the bellows is anchored and connected to the pressure source, the other end to the indicating device. Any change of pressure causes the bellows to expand or contract along the axis in direct proportion to the change of pressure.

Opposed pairs of bellows can be used for differential pressure measurement. For this purpose the bellows are connected end to end with the chambears sealed from each other and the outer ends are connected to their respective pressure sources. The centre of the bellows assembly is connected to the indicating device and the movement at this point is proportional to the difference of pressure in the two bellows. If one of the pair is evacuated and sealed then using this as a reference pressure the differential bellows can be used to measure absolute pressure.

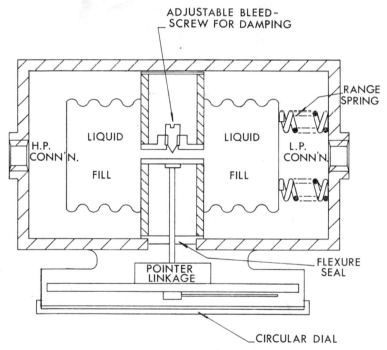

Fig. 10.10. Liquid filled bellows type differential pressure instrument.

A development of the bellows unit for differential pressure measurement is the liquid filled or "Barton type" movement, shown in Fig. 10.10. This unit is basically similar to the opposed bellows system described previously but the plate between the two bellows chambers has a small orifice drilled in it and

24

the unit is completely filled with liquid. The bellows assembly is then enclosed in an outer pressure chamber also divided into two sealed sections with one half of the bellows assembly in each section. These outer sections are the pressure chambers and the differential pressures are applied to the outside of the bellows. When differential pressures are applied, the bellows sensing the higher pressures will contract and the filling liquid, being incompressible will be forced through the orifice into the bellows sensing the low pressure. The movement will continue until the forces produced by the pressure, the liquid movement and the mechanical strain in the bellows are in equilibrium and the resulting displacement of the bellows connecting piece will be transmitted to the indicating device. This type of movement will stand high overload pressures and the orifice, which in production models is usually adjustable, can be set to damp out undesirable pressure oscillations.

(e) *Strain gauge.* Pressure transducers comprise a pressure sensitive element whose deformation is measured by a strain gauge. Thus pressure changes are indicated as changes of electrical resistance. The simplicity of detection is partially off-set by complexity of the indicator.

Modern transducers can be made very small and can be used in locations which are inaccessible for bellows or bourdon tube type pressure instruments. With the development of process control computers and data loggers the use of strain gauge transducers is increasing.

IV. Flow Measurement

A. PROBLEMS

When cryogenic liquids are sufficiently subcooled the problems of measuring flow by most normally accepted methods of industrial flow measurement are no greater than for "normal" fluids. The main difficulty is in metering by orifice plate or other differential producing device where liquid heads at entry to the impulse lines must be measured as gas pressures at near atmospheric temperatures at the indicator or recording instrument. The method used is similar to that used for pressure measurement. That is to run the impulse lines horizontally beyond the point where the liquid–vapour interface forms and connect the measuring instrument as if for gas flow measurement. Care must be taken to ensure that all joints are completely leak tight, otherwise the flow of cold gas leaking through the joint will gradually cool down the pipe until liquid forms and causes measuring errors or even damage to the instrument.

Even with liquids at or near their boiling point, flow-metering with acceptable accuracy is possible using a corner tapped orifice plate, if there is sufficient head of liquid before the orifice. It is essential to prevent boiling before the plate but boiling immediately downstream (at the *Vena Contracta*) does not have any significant effect on accuracy.

The liquid head required before the orifice depends on the flow conditions but, as a general rule, should not be less than 1·5 m. To minimize the risk of flashing the pressure drop across the orifice should be kept to a minimum.

In many orifice metering installations conditions are unsteady due to flow fluctuations, pressure fluctuations or, as previously mentioned, boiling of the liquid. These effects can be minimized by damping the impulse lines. Great care should be taken not to overdamp as this can cause excessive measuring errors. If there is some way of calibrating the metering section under simulated operating conditions the optimum degree of damping can be found experimentally. The presence of solids, e.g. carbon dioxide in the plant can block impulse lines, either partially or completely, causing errors. Reference has also been made to this problem in the section on pressure measurement.

B. TYPES OF MEASURING DEVICE

Types of flow measuring devices come under two main headings, Inferential and Direct. Inferential can be subdivided into differential pressure type and velocity type, the former using orifice plates, venturis etc. as primary elements and the latter, turbine type meters. Direct measuring types mainly employ the principle of volume displacement although methods of mass flow measurement are currently under investigation.

1. *Orifice Plate*

The simplest form of differential producing device and the most widely used in industry generally is the orifice plate, various forms of which are illustrated in Fig. 10.11. The head loss caused by the increase of velocity through the orifice is measured at the pressure tappings and from this can be inferred the flowrate of the fluid in the pipe, flowrate being proportional to the square root of the head loss. For accurate measurement, adequate lengths of straight smooth bore pipe are required before and after the orifice. The upstream length should be between 15 and 40 (or more) pipe diameters according to flowing conditions. The downstream length should in general be not less than 5 diameters.

Various refinements of the basic device have been introduced from time to time. Details of these can be found in the various text books dealing with flow measurement or in British Standard 1042. The British Standard also gives guidance on straight pipe requirements.

Compared with other differential producing devices described later the orifice plate has a comparatively high head loss which is undesirable when measuring flows of cryogenic liquids at or near their boiling point. However, because of its simple form it can be used in cold boxes where space is strictly limited and a trouble free installation is essential for continuous operation of the plant. Orifice metering units for pipes up to two inch bore can be made complete with the necessary upstream and downstream straight pipe lengths and orifice tappings. The complete unit can be welded in position in the cold box. Metering lengths of this type tend to be unwieldy in the larger sizes and orifice carriers for intermediate sizes or orifice flanges for large sizes are normally used.

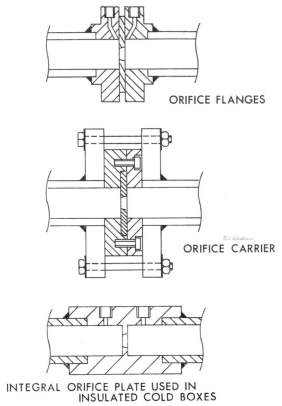

ORIFICE FLANGES

ORIFICE CARRIER

INTEGRAL ORIFICE PLATE USED IN
INSULATED COLD BOXES

Fig. 10.11. Typical orifice plate installations.

2. *Venturi and Dall Tube*

Venturis have a lower head loss and better pressure recovery than orifice plates used for the same flowing conditions. Although they are suitable for metering cold liquids near their boiling point and cold vapour they are little used due to their greater cost and even more because of the long horizontal pipe such as is required to accommodate the meter together with its approach and exit lengths.

The Dall tube is a refinement of the Venturi and it is claimed, has a better recovery than the Venturi, is more compact and requires less straight pipe for installation.

3. *Turbine Meters*

The turbine type meter was used in the U.S. some years ago to meter rocket fuel and has since been developed as an industrial flowmeter. Fluid flowing through the meter impinges on a turbine wheel causing it to rotate, the speed of rotation being proportional to the velocity of the fluid flow and thus to the rate of flow. An electro magnetic pick up is mounted on the outside of the meter body and as each blade of the turbine, usually six in number, passes under the pickup it generates an electrical pulse. The pulses are counted by an electronic counter which displays the total integrated flow or can convert the pulse rate into a flowrate in direct volumetric units.

If a turbine meter is left out of use for any length of time, during plant maintenance, for example, the liquid in the meter and associated pipework will evaporate and the installation will warm up. If safety valves are provided, or the liquid is drained off before it evaporates there is no problem. There is, however, a major problem on start up and that is the effect of cold liquid in a "warm" meter.

Apart from thermal shock, the rapid evaporation of liquid in the warm system will cause the turbine wheel to rotate at a speed far in excess of its maximum permissible value resulting in seized bearings or blade shedding. It is necessary, when starting up, to pass gas slowly through the meter and vent on the downstream side until liquid is observed at the vent. The meter is then sufficiently cooled for normal metering to start.

Stainless steel meters are normally used for cryogenic liquid metering but for oxygen metering this is undesirable on safety grounds, especially where overspeeding may occur. An all bronze meter is currently in use for liquid oxygen with a bar magnet embedded in the rotor to provide the output pulses.

4. *Displacement Meters*

Displacement meters generally work on the principle of rotating pistons or lobes displacing a given volume of liquid for each revolution of the meter. The read out can be by direct drive via suitable linkages and glands, magnetic drive, in which the readout chamber is completely sealed from the metering chamber or remote reading by electro magnetic pulse as in the case of the turbine meter.

Being volumetric devices, positive meters are as susceptible to errors caused by vapour entrainment as other types previously mentioned. Moreover, because of their greater mass and inertia, they are more easily damaged by overspeeding than turbine meters and patience is required when cooling down prior to start up of the metering system.

5. *Mass Flowmeter*

The ideal method of metering cryogenic liquid especially for customer transfer is by mass flowmeter. Although research is being conducted in this field the

TABLE 10.XII. Dielectric Constants of Cryogenic Fluids

Element	NTP gas	Dielectric constant Boiling point Vapour	Liquid
O_2	1·000494	1·00156	1·484
A	1·000517	1·00175	1·532
N_2	1·000548	1·00208	1·433
Ne	1·000123	1·00129	1·174
H_2	1·000254	1·00387	1·228
He	1·000064	1·00618	1·048

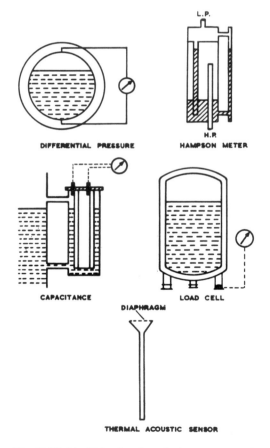

Fig. 10.12. Liquid level and contents measurement.

authors are unaware of any widely accepted method of mass flowmetering of cryogenic liquids.

Amongst types being investigated are a density compensated volume flowmeter and a turbine type in which the velocity of the stream and the torque exerted on turbine blades are measured.

V. Liquid Levels

Many methods of measuring liquid level have been described. The principles on which some operate are illustrated in Fig. 10.12. One measures the differential pressure due to the hydrostatic head of liquid in the vessel. A nonvolatile manometric fluid such as high vacuum oil or dibutyl phthalate is required.

Another type depends on the very considerable difference between the dielectric constants of liquid and vapour as listed in Table 10.XII, a partially immersed parallel plate or concentric cylinder condenser having a capacity varying with the liquid level. For boiling liquids it is advisable to have at least one electrode perforated to allow access of liquid and exit of vapour.

Heat transfer from a hot metal surface to liquid is very much greater than to gas and therefore a vertical wire carrying a current has a higher temperature in the gas phase than in the liquid. The total resistance of the wire can therefore be made a measure of liquid level. If the liquid is helium one can take this to the extreme and use a superconducting metal for the wire so that the immersed section has zero resistance; suitable metals include tantalum, niobium, and lead. However, a large normal resistivity is required to make the gauge sensitive and a superconducting coating of lead on constantan has been found a good combination. In place of a single wire a chain of resistors may be used giving a number of spot levels instead of a continuous indication.

A. INDUSTRIAL LEVEL MEASUREMENT

1. Problems

Cryogenic vessels, whether used for the process or for storage, must be thermally insulated. The insulant may be vacuum, expanded mineral or mineral wool. This requirement for insulation coupled with the necessity to minimize heat inleak is one of the principle problems of cryogenic level measurement.

Other problems are similar to those encountered in flow and pressure measurement, running of differential pressure impulse lines, solid accumulation, etc.

2. Types of Measuring Device

Many of the types of level measuring devices used for conventional liquid level measurement can be used for cryogenic liquids if some minor problems can be overcome. In the case of differential pressure measurement the low boiling point of the liquid is utilized to avoid the necessity for a liquid reference leg.

Some of the more common types of measuring devices are described below. (a) *Differential pressure instrument.* When the differential pressure method is used to measure the level of liquids which boil at above ambient temperature it is advisable to locate the instrument at the same elevation as the base of the vessel. If this is not done the instrument will not indicate zero when the tank is empty and it will be necessary to make allowance for this when calibrating the instrument.

Another possible cause of error is condensation of vapour in the low pressure impulse line. This is overcome by filling the vapour phase impulse line with liquid and using this as a reference head for comparison with the vessel level. With this type of installation a fall of level in the tank produces an increase in differential pressure and the indicator scale, or transmitted output, must be reversed. In cryogenic liquid level measurement the problems described do not apply. Provided the liquid phase impulse line is run horizontally from the base of the vessel to the outside of the lagging case, with a warming loop if necessary, as described in the section on pressure measurement, the liquid–vapour interface will form in the horizontal section and the line outside the lagging case will be vapour filled. Because of the great difference between the cryogenic liquid temperature and ambient, there will be vapour in both the liquid and vapour phase lines outside the cold box and the measuring instrument can be at any elevation with no need for a liquid reference leg. With differential pressure methods of measuring level, particular attention must be given to the size of the impulse lines. If the bore is too small solids (e.g. carbon dioxide) may block the line, as mentioned earlier in the sections on pressure and flow. If the bore is too large there is a possibility of internal circulation in the pipe causing local build up of deposits such as hydrocarbons.

On small transportable cryogenic tanks, such as those used for filling aircraft oxygen systems, space between the inner and outer vessels is limited. There is, therefore a strong temptation to locate differential pressure tappings on liquid decanting and vent lines. This temptation should be resisted if it is intended that the level should be indicated while the vessel is in use, otherwise the flow of liquid or vapour past the measuring point will cause errors.

(b) *Float and displacement instruments.* On large storage tanks use can be made of float-type level measuring instruments. These consist basically of a hollow metal float connected, through a system of pulleys, to a flexible tape, usually metal, calibrated in the required units of liquid level. As the float rises the calibrated tape winds on to a spring loaded drum and the tank liquid level is indicated through a window in the measuring head.

A schematic arrangement of this system is shown in Fig. 10.13.

Normally this type of measurement is suitable only for large tanks where the unavoidable heat inleak through the tape and pulley system is small compared to the total heat capacity of the tank.

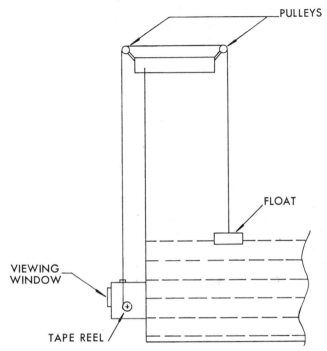

Fig. 10.13. Diagrammatic arrangement of float type tank level indicator.

A variation of the float type level instrument is the displacer type. In this type the movement of the float, which is long and narrow and suspended vertically downwards is restricted by a mechanical linkage connecting it to the measuring head. As the liquid level rises a greater length of float is immersed and the upward force exerted on it increases. This force is transmitted through the linkage and is directly proportional to the change of level in the vessel.

Displacement type instruments are not used for very low temperatures but can be used on LNG and LPG processes where temperatures may in some cases be little below zero Celsius.

(c) *Electrical capacitance.* If the displacer unit mentioned previously had been made of two concentric open ended metal tubes, electrically insulated from each other, an electrical capacitor would be formed with the liquid or vapour in the tank being the dielectric.

As the liquid has a different dielectric constant than the vapour, a rise in liquid level in the concentric tubes will alter the capacitance of the unit, such

24*

capacitance change being indicated on an electrical measuring instrument as a change of level.

Many years ago this principle was employed by Firth Cleveland in their "Pacitor" contents gauge used for aircraft instrumentation and the principle was later developed for use in cryogenic vessels. Although generally satisfactory at the time the dielectric type gauge has now been largely superseded by other methods of liquid level measurement.

A variation of the capacitor type level indicator is the resistance chain in which the rising liquid successively covers a number of electrical resistances connected in series. The change of resistance resulting from the change of temperature can be measured and indicated as a change of liquid level. This method of level measurement can give accurate results but is more suitable for laboratory work than on industrial plant.

(e) *Load cells and weighbeams.* These two types are considered together because they both employ the same principle of mass measurement not liquid volume. The former converts the gravitational force acting on the vessels into an hydraulic pressure or an electrical signal, according to the type of load cell used and the latter balances the gravitational force, via a lever system, against a spring of known rating.

Both types measure the combined mass of liquid and vapour in the tank, this mass being independent of temperature and pressure. Thus, an instrument employing a weighing device as the primary measuring element can be calibrated directly in weight units on a linear scale regardless of the size and shape of the tank. One big disadvantage of using weighing devices for liquid level measurement is the high tare weight compared to the nett weight of the vessel. A double skinned mineral insulated evaporator complete with vaporizing coil, valves etc. can contribute a large proportion of the total gross weight of the liquid filled vessel. When the liquid falls to a low level the tare weight will be many times the nett weight and considerable errors result. Another source of error is wind force on vessels located in the open. The effect of this is much greater when, as is sometimes the case, only one leg of a three-legged vertical vessel is supported by the weighing device, the other two being on solid supports. Connecting pipework can also restrain vessel movement and the effect of this should be minimized by careful design of pipework and location of connections on the vessel.

VI. Analysis

A. APPLICATIONS

One of the most important measurements in an air separation process is quality or composition. Products of air separation are normally sold with a guarantee of purity, such purities being necessary for the user's process. For example, most nitrogen now produced has as little as two volumes per million of impurity. To ensure that high quality of this order is maintained it is necessary to have accurate and reliable measuring instruments.

Although chemical test sets are still used by process operators for spot checks throughout the plant, the development of fully automatic, unattended plants has increased the need for continuous automatic analyzers. In some cases analyzers are fitted with control or alarm devices to adjust the plant or warn the operator of abnormal conditions. In heavily contaminated atmospheres such as are found in the vicinity of oil refineries the accumulation of hydrocarbons in an air separation plant can have serious consequences. The presence of trace quantities of acetylene in the oxygen part of the system can cause an explosion. Consequently, in such atmospheres it is necessary to analyze the air composition at the plant inlet and to check for the presence of hydrocarbons at certain points in the process.

The main analysis measurements in an air separation plant are product impurities, gas and liquid, reflux stream purities and in contaminated atmospheres, hydrocarbon accumulation.

When analyzing for trace elements as in chromatography it is essential that the sample is truly representative. If a liquid sample is taken evaporation prior to entering the detecting element may cause partial separation of component parts resulting in inaccurate readings. This is particularly dangerous when analyzing for acetylene in liquid oxygen where the evaporated oxygen may contain less acetylene than the original liquid sample. The solution of this problem is to ensure "flash" evaporation of the complete sample. Many devices have been tried in attempts to achieve this with varying degrees of success. For commercial reasons most cryogenic plant manufacturers are reluctant to disclose their more recent developments but the most successful method seems to be to draw a small quantity of liquid through a fine bore capillary and expand it into a heated chamber. In this way the liquid sample is vaporized almost instantaneously and the impurities remain in the vapour.

B. TYPES OF ANALYSIS INSTRUMENTS

1. Thermal Conductivity

When a stream of gas is passed over a hot wire some of the heat in the wire will be absorbed by the gas, the amount of heat transferred in this way will depend on the thermal conductivity of the gas. As the wire loses heat and its temperature drops its electrical resistance will decrease and if connected to a constant voltage source the current passing through it will increase. By measuring the current the gas analysis can be inferred.

Thermal conductivity type gas analysis is only suitable for mixtures of two gases with substantially different thermal conductivities. In practice accuracy is limited to tenths of a percent by volume as compared with parts per million for more sophisticated methods of analysis.

2. *Paramagnetic*

There are two basic types of paramagnetic analyzer, "magnetic wind" type and what can be described as "force balance" type. Both types make use of the unique magnetic susceptibility of oxygen gas and the fact that its movement can be influenced by magnetic fields.

(a) *Magnetic wind.* The magnetic wind type of analyzer uses the phenomena of both thermal conductivity and the magnetic susceptibility of oxygen for determination of the gas composition. The basic measuring system is shown diagrammatically in Fig. 10.14, being simply an annular tube with a diametrical connecting tube on which is wound the hot wire element. The centre point

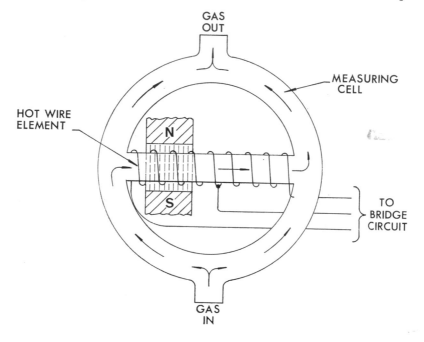

Fig. 10.14. Magnetic wind type oxygen analyzer.

of this element is connected to the measuring instrument and the two outer ends to a Wheatstone bridge circuit. Thus the two halves of the hot wire form two arms of a Wheatstone bridge, the other two arms being fixed resistors. One half of the hot wire element is located in a strong magnetic field.

Referring to Fig. 10.14, the gas to be measured enters the cell and, traversing the annular tube, leaves at the opposite end. Oxygen present in the sample will be drawn into the diametrical tube by the influence of the magnetic field and will be heated by the hot wire element. As the magnetic susceptibility of oxygen is reduced by an increase in temperature the heated oxygen will be displaced by cooler oxygen drawn in from the annular tube. Thus there will be a flow of oxygen along the diametrical tube (from left to right in the diagram).

The heat in the oxygen is absorbed from the first half of the element, therefore this half will be cooler than the second half and its electrical resistance will also be less. The difference of resistance in the two halves is an indication of the amount of oxygen present in the sample.

When using the magnetic wind type analyzer it is essential that the diametrical tube in the measuring cell is truly horizontal, otherwise convection currents through the tube will result in inaccurate readings. Magnetic wind analyzers are fitted with a spirit level to ensure correct installation.

Although, in theory, the cell only detects the presence of oxygen in the sample stream it has been found in practice that, unless the carrier gas is of constant composition and density, accuracy is affected. It is therefore necessary to calibrate the indicating instrument for a known composition of carrier gas.

(b) *Force balance.* In this system a plastic "dumbell" is suspended in the sample stream and is under the influence of opposing electromagnetic and electrostatic fields. The oxygen in the sample is attracted into the magnetic field and tends to displace that half of the dumbell which is between the magnetic poles. An electrostatic force applied to the other half of the dumbell opposes the tendency to motion and equilibrium is reached when the electromagnetic and electrostatic forces are in balance. The current required to produce the electrostatic force is a measure of the oxygen present in the sample stream.

This type of measurement is not so susceptible to errors as the magnetic wind method and the necessity for level mounting is not so important. Both types are restricted to accuracies of about one tenth percent at best.

3. *Galvanic Cell*

This type of instrument is used for detecting trace impurities, one use being to detect oxygen in nitrogen in the range 0 to 10 p.p.m. The measuring cell (described in B.P. No. 707323) consists essentially of a gold or silver cathode according to the analysis requirements, and a cadmium anode in an electrolytic solution.

As the sample stream passes through the cell the oxygen is adsorbed on the cathode and negatively charged hydroxyl ions are produced each having collected one electron from the cathode. In discharging the hydroxyl ions the cadmium anode is oxidized to cadmium hydroxide. The resultant current is proportional to the oxygen present in the sample. The system described uses a silver cathode to measure trace oxygen in nitrogen. Variations of the basic system can be used to measure acetylene, carbon dioxide or hydrogen, the measuring system being generally as described.

Standard ranges are 0 to 10, 0 to 100, and 0 to 1,000 p.p.m. and accuracy is of the order of $\pm 5\%$ of full scale deflection. Because of drift it is necessary to recalibrate at about three day intervals. Facilities provided in the instrument make this a fairly simple operation taking only a few minutes.

4. *Other Methods*

There are various other methods of automatic analysis, among them being gas–liquid chromatography, mass spectrometry and infra-red spectrometry. These are not confined to cryogenics but have a wide application in chemical engineering generally.

VII. Cryostats

To maintain an experimental apparatus at a constant low temperature it is necessary to supply refrigeration continuously to remove the heat entering by radiation, convection or conduction from the surroundings. One common technique is to surround the experimental region with excess refrigeration, for example by a surface at 77°K or 4.2°K and raise the experimental temperature to the required value by electrical heating, varied in response to a temperature signal. This introduces an extra heat load which increases the evaporation rate of the refrigerant but with the ready availability and low price now obtaining this is not of much significance. The refrigerant may be more economically used by a design in which it is circulated through cooling coils in thermal contact with the experiment at a rate varied in response to the temperature signal.

A cryostat of considerable precision of the first type suitable for experiments at temperatures from 80 up to about 273°K was described by Scott (1941) and others and in modified form has been much used by Din and Cockett (1960). It is shown in Fig. 10.15. The refrigerant is contained in a large dewar vessel and in it is immersed a second double walled vessel, filled with a liquid of low freezing point. The gas pressure in the space between the walls of this vessel is varied according to the difference between the temperature of the experimental region and that of the refrigerant. The space is highly evacuated for experiments at higher temperatures but may be filled with gas or even liquid refrigerant for experiments at temperatures near that of the refrigerant. The inner vessel contains a submerged Tufnol tube with hexagonally arranged spacers on which a heater and resistance thermometer are non-inductively wound, the thickness of the spacers being just sufficient to prevent the wires touching the tube so that they are freely suspended in the liquid. A stirrer drives the liquid up the annulus between the Tufnol tube and the vessel wall, over the top of the tube and down its centre thus achieving good temperature distribution. The experimental apparatus is immersed inside the Tufnol tube. Insulating covers of expanded P.V.C. or similar material and connections to a liquid nitrogen level controller complete the cryostat.

Maintenance of close temperature control depends partly on constant refrigerating effect, and this is achieved by maintaining constant the level of the liquid nitrogen. Many level controlling devices have been described in the

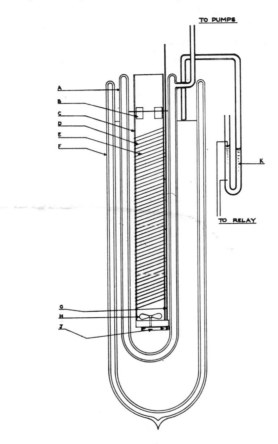

Fig. 10.15. Low temperature cryostat.

literature. In some a small vapour pressure thermometer has its bulb located at the required level; heat conducted down the stem keeps the contained fluid in the vapour phase at some temperature above the normal boiling point and therefore the pressure somewhat above atmospheric. When the refrigerant level rises the bulb is eventually submerged and the fluid liquefies, the pressure falling to 1 atm. This change in pressure can be used to generate a signal that controls the flow either electrically or directly by means of a pressure operated valve. Controllers of this type are available commercially.

Another device uses a thermally conducting dip tube with lower end at the required level and upper end connected to a manometer containing an electrically conducting liquid such as an acid or salt solution. Electrodes placed at the appropriate levels transmit an a.c. signal to control a valve in the liquid transfer line or a pump in the refrigerant container. With this arrangement a third electrode can be used to indicate failure to fill, i.e. usually that the refrigerant supply is exhausted.

Precise temperature control is achieved by connecting the resistance thermometer in a Wheatstone bridge circuit with output going to a phase sensitive detector followed by an amplifier feeding a variable inductor controlling the heating current. The circuitry is quite conventional and all the stages are available as complete units but stability of temperature control requires stability of some essential components, especially those in the bridge network and some effort to achieve this is well justified.

Cryostats up to 15 cm diameter and 100 cm tall have been made of this design and temperature control to better than $\pm 0 \cdot 01°C$ is readily achieved; in smaller cryostats $\pm 0 \cdot 002°C$ can be obtained. In the illustration the liquid nitrogen and cryostat liquid surfaces are shown at about the same level but for tall cryostats the inner liquid may be at a much lower level so long as it overflows the top of the Tufnol tube. This allows a relatively short portion of the experimental apparatus to be immersed but requires that the non-immersed sections be prevented from radiating to the surrounding liquid nitrogen refrigerant, otherwise they will be cooled below the temperature of the cryostat liquid.

Liquids for the inner vessel can be chosen according to the working temperature range required. Carbon tetrachloride can be used down to 255°K ($-18°C$) and it forms a non-flammable eutectic with chloroform useful down to 200°K. A five-component mixture containing 14·5% chloroform, 25·3% methylene chloride, 33·4% ethyl bromide, 10·4% transdichloroethylene and 16·4% trichloroethylene, proposed by Scott can be used from room temperature down to $-130°C$. This mixture hydrolizes readily in the presence of moisture so must be kept dry. Condensation of ice from the atmosphere in it must therefore be prevented otherwise acid products corrode metallic parts of the apparatus whenever the equipment warms up to room temperature. Isopentane is reported to have a freezing point of about 113°K which would make a useful extension to the temperature range coverable but it is difficult to obtain in a sufficient state of purity and most samples the writers have tried freeze at about 143°K. To extend the operating range down to 77°K the most convenient fluid is liquid propane. Pure propane has a triple point of 85·5°K but commercial grades remain liquid a few degrees below this. Like isopentane it is inflammable and every possible precaution should be taken to prevent the accumulation of high concentrations of vapour in the atmosphere. This is no great problem when the cryostat temperature is below about 200°K since the vapour pressure is then below 16 Torr.

The critical temperature of propane is 370°K (97°C) and at room temperature it has a vapour pressure of 10 atm, the normal boiling point being 231°K (−42°C). To transfer it to a cryostat of the type described it must be transferred from a room temperature cylinder at 10 atm pressure through a pressure-reducing valve and a cooling coil immersed in solid carbon dioxide or liquid nitrogen and then through an insulated and pre-cooled tube into the cryostat also previously cooled. At the end of a series of experiments the propane may be returned to the original cylinder by cooling this with liquid nitrogen and syphoning the liquid back into it. Commercial cylinders are not suitable for this operation as, being made of mild steel, they become brittle at the low temperature. A special cylinder must be provided made of a suitable stainless steel or other alloy—see Chapter 6.

For work at temperatures in the region of 4°K the experimental apparatus may be immersed in liquid helium contained in an evacuated glass dewar surrounded by liquid nitrogen. Such dewars are usually made of pyrex or other borosilicate glass which is permeable to helium and therefore the vacuum deteriorates with time due to diffusion of this gas from the atmosphere. Diffusion may also occur from the vapour space over the liquid but this can be minimized by terminating the vacuum space below the level of the liquid nitrogen since permeability decreases with temperature. Provision for repumping the vacuum space must be provided if it is planned to use the dewar for any length of time, or after an appreciable period of storage.

The experimental sample usually requires isolation from the liquid refrigerant so is surrounded by an evacuated metal container, usually of copper or brass. Surfaces facing the vacuum space may be polished or gold plated to reduce heat transfer by radiation. Initial cooling of the specimen has often been hastened by use of a small pressure of helium exchange gas but to remove this later may be time consuming especially if the helium bath is pumped below its normal boiling point. Incomplete pumping, particularly incomplete removal of adsorbed films on the low temperature surfaces, may result in spurious heat capacities rendering specific heat measurements meaningless.

Rose–Innes has (1964) described a simple and versatile design capable of operation over a wide temperature range with economical use of refrigerant.

Most experiments in this temperature range are now carried out in metal cryostats of proprietary design. Figure 10.16 shows a commercially available type, made of stainless steel, equivalent to the glass dewar type already described. The experimental space consists of a wide parallel-sided stainless steel liquid helium container surrounded by a copper shield cooled by liquid nitrogen. The refrigerant liquids are separated and surrounded by a common vacuum space which is pumped hard on cool down by a small pocket of adsorbent attached to the liquid helium container. This design is suitable for immersion of superconducting magnets or other bulky apparatus and for experiments in low temperature calorimetry or on liquid helium. It is available

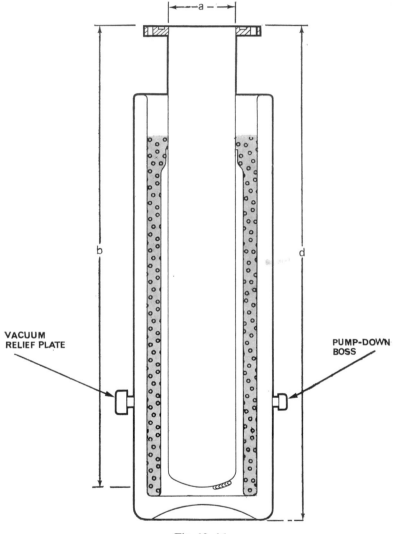

Fig. 10. 16

in a wide range of sizes with internal diameters between 48 and 250 mm and heights between 864 and 1168 mm. Another design specifically intended for superconducting magnets is shown in Fig. 10.17.

Another versatile design is shown in Fig. 10.18, with variable temperature tail for optical experiments. It is fabricated from all-welded polished stainless steel except for the radiation shield which is copper. A 3 l liquid vessel helium surrounded by a liquid nitrogen shield has fittings for attachment of a wide

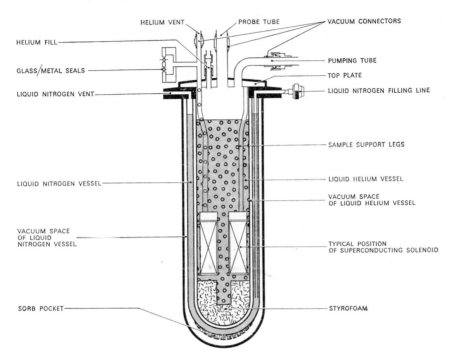

HELIUM VENT   PROBE TUBE   VACUUM CONNECTORS

HELIUM FILL

GLASS/METAL SEALS

LIQUID NITROGEN VENT

PUMPING TUBE

TOP PLATE

LIQUID NITROGEN FILLING LINE

SAMPLE SUPPORT LEGS

LIQUID NITROGEN VESSEL

LIQUID HELIUM VESSEL

VACUUM SPACE OF LIQUID HELIUM VESSEL

VACUUM SPACE OF LIQUID NITROGEN VESSEL

TYPICAL POSITION OF SUPERCONDUCTING SOLENOID

SORB POCKET

STYROFOAM

Fig. 10.17. Liquid helium cryostat designed to contain a superconducting magnet

range of experimental appendages or tails. These may be of small diameter for experiments between the poles of a magnet or may be comparable with the diameter of the helium container. For experiments that can be immersed in the liquid helium the bottom of the vessel may be merely blanked off. It is shown with an open top liquid nitrogen vessel but this can be fitted with an insulating cap to reduce evaporation and cooling of the upper fittings. Standard designs of tails obtainable include those for fixed temperature operation either by direct immersion in liquid helium or clamped to the base of the helium vessel in the vacuum space. This latter arrangement is particularly suitable where a sample is to be subjected to electro magnetic radiation in which case the tails are fitted with optical windows of suitable materials. Windows are often supplied by the user but quartz, mylar, berrylium, silver chloride and sapphire are available. A further modification of this system includes a rotating outer tail allowing the cryostat to rotate relative to the support without impairing the vacuum insulation. The arrangement illustrated includes a heat break and heater between the specimen and the helium bath making possible measurements between about 5 and 100°K. The

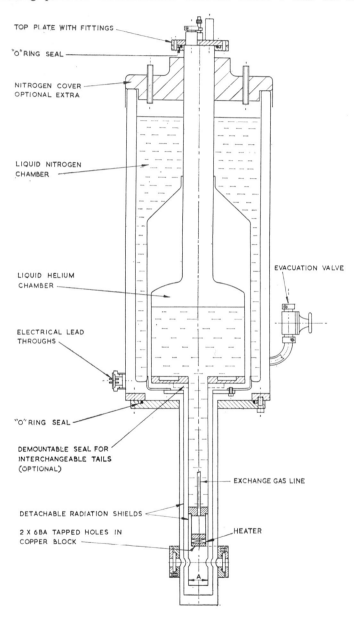

TOP PLATE WITH FITTINGS

"O" RING SEAL

NITROGEN COVER
OPTIONAL EXTRA

LIQUID NITROGEN
CHAMBER

LIQUID HELIUM
CHAMBER

EVACUATION VALVE

ELECTRICAL LEAD
THROUGHS

"O" RING SEAL

DEMOUNTABLE SEAL FOR
INTERCHANGEABLE TAILS
(OPTIONAL)

EXCHANGE GAS LINE

DETACHABLE RADIATION SHIELDS

2 X 6BA TAPPED HOLES IN
COPPER BLOCK

HEATER

A

Fig. 10.18. Helium cryostat for optical experiments.

lowest temperature attainable depends on the thermal dissipation in the experiment and the capacity of the pumping system.

When the experimental apparatus is immersed in liquid helium it is usually placed within an evacuated container allowing experiments over a range of temperatures. Many of the arrangements that have been used for measurements of heat capacities, thermal conductivities and thermo electric properties have been illustrated by White (1961), Hoare *et al.*, (1961) and others; a wide range of variants is possible.

VIII. References

Ambler, E. and Hudson, R. P. (1955). *Rep. Prog. Phys.* 251.

Aston, J. G. (1941). "Temperature, Its Measurement and Control in Science and Industry." Reinhold, New York.

Armstrong, G. T. (1954). *J. Res. N.B.S.* **53**, 623.

Bailey, S. C., Richard, R. T. and Mitchell, G. N. (1969). *Rev. Sci. Instrum.* **40**, 1271.

Barber, C. R. (1950). *J. Sci. Instrum.* **27**, 47.

Barber, C. R. (1955). *J. Sci. Instrum.* **32**, 416.

Barber, C. R. and Horsford, A. (1965). *Metrologia* **1**, 75.

Berman, R. and Huntley, D. J. (1963). *Cryogenics* **3**, 70.

Brewer, J. (1967). Rept. AFOSR No. 67-2795.

Brickwedde, F. G., van Dijk, H., Durieux, N., Clement, J. R. and Logan, J. K. (1960). *J. Res. N.B.S.* **64A**, 1.

Burgess, G. K. (1928). *J. Res. N.B.S.* **1**, 635.

Callendar, H. L. (1887). *Phil. Trans. (London)* **178**, 160.

Clark, A. M., Cockett, A. H. and Eisner, H. S. (1951). *P.R.S.* **A209**, 408.

Clement, J. R., Logan, J. K. and Geoffrey, J. (1955). *Phys. Rev.* **100**, 743.

Clement, J. R. and Quinnel, E. H. (1952). *Rev. Sci. Instrum.* **33**, 213.

Comptes Rendus des Séances du Comite International de la Neuvième Conference Generale des Poids et Mesures, 1948, p. 89.

Din, F. and Cockett, A. H. (1960). "Low Temperature Techniques", p. 92. G. Newnes, London.

Figgins, B. F. (1965). *J. Sci. Instrum.* **42**, 46.

Gabelle, T. H., Morrin, F. J. and Maita, J. P. (1955). *Annexe Bull. Int. Inst. Froid* (Paris) p. 425.

Giauque, W. F., Buffington, R. M. and Schulze, W. A. (1927). *J.A.C.S.* **49**, 2343.

Giaque, W. F. and Eagan, C. (1937). *J. Chem. Phys.* **5**, 45.

Grilly, E. R. (1951). *J.A.C.S.* **73**, 843.

Hall, J. A. (1929). *Phil. Trans. (London).* **A229**, 1.

Hill, J. J. (1968) Private communication.

Hoare, F. E., Jackson, L. C. and Kurti, N. (1961). "Experimental Cryophysics." Butterworth, London.

Hoge, H. J. (1950). *J. Res. N.B.S.* **44**, 321.

Holborn, L. and Otto, J. (1926). *Zeits. F. Physik*, **38**, 366.

Hudson, R. P. (1968). Proc. Int. Conf. on Low Temp. St. Andrews, pp. 501, 506.

Keesom, W. H. (1942). "Helium." Elsevier, Amsterdam.

Knudsen, M. (1927). *Ann. d. Physik* **83**, 797.

Kurti, N. (1968). *Proc. LJ.* 11.

Kurti, N. and Simon, F. E. (1938). *Phil. Mag.* **26**, 849.

Mann, D. B. (1962). NBS Tech Note 154.

Matthews, C. S. and Hurd, G. O. (1956). *Trans. A.K. Ch. E.* **42**, 55.

Mess, K. E., Lubbers, J., Niesen, L. and Huiskamp, W. J. (1968). Proc. Int. Conf. on Low Temp., St. Andrews.

Meyers, C. H. and van Dusen, M. S. (1933). *J. Res. N.B.S.* **10**, 381.

Michels, A. and Wassenaar, T. (1950). *Physica* **16**, 221.

Niesen, L. and Huiskamp, W. J. (1968). *Proc. L.T.*, 11.

Powell, R. L. and Sparks, L. L. (1968). Proc. Int. Conf. on Low Temp., St. Andrews, p. 477.

Preston-Thomas, H. and Kirby, C. G. M. (1968). *Metrologia* **4**, 30.

Preston-Thomas, H. and Bedford, R. E. (1968). *Metrologia* **4**, 14.

Roach, P. R. (1968). *Phys. Rev.* **170**, 213.

Roberts, J. K. and Miller, A. R. (1951). "Heat and Thermodynamics." Blackie, London.

Rose-Innes, A. C. (1964). "Low Temperature Techniques." London Universities Press Ltd.

Ruhemann, M. B. (1937). "Low Temperature Physics." Cambridge University Press.

Rynning, D. F. and Hurd, G. O. (1945). *Trans. A.I.Ch.E.* **41**, 265.

Scott, R. B. (1941). "Temperature, Its Measurement and Control in Science and Industry." Reinhold, New York.

Smith, F. E. (1912). *Phil. Mag.* **24**, 541.

Stall, D. R. (1947). *Ind. Eng. Chem.* **39**, 517.

Stimson, H. F. (1955). "Temperature, Its Measurement and Control in Science and Industry." Reinhold, New York.

van Dijk, H. and Durieux, M. (1955). Conference de Physique des Basses Temperatures. *Annexe Supp. Bull. Int. Inst. Froid* (Paris), p. 95.

van Dusen, M. S. (1925). *J.A.C.S.* **47**, 326.

Wagner, R. R., Bertman, B., Guiffrida, T. S. and van den Berg, W. H. (1968). Proc. Int. Conf. on Low Temp., St. Andrews, p. 472.

Weinstock, H. (1968). Proc. Int. Conf. on Low Temp., St. Andrews, p. 506.

White, G. K. (1961). "Experimental Techniques in Low Temperature Physics." Clarendon Press, Oxford.

White, W. P. (1933). *Rev. Sci. Instrum.* **4**, 142.

Wolfendale, P. C. F. and Firth, J. M. (1968). Proc. Int. Conf. on Cryogenic Engng. Iliffe, London.

Wolfendale, P. C. F. (1969). *J. Sci. Instrum.* (*Series* 2) **2**, 659.

Woolley, W. H., Scott, R. B. and Brickwedde, F. G. (1948). *J. Res. N.B.S.* **41**, 379.

Zimmerman, G. O. and Abeshouse, D. J. (1968). Proc. Int. Conf. on Low Temp., St. Andrews, p. 493.

Chapter 11

Safety

F. J. Edeskuty, R. Reider and K. D. Williamson, Jr.*

*Los Alamos Scientific Laboratory, University of California
Los Alamos, New Mexico, U.S.A.*

* This work was performed under the auspices of the U.S. Atomic Energy Commission.

633

I. Introduction

In the consideration of safety in handling of cryogenic fluids there are two general areas in which hazards can occur. One of these areas is related to the properties of the fluids themselves while the other results from changes in the properties of materials of system components subjected to the low temperature environment. While the latter might be considered secondary in nature, the consequences of these changes are too important to neglect. It is important to remember that the degree of seriousness for a given hazard is related to system volume. In general, hazards are considered to increase as system size increases. This need not, however, always be the case.

Since all cryogenic fluids exist as liquids only at temperatures considerably below ambient, normal storage must account for an unavoidable, inexhaustible heat input available from the environment. Therefore cold systems in general must have suitable vent systems containing provision for pressure relief in order to prevent excessive pressure build-up. Although the critical pressures of the common cryogens are quite low, the pressures necessary to maintain liquid density at ambient temperatures are high (see Table 11.II). The degree to which such a pressure is approached in a sealed cryogenic system depends upon the volume fraction filled and the bursting pressure of the system.

In addition to causing this large pressure build-up, the large volume expansion (typically 700 to 1) accompanying vaporization and warming to room temperature of cryogenic fluids can result in exclusion and/or dilution of air in confined or even partially confined spaces. In the case of oxygen (O_2) an atmosphere which enhances combustion results while for the other cryogens an asphyxiating or toxic environment can be produced. Direct physical contact with either the cryogenic liquid or cold vapour as well as with cold metal surfaces can cause serious tissue damage. Cold metallic surfaces can also condense from the atmosphere an O_2 enriched air mixture which can cause a secondary hazard.

In addition to the foregoing general objections to uncontrolled outleakage of cryogenic fluids, certain fluids present additional hazards. Fluorine is toxic and corrosive while hydrogen (H_2) and low boiling hydrocarbons (in natural gas or petroleum derived cracker gases) can form explosive mixtures over wide ranges of concentration. Carbon monoxide combines a high toxicity hazard with an explosion risk. It is found in low temperature plants separating hydrogen from fuel gas mixtures derived from coal. Since petroleum is now the preferred source of hydrogen, the problem of carbon monoxide will not be considered further. Leakage of air into cryogenic systems is also undesirable. If the leakage is into systems containing liquid hydrogen (LH_2) or liquefied natural gas (LNG) explosive mixtures can result. For all cryogens such leakage can cause line blockage as a result of frozen vapour from H_2O or other gases. Such frozen particles in a system can also be troublesome by causing valve seat or seal erosion or by accumulating undesirable impurities. Such solid impurities do not necessarily leave the system with the cryogen and thus can accumulate over a period of time. For this reason as well as the large liquid to vapour density ratio which proportionally increases any undesirable property of the cryogen, such fluids have been referred to as "risk concentrators". However the degree of risk is usually less than that involved in the use of an equal quantity in a high pressure gas system. In air separation plants there is a tendency for the trace quantities of hydrocarbons present in industrial atmospheres (especially acetylene) or produced by breakdown of compressor lubricant, to accumulate within boiling liquid oxygen at the base of the low pressure column. If precipitation of solid hydrocarbon occurs the risk of explosion is very high. Special precautions regarding the positioning of the air intake, the choice of compressor oil, and the provision of hydrocarbon removal systems, are normally adopted, as described in British Institution of Chemical Engineers (1970).

The low temperatures of the cryogenic fluids can have a profound effect on the properties of the solid materials used in the containing system. The strength of materials is usually not a safety problem at cryogenic temperatures since strength increases as the temperature decreases. However it is advisable to use room temperature strengths in the design of equipment for several reasons. First, most proof testing is done at ambient temperatures; second, the equipment may warm up under stress; third, because of temperature gradients not all of the system will be always at the system's lowest temperature. For example, the top of the inner shell of large, partially full LH_2 dewars under different use conditions can vary in temperature anywhere from almost ambient to $20°K$.

In the design of stress bearing members, consideration must be given to the changes in ductility which can occur. Large losses of ductility do occur at low temperatures for some materials (low temperature embrittlement). Although not universally recognized, hydrogen gas at ambient temperatures can cause embrittlement. Thermal stresses can arise during the cooldown of a system as well as during steady state operation. The physical properties which influence the magnitude of these stresses in a given material are the coefficient of thermal expansion, the thermal conductivity and the elastic modulus. With proper operating conditions these stresses can be kept within safe limits. Finally it should be remembered that the thermal capacity of most solid materials at low temperatures is very small because of the decrease in specific heat as the temperature decreases. For this reason the solid material in a system is not effective in retarding a rapid temperature rise for a given heat input. In general the largest thermal capacity of a cryogenic system resides in the cryogen itself.

II. Properties

A. PROPERTIES OF SOLIDS

The design of safe and reliable cryogenic equipment requires a thorough knowledge of temperature gradients and material properties over the temperature range to be experienced. This is necessary for calculation of thermal stresses and equipment displacement caused by the cooldown to operating temperature. The ultimate and yield strengths of representative materials are given in Chapter 6. In the use of these materials it is customary to incorporate a safety factor to allow for nonuniformities in the material and the presence of stress concentrations. Such factors are incorporated in the so called maximum allowable stress. Typically values for this stress are taken as the lower of either 25% of the room temperature ultimate strength or 62% of the room temperature yield strength. Values for selected materials are given in Table 11.I.

Another property that must be taken into account in the design of low temperature equipment is its retention of ductility at the use temperature. There is no general agreement as to the best method to investigate the onset of low temperature embrittlement. However, methods in use include ratios of notched to un-notched ultimate and yield strengths, impact fracture energies, reduction of area and elongation upon fracture (see Chapter 6). From this standpoint particularly good materials are the aluminium alloys and austenitic stainless steels. In addition 9% nickel steel is in use to temperatures as low as 75°K.

Figure 11.1 illustrates the dependence of ductility upon temperature as determined by fracture energy values. In general such behaviour shows either very little temperature dependence or a very rapid decrease in ductility over a narrow temperature range. Materials in the latter category should not be used for structural applications. The metals with a face-centred cubic crystal structure (Al, Cu, Ni, austenitic stainless steels) are representative of those which retain ductility. Metals with a body-centred cubic crystal structure (low carbon and 400 series stainless steels) represent materials which become embrittled at low temperatures. It should be remembered that some austenitic stainless steels (i.e. AISI 304 stainless steel) can undergo a partial transition to the more brittle martensitic form upon cold working. To compensate for any loss of ductility accompanying this transformation use of room temperature strengths is recommended.

Another mechanism which can lower ductility of a metal is that of hydrogen embrittlement. Though hydrogen can chemically attack many metals at high temperatures, the phenomenon referred to here is most prevalent at room temperatures (Cotterill, 1961). In this case the mechanism is physical rather than chemical and is not yet fully understood. However, the failure in hydrogen service of Bourdon tubes, capillary tubes, and large gas storage vessels previously used in higher pressure nitrogen or helium service leaves little doubt as to the existence of the phenomenon (Mills and Edeskuty, 1956; Chelton, 1964). Many of the early investigations of the phenomena used hydrogen pressurization or cathodic charging of test specimens followed by

TABLE 11.I. Allowable Stress for Selected Materials of Construction (from Fuller and McLagan, 1962)

Material	Allowable stress
304 S.S. (SA–240)	12·9 kN/cm² (1320 kg/cm²)
Monel (SB–127)	12·1 kN/cm² (1230 kg/cm²)
Cu (SB–11)	4·6 kN/cm² (470 kg/cm²)
Al (SB–209 0 temper)	6·9 kN/cm² (700 kg/cm²)

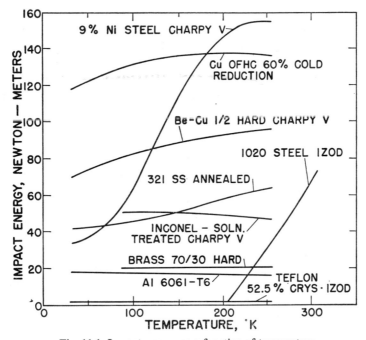

Fig. 11.1. Impact energy as a function of temperature.

removal of the hydrogen source before testing. More recent work (Walter and Chandler, 1969) has determined tensile strengths and ductility of 35 metals for both notched and un-notched specimens while under hydrogen pressures up to 6·9 kN/cm² (700 kg/cm²). Some general conclusions drawn are: the degree of H_2 embrittlement is more severe the higher the pressure but the effect can be evidenced down to 1 atmosphere; hydrogen embrittlement is probably associated with adsorbed H_2 rather than absorbed H_2; reductions of ductility up to 90% and reduction of notched strengths of up to 55% were noted; and fracture initiation occurs after a critical amount of plastic deformation has taken place. In general aluminium alloys, stabilized austenitic stainless steels and copper exhibit negligible embrittlement. Slight embrittlement was noted in nonstabilized 300 series stainless steels, beryllium copper and pure titanium. Severe embrittlement was found in ductile, lower strength steels, pure nickel and titanium based alloys while extreme embrittlement occurred in high strength steels and nickel based alloys (British Institution of Chemical Engineers, 1970). Important factors which influence the susceptibility of a metal to hydrogen embrittlement are its chemical composition, crystal structure, metallurgical history, and the state-of-stress during exposure.

B. PROPERTIES OF FLUIDS

The properties of the more commonly used cryogens which are generally related to their safe handling and use are given in Table 11.II.

Even the highest boiling point cryogen listed in Table 11.II corresponds to a temperature which is the dew point for a water concentration of $\ll 0.01$ v.p.m. (volume parts per million). Therefore, frost formation must always be considered when gases containing any H_2O vapour can come in contact with the low temperature. Furthermore liquid He, H_2, Ne and N_2 are cold enough to condense air. This condensate when in equilibrium with air at 1 atm contains 52% oxygen (Stewart and Johnson, 1961). The density ratios given in column 5 indicate the large volume expansion which occurs between liquid at its normal boiling point and gas at 300°K and 1 atm. This represents the degree of "risk concentration" residing in the liquid phase. It also represents the degree of compression necessary to maintain liquid density of 300°K. The corresponding pressures required to give this compression at ambient temperature are given in columns 6 and 7 represent a practical upper limit of pressure that could be attained if refrigeration is removed from a tightly sealed vessel completely filled with a liquid cryogen. In practice such pressures are seldom reached for two reasons. First, a minimum ullage space of 5–10% of the total volume is provided even in full storage vessels and second, almost all storage vessels will rupture considerably before such pressures are reached.

The latent heats of vaporization, both on a mass and volume basis, are given in columns 10 and 11. Of particular note are the small values of the latter quantity for H_2 and He. The decreased values of the volume latent heat augur increased difficulties in the storage, transfer and handling of the lower boiling cryogens. In addition, as the normal boiling point decreases, there is an increase in the temperature difference driving heat into the system.

TABLE 11.II. Cryogen Properties

Cryogen	Normal boiling point °K	Critical pressure N/cm²	Critical pressure atm	ρ_L/ρ_G (ρ_G at 300°K 1 atm)	Approximate pressure for ρ_L at 300°K kN/cm²	Approximate pressure for ρ_L at 300°K kg/cm²	Specific heat (J/gm K) Saturated [a] liquid	Specific heat (J/gm K) Saturated [a] gas	Heat of vaporization J/gm	Heat of vaporization J/cm³	$\Delta T/H_V$ °K cm³/J
Helium	4·22	22·9	2·3	770	12·4	1300	5·2	6·81	20·2	2·5	118
Hydrogen	20·4	129·0	12·8	865	19·6	2000	9·5	12·0	447·0	31·8	8·8
Neon	27·1	272·0	26·9	1470	—	—	1·82	1·17	85·7	104·0	2·6
Nitrogen	77·3	339·0	33·5	710	29·4	3000	2·05	1·03	199·0	160·0	1·39
Fluorine	85·2	557·0	55·0	975	—	—	1·55	0·81	172·0	260·0	0·83
Argon	87·3	489·0	48·3	865	—	—	1·14	—	163·0	228·0	0·93
Oxygen	90·2	508·0	50·1	880	—	—	1·69	0·91	213·0	243·0	0·86
Methane	111·7	464·0	45·8	650	—	—	3·45	2·09	508·0	215·0	0·88

[a] At 1 atm pressure.

For this reason a useful comparative ratio is the vaporization index given in column 12. This index is the ratio of the temperature difference between 300°K and the normal boiling point to the volume heat of vaporization. The larger this number the greater is the difficulty in storing the cryogenic fluid.

In the purification of gases for subsequent cryogenic use (i.e. liquefaction) one should take note of the enhancement factors which will result in higher concentrations of vapour phase impurities (at low temperatures and high pressures) than those computed from vapour pressure calculations. These enhancement factors (sometimes as large as several million) can be calculated and have also been experimentally determined (Dokoupil et al., 1955; Hiza and Herring, 1965). Such impurity concentrations are usually not a hazard in the gas. However, a change in condition (i.e. pressure, liquefaction) can result in accumulated solids which could be hazardous.

C. SPECIAL CONSIDERATIONS FOR THE CHEMICALLY REACTIVE CRYOGENS

1. Combustibles: H_2 and CH_4 (LNG)

The increasing usage of LNG (approximately 90% CH_4 but composition variable), liquid hydrogen (LH_2), liquid O_2 and liquid fluorine warrants a special consideration of the hazards connected with them. Both LH_2 and LNG are highly combustible and some of the properties relating to the hazards of their combustion are given in Table 11.III. The combustion and detonation ranges give limits of the concentration ranges to be avoided, while the ignition temperature and ignition energy indicate the degree of precaution necessary to prevent unwanted ignition. The ignition temperatures for hydrogen and CH_4 are quite high. The ignition energy for H_2 is very low, however, so that, in most cases of rapid hydrogen release, spontaneous ignition results. This can be initiated from such sources as frictional heat of moving solid particles in the effluent stream and static electricity discharges built up by the rapid discharge of the cryogen.

The range of concentration for combustibility increases for H_2 and CH_4 as the total pressure is increased above normal atmospheric pressure. However, as seen in Fig. 11.2 as the total pressure is decreased the range of combustibility is essentially constant until pressures well below atmospheric are reached. At low pressures the range narrows with a pressure being reached below which the gas will not ignite. For H_2 this pressure is $\simeq 67$ cN/cm² ($\simeq 50$ Torr) and for $CH_4 \simeq 69$ cN/cm² ($\simeq 52$ Torr). The validity of using this pressure as a limit has been questioned, since the phenomena is limited by the quenching distance which increases as pressure decreases (Lewis and von Elbe, 1961). For this reason such an apparent limit may be a function of equipment size. Also the lower limit of pressure which will support combustion is difficult to establish experimentally because of the high spark ignition energies required and the small amount of energy released at low concentration of the reactants.

TABLE 11.III. Combustion Properties of H_2 and CH_4 (gas properties at 1 atm unless otherwise stated)

	H_2	CH_4
Combustion range, volume % in air	4–75	5–15
Combustion range, volume % in O_2	4–96	5–61
Detonation range, volume % in air	20–65	—
Detonation range, volume % in O_2	15–90	—
Ignition temperature in air, °K	858	810
Minimum spark energy to cause ignition in air, mJ	0·02	0·30
Minimum total pressure for combustion (see text)		
cN/m²	67	69
(Torr)	(50)	(52)
Flame temperature, °K	2,323	2,153
Flame velocity, cm/sec	270	38
Flame emissivity	0·10	—
Quenching distance (1 atm), cm	0·06	0·25
Heat of combustion to form H_2O gas and CO_2 gas, kJ/mole	242	802
Liquid electrical resistivity, ohm/cm	$\simeq 4\cdot6 \times 10^{19}$	—
Diffusion coefficient in air at 273°K, cm²/sec	0·634	0·20

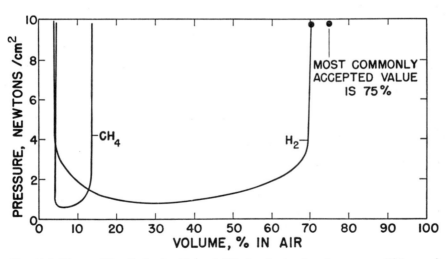

Fig. 11.2. Flammability limits for H_2 and CH_4 in air at reduced pressures. Elston and Laffitte (1947), Lewis and von Elbe (1961).

The very low flame emissivity of burning hydrogen restricts energy transfer from the flame by radiation. While this would seem to decrease the hazard of the flame it must be remembered that the flame is also invisible (if impurities are not present). This combination of properties makes the hydrogen flame very difficult to detect.

The small quenching distance (minimum distance between parallel planes which will just allow a flame to pass without cooling to extinction) for the hydrogen flame complicates the manufacture of "explosive proof" equipment because of the fine tolerances required. The electrical resistivity of liquid hydrogen is so high that it has only recently been determined (Willis, 1966). In general the buildup of an electrical charge within a flowing fluid is a hazard only when its electrical resistance lies between 10^{10} to 10^{15} ohm/cm (Little, 1961). Lower resistivities result in rapid dissipation of any charge to the grounded metallic parts of the transfer system while higher resistivities require a longer residence time for any significant amount of charge formation or dissipation. Liquid hydrogen falls in the latter category. However, the existence of either particulate matter in the pipe or two-phase flow (usually encountered during system cooldown) can greatly enhance the probability of building up electrostatic charges within the flowing fluid. The large diffusion coefficients for hydrogen and methane result in a rapid contamination of the entire enclosure when either of these gases is released into a closed volume. On the other hand the same property results in a rapid mitigation of any hazard when a release takes place outdoors. Since LNG is a mixture (rather than a single component) fractionation occurs upon boiling. Effective purging procedures cannot be carried out unless the coldest spot in the container is warmed to at least 273°K (British Institution of Chemical Engineers, 1970). On the whole, provided proper precautions are taken, the risk involved in the handling of LH_2 and LNG can be similar to that of other commonly handled combustibles such as gasoline.

2. Oxidizing Agents: O_2, O_3 and F_2

Oxygen. The common classification of materials into combustible and non-combustible is based upon experience with atmospheric air. The removal of inert N_2 as well as the presence of highly concentrated (i.e. liquid) oxygen is a sufficient change to cause some non-combustibles to become combustible and some combustibles to become explosive. Included in the former category, particularly when finely divided, are such materials as stainless steels, mild steel, cast iron, aluminium and Teflon. Numerous instances (Reider, 1968) have shown that a material may be considered incombustible or safe only when demonstrated by testing under the conditions of use. Both aluminium and titanium are, under certain conditions, impact sensitive and have exploded in O_2 atmospheres (Vance and Reynales, 1962). Materials which will not normally ignite in the solid form include copper, brass, bronze, silver, nickel, monel, pure asbestos and oil-free silicate-based insulating materials (British Institution of Chemical Engineers, 1970). While some of the materials from both of these lists are used in the construction of O_2 systems, extreme caution must be taken both in the material choice and use. Also important is the elimination of hydrocarbon impurities, such as the oil added to some grades of mineral wool to facilitate handling. It has been recommended that hydrocarbon films must not exceed 0·54 mg/cm² on component surfaces intended for LO_2 service (Ball, 1961). This hazard can be eliminated by proper cleaning techniques which utilize suitable inert oil-free solvents. Following cleaning, the components should be dried with oil-free air or pure inert gas. The cleaned surfaces can be analyzed for hydrocarbon impurities by combination wipe and fluorescence tests while the effluent gases can be analyzed by flame ionization techniques. The recovered liquid cleaning solvent can also be analyzed for impurities. Subsequent use of hydrocarbon based lubricants is not permitted. Lubricants and thread compounds which have been found suitable for LO_2 service include polymers of polychlorotrifluoroethylene, polymers of perfluorocyclic ether, polymers of polytetrafluoroethylene and molybdenum disulphide powder (Vance and Reynales, 1962).

Various tests have been proposed to determine the suitability of materials for O_2 service. Among these is the use of an impact tester employing a 9·04 kg weight dropped from heights of up to 1·1 m onto a sample immersed in LO_2 or LO_2–LN_2 mixtures. The energy input withstood by a given material without visible reaction (exploding or flashing) is a measure of the safety (hazard) with use in LO_2 systems (Key and Gayle, 1964). Such tests have shown that while the dilution effect of LN_2 is important for many materials, the sensitivity at 52% O_2 (equilibrium concentration for air condensing from the atmosphere) approaches that of pure LO_2 (Key and Gayle, 1965).

The determination of the applicability of plastic foams as cryogenic insulations, where some air condensation may occur, is done by an ASTM burning test (American Society for Testing of Materials, 1959). In this test a sample, 5×15 cm in cross-section, is laid flat on a hardware cloth base and exposed to a Bunsen burner flame. The degree of inflammability is determined from the burning rate and the distance of the burn. Classifications are noninflammable (no flame or further consumption once the burner is removed), self-extinguishing (flame goes out before reaching second fiducial mark), and combustible (burns entire specimen). Only the first class should be used where air could condense.

In LO_2 the following safe upper limits have been established for hydrocarbon impurities: propane 2 v.p.m., ethane 10 v.p.m., and ethylene 10 v.p.m. The above are not absolute danger limits but they are useful guidelines for most facilities. Care should also be taken to avoid the settling of the heavy O_2

25

vapour in low areas. In addition gaseous oxygen should never be used as a substitute for air or N_2, and particular care should be taken to avoid the adsorption of O_2 on combustible adsorbents. Similarly only completely inert (fully oxidized) materials should be used in the gelling of liquid O_2.

Ozone (O_3). An allotropic form of oxygen, is a powerful oxidizer even in the gaseous form and can readily be made to detonate in the liquid state. It has a characteristic pungent odour familiar to those who have worked with high voltage devices; the recommended upper limit for atmospheric concentration for working an eight-hour day is 0·1 v.p.m. (Streng, 1960).

Pure O_3 may be kept at 90°K (cooled by liquefied O_2) but the smallest provocation (a spark or fast warming to the boiling point of 161°K) can cause detonation; friction between O_3 crystals in sudden solidification can also cause a detonation.

Fluorine. Organic materials, and even water, are extremely reactive with fluorine. Finely divided materials are quite reactive because of the greater surface exposed for reaction. With most metals, fluorine forms a thin protective film which limits further reaction but if the metals are finely divided the heat of reaction cannot be dissipated quickly and the reaction can become violent. Even finely divided ceramic materials can be ignited in fluorine.

The common materials of construction for use with fluorine at temperatures of 473°K or lower are iron, steel, Monel, nickel, copper, brass and aluminium. These metals form thin protective films by reaction with fluorine.

According to Zabetakis (1967) "liquid fluorine can be handled in clean passivated equipment made of Monel, aluminium, types 304 and 321 stainless steels, copper, and brass. Equipment that contains fluorine must not be bent, flexed, or struck as the protective fluoride film may be removed permitting the fluorine to react violently with the fresh metal surface".

Unsatisfactory materials are those which form volatile reaction products or materials in which the corrosion film becomes powdery or flaky, and thus porous. High silicon alloys are examples of the former where fluorine has a tendency to leach out the silicon by the formation of volatile silicon tetrafluoride. Steel, including the stainless steels, at elevated temperature are examples of the latter.

For joining metals used in fluorine service, welding is the best for steel, nickel or Monel. In some applications silver solder is also satisfactory, but soft solder should be avoided. Gasket materials may be fluoroplastics, aluminium, or soft, annealed copper; all should be cleaned and degreased. If a lubricant is needed for threaded connections, a Teflon dispersion (applied from an aqueous carrier and then dried) or the very thin Teflon tape has been used successfully. The equipment should be free of hydrocarbon oils, organic residues, etc., and should be degreased, preferably before assembly. Care should always be exercised when introducing fluorine into a system, especially during its first contact with fluorine. The surfaces have no protective fluoride film and some reactive deposit, such as water, oil, etc. might inadvertently be present.

Aqueous ammonia and a filter moistened with KI solution are useful in locating F_2 leaks. All work of this nature should be performed in well ventilated work areas. If or when a system is to be opened to the atmosphere for any reason, be sure that the fluorine has been carefully evacuated. It is preferable to refill the system with dry air or nitrogen. When opening a fitting, or flange keep hands away from the connection. In the event that fluorine does issue from the connection, seal it immediately and evacuate the section again, more completely. All equipment containing fluorine or pipe lines containing fluorine should be clearly labelled as to their contents (Farrar, 1969).

III. Hazards

A. FIRES AND EXPLOSIONS

For a combustible cryogen such as LH_2 or LNG to burn it must be mixed with an oxidant (usually air) within appropriate limits, called flammable ranges, and be initiated by an ignition source. Thus the three conditions of the classical fire triangle (fuel, air, ignition) are to be satisfied for a fire or explosion to occur. Normal control measures seek to eliminate two of the three conditions, namely, by avoiding ignition sources and by controlling the release, or accumulation, of the fuel. In some circumstances where large pipes with high flow rates must go through confined spaces or where the free circulation of air is prevented (as in large masses of porous insulation) then one might control the third side of the triangle, namely, exclude air by introducing inerting gas such as nitrogen or argon.

As shown in Table 11.III methane forms flammable mixtures with air over the concentration range 5 to 15 volume % and the minimum value of electrical spark energy requirement for ignition is about 0·3 mJ. Hydrogen also forms flammable mixtures with room air over a wide concentration range, 4 to 75 volume %. The minimum value of electrical spark energy requirement for ignition is 0·02 mJ. High velocity atmospheric releases of hydrogen frequently ignite even in the absence of any apparent source of ignition.

Burning hydrogen exhibits such low emissivity that leaks (audible as well as inaudible) may be aflame and yet not be visible. This non-luminous flame may make it difficult to detect a fire from a small leak. Special devices (ultraviolet, infra-red and television type detectors) have been developed but are not in common use. The use of a broom has been recommended as a physical detection device since the burning bristles held out at a safe distance can be readily seen. But people have received painful, if not serious, burns while searching for leaks or inadvertently finding one. Of course, large leaks may well be audible and this would be a warning.

The burning gas evolving from a leak of combustible cryogenic fluid should not necessarily be extinguished. A continuing gas leak might be more dangerous; therefore, shutting off the supply is an important emergency consideration. Liquefied natural gas authorities insist *"that a major fire should be controlled—not extinguished"* (Dyer and Sommer, 1969).

B. CONTROL OF IGNITION SOURCES

Smoking, open lights and sparking devices should obviously be avoided in areas where hydrogen, methane or other light hydrocarbons might be present. Whenever work is to be performed in areas where the continued absence of combustible gas cannot be confidently assured, the need for doing this work and the methods employed must be scrutinized by the appropriate authority, most likely the operating supervisors.

The use of "non-sparking" tools cannot be relied upon to avoid ignition of inflammable gas. They have little safety value compared with the elimination of fuel wherever tools are used (Kerr, 1962).

As aluminium painted iron surfaces can readily spark when struck, one should avoid such thermite combinations in closed areas.

The U.S. National Fire Code, Volume 5, "Electrical", one of the ten volumes published by the National Fire Protection Association, covers the safe installation and use of electrical equipment (National Fire Protection Association, 1968). Articles 500–517 provide standards for the use of electrical equipment in hazardous areas.

The term "explosion proof" equipment is used rather loosely in practice. By this term manufacturers intend encased equipment suitable to prevent the spread of any explosion which might originate within it.

For atmospheres containing methane, Group D (Class 1) fixtures are specified; for hydrogen, Group B (Class 1) are specified. Because of the high explosion pressure and small quenching distance, approved equipment for hydrogen is difficult to build and therefore expensive when available. Some ingenuity may be employed to control the electrical sparking problem without the necessity of the more expensive devices approved for hazardous areas, particularly in hydrogen service.

If electrical devices spark with less energy than that necessary for the ignition of a gas–air mixture they are "intrinsically safe" and may be used in the hazardous zone. Such devices are being developed at an increasing rate (McKinney, 1967; Weismantel, 1969; Redding, 1969; Hickes, 1969; National Fire Protection Association, 1967a).

An alternate to making equipment intrinsically safe or explosion proof is to place the equipment in an enclosure purged above ambient pressure with an inert gas such as nitrogen or with air from an uncontaminated source (National Fire Protection Association, 1967b).

Locating electrical equipment outside of hazardous areas is a good choice if it can be done. Light fixtures, switches, outlets and instruments are examples of devices that can often be located in a less hazardous or non-hazardous environment.

C. STATIC ELECTRICITY

Because of their low ignition energies any flammable mixture of hydrogen–air or methane–air (see Table 11.III) can be spark-ignited by commonly occurring electrostatic discharges. If a person with a capacitance of 300 pF were charged to 10,000 V then the individual would have a charge energy of 0·015 J, or 15 mJ (United States Bureau of Mines, 1963). Ungrounded pieces of equipment, wiring or structures may acquire much larger energies than 15 mJ and precautions should be taken to prevent such charge accumulation where combustible gas–air mixtures may prevail.

Grounding (or earthing) is the most effective way of preventing static charge accumulation. All electrical equipment and all equipment containing combustible cryogens, or their evolved gases, including fill lines and vent lines, should be grounded. In transfer operations grounding all equipment to a common potential is important.

The American Petroleum Institute (1968) standard on liquefied natural gas (methane) deals with this problem as follows: "Static protection is not required when tank cars, tank vehicles, or marine equipment are loaded or unloaded by conductive or nonconductive hose, flexible metallic tubing, or pipe connections through or from tight (top or bottom) outlets because no gap exists over which a spark can occur."

Where conductive floors are used the floor should also be grounded. Normal resistances of such floors are sufficiently high, usually more than 100,000 ohms, so that power voltages are not a risk. Since normal good practice requires containment of the gas it is not certain that special provision for grounding individuals serves any useful purpose.

Humidification may make a contribution to safety in avoiding static sparking by improving conductivity in the environment; relative humidity over 70% is particularly effective in preventing buildup of static charges (Forrest, 1953; Cooper, 1953). The Safety Manual of the United States Army Material Command (United States Department of the Army, 1964) states that "Humidification for preventing static electrical accumulations and subsequent discharges is usually effective if the relative humidity is above 60%."

Concern for static electricity may be extended to include animal life. One may find in several authorities (United States National Aeronautics and Space Administration, 1968, and United States National Bureau of Standards, 1960) under the heading "Static Electricity" a warning to keep furred animals out of hydrogen areas. It is likely that the furred animals present other risks more dangerous than a potential ignition source. No animals, furred or otherwise, belong in an establishment handling dangerous commodities.

D. DEFLAGRATION AND DETONATION

Ignition of combustible gas–air mixtures by the previously discussed sources usually results in ordinary reactions of combustion or deflagrations. These reactions continue through that portion of the unburned mixture within the appropriate range, at significantly less than sonic speeds. However, under some circumstances, usually of confinement, a deflagration may change into a detonation wherein the reaction propagates into the unburned mixture at supersonic rates. If combustible gas–air mixtures can occur in confined spaces the destructive effect of explosions can be moderated by providing pressure relief of sufficient area and ease of opening. The suggested relief area is 10–30 $m^2/100$ m^3 of room volume (National Fire Protection Association, 1967c). The progression from deflagration to detonation will be interrupted as the products of combustion and the unburned mixture expand into free space. Where possible the introduction of an inert diluent or sufficient quantities of diluting air can also minimize the possibility of a destructive explosion.

The energy available in combustible cryogens and in their gas phases is large when related to the energy available in condensed chemical explosives.[†] Nevertheless, in a combustible gas–air explosion not all the available heat can be converted into a pressure wave. In actual events the rate of energy release bears largely on the effect produced. Mixing is not always uniform nor within an approximately stoichiometric range so only a small percentage (probably less than 10%) of the available fuel is usually involved at a given instant.

There is some evidence that accidental hydrogen explosions in free space can do damage but this damage has been observed only in light structures. Reider et al. (1965) and others described an experiment wherein hydrogen was released into the air at the rate of 54 kg/sec; inadvertent ignition yielded a pressure wave equivalent to the pressure presented by a wind of approximately 250 km/hr. The R–38 (hydrogen lift) airship disaster over Hull, England, in 1921 is reported to have shattered windows over a 3 km radius.

Cassutt et al. (1960) and others were able to get a full explosive yield in stoichiometric hydrogen–air mixtures in a 1·5 m diameter balloon only

---

† Methane has 12 times the energy of an equal weight of trinitrotoluene; hydrogen has available 29 fold the energy of equal TNT weight. These ratios are this high because the weight of oxidant (air) is not included with the methane and hydrogen while the TNT molecules already contain oxygen.

when using a strong initiating source of 2 g of pentolite. Using a blasting cap initiator, which is approximately 0·5 g of explosive, reduced the yield 95%, while flame sources, sparks, and hot wires gave only combustion of the gases with no measurable pressures.

E. NUCLEAR PHYSICS PROBLEMS

There are an increasing number of experiments being undertaken which involve the use of superconducting magnets and glass or thin metal windows. In the case of the magnets, provision must be made to handle the entire circulating current in normal conductors should the superconductor go normal. This is usually done by the use of a composite wire containing both superconductor and low resistance normal conductor. Where windows are used, provisions must be made to shut the experiment down safely in the event any or all windows fail. Frequently nuclear radiation damage is a factor in such failure.

The use of liquefied nitrogen around high nuclear radiation fields has resulted in a unique hazard. The radiative generation of ozone or nitrogen oxides has been suggested as the mechanism for explosions that occur after the evaporation of the nitrogen (Chen and Struss, 1969). The most significant control measure is the avoidance of contamination of the nitrogen with oxygen condensed from the atmosphere. Back diffusion of air into the cryostat should be positively prevented either by operating with a continuous flow of uncontaminated LN_2 or with a positive gauge pressure. The continual replenishment of the original supply may lead to an accumulation of impurities and should be avoided. The liquid nitrogen supply used around radiation fields should have a maximum allowable concentration of 20 v.p.m. O_2 and 5 v.p.m. oxidizable hydrocarbons (Carter, 1969).

F. SPILLS OF LIQUID HYDROGEN

In the case of a large spill of LH_2, ignition can be expected even if the obvious ignition sources are absent. Energy from rending metal in the case of vessel failure or dislodged particulate matter can cause ignition. In the combustion of H_2 in large spills the phenomena to be protected against are flame radiation, blast and shrapnel.

Spills of liquid hydrogen from a few litres to 20,000 litres have been investigated (Little, 1959; Zabetakis and Burgess, 1960). These tests investigated the dispersion rate of the spilled fluid and resultant vapour, the effect of confinement upon the production of a detonation and the characteristics of the combustion.

For spills with unconfined volumes (less than 2 external walls plus ground) the tests did not produce a detonation unless initiated by a strong detonator (i.e. explosive). When in the open with no wind the flame burns directly above the spilled area. The presence of wind will deflect the flame thus increasing the hazardous area. However, burning duration of a hydrogen flame is only 2–5 % of that of an equal volume of hydrocarbon. The experiments showed that one could not visually determine the limits of the combustible cloud from observation of the condensing water vapour. Figure 11.3 shows the extent of the flammable mixture as a function of height and time for a small hydrogen spill. The attenuation of radiation from a hydrogen flame is greatly enhanced by the presence of H_2O vapour in the air hence this is a factor in selecting the spacing of adjacent LH_2 dewars. The intensity of radiation, I, at a distance, r (m) from the flame is given as a function of initial flame intensity, I_0, and the volume percent water vapour, w (Zabetakis and Burgess, 1960)

$$I = I_0 e^{-0.0492wr}$$

Spill tests in confined areas (exceeding two walls and the ground) have shown detonations to result from burning hydrogen–air mixtures. For this reason

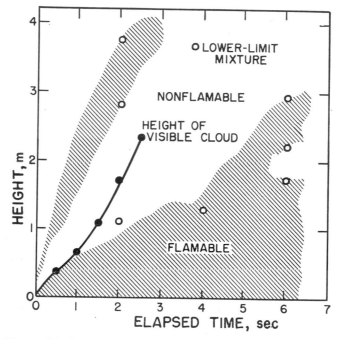

Fig. 11.3. Extent of the flammable mixtures and height of the visible cloud formed after the rapid spillage of 3 l of liquid hydrogen on a dry macadam surface in a quiescent air atmosphere at 288°K (Zabetakis and Burgess, 1960).

full height protection walls are not to be used around liquid hydrogen dewars. Instead the lowest possible dikes consistent with complete containment of the spilled fluid are desirable. If a safer area exists into which the fluid can be channelled, this would be the preferable solution. Crushed rock should be used for the lining of the confinement area in order to promote rapid evaporation of the spilled fluid. These factors plus the degree of approach to the maximum energy yield from the combustion must be considered in the preparation of quantity–distance charts which will be discussed later.

G. COLD INJURY

All cryogenic fluids exist at temperatures low enough to damage tissue. Surfaces cooled by such fluids, or their evolved gases, may also be cold enough to bring tissue to a point below 0° centigrade (273°K) where damage occurs. Such cooling by the fluids does not take place immediately because the blood supply to tissue acts as a heat source and the contact of a cryogen with warm tissue creates a gas film which is not a good conductor of heat (touching cold metal can result in cold damage to tissue more rapidly). This time delay usually permits successful measures to be taken if the exposed part of the body is splashed by a cryogenic fluid. Two measures make significant contributions to safety: (1) the ability to elude the continuous contact of the cryogen by shutting off the flow or escaping the area (or both), (2) immediately flooding the tissue or clothing with water. The enormous heat capacity of water, its harmlessness and its ready availability all combine to make water an important safety contribution to cryogenic operations. So universal is the regard for this safety measure that cryogenic operations, except for routine, well-proven procedures, should not proceed without its ready presence.

An internationally known major producer and handler of cryogenic fluids for over thirty years states, "our personnel injury experience from 'cold' burns has been practically nil". This authority compares liquid nitrogen to hot water (71°C, 344°K) in its body-contact effect (Neary, 1969). Momentary or very brief contact with a small amount of cryogen will not cause tissue damage but prolonged contact will "burn". While tissue damage may be virtually unknown with such cryogens as helium and hydrogen it is more frequently seen among those cryogenic fluids with higher heat capacity. Liquefied oxygen spills have produced belated accidents from fire burns because of smoking and stray sparks on oxygen enriched clothing even after the individual has left the environment of the spill.

The treatment of cryogenic burns is simply to restore tissue to normal body temperature (37°C or 310°K) as rapidly as possible followed by protection of the injured tissue from further damage and infection (Rosenberg, 1969). Rapid warming of the affected part is best achieved by using water at 42°C (315°K) (Meryman, 1957). Under no circumstances should the water be over 44°C (317°K); nor should the frozen part be rubbed either before or after rewarming. A patient should not smoke or drink alcohol (Washburn, 1962; Mills, 1962).

25*

Safety glasses should be generally worn when working with laboratory and industrial cryogenic and pressurized, or potentially pressurized, systems. Where splashing is a real possibility, face shields may be indicated. One should not look into an open cryogenic system without eye protection; even small mechanical or thermal disturbances have caused dangerous ejection of fluid. Personnel may require special clothing, gloves and footwear. Gloves should be clean and dry; loose fitting gloves are preferred so they may be removed quickly if splashed or spilled upon.

H. ASPHYXIATION AND TOXICITY

Cryogenic fluids released or spilled in a confined area can rapidly alter the air composition, by displacement or contamination, and introduce an asphyxiating or toxic risk.

While breathing air is normally 21% O_2 with a partial pressure of about 0.2 atm at sea level the human can tolerate large quantities of contaminant gases if these gases are non-reactive and only slightly soluble in the blood. Thus, with physiologically inert gases such as helium, methane, hydrogen, nitrogen, argon and neon, oxygen concentrations (at 1 atm total pressure) as low as 13 volume $\%$ can be tolerated by healthy subjects. However, the minimum concentration to oxygen necessary to support humans is also related to the activity and health of the individual. With an increase in activity there is an increase demand for oxygen (Zabetakis, 1967).

If the oxygen content of the air falls slowly and the individual is at rest, symptoms ensue gradually. If these symptoms arise slowly enough an individual's difficulties may be masked by a state of "euphoria" giving the victim a sense of security and well being which is entirely false and may preclude self-rescue (see Table 11.IV).

TABLE 11.IV. Symptoms of O_2 Deficiency (from Johnstone and Miller, 1960)

$O_2\%$ at 1 atm total pressure	At rest, signs and symptoms of O_2 deficiency
12–14	Respiration deeper, pulse faster, coordination poor
10–12	Giddiness, poor judgment, lips blue
8–10	Nausea, vomiting, unconsciousness, ashen face
6–8	8 min, 100% die; 6 min, 50% die and 50% recover with treatment; 4–5 min, all recover with treatment
4	Coma in 40 sec, convulsions, respiration ceases, death

When the diluent gas is carbon dioxide, oxygen concentrations below about 19 volume $\%$ can be tolerated only for a few minutes (19 volume $\%$ oxygen concentration is found in a nine volume $\%$ CO_2–air mixture). Thus, although carbon dioxide is a physiological necessity in the normal breathing cycle because it stimulates breathing, the effect of an excessive amount on the breathing rate can affect the ability of the body to rid itself of the carbon dioxide in the blood (Zabetakis, 1967).

Although oxygen is a physiological necessity for human beings, there is a toxic potency in an environment superabundant in oxygen. Oxygen tolerance is virtually unlimited with enrichment of air up to an oxygen concentration of 60 volume % at total pressures of 1 atm. At higher oxygen concentrations difficulties appear after periods of several hours to several days and depend on a variety of factors such as activity, moisture, diluents, oxygen partial pressure, and so on. Having such a high oxygen concentration for a long enough time would only be expected in a planned experiment (Roth, 1965).

Fluorine is the most toxic gas of cryogenic interest; it is also the most powerful oxidizer of all the elements. All organic materials are extremely reactive with fluorine; contact with skin may produce burns which can be painful and difficult to heal. At concentrations of 50 v.p.m. of fluorine breathing is reported to be impossible without respiratory equipment; skin irritation occurs at about 100 v.p.m. The threshold limit value for fluorine is 0·1 v.p.m. based on average exposure for a normal 40 hour work week. Despite its extraordinary toxicity and reactivity it has been reported that fluorine has been handled on a large scale for more than twenty-two years without a disabling injury, attesting to the fact that good safety procedures can minimize risk in handling hazardous materials (Farrar, 1969).

I. AIR CONDENSATION AND OXYGEN ENRICHMENT

Liquefied helium, hydrogen, neon and nitrogen are capable of condensing (and solidifying in the case of the first three) air that may come in contact with the fluids. The solidified air can plug vent paths and jeopardize vessels because of pressure build-up; solid air, probably oxygen enriched, in liquefied hydrogen can present an explosion hazard. Whenever possible, liquefied helium and hydrogen should be stored, transferred, and otherwise handled only in closed systems pressurized to greater than atmospheric pressure to prevent backflow of air; relief devices should be designed to prevent back leakage of air if it is necessary to subcool the liquid by drawing a partial vacuum on it (Neary, 1963).

Air coming in contact with a surface cooled below 82°K will partially condense with the condensate containing approximately 52% O_2 and will continue to condense in this fashion as long as air with 21% O_2 is supplied to the surface. Such an oxygen enriched mixture will significantly enhance burning rates of combustibles and permit some materials to become combustible that might not qualify as such in normal air. Provision should be made either to avoid the formation of such condensate or to ensure that the run-off does not cause any hazard. This run-off should not be permitted to contact combustibles or components which may become embrittled or cease functioning at low temperature.

When vacuum insulation is not required, applied external insulation is frequently used. These are generally poured-in-place foams with densities from 0·03 to 0·1 gm/cm³. Many foams such as polyurethane and polyvinyl chloride exhibit toxic properties during their formation and appropriate precautions for their handling should be taken. Since these are cellular structures, it may be possible for oxygen enriched air to accumulate in them, although most foams are closed-cell which are less susceptible to penetration than open-cell foams. It is usually desirable to provide a barrier against the infusion of air as it might also accumulate behind the foam adjacent to the pipe and expand when the pipe warms up (Neary, 1963). A foam of proven non-combustibility should be selected not only because of the enriched air possibility but also because it is always desirable to reduce the quantity of combustibles present in cryogenic facilities. These facilities are usually so expensive that a fire of just a few pounds of extraneous combustibles can do a disproportionate amount of damage and the combustion products can be a toxic hazard. Insulating materials of glass fibres or mineral wool with desirable low combustibility characteristics are also available. However care must be taken to ensure that no oil was added to these materials to facilitate handling.

IV. Design and Operation

A. STORAGE AND TRANSFER

In the storage and transfer of cryogenic fluids the safety objectives are to get the fluids to the use or storage point in a pure state, to effect any disposal in a safe manner and to prevent any damage to the system.

For cases where contamination of the cryogen is hazardous it is necessary to remove undesirable gases from the system by purging. For portions of the system with large length to diameter ratios ($\ll 1$) a flow-through purge is usually most effective. This portion of the system must be equipped with suitable purge-gas introduction and vent ports, a purge-gas supply system, and gas-sample ports. A gas-analysis capability should also be available. For components with small length to diameter ratios or in which the entire volume cannot be swept with purge gas, purging is effected by either evacuation and backfill to 1 atm or pressurization to a few atmospheres followed by venting to 1 atm. The previous steps or combinations of them are repeated as required to obtain the desired purity. One should not rely upon computations or intuition to determine the extent of purge effected. The inevitable existence of dead-ends, adsorbed gas and incomplete mixing make purging less than 100% efficient even for simple systems. For this reason gas analysis is required for verification of purge adequacy. Also gas samples should be taken from locations where purging is least likely to be complete. Gas analysis is usually accomplished by means of a mass spectrometer or a gas chromatograph.

It is frequently difficult to prevent all leakage paths in a large and complicated system. It is usually better to maintain the system at above atmospheric pressure so that if leakage occurs it will be out-leakage which is usually tolerable. To prevent internal leakage between parts of a given system additional precautions may be necessary. One form of isolation is obtained by a "double block and bleed" which consists of two valves in series with an open side valve between them leading directly to the atmosphere. An application for this is in the isolation of stored gaseous hydrogen from its inerted discharge lines. For the long term storage of cryogenic liquids (e.g. LH_2) a more sophisticated procedure is required to isolate the liquid. To minimize heat leak into the storage vessel the liquid shut-off valves are almost always located downstream of a gas-trapped section of line. If leakage from the shut-off valves were vented continuously the gas trap would be lost, the valve would cool and leakage would increase. To prevent this a "He block" is used. This consists of two valves in series with the included section of line maintained under He-gas pressure slightly greater than the dewar storage pressure. Thus any valve leakage is He gas into the dewar, rather than LH_2 leaking out. (In the case of 2,000,000 l LH_2 storage dewars at the Nuclear Rocket Development Station (NRDS) in Nevada, the liquid is stored at 0·5 atm gauge pressure and the "He block" is maintained at 0·6 atm gauge. In this system the He block pressure and gas purity are monitored periodically.)

When the pressure in a dewar is significantly above atmospheric pressure, the venting takes place through a pressure-relief valve or an area-restricted vent-line, either one of which will prevent backflow of air into the dewar. Such a backflow in the case of LH_2 and LHe storage would be particularly undesirable because air exerts essentially zero vapour pressure at these temperatures and would therefore continue to accumulate. This particular hazard is usually greater for the small laboratory sized dewar which is not fitted with a sophisticated vent system. Such dewars when not protected in any way from back diffusion of air and its accompanying moisture have exploded from pressure build-up resulting from plugged vents. During storage, small dewars should be connected to a safely purged vent system or a so-called Bunsen-valve (slit rubber-hose). Following an experiment, hazardous gas should be purged from the system where possible in order to obtain a safe standby condition. The same purge techniques discussed above are applicable.

As an additional precaution in the long term storage of liquid cryogens, periodic warm-up and analysis of the dewar contents is desirable. Each warm-up should bring the coldest spot in the dewar interior to a high enough temperature that the suspected impurity is above its normal boiling point. This is necessary to prevent excessive accumulations of solid air which has no way of being detected or removed under normal cryogen storage conditions. The period between such warm-ups is a function of the individual system and its use. Theoretically, it should be possible by careful control of operating procedures to obviate the requirement for such warm-ups. However, such a procedure is a positive protection against unsuspected in-leakage, inadequacies in procedures and undetected procedural errors. [At NRDS the warm-up of the 2,000,000 l LH$_2$ dewars has been performed annually for operational reasons. This has proved to be more frequent than required by safety considerations since in all cases only a few p.p.m. of O$_2$ ($\simeq 2$ g total accumulation) has been detected.]

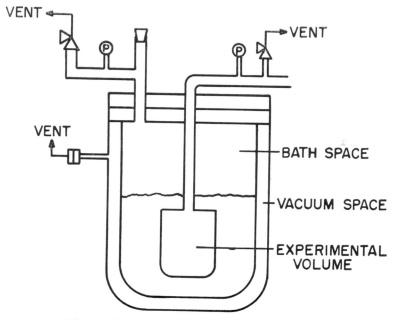

Fig. 11.4. Spaces requiring pressure relief protection.

Automatic pressure relief must be provided in several situations (see Fig. 11.4). All isolatable spaces which can hold liquefied cryogen (including sections of transfer line between valves in series) must be provided with relief valves of sufficient size to accommodate at safe pressure the maximum gas evolution rate that could occur. (This usually occurs upon the complete loss of insulating vacuum). For long term storage in large dewars the liquid space should be protected by duplicate pressure relief valves. Secondly, any spaces containing a gas or liquid refrigerated by another source (e.g. cryogen bath or refrigerator) must likewise be protected. One typical problem is given as an example; argon was being purified in a small glass system by immersion in a liquefied nitrogen trap. The argon condensed in the system which subsequently was sealed off. When the nitrogen trap was removed the argon evaporated causing enough pressure to rupture the glass system. The experimenter who had not realized that argon as well as impurities condensed at liquefied nitrogen temperatures suffered an eye injury. Thirdly, each vacuum insulation space should be individually protected. For the lower boiling cryogens which can cryopump air, protecting the vacuum insulation space is particularly important because large quantities of undetected air could accumulate over the storage period. Frequently this pressure relief valve can be made integral with the evacuation valves thus requiring only one fitting.

A valuable and necessary adjunct to any transfer and/or storage system is adequate instrumentation. The requirements of course vary from system to system but in general should include pressure measurements in storage spaces, insulation spaces (vacuum) and in the transfer lines; temperature measurements in the storage space (to facilitate warm-up analysis); liquid level measurements in storage dewars; flow measurements on both the liquid and purge systems (to provide reproducible purge flows) and combustible gas detectors where necessary. Since safety requirements can vary from one system to the next, additional diagnostic instrumentation which facilitates a gaining of a better understanding of the operation of the system is a valuable asset.

B. STRESSES

For safe operation of a cryogenic system consideration must be given to the stresses which arise from thermal contraction and temperature gradients which occur during the cool-down. The magnitude of such stresses (which are a function of thermal conductivity, modulus of elasticity and coefficient of thermal expansion of the metals involved) can be held at acceptable levels by proper procedures.

In well insulated systems at operating temperatures usually the only significant stress problems result from the axial thermal contraction of the cooled line. Obviously these stresses should be evaluated during the design of any proposed cryogenic system. The magnitude of this problem is relatively independent of the particular cryogen used since the coefficient of thermal contraction decreases as the temperature is lowered with most of the contraction being attained by the time LN_2 temperatures have been reached. In theory such stresses can be computed easily by standard techniques of mechanics (Crocker and King, 1967). However in most large scale systems the actual configuration of the pipe, and its attached lines, supports, and anchors sufficiently complicate the problem that a computer solution is required.

There are three methods to lower the thermal contraction stresses. For vacuum jacketed lines a bellows is normally placed in the outer line to allow it to conform to the inner line contraction. Stresses accompanying the contration of the inner line are then held to an acceptable level of providing sufficient flexibility. In many systems the required routing of the pipe will provide sufficient flexibility; however, if such is not the case, additional U-bends, etc. must be provided. A second method is to provide bellows in the inner line. This is less commonly used since it requires extra inner line joints and complicates the repair of a leaking bellows. Another possibility is the use of slip joints on the inner line but in this case protection against potential leakage into the vacuum space presents a further complication.

During cool-down and in systems with large heat fluxes (e.g. heat exchangers and non-insulated lines) radial and circumferential stresses can also occur. Where the temperature gradients occur during steady-state operation they must be considered in the system design. Stresses due to transient gradients can be minimized by proper operating procedures.

Radial temperature gradients result during cool-down because the inner surface of the pipe cools first. The faster the coolant fluid flows through the pipe the greater is the heat transfer and a higher stress results. In addition, the magnitude of the thermal stress depends upon the shape of the temperature gradient as well as the total temperature difference across the pipe wall. In general pipe walls are sufficiently thin so that such stresses are not a problem. However, with rapid cooling, large thermal stresses can be built up in flanges and other thick walled components. A theoretical analysis of this problem has been carried out (Novak, 1970) and some results are presented in Fig. 11.5. The results obtained are highly dependent upon the heat transfer correlation used. The maximum flow limits presented in Fig. 11.5 use the most conservative correlation (Dittus and Boelter, 1930). If these curves are used, stresses due to radial temperature gradients during cool-down should not exceed allowable room temperature thermal stresses.

Circumferential temperature gradients can also cause additional stresses. Such gradients can result during the cool-down process from stratified flow in horizontal pipes which permits the bottom of the pipe to cool faster than the top. The differential thermal contraction will then cause the pipe to bow and the resulting stresses will depend upon the shape of the temperature gradient and the manner in which the pipe is restrained. These stresses can either be held to acceptable levels by maintaining sufficient piping flexibility (Fleider et al., 1960) or be eliminated by avoiding stratified flow during the bulk of the vaporization process during line cool-down (Bronson et al., 1962). While not firmly established as being applicable, two-phase flow correlations exist which predict the flow patterns expected. Based upon one of these (Baker, 1954), one can estimate minimum flowrates which are high enough to avoid stratified flow during at least 95% of the vaporization process (Fig. 11.6). Thus, it is seen that the range of permissible flowrates for safe cool-down of a given system lies between the limits given in Figs. 11.5 and 11.6.

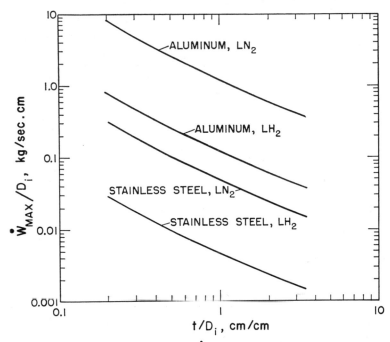

Fig. 11.5. Maximum cool-down flow rates ($\dot{W}$) of cryogen to keep thermal stresses below maximum allowable thermal stress for a pipe of wall thickness t and inside diameter D_i.

In case of evacuated-powder insulated, large cryogenic storage vessels (> 100,000 l) consideration should be given to the possibility of the build-up of forces within the powder in the insulation space (Novak, 1966; Leonhardt, 1970). The occurrence of such forces has been known to break the support rods and rupture the bellows connecting the inner and outer shells of a vessel. These forces can build up as a result of the insulating powder spilling into the void created when the inner shell cools to operating temperatures. When the vessel is rewarmed the powder is compressed rather than flowing back to its original position. Repetition of this process can result in damaging stresses. This effect may also occur in large, non-evacuated, powder insulated storage tanks for O_2, N_2, and LNG. There are a number of ways to alleviate this problem, such as wrapping the outer surface of the inner shell with a resilient material like a fibreglass batting, or supporting of the inner shell so that it can move in such a way as to relieve any compressive stresses (attached pipes must be flexible enough to permit this). In any case the vessel should be made as symmetrical as possible to minimize these forces in all directions except the vertical.

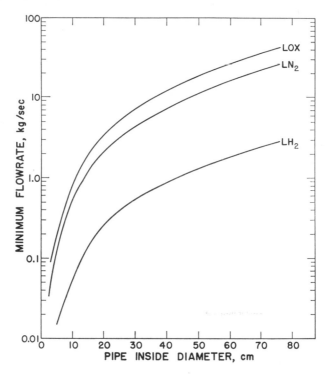

Fig. 11.6. Minimum flow rate for which is predicted non-stratified flow for pipeline fluid qualities below 95% (Edeskuty and Williamson, 1971).

C. DISPOSAL

The disposal of cryogens should take place in such a way that significant atmospheric accumulations of the cryogen cannot occur. Usually this would be from a vent located at a high point in the system. Depressed areas such as pits which could trap cold vapour and create an asphyxiating (or O_2 enriched in the case of O_2 venting) atmosphere should be avoided near the venting area. Precautions should also be taken to prevent any damage to personnel or equipment from the cold stream being discharged. Under some circumstances this may involve the insulation of the disposal line to prevent frostbite, condensation of an O_2 enriched liquid air, or embrittlement of adjacent equipment.

Disposal of combustible cryogens can be effected by either venting directly to the atmosphere or by burning. The total quantity to be vented will influence the disposal method; however for low flowrates (e.g. <6 l/sec of liquid hydrogen) where the vent stack is isolated from ignition sources (> 20 m) such disposal could be directly to the atmosphere without burning. Prior to letting combustible gas into the stack it should be purged of O_2 (i.e. air) to prevent the formation of explosive mixtures within the stack. Since in such a disposal system burning is not desirable and since ignition can result from electrical discharges during thunderstorms, provisions should be made for inert gas purging to extinguish any flame. The quantity and flowrate of purge gas available should be sufficient to extinguish the flame and to cool any hot surfaces so that reignition will not occur. For high flowrates (e.g. > 60 l/sec of liquid hydrogen), or where sufficient isolation from ignition sources is not possible, disposal of combustible cryogens is effected by burning. Frequently it is advantageous to design the external piping of a dewar so that the liquid can be disposed of through the normal gas venting system. Burning is normally carried out by use of a flare stack. Burn-ponds in which the gas is released under water and allowed to flow up to burn on the water surface have also been used. In both cases the system must be designed to prevent blow-out of the flame as well as back diffusion of air. The blow-out can be prevented with a continuous ignition system. The dynamics of this combustion have been studied and it is possible to predict the necessary flowrates and exhaust velocities to maintain a stable flame (Zabetakis, 1967). In the case of a burn-pond back diffusion is prevented by the water seal while for the flare stack a molecular seal is used. For intermediate flowrates the disposal method used will be dictated by the individual system and the degree of isolation from surroundings. For large storage dewars containing combustible cryogens it is advisable to isolate each dewar as much as possible from common vent lines.

Massive spills will probably be accompanied by ignition of the cryogen. Damage from such spills can be minimized by either diverting the spilled liquid by means of ditches to a safe burning area or if this is not possible by containing it within a dike around the base of the dewar. In either case the walls of the ditches or dike should be only high enough to contain the liquid so as to minimize the possibility of their converting the deflagration to detonation.

At NRDS disposal of hydrogen has followed a slightly different philosophy. During stand-by conditions venting rates are small and during this time each dewar is vented individually through a non-flared local stack. On run days disposal quantities vary from low flow dewar boil-off to high liquid hydrogen flowrates (up to 1,500 l/sec of LH_2) all of which are burned in a common flare stack.

D. VENTILATING AND INERTING

Where combustible cryogens must be used in confined spaces it is necessary to prevent the formation of a combustible atmosphere should leakage occur.

Where the possible leakage quantity is not too large adequate ventilation will suffice (30–60 room air changes/hr). Where leakage could be large enough that a practical ventilation system cannot maintain the room atmosphere below the lower explosive limit, the room can be made safe from ignition by an inflow of gaseous nitrogen which maintains an oxygen concentration below 4%. For a well sealed room the initial inerting can be accomplished with a volume of N_2 equal to 1·7 times that of the room (Boyer et al., 1965). Additional gas must be added as required. With such a system large leaks resulting in hydrogen concentrations in excess of 10% have caused no problems (Edeskuty, 1969). To be effective, the oxygen concentration must be continuously monitored to prevent combustion during a run. To insure a breathable atmosphere for personnel re-entry after a run, de-inerting is required unless self-contained air breathing equipment is used. Before the entry of personnel into previously inerted rooms, tanks or other confined structures, oxygen concentration monitoring precautions should be taken.

E. HAZARDOUS LEAKAGE MONITORING

In the handling of combustible cryogens in confined spaces a suitable gas detection system must be used. In general single point detection should not be relied upon. The detectors should be located where the escaping gas will be most quickly detected (i.e. close to possible leakage points, high point locations for hydrogen, etc.). Examples of monitors used for H_2 detectors include thermal conductivity analyzers and combustion analyzers (to use this type in inert atmospheres requires an auxiliary stream of air). Both of these can either have the analyzer at the sampling point (diffusion head) or can have the gas to be analyzed aspirated to a remote central location for analysis. Loss of response time is an important consideration in the latter and in all cases the response time of a given system should be determined after installation. The equipment should be maintained so that the number of false alarms is minimized. Also personnel should be familiar with well established procedures which dictate the proper response to an alarm.

F. TRANSPORT

Highway transport of cryogens involves the use of containers ranging from a few to 50,000 l. Railroad cars with a capacity as large as 130,000 l have been built, and there are barges that carry 350,000 l of liquid oxygen and 10^6 l of liquid hydrogen. Liquefied natural gas ocean tankers of 7×10^7 l are under construction (Culbertson, 1970). Air shipments up to 500 l as well as 38,000 l tank truck shipments of liquid helium are now routine. The rapid transportation of cryogenic fluids particularly in the case of land and air travel accentuates the problem of air inflow to the container mentioned previously. This problem can be eliminated by maintaining the liquid at a gauge pressure greater than any pressure variation expected because of altitude and weather changes during transport. Occasionally plans go astray and a dewar is discovered to be plugged. Action to be taken under this circumstance depends upon the known history of the dewar. If the age of the plug and the heat leaks to the dewar can be estimated, internal pressure can be estimated. Since for most liquid helium transport dewars heat leak down the neck is minimized by utilization of refrigeration in the effluent gas, a vent tube plug destroys this source of refrigeration resulting in a larger heat leak by factors as large as 10 (Scott, 1964). For a completely filled 25 l liquid helium dewar self-pressurization rates of $2 \cdot 0$ N/cm^2 hr ($0 \cdot 2$ atm/hr) have been reported. For either incompletely filled dewars or larger dewars lower rates would occur. If the pressure is not excessive the plug can be drilled out or melted by a warm He gas stream or copper (high thermal conductivity) rod. In cases where excessive or unknown pressures exist there is no completely safe method of disposal. In one instance a LHe dewar in this condition was taken to a remote location and upon being shot rocketed into the air approximately 30 m.

During shipment of large dewars the pressure rise to the relief valve setting is usually more rapid when the dewar is at rest than when in motion. This is due to the fact that during motion the contents are constantly agitated and thus kept at temperature equilibrium and consequently the heat input is distributed uniformly throughout the cryogen. While at rest the liquid quickly stratifies and most of the incoming heat is deposited in a warm stratified layer at the top of the liquid and in the gas ullage space. Because of the stirring action during shipment and low heat leak of most transport dewars such shipments can frequently be made without venting. However, provision must be made for venting inflammable gases in a safe manner.

In the case of the LNG tankers the boil-off gases are burned in the ships' engines. For highway transport of liquid hydrogen the driver vents the truck as required in isolated areas. For railroad cars carrying liquid hydrogen a fan provides an air supply sufficient to dilute the venting hydrogen-gas below the lower explosive limit.

G. DRIVER TRAINING

The person accompanying and responsible for the shipment of a cryogen should be aware of the nature and consequences of a mishap involving his load. He should receive training in emergency action required and be provided with specific instructions. These instructions should include fire and first aid measures, loading and travel requirements as necessary, and sources for additional aid.

H. LABELLING

Containers and pipelines should be labelled with the common name of the contents. Shipping containers that might meet with an emergency evoking response by public (and perhaps not fully aware protective forces) should also be labelled with the nature of the risk—inflammable, toxic, corrosive. Jargon should be avoided even within a given establishment; if colour codes or numerical or letter codes must be used they should be accompanied by appropriate legend. Code designations which may be misunderstood should be avoided.

I. SAFETY TRAINING

Every employee involved with cryogens must be conscious of misadventure potential and therefore safety-minded. This consciousness can and should be developed by regular and specific training including cryogenics as well as other safety subjects.

Circumstances of actual and near accidents should be made known; incidents may be found in the literature. Safety papers may be assembled in topical collections and made available to employees at all technical levels to enhance their knowledge of and interest in the safe use of cryogens.

1. *Operating Procedures*

The operation of cryogenic facilities should proceed within the framework of a set of safety rules or standard operating procedures written specifically for that facility. In general these would include items previously discussed such as pre-run and post-run purging, ignition control, pressure relief, etc. In addition each individual experiment should have its own safety considerations which include specific safety features required for that experiment. Use of written checklists for normal operation and maintenance of equipment is encouraged. There should also be contingency procedures which cover any anticipated emergency conditions. These procedures may be developed by the personnel involved but should be approved by competent higher authority. Deviations from agreed upon rules and regulations should be permitted only upon approval by this higher authority.

Procedures should be reviewed at intervals determined to be appropriate on the basis of frequency of use, change in programs or facilities, development of experience or new information, and personnel changes. It has been found useful for persons outside the chain of responsibility, such as safety or consulting authorities, to contribute to these reviews. Significant procedural or facility changes should be carefully reviewed for possibly adverse influences on safety.

2. *Emergency Plans*

An understanding of the principal hazards and procedures to be followed in an emergency by the personnel required to work in a hazardous area is generally accepted to be of major import from the standpoint of safety.

Elaborate emergency plans and emergency equipment that bear little resemblance to credible accidents should be avoided for fear that they could detract from the important elements of emergencies.

It has been common practice in the potentially high hazard technologies to go to considerable effort to acquaint employees with the consequences of misadventure. With a reasonable knowledge of such consequences the employee can be additionally motivated toward safety from the natural interest of self-preservation. Not only is a reasonable understanding of risk fundamental to the enhancement of self-motivation for safety but it relieves uneasiness about new risks and permits a more rational approach to emergency situations and to safety improvements.

Fire brigades or other personnel who might be called upon to respond to an emergency should also be specifically trained and made aware of the unique problems of the cryogenic facility. If they represent an independent authority they should be encouraged to make their own plans for emergency response, subject to technical counsel or approval. Significant changes in the facility or the cryogenic operations require a re-evaluation of emergency plans.

V. Codes and Regulations (In the U.S.A.)

The field of cryogenics is relatively new and therefore few codes have been written for or make mention of cryogenic fluids. However, existing codes do cover vessel construction, electrical equipment, some phases of facility design, and the transport of cryogens. Laws are required to make use of these codes mandatory.

As a rule cryogenic storage vessels are designed, built and tested according to the provision of the American Society of Mechanical Engineers' Boiler and Pressure Vessel Code, Section III (1965) or Section VIII (1968). (In general, Section III allows one to use a somewhat higher allowable stress than Section VIII but requires a more sophisticated stress analysis.) This code covers material selection, protection against excess pressure by relief devices, design and fabrication criteria, and gives test procedures to be carried out before and after fabrication. Materials must be suitable for the use temperature and pressure and must retain ductility at the use temperature. Chemical composition and mechanical properties must be determined for the actual materials of construction. A final pressure test according to Section VIII requires either a hydrostatic test at 1·5–2 times the use pressure or a pneumatic test at 1·25–2 times the use pressure. The pneumatic test is more commonly used for cryogenic equipment since water is difficult to eliminate, and H_2 and He vessels are not normally constructed to handle the weight of the higher density water. Although not required by the code, it is desirable to subject the vessel to a final cold test pressure test while filled with cryogen.

The National Electrical Code (National Fire Protection Association, 1968) covers the design of electrical equipment to be used where combustible gases can accumulate in flammable concentrations. Hydrogen and methane are covered by this code.

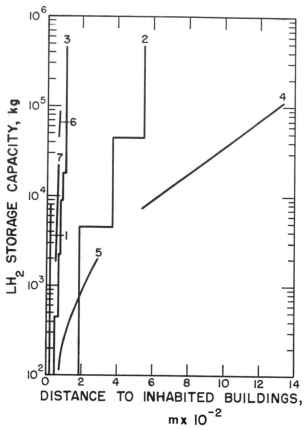

Fig. 11.7. Quantity–distance requirements for storage of LH_2 relative to inhabited buildings (Edeskuty, 1969). 1.—Compressed Gas Association (1966b); 2.—U.S. Department of Defense (1964); 3.—Zabetakis *et al.* (1961); 4.—U.S. Department of the Army (1964); 5.—Zabetakis, *et al.* (1961) (no water vapour); 6.—Zabetakis, *et al.* (1961) (2% absolute humidity); 7.—A. D. Little, Inc. (1960).

The Compressed Gas Association has developed a standard specifically written for quantities of liquid hydrogen from 150 to 110,000 l entitled "Standard for Liquefied H_2 Systems at Consumer Sites" (Compressed Gas Association, 1966b). Among its provisions are requirements for the use of Section VIII of the ASME Boiler and Pressure Vessel Code and the National Electrical Code, fire protection and operating instructions and storage distance requirements (see Fig. 11.7 and 11.8). It covers the design of systems, the location of storage vessels and the design of LH_2 vaporizers.

Recently standards have been developed to cover the use of LNG in the U.S.A. The American Petroleum Institute has developed API 2510-A, "Design and Construction of LNG Installations at Petroleum Terminals, Natural Gas Processing Plants, Refineries, and Other Industrial Plants" (American Petroleum Institute, 1968), while the National Fire Protection Association has developed standard NFPA 59A, "Liquefied Natural Gas at

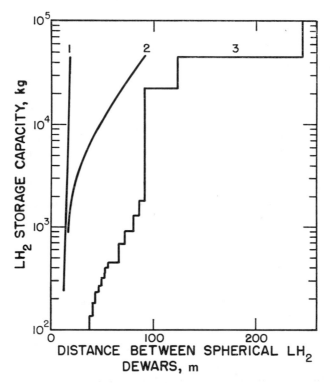

Fig. 11.8. Quantity–distance requirements for storage of LH_2 relative to inter-dewar spacing (Edeskuty, 1969). 1.—Baily et al. (1959); 2.—Zabetakis et al. (1961); 3.—U.S. Department of the Army (1964).

Utility Gas Plants" (National Fire Protection Association, 1969). Currently an attempt is being made to combine these two standards into one covering all LNG facilities (Dyer and Sommer, 1969). Both of these standards cover diking, tank spacing and the construction of LNG tanks with API 2510-A requiring compliance with API Standard 620 and NFPA 59A requiring compliance with Section VIII of the ASME Boiler and Pressure Vessel Code.

In addition to the above codes a number of regulations have been established by various organizations for their own use. One regulation of this type deals with the separation of storage dewars and the separation of storage dewars from inhabited buildings. There are several considerations important in the determination of safe quantity-distance relations. These include the likelihood and magnitude of a catastrophic dewar failure followed by an explosive reaction of its contents with air. The consequences of such a failure depend upon the total energy released and the rate of energy release. It is very unlikely at any given moment, that a large fraction of the stored hydrogen will be in a stoichiometric hydrogen–oxygen mixture. Thus only a fraction of the maximum total energy release would ever be realized.

The rate of energy release is dependent upon the type of combustion—deflagration against detonation. With proper precautions the possibility of a detonation can be minimized thus lowering the rate of energy release. The quantity-distance relationship presented in Figs. 11.7 and 11.8 exhibit a wide spread as a result of different interpretations concerning the consequences of a given failure. The spacing requirements between storage dewars (Fig. 11.8) seem somewhat artificial since there are no apparent safety reasons to limit the size of a single storage dewar. Statistics have shown that the number of accidents involving dewars used as storage vessels is considerably less than the number of accidents involving dewars used in experiments. This can be attributed to the fact that tanks being used in experiments frequently have to meet special requirements that increase the opportunity for operational error. Where tanks serve both as storage- and run-vessels, distance for exclusion of personnel may depend upon tank status.

Inter-state shipments of liquid hydrogen and liquid oxygen in the United States of America are controlled by federal government regulations. Similar regulations for intra-state shipment are possible but in general do not exist specifically for cryogens. Federal regulations (United States Department of Transportation, 1969) specify detailed construction requirements for railroad storage tank cars while highway shipments are controlled by Interstate Commerce Commission permit (McKinley, 1962). While there are no restrictions imposed upon the seller of liquid hydrogen, the producers have assumed the responsibility of delivering only to qualified users. Transportation of cryogenic fluids has been considered to come under the purview of the Bureau of Explosives Tariff No. 13 (McKinley, 1962). Purity requirements are imposed by the user and in the case of H_2 (for which a minimum of 95% parahydrogen is specified) the United States National Aeronautical and Space Administration has limited total impurities to 5 v.p.m. with oxidizing impurities being limited to 1 v.p.m. A typical analysis for liquid hydrogen currently produced under these specifications shows 97·6% parahydrogen, 1·16 v.p.m. total impurities and 0·57 v.p.m. oxidizing impurities.

VI. Conclusions

Where hazardous materials and expensive facilities are used the importance of a high standard of housekeeping and a thoughtful arrangement cannot be over-emphasized. Cleanliness and an uncluttered work place will improve the efficiency and safety of operations.

Combustibles, even in modest quantities, should be avoided in high-value or potentially dangerous areas to prevent small fires that may be compounded into major losses (Grumer *et al.*, 1967; United States Atomic Energy Commission, 1966). In one instance, the combustion of a few pounds of combustible foamed insulation jeopardized thousands of dollars-worth of expensive cryogenic valves and controls.

Optimum conditions for safe operation occur when safety has been considered in the design, construction and checkout as well as in the operation of a cryogenic system. Occasionally deviations from the normal safety procedures are either desirable or unavoidable. Such deviations are permissible only if the decision is made by the proper authorities. Such steps can only be taken when all concerned are fully aware of the consequences and understand the alternate procedures to be taken. Also, it should be remembered that the maintenance of safety is a continuing concern. Independent periodic safety reviews and continual safety training of personnel are desirable. Because of the rapid expansion of cryogenic technology, effort must be expended on safety technology to keep pace. Where knowledge is inadequate, either one must take risks or build in excess safeguards. To do the former is usually disastrous, while to do the latter is always expensive, and may involve a hidden danger because it reduces respect for the safety precautions which are necessary.

VII. Literature Sources

The following is a list of representative literature which contains further information pertaining to the safe handling of cryogenic fluids and equipment.

"Advances in Cryogenic Engineering", (Ed. K. D. Timmerhaus) Vol. 1–16, Plenum Press, New York. (Issued yearly.)

"Safety with Cryogenic Fluids" (1967). Zabetakis, N. G., Plenum Press, New York.

"Technology and Uses of Liquid Hydrogen" (1964). (R. B. Scott, W. H. Denton and C. M. Nicholls, eds.) Pergamon Press, New York.

"Applied Cryogenic Engineering" (1962). (eds. R. W. Vance and W. M. Duke.) J. Wiley and Sons, New York.

"Cryogenic Technology" (1963). (R. W. Vance, ed.) J. Wiley and Sons, New York.

"Technology of Liquid Helium" (1968). (R. H. Kropschot, B. W.

Birmingham and D. B. Mann, eds.) National Bureau of Standards Monograph #111, U.S. Govt. Printing Office.

"Cryogenic Safety Manual" (1970). British Cryogenics Council, Institution of Chemical Engineers, London.

"Progress in Cryogenics" (K. Mendelssohn, ed.) Vols. 1–4. Heywood and Co. Ltd., London.

Several periodicals including *Cryogenics, Cryogenics and Industrial Gases* (formerly *Cryogenic Engineering News*), *Bulletin of the International Institute of Refrigeration, Cryogenic Information Report* and *Cryogenic Technology*.

Various safety manuals issued by governmental and industrial organizations.

VIII. References

A. D. Little, Inc. (1959). Report No. C-61092, Cambridge, Mass.

A. D. Little, Inc. (1960). Report No. C-61092, Cambridge, Mass.

A. D. Little, Inc. (1961). Report No. C-61092, Cambridge, Mass.

American Petroleum Institute. (1968). API No. 2510-A. New York.

American Society of Mechanical Engineers (1965). ASME Boiler and Pressure Vessel Code, Section III. New York.

American Society of Mechanical Engineers (1968). ASME Boiler and Pressure Vessel Code, Section VIII. New York.

American Society for Testing of Materials (1959). ASTM-D-1692-59T. Philadelphia.

Baker, O. (1954). *Oil and Gas J.* **53,** 185.

Ball, W. L. (1961). *In* "Proceedings of the A.I.Ch.E. Air and NH_3 Plant Symposium". American Institute of Chemical Engineers, New York.

Bailey, B. M., Benedict, D. C., Byrnes, R. W., Campbell, C. R., Fowle, A. A., Moore, Jr., R. W., Pertalozzi, W. G., Richter, E. G., and Schulte, C. A. (1959), WADC-TR-59-386, Air Development Center, Wright–Patterson Air Force Base, Ohio.

Boyer, K., Otway, H. and Parker, R. C. (1965). *In* "Advances in Cryogenic Engineering" (K. D. Timmerhaus, ed.), Vol. 10, p. 273. Plenum Press, New York.

British Institution of Chemical Engineers (1970). "Cryogenic Safety Manual." Safety Panel of British Cryogenics Council.

Bronson, J. C., Edeskuty, F. J., Fretwell, J. H., Hammel, E. F., Keller, W. E., Meier, K. L., Schuch, A. F. and Willis, W. L. (1962). *In* "Advances in Cryogenic Engineering" (K. D. Timmerhaus, ed.), Vol. 7, p. 198. Plenum Press, New York.

Carter, H. G. (1969). Private communication.

Cassutt, L. II., Maddocks, F. E. and Sawyer, W. A. (1960). *In* "Advances in Cryogenic Engineering" (K. D. Timmerhaus, ed.), Vol. 5, p. 55. Plenum Press, New York.

Chelton, D. B. (1964). *In* "Technology and Uses of Liquid Hydrogen" (R. B. Scott, W. H. Denton and C. M. Nicholls, eds.). Pergamon Press, Oxford.

Chen, C. W. and Struss, R. G. (1969). *Cryogenics* **9,** 131.

Compressed Gas Association (1966a). *In* "Oxygen Deficient Atmospheres". Safety Bulletin SB-2. New York.

Compressed Gas Association (1966b). *In* "Standard for Liquefied Hydrogen Systems at Consumer Sites". Pamphlet G-5.2. New York.

Cooper, W. F. (1953). *Brit. J. Appl. Phys.* Supplement No. 2, S-73.

Cotterill, P. (1961). *In* "Progress in Materials Science", Vol. 9, No. 4, p. 231. Pergamon Press, Oxford.

Crocker, S. and King, R. C. (1967). "Piping Handbook." McGraw–Hill, New York.

Culbertson, W. L. (1970). *In* "Advances in Cryogenic Engineering' (Ed. K. D. Timmerhaus), Vol. 15. Plenum Press, New York.

Dittus, F. W. and Boelter, L. M. K. (1930). Publications in Engineering **2**, No. 13, 443. University of California (Berkeley).

Dokoupil, Z., Van Saest, G. and Swenker, M. D. P. (1966). *Appl. Sci. Res.* **A5**, 182.

Dyer, A. F. and Sommer, E. C. (1969). *In* "Proceedings of the Conference on Liquefied Natural Gas." Bulletin of the International Institute of Refrigeration.

Edeskuty, F. J. (1969). *In* "Proceedings of the XIIth International Congress of Refrigeration", Vol. I, p. 283. Graficas Reunidas, S.A., Hermiosilla, Madrid.

Edeskuty, F. J. and Williamson, K. D. (1971). *In* "Cryogenic Engineering Handbook". J. Wiley and Sons, New York. To be published.

Elston, J. and Laffitte, P. (1947). *Compte Rendue* **225**, 1313.

Farrar, Jr., R. L. (1969). Report K-L-6224, Union Carbide Corporation, Nuclear Division, Oak Ridge.

Flieder, W. G., Smith, W. J. and Wetmore, K. R. (1960). *In* "Advances in Cryogenic Engineering" (K. D. Timmerhaus, ed.), Vol. 5, p. 111. Plenum Press, New York.

Forrest, J. S. (1953). *Brit. J. Appl. Phys.* Supplement No. 2, S-38.

Fuller, P. D. and McLagan, J. N. (1962). *In* "Applied Cryogenic Engineering" (R. W. Vance and W. M. Duke, eds.), ch. 9. J. Wiley and Sons, New York.

Grumer, J., Strasser, A. and Van Meter, R. A. (1967). Principles for Safe Handling of Liquid Hydrogen. U.S. Department of the Interior, Bureau of Mines, Explosives Research Center, Pittsburgh.

Hickes, W. F. (1967). *Chem. Eng.* **76**, 139.

Hiza, M. J. and Herring, R. N. (1965). *In* "International Advances in Cryogenic Engineering" (K. D. Timmerhaus, ed.), Vol. 10, p. 182. Plenum Press, New York.

Johnstone, R. T. and Miller, S. E. (1960). *In* "Occupational Diseases and Industrial Medicine", p. 106. W. B. Saunders Company, Philadelphia.

Kerr, E. C. (1962). "Liquid Hydrogen, A Guide for the Safe Handling and Storage of Liquid Hydrogen at LASL Facilities." Los Alamos Scientific Laboratory, Los Alamos, N.M.

Key, C. F. and Gayle, J. B. (1964). NASA-TM-X-53144. Marshall Space Flight Center, Huntsville.

Key, C. F. and Gayle, J. B. (1965). NASA-TM-X-53208. Marshall Space Flight Center, Huntsville.

Leonhardt, E. H. (1970). *In* "Advances in Cryogenic Engineering, Vol. 15. Plenum Press, New York. To be published.

Lewis, B. and von Elbe, G. (1961). *In* "Combustion Flames and Explosions of Gases". Academic Press, New York.

McKinley, C. (1962). *In* "Applied Cryogenic Engineering" (R. W. Vance and W. M. Duke, eds.). J. Wiley and Sons, New York.

McKinney, A. H. (1967). *In* "Electrical Safety Abstracts", Third Edition. Instrument Society of America.

Meryman, H. T. (1957). *Physiol. Rev.* **37**, 233.

Mills, R. L. and Edeskuty, F. J. (1956). *Chem. Eng. Prog.* **52**, 477.

Mills, Jr., W. J. (1962). U.S. Naval Research Reviews.

National Fire Protection Association (1967a). *In* "Intrinsically Safe Process Control Equipment for Use in Hazardous Locations", NFPA 493-T, Boston.

National Fire Protection Association (1967b). *In* "Standard for Purged Enclosures for Electrical Equipment in Hazardous Locations", NFPA 496, Boston.

National Fire Protection Association (1967c). *In* "Guide for Explosion Venting", NFPA 68-1954, Boston.

National Fire Protection Association (1968). National Fire Code, Vol. 5.

National Fire Protection Association (1969). *In* "Liquefied Natural Gas at Utility Gas Plants", NFPA 59A, Boston.

Neary, R. M. (1963). *In* "Transactions National Safety Congress", vol. 5. (Chemical and Fertilizer Industries.) Chicago.

Neary, R. M. (1969). March Safety News-letter. Chemical Section, National Safety Council, Chicago.

Novak, J. K. (1966). Report LA-DC-7158. Los Alamos Scientific Laboratory.

Novak, J. K. (1970) *In* "Advances in Cryogenic Engineering" K. D. Timmerhaus, ed.), Vol. 15. Plenum Press, New York.

Redding, R. J. (1969). *Chem. Eng.* **76**, 137.

Reider, R., Otway, H. J. and Knight, H. T. (1965). *Pyrodynamics* **2**, 249.

Reider, R. (1968). *In* "Proceedings of the Second International Cryogenic Engineering Conference", p. 154. Iliffe Science and Technology Publications, Ltd., Guildford.

Rosenberg, J. C. (1969). *Cryogenic Eng. News* **4**, No. 3, 21.

Roth, E. M. (1964). NASA SP-47, NASA, Washington.

Scott, L. E. (1964). *In* "Advances in Cryogenic Engineering", (K. D. Timmerhaus, ed.), Vol. 9, p. 379. Plenum Press, New York.

Stewart, R. B. and Johnson, V. J. (1961). N.B.S. WADD Technical Report 60–56, Part IV, Boulder.

Streng, A. G. (1960). *Explosivstoffe* **8**, 225.

United States Atomic Energy Commission (1966). TID-22594, Cambridge, Mass.

United States Bureau of Mines (1963). Circular 8137, Pittsburg.

U.S. Department of the Army (1964). U.S. Army Material Command Safety Manual, No. 385-224 (formerly Ordnance Safety Manual ORD M7-224).

U.S. Department of Defense (1964). Instruction No. 4145.21.

United States Department of Transportation (1969). Hazardous Materials Regulations Tariff No. 23, Washington.

U.S. National Aeronautics and Space Administration (1968). Technical Memorandum X-52-454, Washington.

U.S. National Bureau of Standards (1960). Memorandum Report No. CMF-4, Boulder.

Vance, R. W. and Reynales, C. H. (1962). *In* "Applied Cryogenic Engineering" (R. W. Vance and W. M. Duke, eds). J. Wiley and Sons, New York.

Walter, R. J. and Chandler, W. T. (1969). Report R-7780-1. Rocketdyne Division of North American Rockwell Corporation, Canogn Park.

Washburn, B. (1962). *New England J. Med.* **266**, 974.

Weismantel, G. E. (1969). *Chem. Eng.* **76**, 132.

Willis, W. L. (1966). *Cryogenics* **6**, 279.

Zabetakis, M. G. and Burgess, D. S. (1960). Bureau of Mines. WADD-TR-60-141, Pittsburg.

Zabetakis, M. G., Furno, A. L. and Martindill, G. H. (1961). *In* "Advances in Cryogenic Engineering" (K. D. Timmerhaus, ed.), Vol. 6, p. 185. Plenum Press, New York.

Zabetakis, M. G. (1967). *In* "Safety with Cryogenic Fluids." Plenum Press, New York.

Chapter 12

Thermophysical Data for Cryogenic Materials

R. M. GIBBONS

The Gas Council, London Research Station,
London, England

I. Introduction

The purpose of this chapter is to provide a convenient and usable source of cryogenic data for the properties of solids and single component cryogenic fluids. As far as possible the values presented are based on experimental data but in some cases it has been necessary to use predictive methods to supplement the experimental data. Naturally these predicted values are less reliable and this is indicated where these methods are used. Within the limited space available it is not possible to present data with great accuracy but the sources selected are thought to be the best available and should be referred to where greater accuracy is required.

The chapter is divided into sections on the properties of solids and fluids followed by a short discussion of some modern methods of prediction of the properties pure fluids and their mixtures. Section II provides graphs of the thermal and electrical conductivities, the heat capacities and coefficients of thermal expansion for the common metals, alloys and other substances used in cryogenic hardware. Section III contains small $T–S$ or $P–H$ diagrams for each of twelve cryogenic liquids; diagrams for the vapour pressure and saturated liquid densities of these fluids are also included in this section. Section IV provides graphs of the thermal conductivity and viscosity of the same fluids as in Section III. The general accuracy of the experimental data for transport properties is much lower than for thermodynamic properties. Also it has been necessary to use equations and the principle of corresponding states extensively to supplement the experimental data for viscosity and thermal conductivity.

Section V provides a short introduction to some methods for the prediction of the thermodynamic properties of cryogenic fluids. The discussion is confined to the Benedict–Webb–Rubin equation and the principle of corresponding states and its application to the calculation of PVT data using generalized correlations for the second and third virial coefficients. The extension of these methods to the calculation of the properties of mixtures, including the calculation of vapour–liquid equilibria, is also briefly discussed and a range of references is given.

It has not been possible to use the same set of units for all the diagrams but preference has been given to c.g.s. units. For the convenience of the reader a table of conversion factors has been included at the end of this chapter (page 736).

II. The Properties of Cryogenic Solids

The data in this section are all taken from V. Johnson et al. (1960) except for specific heat data which are taken from Corruccini (1957). There is an excellent review of the thermal properties of solids at low temperatures by Dillard and Timmerhaus (1968) which lists approximately 300 references for a wide range of substances. The theory and method of predicting the properties of solids are also discussed by these authors.

A. THERMAL EXPANSION OF SOLIDS

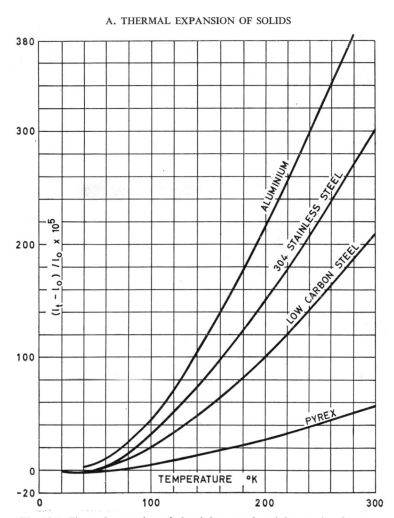

Fig. 12.1. Thermal expansion of aluminium, steel, stainless steel and pyrex.

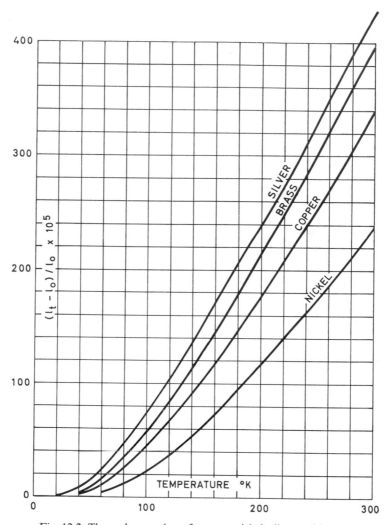

Fig. 12.2. Thermal expansion of copper, nickel, silver and brass.

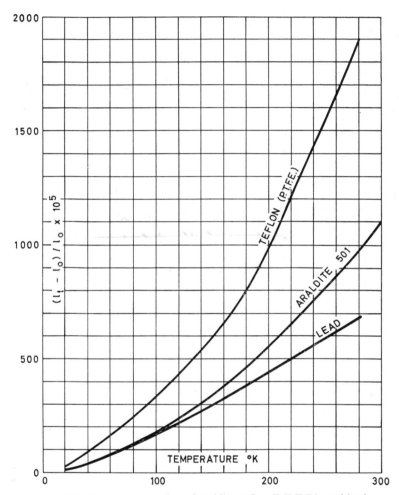

Fig. 12.3. Thermal expansion of araldite, teflon (P.T.F.E.), and lead.

B. THERMAL CONDUCTIVITY OF SOLIDS

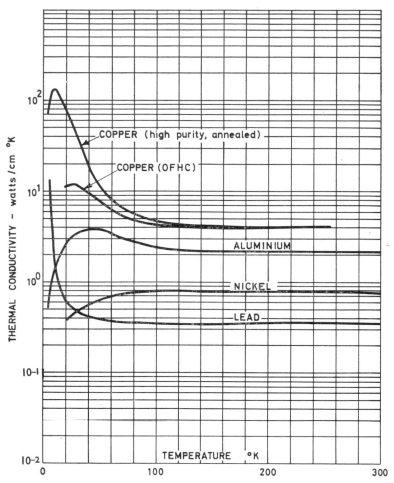

Fig. 12.4. Thermal conductivity of copper, aluminium, nickel and lead.

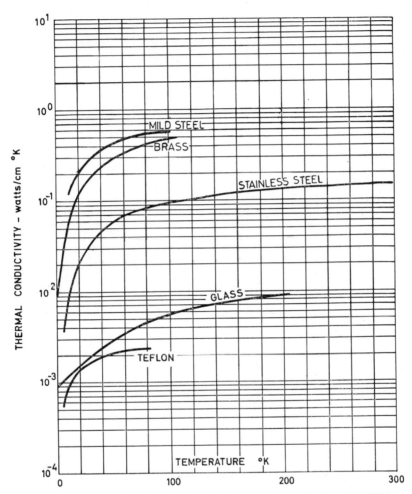

Fig. 12.5. Thermal conductivity of steels, brass, glass and teflon (P.T.F.E.).

C. HEAT CAPACITY OF SOLIDS

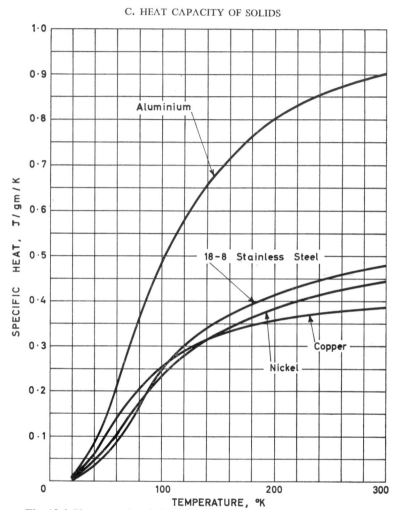

Fig. 12.6. Heat capacity of aluminium, copper, nickel and stainless steel.

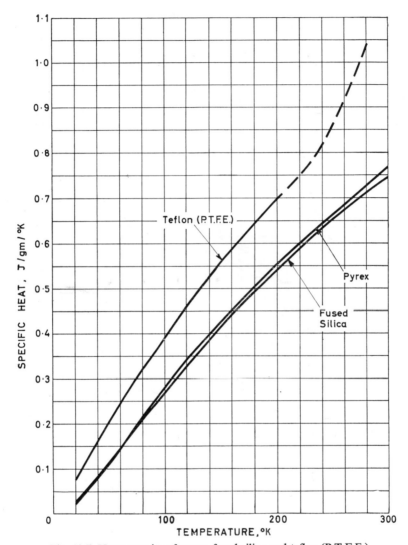

Fig. 12.7. Heat capacity of pyrex, fused silica and teflon (P.T.F.E.)

D. ELECTRICAL RESISTIVITY

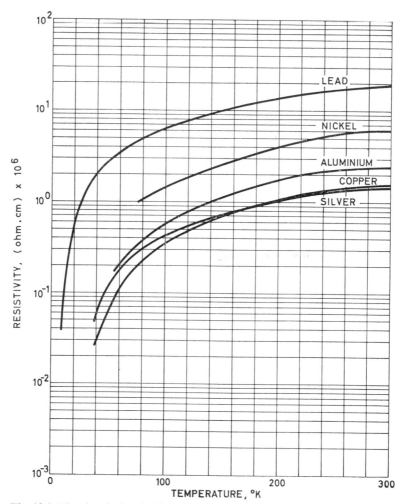

Fig. 12.8. The electrical resistivity of copper, silver, aluminium, lead and nickel.

III. The Thermodynamic Properties of Cryogenic Fluids

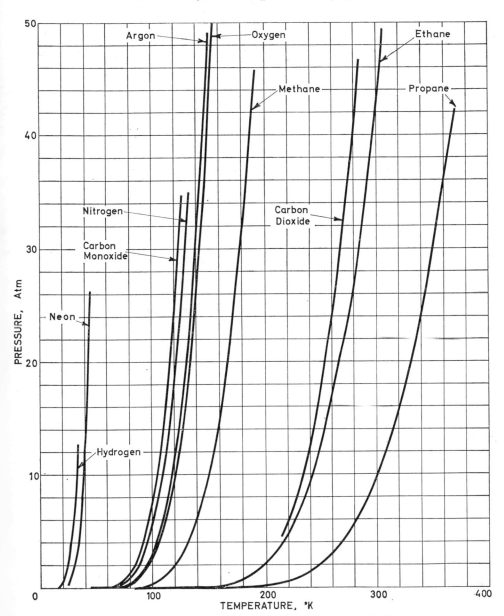

Fig. 12.9. The vapour pressure for several cryogenic fluids.

26*

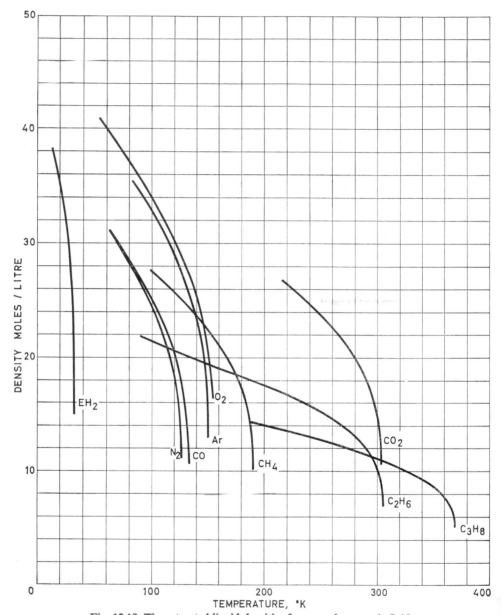

Fig. 12.10. The saturated liquid densities for several cryogenic fluids.

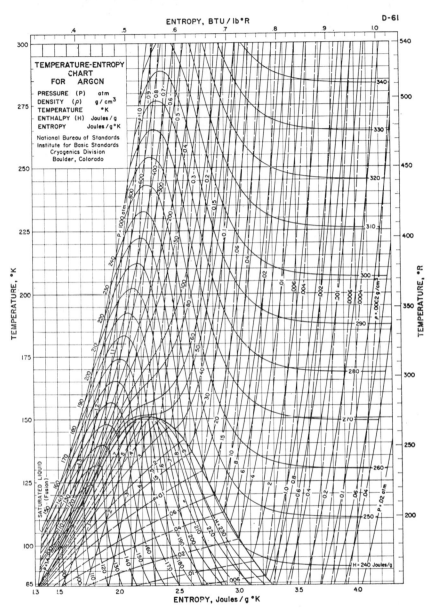

Fig. 12.11. *T–S* Chart for argon.
(Reproduced with permission from Gosman *et al.*, 1969),

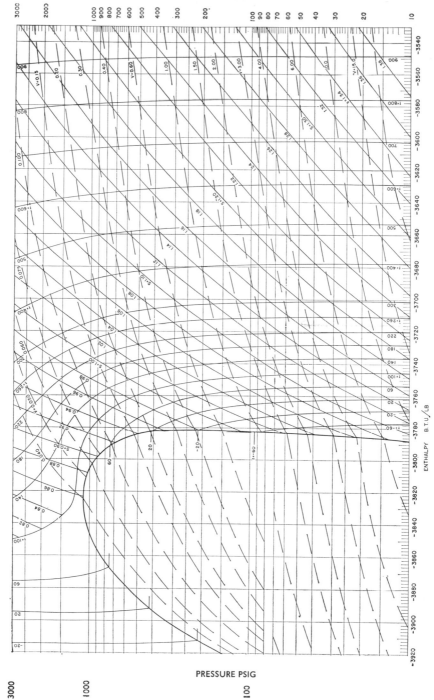

Fig. 12.12. *P–H* chart for carbon dioxide. (Reproduced with permission from Canjar *et al.*, 1966).

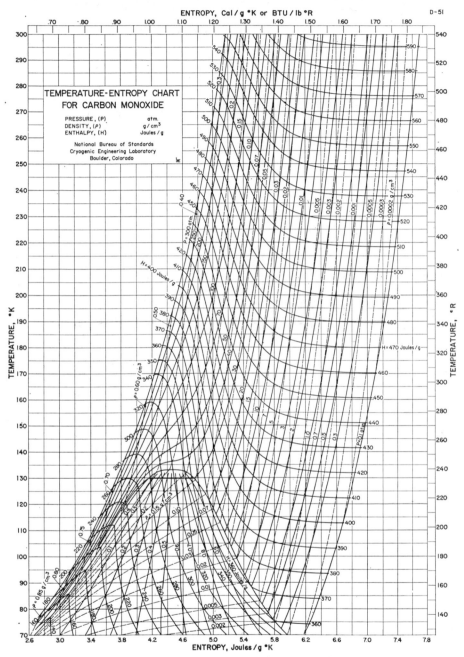

Fig. 12.13. *T–S* chart for carbon monoxide.
(Reproduced with permission from Hust and Stewart (1963)).

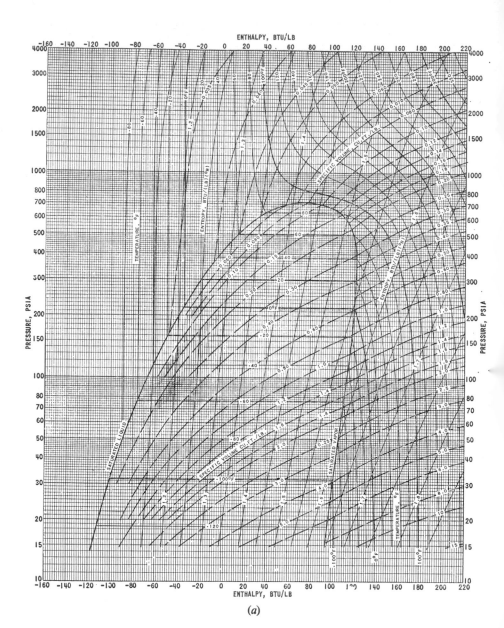

(a)

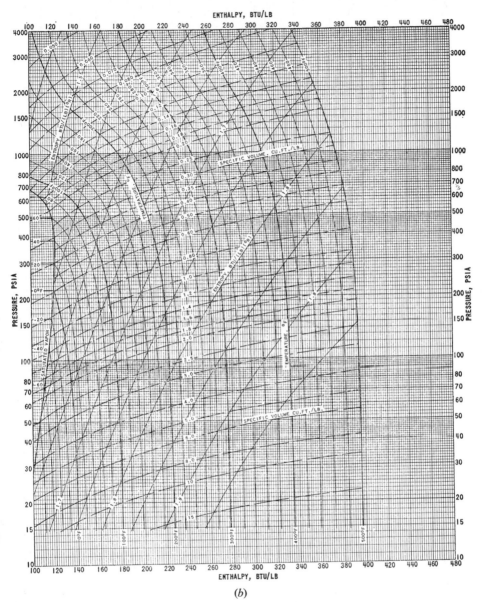

(b)

Fig. 12.14. *P–H* chart for ethane.
(Reproduced with permission from Edmister, W. C. (1961). "Applied Hydrocarbon
Thermodynamics", Gulf Publishing Co., Houston.)

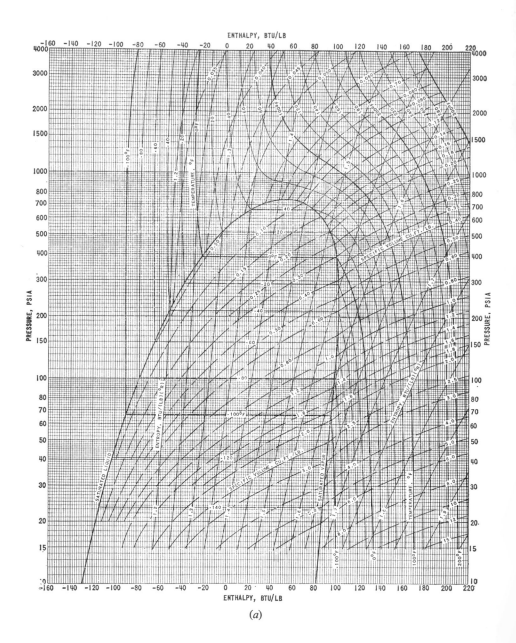

(a)

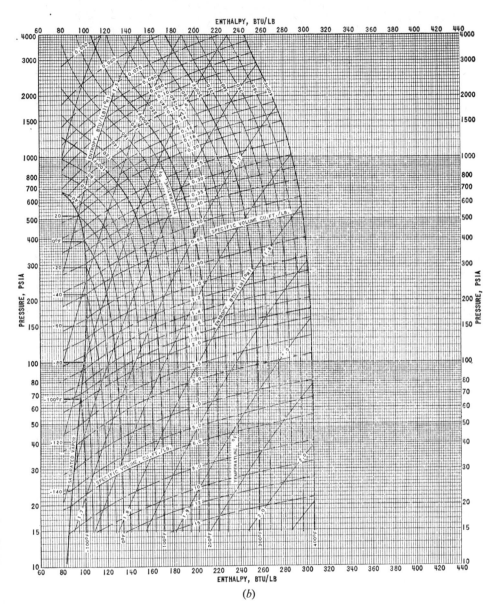

(b)

Fig. 12.15. *P–H* chart for ethylene.
(Reproduced with permission from Edmister, W. C. (1961). "Applied Hydrocarbon Thermodynamics", Gulf Publishing Co., Houston.)

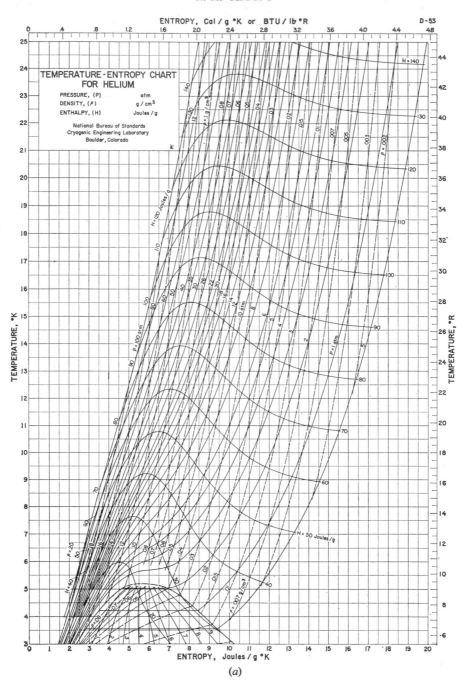

(a)

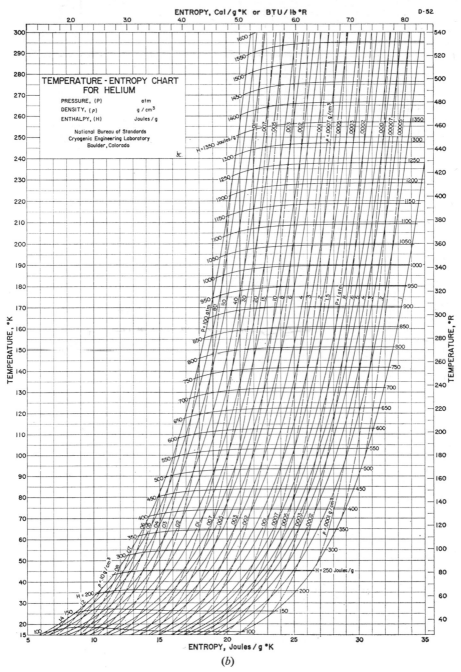

Fig. 12.16. *T–S* chart for helium-4.
(Reproduced with permission from Mann, 1962.)

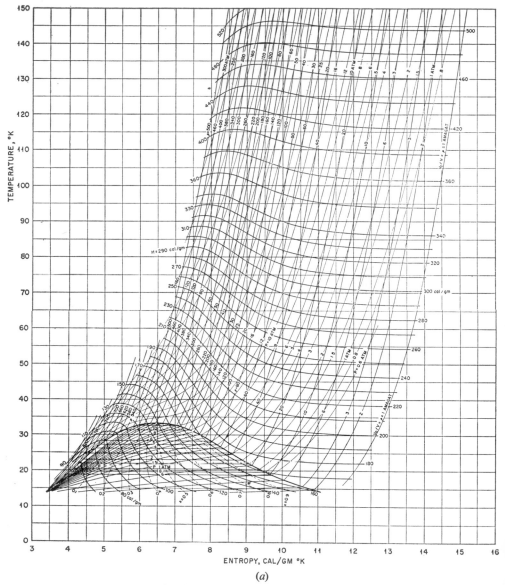

(a)

Fig. 12.17. *T–S* chart for n-hydrogen.
(Reproduced with permission from Wooley *et al.*, 1948.)

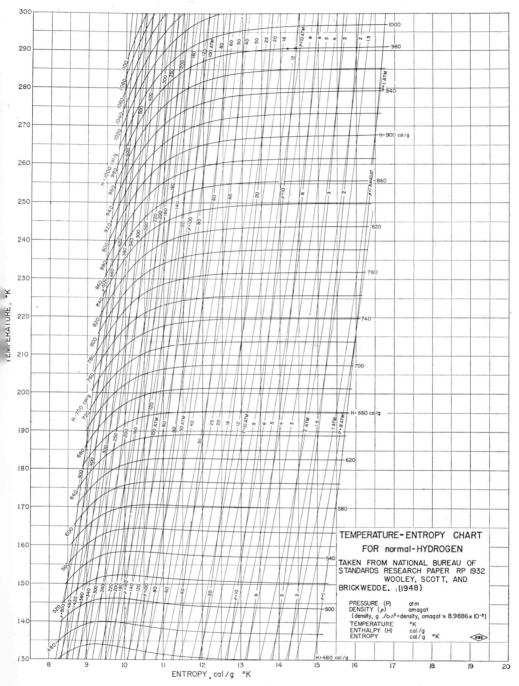

TEMPERATURE-ENTROPY CHART
FOR normal-HYDROGEN

TAKEN FROM NATIONAL BUREAU OF
STANDARDS RESEARCH PAPER RP 1932
WOOLEY, SCOTT, AND
BRICKWEDDE. (1948)

PRESSURE (P)	atm
DENSITY (ρ)	amagat
(density, g /cm³ =density, amagat × 8.9886 × 10⁻⁵)	
TEMPERATURE	°K
ENTHALPY (H)	cal/g
ENTROPY	cal/g °K

(b)

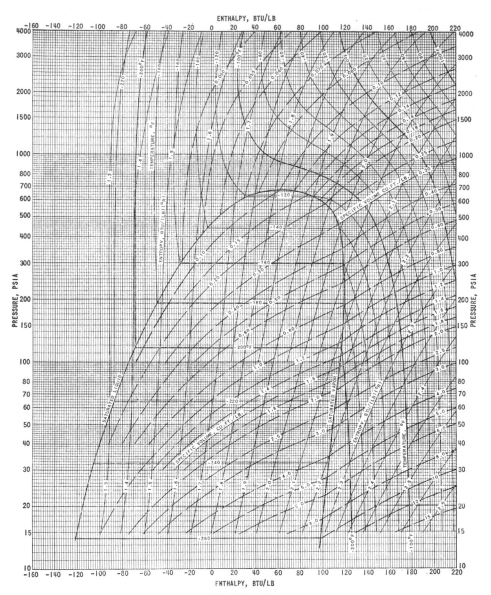

Fig. 12.18 (a).

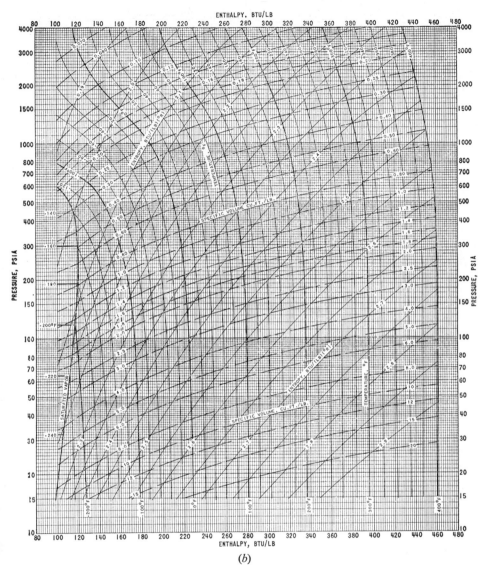

(b)

Fig. 12.18. *P–H* chart for methane.
(Reproduced with permission from Edmister, W. C. (1961). "Applied Hydrocarbon Thermodynamics", Gulf Publishing Co., Houston.)

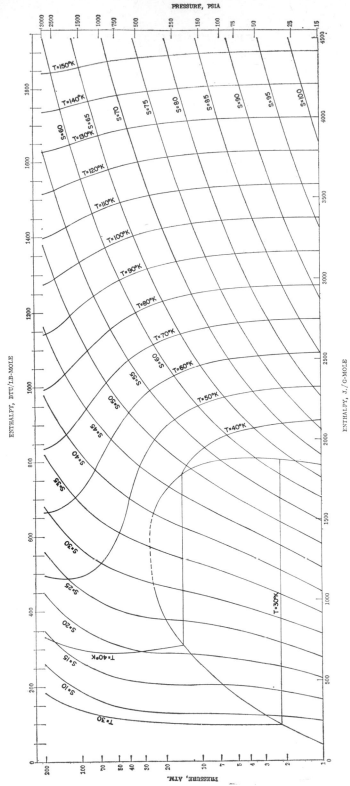

PRESSURE, PSIA

ENTHALPY, BTU/LB-MOLE

ENTHALPY, J./G-MOLE

PRESSURE, ATM.

Fig. 12.19 (a).

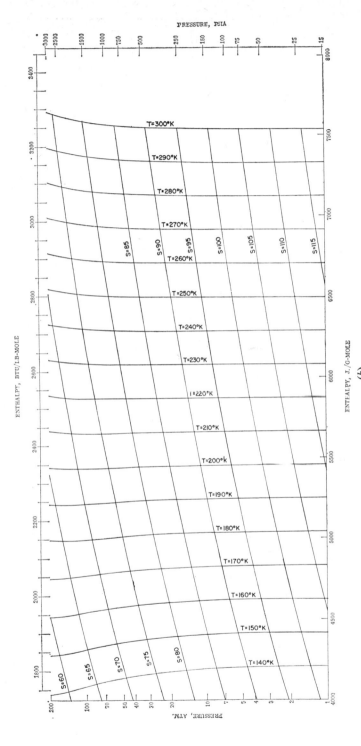

Fig. 12.19. *P–H* chart for neon.

(Reproduced with permission of the Air Force Materials Research Laboratory, Wright Patterson Air Force Base, U.S.A.; Gibbons and Keubler, 1968.)

(b)

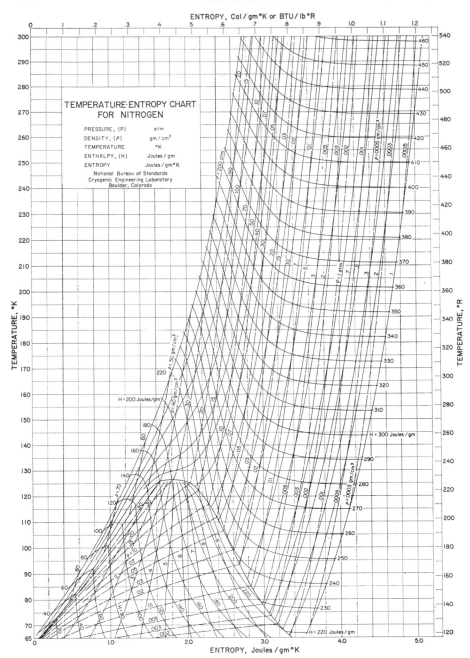

Fig. 12.20. *T–S* chart for nitrogen.
(Reproduced with permission from Strobridge, 1962.)

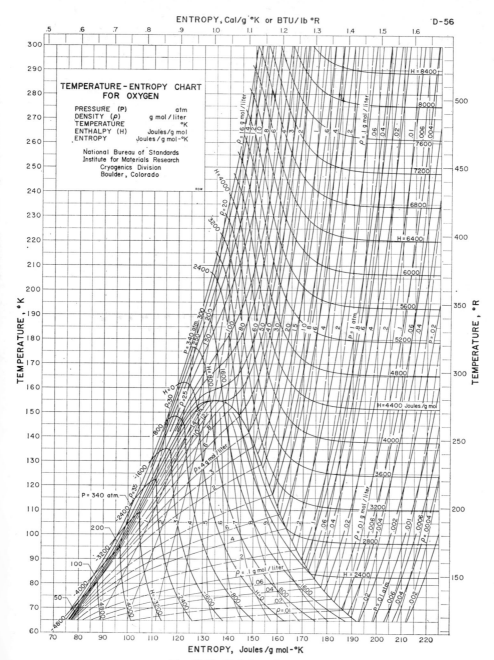

Fig. 12.21. *T–S* chart for oxygen.
(Reproduced with permission from Stewart, 1966).

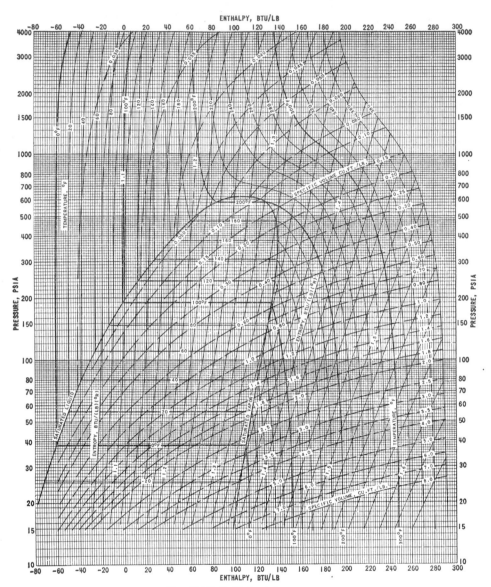

Fig. 12.22. *P–H* chart for propane.
(Reproduced with permission from Edmister, W. C. (1961). "Applied Hydrocarbon
Thermodynamics", Gulf Publishing Co., Houston.)

IV. The Transport Properties of Cryogenic Fluids

The data available for the viscosity and thermal conductivity are much sparser than is the case for the thermodynamic properties of cryogenic fluids. Furthermore, there is substantial scatter in the experimental data for viscosity and thermal conductivity of a single substance reported by different authors for the same sets of conditions. The experimental method used, particularly for viscosity measurements, appears to be important in this respect. As far as possible experimental data were used in preparing graphs of the viscosity and the thermal conductivity. In many cases, however, it has been found necessary to supplement the experimental data by use of correlations based on data and the principle of corresponding states. Values for the thermal conductivity of argon, helium, nitrogen (Gibbons and Keubler, 1968), n-hydrogen (Rogers et al., 1962) and methane (Manni and Vennart, 1969), the viscosity of argon, helium (Gibbons and Keubler, 1968) and n-hydrogen (Rogers et al., 1962) were obtained from equations based on the experimental data for each substance. For the remaining substances generalized correlations developed for inert gases by Gibbons and the more general methods of Stiel and Thodos (1964) and of Jossi et al. (1962) were used.

Both these types of correlations calculate the excess viscosity or thermal conductivity. The excess quantities, defined as the property at a given temperature minus the value of that property at a low pressure and the same temperature, vary only slowly, if at all, with temperature. The correlations of Gibbons are based on the equations which describe the data for argon and the principle of corresponding states. The reduction parameters chosen were based on the Lennard Jones potential but were adjusted to obtain the best agreement between the calculated and experimental data for each inert gas. Further details are given by Gibbons and Keubler, (1968).

The correlation of Stiel and Thodos (1964) for the thermal conductivity of non-polar gases is based on the observation that the excess thermal conductivity, $k - k_0$, is independent of the temperature but depends on the density and dimensional constants for each substance, which are taken to be the molecular weight and the critical constants. A dimensional analysis of the excess thermal conductivity leads to the following expression.

$$\lambda(k - k_0) = f(d_R)/(Z_c^5) \tag{12.1}$$

where k_0 is the thermal conductivity at 1 atm at temperature T, Z_c is the critical compressibility for the gas and $\lambda = T_c^{\frac{1}{6}} M^{\frac{1}{2}} P_c^{\frac{2}{3}}$ .The function, $f(d_R)$ was obtained from analysis of the experimental values of the quantity $(k - k_0)Z_c^5$ which should be the same for all substances. Stiel and Thodos (1964) were unable to obtain a single analytical expression for this function, $f(d_R)$, which was applicable over the whole range of density. However, they did provide a tabular set of values and three expressions applicable over restricted ranges of reduced density.

The correlation of Jossi *et al.* (1964) for the viscosity of non-polar gases is based on the independence of the excess viscosity, $\Delta\mu_0$, with respect to temperature. In fact the excess viscosity is somewhat dependent on temperature at higher densities but the predicted values from this correlation appear to agree reasonably well with the experimental data. The dimensional analysis of the excess viscosity leads to the expression,

$$\xi(\mu - \mu_0) = f(d_R) \tag{12.2}$$

where $\xi = M^{-\frac{1}{2}}p^{-\frac{2}{3}}T_c^{+\frac{1}{6}}$, μ_0 is the viscosity at 1 atm and temperature T, and $f(d_R)$ is a function of the reduced density which has to be obtained from analysis of the experimental data. Jossi *et al.* (1962) were unable to obtain an analytical representation of this function over the entire range of density but did provide a tabulation of values of the quantity $f(d_R)$ as a function of density.

A. THE THERMAL CONDUCTIVITY OF CRYOGENIC FLUIDS

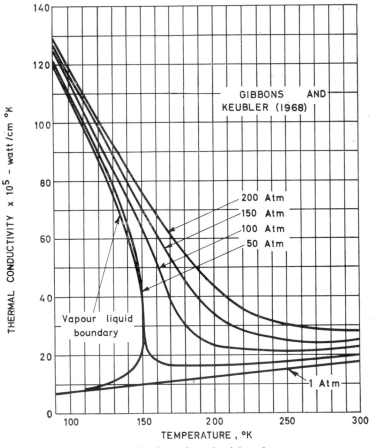

Fig. 12.23. The thermal conductivity of argon.

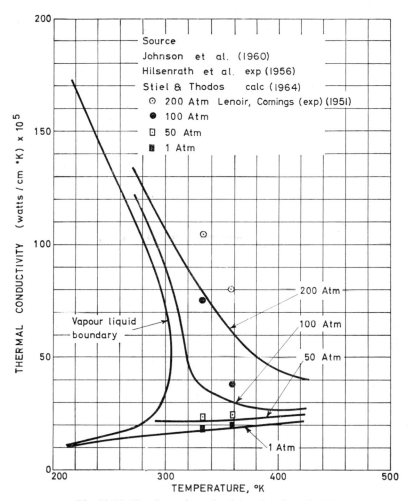

Fig. 12.24. The thermal conductivity of carbon dioxide.

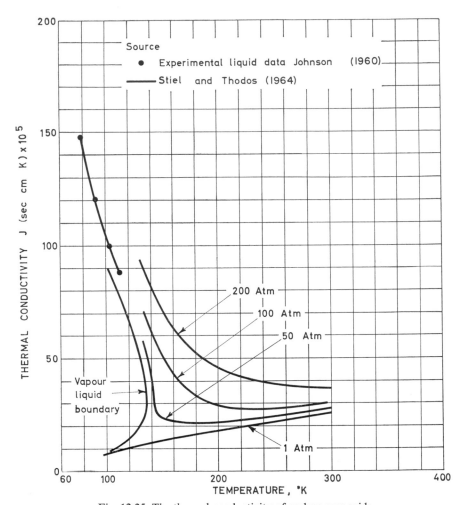

Fig. 12.25. The thermal conductivity of carbon monoxide.

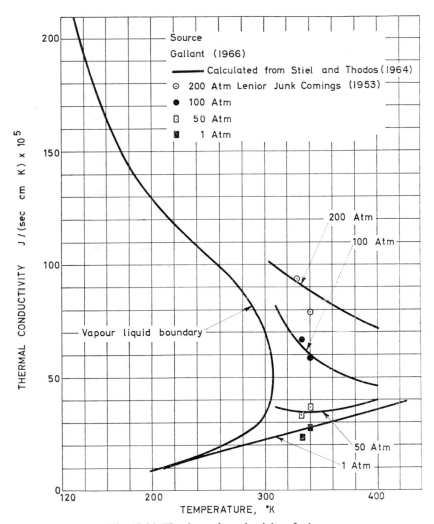

Fig. 12.26. The thermal conductivity of ethane.

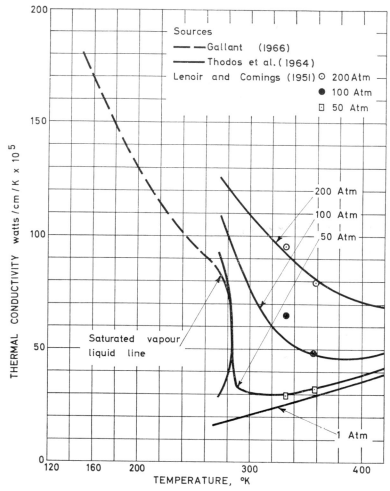

Fig. 12.27. The thermal conductivity of ethylene.

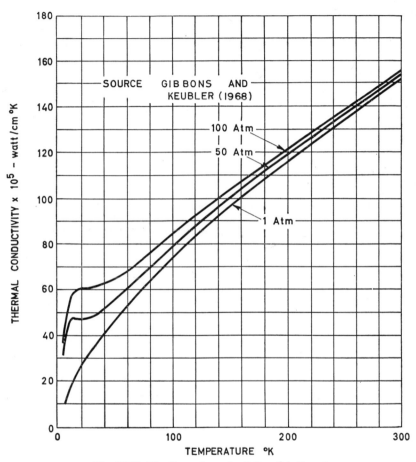

Fig. 12.28. The thermal conductivity of helium-4.

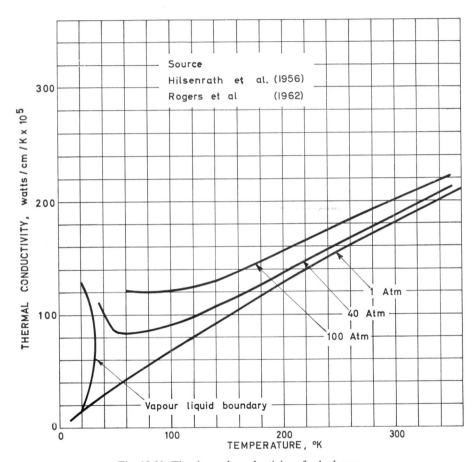

Fig. 12.29. The thermal conductivity of n-hydrogen.

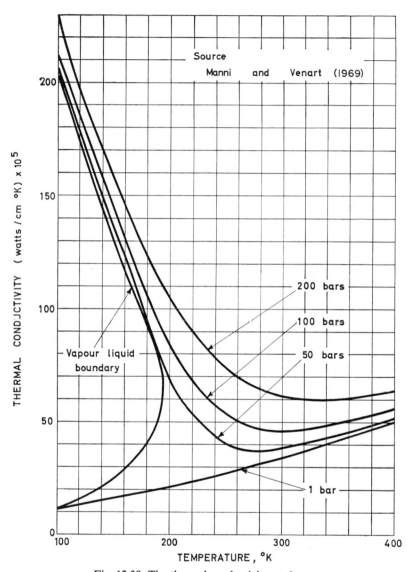

Fig. 12.30. The thermal conductivity methane.

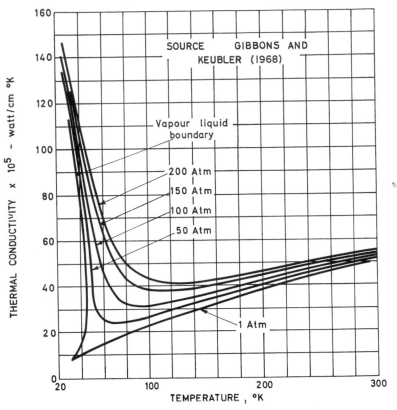

Fig. 12.31. The thermal conductivity of neon.

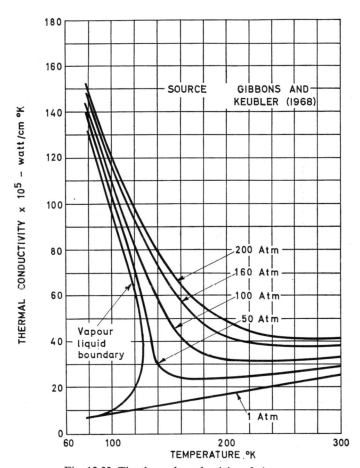

Fig. 12.32. The thermal conductivity of nitrogen.

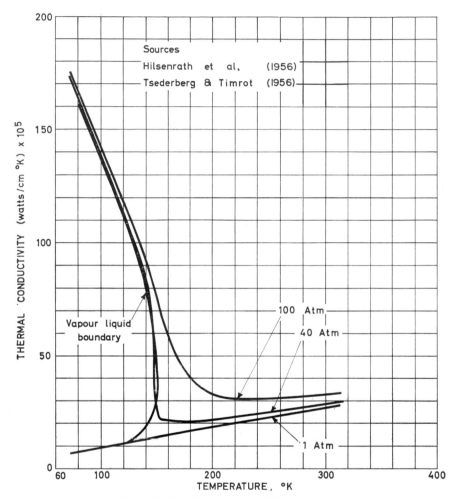

Fig. 12.33. The thermal conductivity of oxygen.

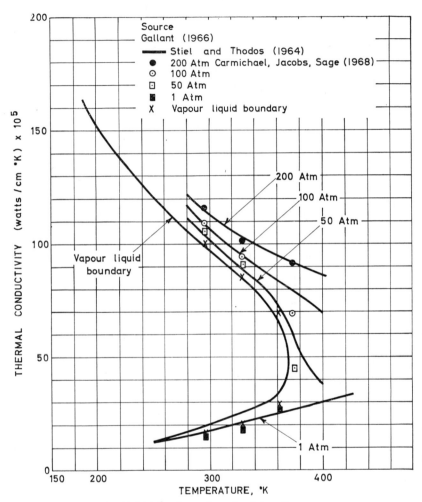

Fig. 12.34. The thermal conductivity of propane.

B. THE VISCOSITY OF CRYOGENIC FLUIDS

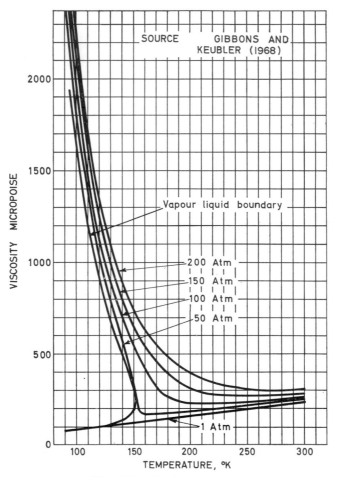

Fig. 12.35. The viscosity of argon.

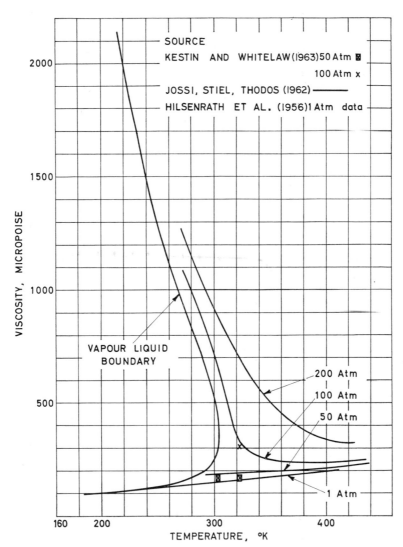

Fig. 12.36. The viscosity of carbon dioxide.

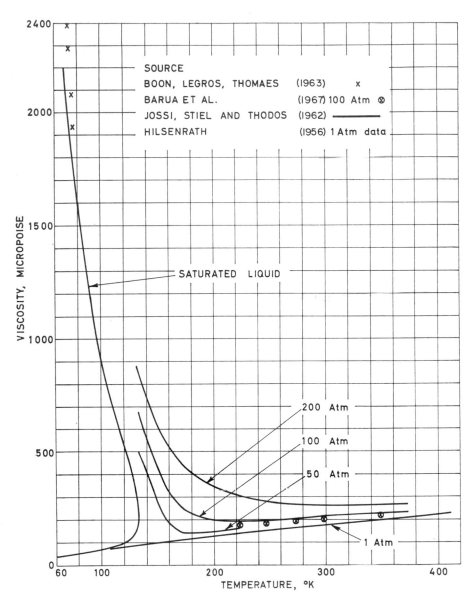

Fig. 12.37. The viscosity of carbon monoxide.

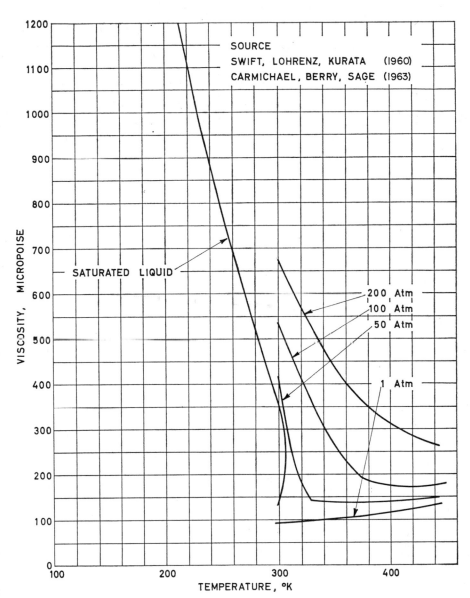

Fig. 12.38. The viscosity of ethane.

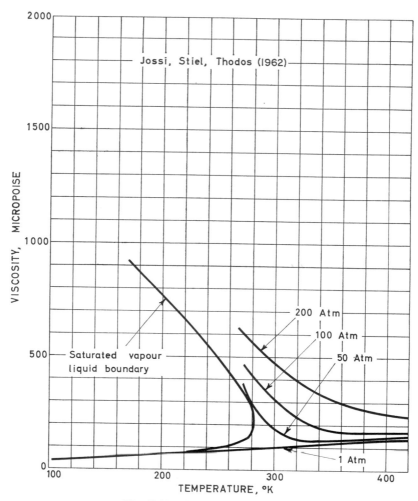

Fig. 12.39. The viscosity of ethylene.

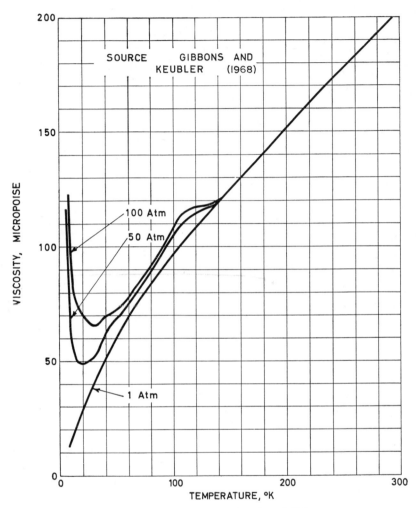

Fig. 12.40. The viscosity of He⁴.

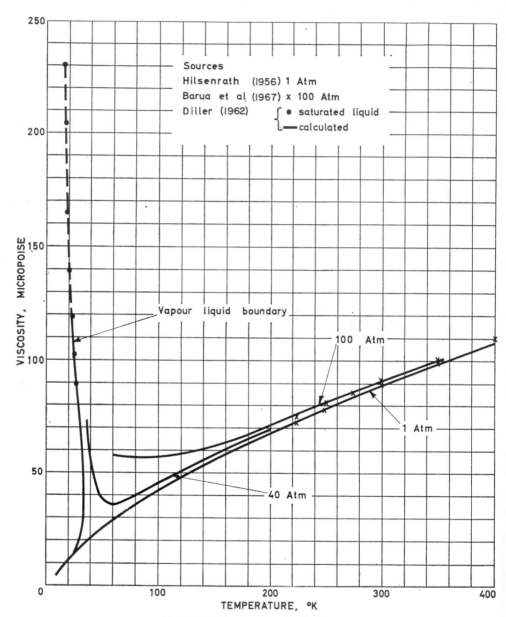

Fig. 12.41. The viscosity of n-hydrogen.

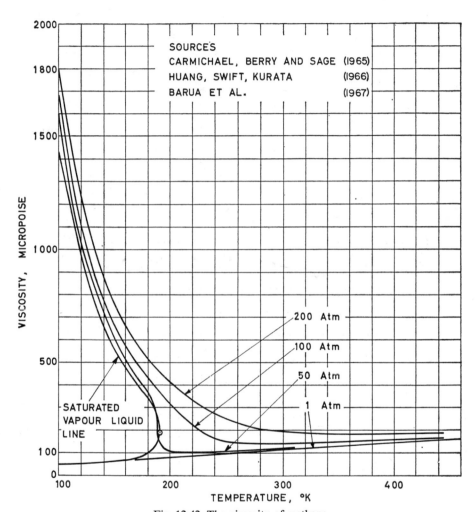

Fig. 12.42. The viscosity of methane.

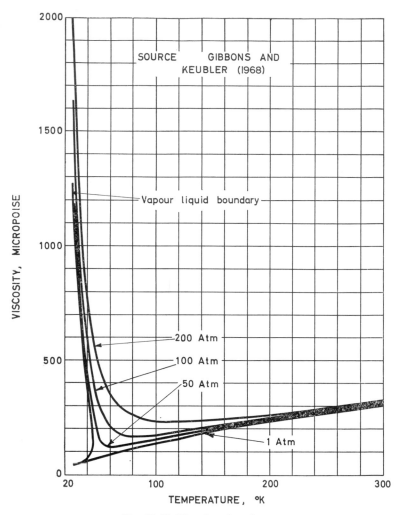

Fig. 12.43. The viscosity of neon.

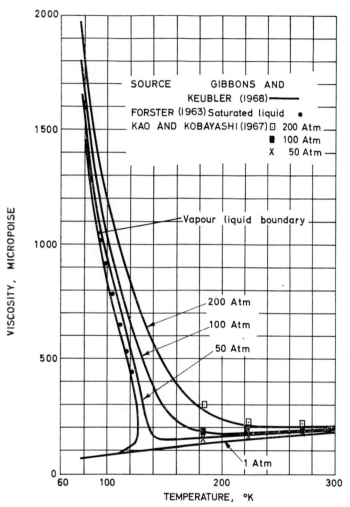

Fig. 12.44. The viscosity of nitrogen.

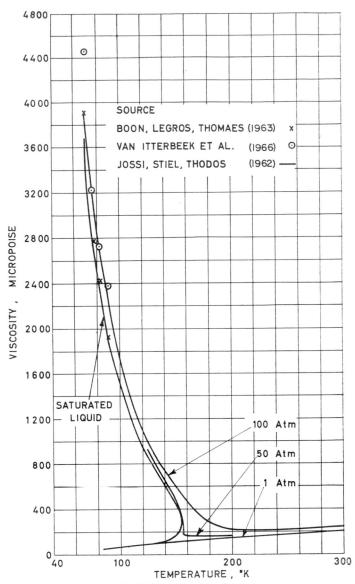

Fig. 12.45. The viscosity of oxygen.

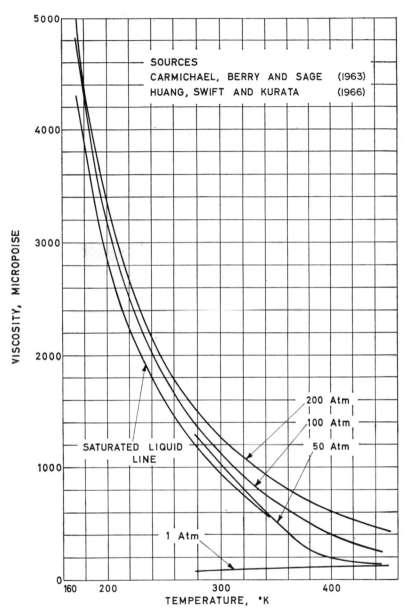

Fig. 12.46. The viscosity of propane.

V. The Calculation of Thermodynamic Properties

There are two main approaches to calculating, as opposed to measuring, thermodynamic properties. The first approach is to fit an equation of state to the PVT data of the substance. The second approach is to use the principle of corresponding states (CSP) and an equation for the PVT surface of another substance. Both methods are concerned with calculation of the difference of a property from the value of that property for an ideal gas under the same conditions of temperature and pressure. The properties of each substance in the ideal gas state are therefore required as a function of temperature. In most cases these ideal properties can be calculated from statistical mechanical formulae and spectroscopic data and have been tabulated for a wide range of substances (American Petroleum Institute, 1945). The formulae relating the entropy, enthalpy and fugacity can be derived from the fundamental thermodynamic relationships and are are given explicitly by equations (12.3–12.5).

$$S = S_0(T_0) + \int_V^\infty \left(\frac{R}{V} - \left(\frac{\partial p}{\partial T} \right)_V \right) dV + \int_{T_0}^T C_p^\circ \frac{dT}{T} + R \ln \left(\frac{pV}{RT} \right) \qquad (12.3)$$

$$H = H_0(T_0) + \int_V^\infty \left(p - T \left(\frac{\partial p}{\partial T} \right)_V \right) dV + \int_{T_0}^T C_p^\circ \, dT + pV - RT \qquad (12.4)$$

$$F = F_0 - \int_V^\infty \left(p - \frac{RT}{V} \right) dV + R \ln \left(\frac{RT}{V} \right) \qquad (12.5)$$

A summary will be given of the calculation of these properties by the Benedict–Webb–Rubin (1951) equation, and, more sketchily, from (CSP), for both pure substances and mixtures.

A. THE BWR EQUATION

The BWR equation is a flexible 8 constant equation which was developed empirically by Benedict *et al.* (1951) to describe the PVT data for the light hydrocarbons. Its form is

$$p = RT/V + [RT(B_0 + b/V) - (A_0 + a/V - a\alpha/V^2) -$$
$$(C_0 - c/V(1 + \gamma/V^2) e^{(-\gamma/V^2)})T^2]/V^2 \qquad (12.6)$$

Constants for a selection of substances are shown in Table 12.I (Haselden and Sood, 1970). Various schemes for generalizing these parameters have been suggested by Orye (1969), Griskey and Canjar (1963), Canjar *et al.* (1955) and Kaufman (1968). However, this seems an inherently dangerous procedure inasmuch as the calculated pressure is the result of a sum of positive and negative terms, each of which may be much larger than the resultant pressure. Small errors in these terms can produce large errors in the calculated pressures.

Attempts have been made to improve the BWR equation particularly for the prediction of the fugacity and hence the vapour liquid boundary, which is

TABLE 12.I. BWR Coefficients in the Units litre atm g mole $^\circ$K.

Compound	B_0	A_0	C_0	b	a	c	$\alpha \times 10^3$	$\gamma \times 10^2$
Nitrogen[a]	0·05054310	1·481130	2676·423	0·003066334	0·01471387	428·6965	0·1139511	0·587570
Nitrogen	0·0458013	1·19257	5889·4	0·00198165	0·0149013	548·110	0·291569	0·750042
Carbon Monoxide	0·0400	1·03115	11240·0	0·00263158	0·03665	1040·0	0·1350	0·6000
Argon	0·022282597	0·823417	13141·25	0·00215289	0·0288358	798·2437	0·03558895	0·23382711
Oxygen[a]	0·03928469	1·516162	2585·949	0·002380833	0·01585147	508·6938	0·04641719	0·39
Methane	0·042600	1·8550	22570·0	0·00338004	0·04940	2545·0	0·124359	0·60
Ethylene	0·0556833	3·33958	131140·0	0·008600	0·259000	21120·0	0·178000	0·923000
Carbon Dioxide	0·0448849	2·51606	147443·0	0·0041239	0·136814	14918·0	0·084663197	0·52533158
Ethane	0·0627724	4·1556	179592·0	0·0111220	0·345160	32767·0	0·243389	1·18000
Propylene	0·0850647	6·11220	439182·0	0·0187059	0·774056	102611·0	0·455696	1·82900
Propane	0·0973130	6·87225	508256·0	0·0225000	0·947700	129000·0	0·607175	2·2000
iso-Butane	0·137544	10·23264	849943·0	0·0424352	1·93763	286010·0	1·07408	3·40000
1-Butene	0·116025	8·95325	927280·0	0·0348156	1·69270	274920·0	0·910889	2·95945
n-Butane	0·124361	10·0847	992830·0	0·039983	1·88231	316400·0	1·10132	3·40000
iso-Pentane	0·160053	12·7959	1746320·0	0·0668120	3·75620	695000·0	1·7000	4·63000
n-Pentane	0·156751	12·1794	2121210·0	0·0668120	4·07480	824170·0	1·81000	4·75000

[a] Values from McCracken and Mullins.

obtained from the conditions

$$f_l = f_g \text{ and } p_l = p_g,$$

where f is the fugacity, p the pressure and the subscripts l and g refer to the vapour and liquid respectively. At low temperatures agreement of calculated values with the vapour–liquid boundary data is improved by making C_0 temperature dependent. A successful form for this temperature dependence of C_0 is given by

$$C_0 = A_0 + A_1\phi + A_2\phi^2 + A_3\phi^3 + A_4\phi^4 \tag{12.7}$$

where

$$\phi = (T - T_c)/T_c$$

Values for the constants A_0, A_1, A_2, A_3 and A_4 are given in Table 12.II.

The BWR equation can be extended to mixtures by defining mixing rules relating the constants of the equation to the composition of the mixture. The forms suggested 12.7 are:

$$A_0 = [\Sigma_i x_i (A_0)_i^{\frac{1}{2}}]^2 \qquad\qquad a = [\Sigma_i x_i a_i^{\frac{1}{3}}]^3$$
$$B_0 = \tfrac{1}{8}[\Sigma_{i,j} x_i x_j [(B_0)_i^{\frac{1}{3}} + (B_0)_j^{\frac{1}{3}}]]^3 \qquad b = [\Sigma_i x_i b_i^{\frac{1}{3}}]^3$$
$$C_0 = [\Sigma_i x_i (C_0)_i^{\frac{1}{2}}]^2 \qquad\qquad c = [\Sigma_i x_i c_i^{\frac{1}{3}}]^3$$
$$\alpha = [\Sigma_i x_i \alpha_i^{\frac{1}{3}}]^3 \qquad\qquad \gamma = [\Sigma_i x_i \gamma_i^{\frac{1}{2}}]^2$$

These combining rules are empirical and generally give best results for close boiling mixtures.

All other thermodynamic properties of a substance can be obtained from the equation of state using the standard thermodynamic relationships. The explicit relationships for the entropy, enthalpy and fugacity obtained from equation 12.7 are

$$S = \Sigma_i x_i (S_i^0 - R \ln (x_i RT/V) - (RB_0 + 2C_0/T^3)/V - bR/2V^2 + \\ 2c((1 - e^{-\gamma/V^2})/\gamma - e^{-\gamma/V^2}/2V^2)/T^3 \tag{12.9}$$

$$H = \Sigma_i x_i H_i^0 + (B_0 RT - 2A_0 - 4C_0/T^2)/V + (2bRT - 3a)/2V^2 + \\ 6a\alpha/5V^5 + c[3(1 - e^{-\gamma/V^2})/\gamma - e^{-\gamma/V^2}/2V^2 + (\gamma e^{-\gamma/V^2})/V^4]/T^2 \tag{12.10}$$

$$RT \ln f_i = RT \ln (x_i RT/V) + [(RT/4)\Sigma_j ((B_0)_j^{\frac{1}{3}} + (B_0)_i^{\frac{1}{3}})^3 - 2A_0(A_0)_i^{\frac{1}{2}} - \\ 2[C_0(C_0)_i]^{\frac{1}{2}}/T^2]/V + 3[RT(b^2 b_i)^{\frac{1}{3}} - (a^2 a_i)^{\frac{1}{3}}]/2V^2 + \\ 3[a(\alpha^2 \alpha_i)^{\frac{1}{3}} + \alpha(a^2 a_i)^{\frac{1}{3}}]/5V^5 + [1 - e^{-\gamma/V^2}] \times \\ [3(c^2 c_i)^{\frac{1}{3}} - 2c(\gamma_i/\gamma)^{\frac{1}{2}}]/\gamma T^2 - \\ e^{-\gamma/V^2}[(3(c^2 c_i)^{\frac{1}{3}}/2 - 4c(\gamma_i/\gamma)^{\frac{1}{2}})/2V^2 T^2 - c(\gamma\gamma_i)^{\frac{1}{2}}/T^2 V^4] \tag{12.11}$$

If the constant C_0 is allowed to be temperature dependent as in equation (12.7) the following equations for the entropy, enthalpy and fugacity are obtained.

TABLE 12.II. Polynomial Coefficients for C_0 Adjustment in Units litre atm g mole °K.

Compound	Temperature range, °K	A_0	A_1	A_2	A_3	A_4
Nitrogen[a]	Not given	1·00000	$-1·689605 \times 10^{-1}$	3·606153	$-8·621769$	5·339828
Nitrogen	63 to 126	0·9997103	$-2·526856 \times 10^{-3}$	$4·381323 \times 10^{-2}$	$-4·726279 \times 10^{-2}$	$9·256033 \times 10^{-3}$
Carbon monoxide	64 to 133	1·018458	$-2·638971$	9·586363	$-14·755768$	7·603512
Argon	70 to 151	0·9987691	$-0·7174757$	2·651036	$-6·601487$	5·3088647
Oxygen[a]	Not given	1·00000	0·8689789	0·9792821	$-1·248256$	$-2·175794$
Methane	90 to 191	0·9979814	$-3·320010 \times 10^{-2}$	$2·962412 \times 10^{-1}$	$-1·932208$	1·1346987
Ethylene	124 to 280	0·9947091	0·2462640	$-1·411671$	2·881806	$-2·871082$
Carbon dioxide	139 to 304	1·071528	$-2·413171$	20·329057	46·96394	32·152077
Ethane	136 to 305	1·000434	$0·2716179 \times 10^{-1}$	$-4·379819 \times 10^{-2}$	$-0·4433283$	$-0·2001602$
Propylene	166 to 363	0·9858207	0·3294600	$-1·7105199$	3·5640557	$-34·42999$
Propane	145 to 370	0·9965831	$1·574242 \times 10^{-2}$	0·2005013	$-0·8345010$	$-2·544982 \times 10^{-2}$
iso-butane	183 to 408	0·9997103	$-8·168955 \times 10^{-3}$	0·4579070	$-1·596897$	1·0110428
1-butene	216 to 398	1·006835	$0·8245186 \times 10^{-1}$	$-0·3141616$	$4·366620 \times 10^{-3}$	$0·430830 \times 10^{-4}$
n-butane	195 to 403	0·9319580	1·192142	$-2·6581867$	17·290853	$-28·408991$
iso-pentane	231 to 461	0·992118	$-0·1386477 \times 10^{-1}$	0·5157715	$-1·430737$	0·5618867
n-pentane	224 to 470	0·9972927	$0·3197312 \times 10^{-1}$	$-0·5942343 \times 10^{-1}$	0·284999	$-1·1674545$

[a] Values from McCracken and Mullins.

$$S = S_{\text{BWR}} + C_0(A_1 + 2A_2\phi + 3A_3\phi^2 + 4A_4\phi^3)/(T_cT^2V) \qquad (12.12)$$

$$H = H_{\text{BWR}} + C_0(A_1 + 2A_2\phi + 3A_3\phi^2 + 4A_4\phi^3)/(T_cTV) \qquad (12.13)$$

$$F = F_{\text{BWR}} \qquad (12.14)$$

B. THE PRINCIPLE OF CORRESPONDING STATES

The principle of corresponding states (CSP) was originally suggested by Van der Waals and has since been restated in terms of statistical mechanical quantities by Pitzer (1939). Leland and Chappelear (1968) give a complete discussion of modern developments of (CSP). This survey is confined to some practical applications of (CSP) using constants obtained from macroscopic data.

(CSP) states that the reduced properties of substances are the same for all reduced conditions. Pitzer has shown by dimensional analysis of the canonical ensemble partition function that (CSP) will be obeyed by molecules satisfying the following conditions.

1. The canonical ensemble partition function is separable into independent translational, vibrational, rotational and configurational states.
2. The internal energy states are independent of the density.
3. Maxwell Boltzmann statistics apply; this excludes quantum effects.
4. The potential energy is a function of an energy parameter and the reduced distances between the particles. It is not necessarily dependent on pairwise addivity; the potential functions need only be conformable.

These assumptions lead to a reduced equation of state in terms of molecular constants of the general form

$$Z = F(kT/\varepsilon, V/N\sigma^3) \qquad (12.15)$$

where ε is the intermolecular energy parameter and σ the molecular diameter. This equation can be rewritten in terms of macroscopic reduction constants by relating ε and σ to the experimental critical temperature and volume leading to the equation,

$$Z = F(T/T_c, V/V_c). \qquad (12.16)$$

This two parameter form of (CSP) is only accurate for the inert gases, argon, krypton and xenon, as these are the only substances which satisfy the restrictive basic assumptions.

(CSP) can be reformulated to take account of departures from the basic assumptions by the inclusion of more molecular parameters which are related to the molecular interaction energy and/or the statistics followed by the system of particles. For most practical applications, however, additional parmeters based on deviations of macroscopic properties have proved more successful. Two of the more successful types of macroscopic parameters are the acentric factor of Pitzer and Curl (1957) and the shape factors of Leach et al. (1966, 1968).

TABLE 12.III. Critical Constants and Accentric Factors for Some Selected Substances

Substance	P_C atm	V_C cm³/g mde	T_C °K	w Accentric factor	M Molecular weight
Ammonia	111·30	72·5	405·50	0·25	17·03
Argon	48·0	75·2	150·72	0·0	39·95
Benzene	48·6	256·60	562·00	0·211	78·11
n-Butane	37·47	256·1	425·17	0·200	58·12
iso-Butane	36·00	262·99	408·13	0·1918	58·12
Carbon Dioxide	72·80	94·0	304·16	0·225	44·01
Carbon Monoxide	34.53	93·1	132·90	0·049	28·01
Ethane	48·20	148·00	305·40	0·105	30·07
Ethylene	50·50	124·00	283·06	0·097	28·05
Helium	2·26	58·18	5·22	—	4·003
n-Hexane	29·9	367·66	507·9	0·298	86·18
Hydrogen	12·80	65·03	33·19	—	2·016
Krypton	54·18	92·20	209·39	0·0	83·8
Methane	45·80	99·0	190·66	0·013	16·04
Neon	26·19	41·79	44·40	0·0	20·18
Nitrogen	33·54	90·1	126·26	0·040	28·01
Oxygen	50·14	78·0	154·78	0·021	32·00
iso-Pentane	32·92	307·75	460·95	0·206	72·15
n-Pentane	33·03	310·72	470·35	0·252	72·15
Propane	42·01	200·0	369·97	0·152	44·09
Sulphur Dioxide	77·8	122·0	430·70	—	64·06
Toluene	40·0	331·1	594·0	0·241	92·13
Water	218·3	55·2	647·4	0·344	18·015

Pitzer defined an acentric factor, w, based on the deviation of the vapour pressure of a substance from that of argon at a reduced temperature of 0·7. The relationship for w is,

$$w = -1·0 - \ln P_R \text{ at } T_R = 0·7. \qquad (12.17)$$

w is a measure of the departure from spherical symmetry of a molecule and increases with polarity or elongation of the molecule. Tables of w, and critical constants, for most cryogenic fluids are shown in Table 12.III. (CSP) can be extended formally to include the acentric factor by writing

$$Z = F(T_R, V_R, w) = F(T_R, V_R, 0) + w F^1(T_R, V_R, 0) \qquad (12.18)$$

Graphs of Z as a function of P_R and T_R have been prepared with corrections as a function of w. Perhaps its greatest success is its use in an accurate equation for the second virial coefficient, B, developed by Pitzer and Curl (1957), which is applicable to a wide range of substances and is given by

$$BPc/RT_c = (0\cdot1445+0\cdot073\,w)-(0\cdot33-0\cdot46w\,)/T_R-$$

$$(0\cdot1385+0\cdot5\,w)/T_R^2-(0\cdot0121+0\cdot097\,w)/T_R^3-$$

$$0\cdot0073\,w/T_R^8 \qquad (12.19)$$

Gunn, Chueh and Prausnitz (1966) have extended this approach to the quantum gases hydrogen and helium by defining temperature dependent reduction factors,

$$T_c = T_c^\circ(1\cdot0/(1\cdot0+c_1/MT))$$

$$V_c = V_c^\circ(1\cdot0/(1\cdot0+c_2/MT)) \qquad (12.20)$$

$$c_1 = 21\cdot8^\circ\,K, \qquad c_2 = -9\cdot91^\circ\,K$$

where T_c° and V_c° are the experimental values of the critical constants.

An expression for the reduced third virial coefficient has been obtained by Chueh and Prausnitz (1967).

$$C/V_c^2 = (0\cdot232/T_R^{0\cdot25}+0\cdot466/T_R^5)(1\cdot0- \exp(1\cdot0-1\cdot89T_R^2))+$$

$$D \exp -(2\cdot49-2\cdot3T_R+2\cdot7T_R^2) \qquad (12.21)$$

D is a parameter specific for each substance which could not be related to the accentric factor or any other simple parameter; it is important only at reduced temperature of two or less. The expression is accurate for most non-polar gases and for quantum gases using the expressions from equation (12.20) for the reduction of parameters. The second and third virial coefficients defined by equations (12.19) and (12.21) when substituted into the virial series provide a means of calculating the PVT surface of simple substances. These expressions can be used above the critical temperature to calculate the density for a given temperature and pressure with an accuracy of 1% at pressures of less than 100 atmospheres or densities less than the critical.

Other properties which can be accurately calculated by three parameter forms of (CSP), such as saturated vapour and liquid densities, the vapour pressure and latent heat are discussed by Leland and Chappelear (1968).

The simple two parameter form of (CSP) assumes that all substances have the same reduced properties for the same reduced conditions. Alternatively it can be restated as follows in terms of the properties of individual substances. Given the reduced properties of a reference substance there are two parameters independent of temperature and density, which will make the reduced properties of a second substance identical with those of the reference substance for all reduced conditions. In general this is not the case because of deviations.

These deviations can be allowed for in an empirical way by assuming the reduction parameters depend on the reduced temperature and density; such reduction parameters are commonly called "shape factors".

Values of the "shape factors" have been given by Leach, Leland and Chappelear for light hydrocarbons systems, and air. To obtain the properties of the reference substance these authors used the equation of Vennix and Kobayashi (1969), which accurately describes the properties of methane. These "shape factors" depend on the acentric factor and the reduced temperature and density.

C. THE PROPERTIES OF MIXTURES

The properties of mixtures are outside the scope of this chapter but no discussion of thermodynamic data for cryogenic processes would be complete without at least some mention of the properties of mixtures. In the space available it is only possible to give a short list of useful bibliographies of existing data for vapour liquid equilibria compressibility and enthalpy data for mixtures (Hala *et al.*, 1957; Muckerloy, 1962; Ruheman and Harmens, 1967; Reid, 1968; Clarke *et al.*, 1966), and to review briefly methods of calculating these properties of mixtures (Prausnitz *et al.*, 1967).

The calculations of the properties of mixtures are based on the methods developed for the calculation of the properties of pure components. They fall into two categories, those based on equations of state and those based on (CSP). The central problem in both methods is to relate the parameters for the mixture to those for the pure substances. Once this has been done the properties of the mixture in a single phase can be obtained from standard thermodynamic relationships. For the calculation of vapour–liquid equilibria in an n component mixture it is necessary, in addition, to find a solution to the set of equations defining the equilibrium, that is

$$f_i^l = f_i^g \qquad \sum x_i = 1 \qquad \sum y_i = 1 \qquad (12.22)$$

where f_i is the fugacity of component i. The fugacity of components in the vapour phase can be calculated using either the truncated virial series (Chueh and Prausnitz, 1967) or some empirical equation of state such as the BWR equation (Benedict *et al.*, 1951) or, as used by Prausnitz *et al.* (1967), Chueh and Prausnitz (1968) and Wilson (1963), the Redlich–Kwong equation. The properties of the liquid phase are more difficult to represent accurately. Chao and Seader (1963) used the Scatchard–Hildebrande theory to describe liquid phase properties of the substances from the basic assumptions. Prausnitz and his co-workers have developed methods for the modified Wohl equation (Prausnitz *et al.*, 1967) or a modified Wilson equation for the activity coefficient (Chueh and Prausnitz, 1968). Other equations of state have been used for the liquid phase including the BWR (Benedict *et al.*, 1951) and Redlich–Kwong equations (Wilson, 1963).

28*

(CSP) as developed by Leland *et al.* (1968, 1969) and Rowlinson (1969) has also been developed for the calculation of vapour–liquid equilibria. As for equations of state the problem lies in the choice of mixing rules for the reduction parameters of the mixture.

TABLE 12.IV. Conversion Factors for Some Quantities between British, cgs and SI Units

Quantity	To convert from	To	Multiply by
Density	lb/ft³	kg/m³	16·0185
	g/l	kg/m³	1·0
Energy	erg	Joule	10^{-1}
	B.T.U.	Joule	1055·06
	B.T.U.	k Joule	1·05506
	Cal	Joule	4·1868
	atm. l	Joule	101·328
Energy, specific	B.T.U./lb	Joule/gm	2·326
		kJ/kg	2·326
Entropy, specific	B.T.U./lb/°R	Joule/kg/K	4186·8
		kJ/kg/K	4·1868
Heat flow	B.T.U./hr	W	0·293071
Mass	lb.	kg	0·453592
Pressure	p.s.i.	atm	0·0680457
	atm	bar	0·986923
	p.s.i.	bar	0·0689476
	N/m²	bar	10^{-5}
Thermal conductivity	B.T.U. ft/ft² hr°R	W/m°K	1·73073
	B.T.U./m/ft² hr°R	W/M°K	0·144228
Viscosity	poise	micropoise	10^6
	centipoise	lb. sec/ft	$2·0885410^{-5}$
	lb. sec/ft	centipoise	$4·7880310^{-5}$
Volume	ft³	m³	0·0283168
	ft³	l (dm³)	28·3168
	in.³	cm³	16·3871
Volume, specific	ft³/lb.	1(dm³)/kg	62·428
		cm³/g	62·428
	in./lb.	cg³/g	0·0361273
Volume, rate of flow	ft³/sec	m³/sec	0·0283168
	gal/hr	l(dm³)/hr	4·53592

VI. Nomenclature

C_V Specific heat of constant volume
C_p Specific heat at constant pressure
d Density
d_R d/d_c = reduced density
x_i Mole fraction of component i in the liquid phase
y_i Mole fraction of component i in the vapour phase

H	Enthalpy
F	Free energy
f	Fugacity
K_0	Thermal conductivity at 1 atm
ΔK_0	Excess thermal conductivity
K	Thermal conductivity
k	Boltzmann's constant
P	Pressure
P_R	$P/P_C =$ reduced pressure
S	Entropy
T	Temperature
T_R	$T/T_c =$ reduced temperature
V	Volume
V_R	$V/V_c =$ reduced volume
μ_0	Viscosity at 1 atm
$\Delta\mu_0$	Excess viscosity = viscosity $-\mu_0$
μ	Viscosity

Superscripts

$\circ$	used with thermodynamic properties to denote the ideal gas value of the property
i	denotes the quantity is a property of component i
c	denotes the critical value of a property.

VII. References

American Petroleum Institute (1945). Project 44, Texas A and M University.

Barua, A. K., Afzal, M. A., Flynn, G. P. and Ross, J. (1967). *J. Chem. Phys.* **41**, 374.

Benedict, M., Webb, G. B. and Rubin, L. C. (1951). *Chem. Eng. Progr.* **47**, 419, 449.

Boon, J. P., Legros, J. C. and Thomaes, F. (1963). *Physica* **24**, 335.

Canjar, L. N., Pollock, E. K. and Lee, W. E. (1966). *Hydrocarbon Process* **45**, 139.

Canjar, L. N., Smith, R. F., Voliantis, E., Galluzzo, J. G. and Cabarcos, M. (1955). *Ind. Eng. Chem.* **47**, 1028.

Carmichael, L. J., Berry, V. and Sage, B. H. (1963). *J. Chem. Eng. Data* **8**, 94.

Carmichael, L. J., Berry, V. and Sage, B. H. (1965). *J. Chem. Eng. Data* **10**, 57.

Carmichael, L. J., Jacobs, J. and Sage, B. H. (1968). *J. Chem. Eng. Data* **13**, 40.

Chao, K. C. and Seader, J. D. (1963). *A.I.Ch.E.* **1**, 598.

Chueh, P. L. and Prausnitz, J. M. (1968). "Computer Calculations for High Pressure Vapour Liquid Equilibria." Prentice Hall, New Jersey.

Chueh, P. L. and Prausnitz, J. M. (1967). *A.I.Ch.E.* **13**, 896.

Clarke, R., Hyman, F. and Wilson, G. (1966). Air Materials Research Laboratory, TR-AFML-TR-66, 136.

Corruccini, R. J. (1957). *Chem. Eng. Progr.* **53**, 262, 342, 397.

Dillard, D. S. and Timmerhaus, K. D. (1968). *C.E.P., Symp. Series* **64**, 1.

Diller, D. E. (1962). *J. Chem. Phys.* **42**, 2089.

Edmister, W. C. (1961). "Applied Hydrocarbon Thermodynamics." Gulf Publishing Co., Houston.

Forster, S. (1963). *Cryogenics* **3**, 176.

Gallant, R. (1966). *Hydrocarbon Process* **45**, 113.

Gibbons, R. M. and Keubler, G. P. (1968). Wright Patterson Air Force Base TR AFML-68.

Gosman, A. L., Hust, J. G. and McCarty, R. D. (1969). NSRDS-NBS 27.

Griskey, R. W. and Canjar, L. N. (1963). *A.I.Ch.E.* **9**, 182.

Gunn, R. D., Chueh, P. L. and Prausnitz, J. M. (1966). *A.I.Ch.E.* **12**, 937.

Hala, E., Pick, J., Fried, V. and Vilim, O. (1957). "Vapour Liquid Equilibria." Pergamon Press, London.

Haselden, G. G. and Sood, S. K. (1970). *A.I.Ch.E.* **16**, 891.

Hilsenrath, J. *et al.* (1956). N.B.S. Circular 564.

Huang, E. T. S., Swift, G. W. and Kurata, F. (1966). *A.I.Ch.E.* **12**, 932.

Hust, J. G. and Stewart, R. B. (1963). NBS TN 202.

Johnson, V. (1960). WADD Technical Report 60-50.

Jossi, J. A., Stiel, L. I. and Thodos, G. (1962). *A.I.Ch.E.* **8**, 59.

Kao, J. T. and Kobayashi, R. (1967). *J. Chem. Phys.* **47**, 2836.

Kaufman, T. (1968). *Ind. Chem. Fund* **7**, 115.

Kestin, J. and Whitelaw, J. H. (1963). *Physics* **29**, 335.

Leach, J., Leland, T. and Chappelear, P. (1966). *Proc. NGPA* **45**, 8.

Leach, J., Leland, T., and Chappelear, P. (1968). *A.I.Ch.E.* **14**, 568.

Leland, T. and Chappelear, P. (1968). *Ind. Eng. Chem.* **60**, 15.

Leland, T. W., Rowlinson, J. S. and Sather, G. A. (1968). *T.F.S.* **64**, 1447.

Leland, T. W., Rowlinson, J. S., Sather, G. A. and Watson, I. D. (1969). *T.F.S.* **65**, 2034.

Lenoir, J. M. and Comings, E. W. (1951). *Chem. Eng. Prog.* **47**, 221.

Lenoir, J. M., Junk, W. A. and Comings, E. W. (1953). *Chem. Eng. Prog.* **45**, 539.

Mann, D. (1962). NBS TN 154.

Manni, N. and Venart, J. E. S. (1969). Proceedings 1st International Conference on LNG.

Muckleroy, J. S. (1962). "Bibliography on Hydrocarbons." NGPA.

Orye, R. V. (1969). *Ind. Chem. Eng. Proc. Design and Dev.* **8**, 579.

Pitzer, K. E. (1939). *J. Chem. Phys.* **7**, 583.

Pitzer, K. E. and Curl, R. F. (1957). *J.A.C.S.* **77**, 3427.

Prausnitz, J. M., Orye, R. V., O'Connell, J. P. and Eckert, C. A. (1967). "Calculations for Multicomponent Vapour Liquid Equilibria." Prentice Hall. New Jersey.

Reid, R. C. (1968). *C.E.P. Monograph Series* **64**, No. 5.

Rogers, J. D., Zeigler, R. K. and McWilliams, P. (1962). *J. Chem. Eng. Data* **1**, 179.

Rowlinson, J. S. (1969). International Conference on Distillation, Brighton.

Ruheman, M. and Harmens, A. (1967). *Chem. Eng.* 254.

Stewart, R. B. (1966). Ph.D. Thesis, University of Iowa.

Stiel, L. I. and Thodos, G. (1964). *A.I.Ch.E.* **10**, 26.

Strobridge, T. R. (1962). NBS TN 129.

Swift, G. W., Lohrenz, J. and Kurata, F. (1960). *A.I.Ch.E.* **6**, 415.

Tsederberg, N. V. and Timrot, D. L. (1956). *Sov. Phys. Tech. Phys.* **1**, 1791.

Tsederberg, N. V. and Timrot, D. L. (1956). *J. Tech. Phys.* (*U.S.S.R.*) **26**, 1849.

Van Itterbeck, A., Helleman, J., Zink, H. and van Canteren, T. (1966). *Physica* **32**, 2171.

Vennix, O. and Kobayashi, R. (1969). *A.I.Ch.E.* **15**, 962.

Wilson, G. M. (1963). "Advances in Cryogenic Engineering", Vol. 9. Plenum Press, New York.

Wooley, H. W., Scott, R. B. and Brickewedde, F. G. (1948). *N.B.S. J. Res.* **41**, 379.

Author Index

The numbers in *italics* refer to pages in the References at the end of each Chapter.

A

Abel, W. R., 83, *89*
Abeshouse, D. J., 607, *632*
Abrikosov, A. A., 506, *568*
Acrivos, A., 392, *400*
Adams, 391, *400*
Afzal, M. A., 718, 722, 723, *737*
Alers, G. A., 502, *568*
Allen, J. F., 35, *89, 307*
Ambler, E., 38, *89*, 607, *631*
American Petroleum Institute, 647, 667, *670*, 728, *737*
American Society for Testing of Materials, 643, *670*
American Society of Mechanical Engineers, 665, *670*
Amundson, N. R., 392, *400*
Anderson, P. W., 524, *569*
Andres, K., 496, *570*
Appleton, A. D., 568, *569*
Aris, R., 392, *400*
Armstrong, G. T., 590, *631*
Arp, V., 302, 303, *307*
Arpaci, V. S., 96, 97, 100, 102, 144, *192*
Arthur D. Little, Inc., 145, 146, *192*, 642, 649, 666, *670*
ASME, 262, 263, 269, 271, 272, 273, 274, *307*, 313, 336, 357, 359, 361, 362, 363, *373*
Aston, J. G., 603, *631*
Astruc, J. M., 126, 129, *192*

B

Baily, B., 103, *193*
Bailey, B. M., 667, *670*
Bailey, S. C., 603, *631*
Baker, C. K., 204, *234*
Baker, O., 142, *193*, 269, *307*, 658, *670*
Baldus, W., 74, *89*
Ball, W. L., 643, *670*
Balzhiser, R. E., 122, *193*
Banchero, J. T., 131, *193*
Barakat, H. Z., 101, 102, 103, 118, *193, 194*
Barber, C. R., 586, 596, *631*
Barber, W. R., 50, 53, *89*
Bardeen, J., 503, *569*
Barker, G. E., 131, *193*
Barker, J. J., 173, *193*
Barnes, C. B., 86, *89*
Barnett, D. O., *193*
Barratt, P. B., *570*
Barron, R. F., 144, 146, *193*
Barry, H. M., 376, 393, *400*
Bartlit, J. R., 105, *193*, 277, 289, 290, *307*
Barua, A. K., 718, 722, 723, *737*
Basmadjian, D., 386, *400*
Bauer, E., 276, 285, 286, 287, *309*
Bean, C. P., 527, 531, 532, *569*, 571
Beasley, S. A., 405, *488*
Becker, H., 462, *489*
Bedford, R. E., 586, *632*
Beher, J. T., 222, *234*
Benedict, D. C., 667, *670*
Benedict, M., 728, 735, *737*
Benser, W. A., 112, *193*
Benz, M. G., 500, *569*
Berben, T. J., 515, *569*
Berenson, P. J., 127, 130, 131, 132, 141, 142, *193, 196, 197*
Berényi, L., 383, 384, 385, *400*
Bergles, A. E., 133, *193, 195*
Berman, R., 603, *631*
Bernstein, B. A., 233, *234*
Berry, V., 719, 723, 727, *737*
Bertman, B., 607, *632*
Betts, D. S., 82, *89*
Bewilogua, L., 126, *193*
Birkholz, U., 34, *89*
Birmingham B. W., 65, *90*, 405, 439, *489*

739

Subject Index

751